S0-DTC-355

Man and the Natural World

AN INTRODUCTION TO LIFE SCIENCE

MAN AND THE NATURAL WORLD

COLEMAN J. GOIN ■ OLIVE B. GOIN
MUSEUM OF NORTHERN ARIZONA

SECOND EDITION

Man and the Natural World

AN INTRODUCTION TO LIFE SCIENCE

Macmillan Publishing Co., Inc.
NEW YORK
Collier Macmillan Publishers
LONDON

Printed in the United States of America

Macmillan Publishing Co., Inc.
866 Third Avenue, New York, New York 10022

Collier-Macmillan Canada, Ltd.

Library of Congress Cataloging in Publication Data

Goin, Coleman Jett (date)
Man and the natural world.

Includes bibliographies.
1. Biology. I. Goin, Olive Bown, joint author.
II. Title.
QH308.2.G64 1975 574 73-19043
ISBN 0-02-344240-9

Printing: 1 2 3 4 5 6 7 8 Year: 5 6 7 8 9 0

to Lewis Berner

staunch advocate of liberal education

Preface

TO THE SECOND EDITION

Rapid advances in biological knowledge in the past few years have made it necessary for us to revise this book. We have not changed the basic philosophy and organization of the first edition, but we have incorporated new information into almost every chapter. Some of the most exciting advances in human biology include the characterization of the B-cells and T-cells of the immune system, the elucidation of the role of cyclic AMP as the mediator of the activity of many hormones, the demonstration that the two halves of the brain differ in function, and suggestions about the nature of the engram, the memory trace in the brain. We have expanded the discussion of contraceptives in the section on reproduction and added a brief account of aging. The genetics section now includes a discussion of the multiple copies of many genes found in higher organisms. On the other hand, the Jacobs-Monod model of gene regulation has been found to apply only to bacteria and has been dropped. We have added an account of continental drift and its effect on the distribution of plants and animals to the section on evolution, as well as a consideration of the possible evolutionary role of changes in amount of DNA per nucleus. We have modified the discussion of human evolution to bring it into accord with the latest findings of anthropology. We have rewritten the chapters on behavior to include accounts of learning and of primate societies. Finally, we have inserted metric measurements and added an appendix on the metric system.

No people ever had better friends and colleagues. We are grateful to Miss Pam Lungé for her assistance with the drawings. Our good friends, Dr. and Mrs. Edwin C. Colbert of the Museum of Northern Arizona, have helped us in many ways. Margaret Colbert made the delightful drawings of the squirrel and hawk in Chapter 31 and Ned Colbert allowed us to use his illustrations on continental drift. Dr. Thomas Emmel of the University of Florida provided the photograph of the butterfly and sunflower in Chapter 30. We thank E. P. Dutton & Co. and W. W. Norton & Company for permission to use these illustrations.

Dr. William Lipe, assistant director of the Museum of Northern Arizona, and Dr. Alexander Lindsay and Miss Dana Hartman, of the Department of Anthropology of the museum, all helped most generously with the section on the evolution of man.

Dr. Mildred Griffith of the University of Florida kindly provided titles for the suggested readings.

Mrs. Sally Carothers assisted with the typing. Mrs. Deloris Douglas not only did most of the typing, but she also gave cheerful, ready help in many other ways, numbering pages, xeroxing, and assisting in all the multitudinous chores involved in getting a manuscript ready for press.

Above all we are indebted to Dr. Edward B. Danson, director of the Museum of Northern Arizona, and to our colleagues there, especially in the Department of Biology. They rallied round in a time of crisis, and without their help this book would not have been finished.

C. J. G.
O. B. G.

Preface

TO THE FIRST EDITION

This book was written with two firm convictions. The first is that the general student is just as intelligent, just as responsive, as the preprofessional student, and the second is that such universal biological phenomena as respiration, osmosis, and energy exchange are more interesting, more understandable, and more thought-provoking for the general student when examined in the context of his own body and life.

We believe that the general student, to be a well-informed citizen, should not only have an understanding of his own body, his own reproductive processes, and his own inheritance, but should also have a sufficient biological background to comprehend the problem of population control, the genetic effects of radiation, the implications of pollution, and the basic concepts of behavior. On the other hand, we do not feel that such technical details as the anatomy of a clam or the whole series of reactions involved in the tricarboxylic acid cycle are a necessary part of the core of knowledge of a banker, lawyer, merchant, or housewife.

It is obvious that to prepare such a book we have had to make many decisions as to what to include and what to leave out. While it is hardly to be expected that anyone will agree with us entirely regarding these decisions, in every case we have decided on the basis that this book was written for the student who wants to become an educated, intelligent citizen.

The preparation of a book of this kind would be impossible if it were not for the fine cooperation of others. In many ways, this portion of the manuscript is the most difficult of all to prepare, for we have to list those to whom we are grateful in some order while, in fact, they all share first place in our gratitude. Dr. M. Graham Netting made the extensive photographic file of the Carnegie

Museum available to us, and other members of the Carnegie staff, Drs. J. K. Doutt, L. K. Henry, and Clarence J. McCoy, also allowed us to use personal prints of their own. Mr. William C. Gamble, president, and Mr. Henry Gresham of Wards Natural Science Establishment, Inc., were most generous in providing us with illustrative material. Government photographs were made available to us by the Armed Forces Institute of Pathology and the United States Department of Agriculture. The W. Atlee Burpee Company gave permission to use photographs of theirs. Special illustrative material was prepared for or presented to us by Drs. Archie Carr, H. E. Huxley, William S. Justice, F. Wayne King, George B. Rabb, Paul Smith, E. G. F. Sauer, and W. C. Thomas, Sr. Dr. Glenn Jepsen gave us permission to reprint part of his essay on dinosaur extinction, and Dr. and Mrs. E. H. Colbert had copies of Mrs. Colbert's beautiful Paleozoic and Mesozoic panoramas prepared for us. The following publishers granted permission to copy or redraw certain figures of which they own the copyright: Scientific American, D. C. Heath and Company, W. H. Freeman Company, McGraw-Hill Book Company, Inc., Prentice-Hall Inc., W. B. Saunders, American Genetic Association, S. Karger–A. G. Basel, Cold Spring Harbor Laboratory, Charles C Thomas, Publisher, and Random House Publishing Company. All illustrations are acknowledged in the legends of the particular figure or plate. Mr. Paul Laessle gave us the benefit of his advice and assistance in the design of some of the text figures. The many fine drawings were prepared for this text by Mr. Peter Loewer. Drs. David Anthony, R. E. Perry, Mildred Griffith, and E. G. F. Sauer read portions of the manuscript to our profit. The preparation of the manuscript was made easier by the care and interest devoted to it by our three typists, Mrs. Phyllis Durell, Miss Janice Johnson, and Mrs. Sandra Smith. Finally, Mr. William Eastman, editor, and the staff of Macmillan have been more helpful to us than words can express. Despite all of this cooperation, this book surely contains its share of errors. These are our own.

C. J. G.
O. B. G.

Contents

III INHERITANCE—THE MECHANISM OF VARIATION

IV THE DIVERSITY OF LIFE

V THE HISTORY OF LIFE

VI THE WORLD TODAY

An Open Letter

TO THE STUDENT

Why are you studying biology? Perhaps it is because so much that is biological has appeared in newspapers and magazines lately—RNA as a "memory molecule," the rejection reaction in organ transplants, DNA and its synthesis in a test tube. Your curiosity may have been roused and you decided you wanted to find out more about these topics. Perhaps someone in authority decreed that you must take at least one basic science course and biology happened to fit into your schedule better than chemistry or physics. Perhaps the authority even required that you take biology. You may be taking it for any one of a number of different reasons, but you are not planning to major in it. You do not expect to become a professional biologist or to go into one of the related fields such as medicine or agriculture. We have written this book for you.

We have written it because we believe that some knowledge of biology should be part of the mental equipment of every educated person. We believe you should know something of the structure and functioning of your own body. You should know it not just for practical reasons, although surely it is one of the most practical things a man can know. But it is very much more. Self-knowledge is said to be the beginning of wisdom. You cannot know yourself without knowing your body.

You also need a knowledge of biology because of your obligations as a citizen. You, and the other members of your generation, face problems as pressing, as terrible, as any generation has ever faced. The population explosion, the destruction of the environment, the effects of atomic radiation, race relations—these are all basically biological problems. We cannot offer solutions to these problems, but we hope that a knowledge of biology will provide you with information that will help you in your search for solutions.

Finally, we believe that one primary objective of your education should be to give you a conception of the universe and your place in it. The great contribution of biology to philosophy has been the idea of evolution. You cannot really understand yourself and the world about you unless you have some idea of how both came to be that way.

For all these reasons we think that you should study biology. But we do not believe that you should study the same biology as the student who is planning to major in it. We do not feel that a detailed knowledge of the anatomy of a crayfish or of the chemical stages by which an organism changes sugar into carbon dioxide and water would contribute much to your education. This does not mean that you will be offered a watered-down, simplified biology. We have omitted, or passed over lightly, some topics that are usually covered in great detail in introductory biology courses. But we have treated other topics, such as the structure and functioning of the human body, in far greater detail.

This book is not only for you, it is about you—what you are, how you evolved, your environment, and how you both affect and are affected by it. We hope you will enjoy it and find it profitable.

CHAPTER **1**

Introduction—The Organism

Biology is the science of life. The trouble with this statement is that not even biologists can agree on a mutually satisfactory definition of life. It is too diverse, its manifestations too various, to be condensed within a single brief statement. We do not attempt to define life here. We simply say that for our purposes we consider life to be the sum total of the activities of organisms and then try to indicate what it is that makes something an organism. A tree, a bacterium, and a man are all organisms. What do they have in common that allows us to lump them together? How do they differ from a rock, a cloud, or a river?

THE ORGANISM

Like the rock and cloud and river, an organism is composed of molecules, some of which are the same as the molecules found in the nonliving environment. Like the cloud and river, a man contains many molecules of water. The molecules that organisms share with the nonorganic world are usually relatively simple ones, composed of a few atoms. But all organisms, even the simplest ones, also contain certain very large and highly complex molecules comprising thousands of atoms arranged in an intricate, orderly, very precise pattern. These molecules are not found in the nonorganic world—they are present only in organisms and in the products of organisms. An organism, then, is a chemical mixture containing certain complex molecules that are peculiar to the living world.

A mixture of chemicals, though, does not make an organism. If you took all the chemical constituents of a dog in exactly the right proportions and dumped them together in a bowl they would not make a dog. An essential aspect of an organism is the way in which the various component molecules are arranged. If the arrangement is destroyed, the organism is destroyed, though all the parts are still there. This arrangement of molecules maintains itself as a discrete entity, occupying a given space, sharply delimited from the world around it. And it maintains itself in time. The constituent molecules and atoms may change, but the arrangement persists, for a few hours, a few days, or many years.

Still, nothing has been said that will allow you to distinguish between a dog and a wooden table. The table is an entity existing in space and time. It contains highly complex molecules (they were once part of a living organism, a tree). These molecules are arranged in an orderly way that you cannot destroy without destroying the table. Yet the table is not an organism. Why? The answer lies in chemical activity. The various molecules that make up an organism are constantly interacting with one another. They separate and recombine, their constituent atoms are shifted in position. Large molecules are broken apart into smaller units and the parts put together in a different way to form new molecules. The chemical reactions that take place within an organism are like those that take place in the nonliving world. But the rate at which they take place is very different. Remember that the molecules in a mixture are in a state of constant motion, and that for two molecules to interact they must come in contact with each other. The rate of reaction can be increased if the concentration of the molecules is increased, because the more there are, the more chance there is that they will come in contact. The same thing can be done by putting pressure on the mixture, because this pushes the molecules closer together and in effect increases the concentration. Or you can heat the mixture and thereby speed up the rate at which the molecules are moving and so increase the chances of collision. Given the concentration of the various molecules in an organism, and the conditions of temperature and pressure under which organisms live, most of their chemical reactions should take place so slowly as to be imperceptible. In point of fact, they take place very rapidly. This is made possible by the presence of certain complex molecules called *enzymes,* which have the ability to alter the rate of reactions. All the various chemical activities of an organism are lumped together under the term *metabolism.* Here is a major difference between the table and the dog—the dog is carrying on metabolic activities, the table is not.

Characteristics of Organisms

As a result of its metabolic activities, an organism exhibits three characteristics not shown by nonliving things. These are growth, responsiveness, and reproduction.

Growth. When biologists speak of the growth of an organism, they generally have in mind more than mere increase in size. The organism is able to take in substances from the environment. It can break down some of the molecules it acquires in this way to release the energy stored in their chemical bonds. By this means it provides itself with the energy it needs for its activities. But it can also reorganize some of the molecules after its own particular pattern and build them into its substance. It may use the reconstituted molecules simply for repair, to replace other molecules that have been destroyed by the wear and tear of existence. But if all the reconstituted molecules are not needed for repair, the organism will grow. A crystal can grow by attracting to itself similar molecules from the environment; it grows from the outside. But only an organism can take in and remake molecules after its own pattern and grow, as it were, from the inside.

Responsiveness. Responsiveness means that an organism is able to react to changes in the environment. Now in a sense reactivity is one of the basic laws of matter. Suppose a gust of wind comes in an open window by the desk where you are studying, picks up a sheet of paper, and wafts it to the floor. Given the weight and shape of that particular piece of paper, how far it travels is determined by the energy imparted to it by the moving particles of air. You look at the paper, bend down and pick it up, and then get up and close the window. The energy you expend is out of proportion to the energy of the stimulus you received; it comes from within you. On the other hand, you may decide to do nothing at all. The response of an organism to a stimulus is determined more by the organism than by the stimulus. Furthermore, the responses of an organism are by and large adaptive. They serve either to protect the organism from destructive forces in the environment or to enable it to take what it needs from the environment. When biologists speak of responsiveness they are referring to these adaptive, organism-determined reactions.

Reproduction. Not all the material an organism takes in from the environment is used for its maintenance and growth. Some of this material usually goes into the production of offspring, of other discrete organisms. Because of this, all the living matter of the world is not gathered into one huge superorganism but is divided into untold billions of separate organisms. Moreover, offspring not only resemble their parents they also tend to differ from them. These differences have provided the raw material for evolution, for the development of well over a million different kinds of organisms, of different ways of meeting the problems of existence.

Chemical Constituents of Organisms

A little over a hundred years ago the word *protoplasm* came into use to denote the material of which organisms are composed. It has been defined simply as "the living stuff." In those days little was known of the structural organization or chemical composition of protoplasm. Today biologists recognize that in a chemical sense the term is imprecise. Protoplasm is a complex mixture that differs from one individual to another, from time to time in the same individual, and even from one part of the body to another. Yet in spite of these differences, there is an overriding, remarkable similarity in the chemical constituents and organization of the protoplasm of the most diverse organisms. So biologists continue to use the word, recognizing the unity that underlies the diversity.

There are in protoplasm approximately twenty different elements, although many of them are present only in minimal amounts. Oxygen makes up about 62 per cent by weight of protoplasm, carbon about 18 per cent, hydrogen 10 per cent, and nitrogen 3 per cent. Calcium, iron, potassium, phosphorus, sulfur, and so on are present in smaller amounts. Usually the elements are combined into compounds.

Inorganic Compounds. Inorganic compounds are compounds that do not contain carbon. The most important inorganic compound in protoplasm is water, which makes up about two thirds of the weight of the human body. Many of the other constituents of protoplasm are dissolved or suspended in water; salts are ionized by it and become chemically reactive; and the water molecules themselves enter into many of the chemical activities of protoplasm.

Organic Compounds. Originally the term *organic chemistry* referred to the chemistry of substances found in living things. Now it is used for the study of carbon compounds whether or not they occur in organisms. Nylon, for example, is a man-made carbon compound that is not present in nature. Many simple carbon compounds are found in living beings. Thus, carbonic acid is an important constituent of blood.

Biochemical Compounds. Some large and complex organic compounds are characteristic of all organisms and are not found in nature except in organisms. They fall into four major groups: carbohydrates, lipids, proteins, and nucleic acids. The basic building blocks of these molecules are shown in Appendix A. Students who have not had any organic chemistry should read this appendix before going further.

CARBOHYDRATES. The basic carbohydrates of organisms are called *simple sugars*, or *monosaccharides*. A simple sugar is one that cannot be broken down

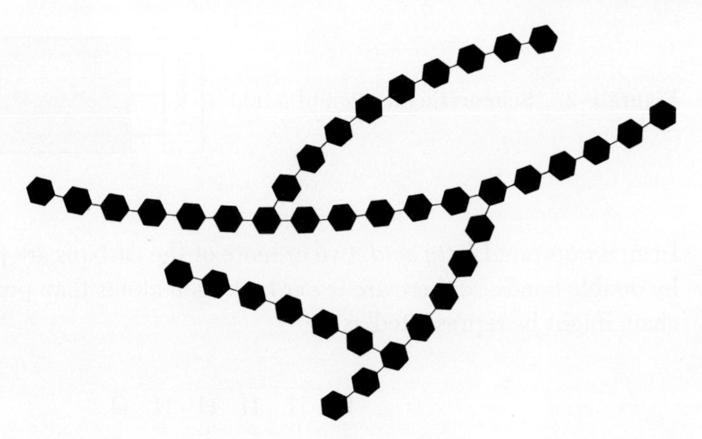

Figure 1–1. Diagram of part of a starch molecule. It is a branched chain of linked sugar molecules.

to form smaller sugar molecules. A simple sugar can be oxidized to furnish energy for the organism or can be used in building other organic molecules. When two simple sugars come together in an organism, they may combine by condensation to form a double sugar, or *disaccharide*. This process of adding sugars can be continued indefinitely until a very long, sometimes branching chain is formed. Such chains of similar units are called *polymers*. Starch, which is made by plants, and *glycogen*, the so-called animal starch, are *polysaccharides*, polymers of simple sugar molecules. They are storage molecules, holding sugars in reserve until they are needed by the organism. As they were built up by the removal of water, they can be broken down by the addition of water (*hydrolysis*). Carbohydrates are also important structural units of organisms, particularly of plants, which are made up largely of the carbohydrate cellulose. Figure 1–1 is a schematic diagram of part of a starch molecule.

LIPIDS. Lipids, like carbohydrates, contain carbon, hydrogen, and oxygen, but they differ in having less oxygen for the amount of hydrogen. Lipids are only very slightly soluble in water. Structurally they are a heterogeneous group, including fats, phospholipids, waxes, and steroids.

The building blocks of a fat molecule are three fatty acid chains and one molecule of glycerol. The fatty acids differ in length and also in whether they are saturated or unsaturated. A *saturated fatty acid* has as many hydrogen atoms as possible. A part of the chain can be pictured as:

```
    H  H  H  H  H
    |  |  |  |  |
H—C—C—C—C—C—
    |  |  |  |  |
    H  H  H  H  H
```

Figure 1–2. Schematic diagram of a fat.

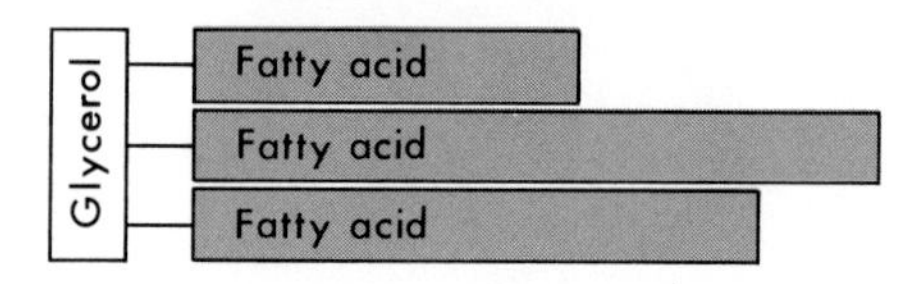

In an *unsaturated fatty acid,* two or more of the carbons are joined to each other by double bonds, so there are fewer hydrogen atoms than possible. A part of the chain might be represented as:

```
  H   H   H   H   H   H
  |   |   |   |   |   |
—C—C=C—C=C—C—
  |                   |
  H                   H
```

Just as the simple sugars are bonded together by the removal of water, so are fatty acids bonded to glycerol. Figure 1–2 is a schematic diagram of a fat. Highly unsaturated fats, those that have many double-bonded carbons, are liquid at room temperature and are usually called oils. Like carbohydrates, fats serve for energy storage. Because they contain proportionately more hydrogen than carbohydrates do, fats release more energy when oxidized.

Phospholipids are fats in which one fatty acid chain is replaced by a phosphoric acid group attached to a nitrogen-containing group. Many structures of the organism are composed partly of phospholipids.

Waxes are made of fatty acids joined to an alcohol larger than glycerol. Plant waxes form a protective covering for the leaves, stems, and fruits; they are used by man in the manufacture of many of the better grade shoe, floor, furniture, and automobile polishes.

Steroids are entirely different structurally from the other lipids. A steroid consists of four interlocked carbon rings with a short side chain. *Cholesterol* is a steroid that is an essential structural unit of the body. But when the diet is high in saturated fats, cholesterol tends to accumulate in the walls of blood vessels, making them narrow and reducing the blood supply to the various parts of the body. This is why most biologists believe that unsaturated plant fats are better as food for man than the more saturated animal fats that most people prefer.

PROTEINS. Proteins are polymers of a group of organic acids called *amino acids* (see Figure 1–3). Each amino acid is built around a carbon atom. One of the four bonds of this carbon atom is attached to a hydrogen atom, one to an amino group, and one to a carboxyl group. To the fourth bond is attached a side

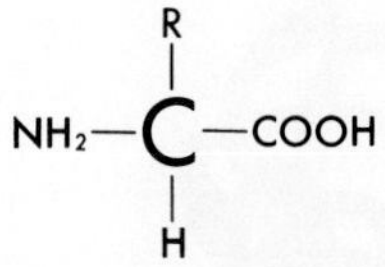

$$NH_2-\underset{\displaystyle H}{\overset{\displaystyle R}{\overset{|}{\underset{|}{C}}}}-COOH$$

Figure 1–3. The structure of an amino acid. R stands for any one of a number of possible side chains.

chain, which varies from one amino acid to another. As with sugars and fats, amino acids are linked together by condensation, the carboxyl group of one amino acid donating the hydroxyl and the amino group of another donating the hydrogen to form the water molecule. The bond that forms between the carbon of the first amino acid and the nitrogen of the second amino acid is called a *peptide bond,* and chains of amino acids held together by peptide bonds are known as *polypeptides.* Proteins consist of one or more long polypeptide chains, comprising from fifty to several thousand amino acids. Twenty different kinds of amino acids are commonly found in proteins. There is some constraint in the order in which they are joined. An amino acid with a short side chain is more likely to be found next to another amino acid having a small chain than it is to be next to an amino acid with a long chain. Still, the permutations and combinations of amino acids make possible an incredibly large number of different proteins—no one knows how many actually exist.

The sequence of amino acids is not all there is to protein structure. Chemical bonds may form between various parts of the polypeptide chains. One of the important amino acids is *cysteine,* in which the side chain has a sulfhydral (SH) group. If two of these groups are brought together and the hydrogens removed by oxidation, a covalent bond forms between the two sulfurs. If both cysteines are in the same polypeptide chain, a loop forms in the chain. If they are in different chains, the two chains are held together by the disulfide bond.

A hydrogen bond may form between the amino end of one amino acid and the carboxyl end of another amino acid further along the chain. Frequently, a single chain is twisted around itself into a helical shape like a coiled spring with the loops of the coils held together by such hydrogen bonds. Several polypeptide chains may twist together to form a ropelike molecule, the strands being linked by a series of hydrogen bonds between the chains. Protein molecules such as these form long, linear structures that are relatively insoluble in water and are usually not very reactive chemically. They are called *fibrous proteins* and are important structural parts of the body. Some of these proteins are apparently able to carry on enzymatic activity.

Usually, the polypeptide chain is folded and convoluted to form a rounded, tangled mass, held in shape by various kinds of chemical bonds between the parts of the chain. Sometimes several chains are folded together. Proteins that take a rounded shape are called *globular proteins.* They are soluble

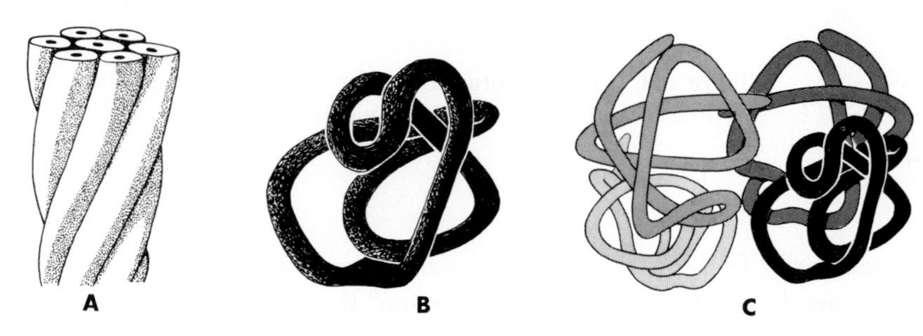

Figure 1–4. Diagram of protein structures. A: A fibrous protein. Each strand represents a coiled polypeptide chain. B: A globular protein. C: A globular protein formed of several polypeptide chains.

and usually very reactive chemically. Most enzymes are globular proteins. Figure 1–4 shows schematic diagrams of a fiber formed of fibrous protein strands, of a globular protein, and of a protein formed of several linked polypeptide chains.

NUCLEIC ACIDS. The nucleic acids, familiarly known as DNA (*deoxyribonucleic acid*) and RNA (*ribonucleic acid*), are polymers of nucleotides. They determine what proteins an organism is able to make and hence what chemical reactions it can carry on. Moreover, DNA is the substance that passes the ability to synthesize the same proteins from parent to offspring. We discuss how nucleic acids are formed and how they function in Chapter 16.

SOME OTHER IMPORTANT MOLECULES. In order to carry on their metabolic activities, all organisms need a source of energy. Green plants use the energy of sunlight; animals, the chemical bond energy in the food they eat. In neither case, though, is the energy used directly. There are, in all organisms, molecules of *adenosine diphosphate* (ADP). Each consists of a molecule of adenine (Figure A–9) joined to a five-carbon sugar molecule, which is joined to two phosphate groups. By a complicated series of steps (see Chapter 6), energy is used to join a third phosphate group (indicated as Ⓟ) to ADP to produce *adenosine triphosphate*, ATP. When energy is needed, this bond is broken and the energy is released. The resultant ADP can then be recombined with another phosphate group and so used over and over again. ATP is thus a molecule that serves for temporary storage, providing a readily available source of energy that can be released when it is needed and in the amount needed. The source of energy may be different, but so far as is known, all organisms, from bacterium to tree to man, make use of the ATP-ADP mechanism. Figure 1–5 shows the cycle as it occurs in relation to the oxidation of a sugar molecule (glucose).

Another important molecule is *cyclic adenosine monophosphate* (cyclic

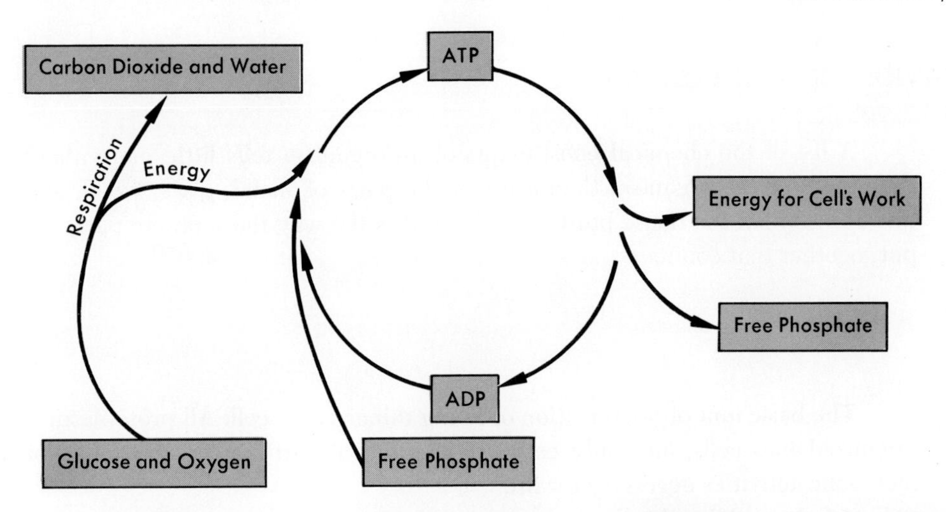

Figure 1–5. The ATP–ADP cycle in the temporary storage and release of energy.

AMP), which can be formed from ATP. The cell does not carry on all the processes it is capable of simultaneously. Many of its enzymes are present in an inactive state but they can be activated by cyclic AMP. In turn, the formation of cyclic AMP can be regulated by hormones (see Chapter 10).

A great many of the chemical reactions of organisms are *oxidation-reductions.* They involve the removal of hydrogen, or of electrons, from one substance (oxidation) and the addition of them to another (reduction). There are in the organism several types of molecules called *hydrogen acceptors.* They pick up hydrogen and hold it in temporary storage. When hydrogen is needed for a reduction reaction, they release it, thus becoming hydrogen donors. Their full names are rather formidable and they are usually indicated simply by their initials. Three of the most important hydrogen acceptors are

NADP—nicotinamide adenine dinucleotide phosphate.
NAD—nicotinamide adenine dinucleotide.
FAD—flavin-adenine dinucleotide.

Structurally these molecules are related to ATP; like it they can be used over and over again, and like ATP, they may be thought of as temporary storage molecules.

There are also electron acceptor-donor molecules in the organism, notably the *cytochromes,* which are proteins joined to pigment molecules.

ORGANIZATION OF THE INDIVIDUAL

A list of the chemical constituents of an organism tells little more about the nature of the organism than a list of the parts of metal, glass, wood, and brick tells about the house built from them. It is the way the separate parts are put together that counts.

The basic unit of organization of living things is the cell. All protoplasm is organized into cells, and only cells are capable of carrying on the range of metabolic activities necessary for life.

A few specialized cells, such as the yolk of an egg, are large, but most can be seen only with a microscope. For this reason, knowledge of cells had to await the development of the microscope. The first known observations on cellular structure were made by Robert Hooke in 1665. With a primitive microscope, Hooke studied the structure of cork and other plant tissues. He saw that cork is divided into small spaces that look like little rooms, which he accordingly called *cells*. Later in the seventeenth century, the Italian Malpighi made additional microscopic studies of insects, plants, and human tissues. He also remarked on the "repeated vesicles" that seemed to make up many of the tissues he examined. With additional refinements of the microscope, knowledge of cellular structure developed until, in 1839, Schwann announced the hypothesis that all organized bodies are composed of essentially similar parts—namely cells. This original hypothesis has been modified and extended and is no longer thought of as an hypothesis, but rather as the cell principle: "All living things are composed of cells and cell products."

Individual cells differ enormously from one another, not only from organism to organism but also from one part of the body to another. There is no such thing as a typical cell. But there are certain features that are common to all, or to most, cells. Figure 1–6 is a composite picture of an animal cell showing these features. The cell is bounded by the *cell* or *plasma membrane*. Within the membrane is the *cytoplasm* in which are suspended a number of structures. The largest and most conspicuous of these, in most cells, is the *nucleus*. Other, smaller structures, called *organelles*, are scattered through the cytoplasm. And here the facile analogy of cell and building breaks down. For the parts of the cell are much more than structural units. They are dynamic, functional centers, cook and kitchen combined, dining room and diners. These organelles of the cell may move around and change in shape; they sometimes appear to break

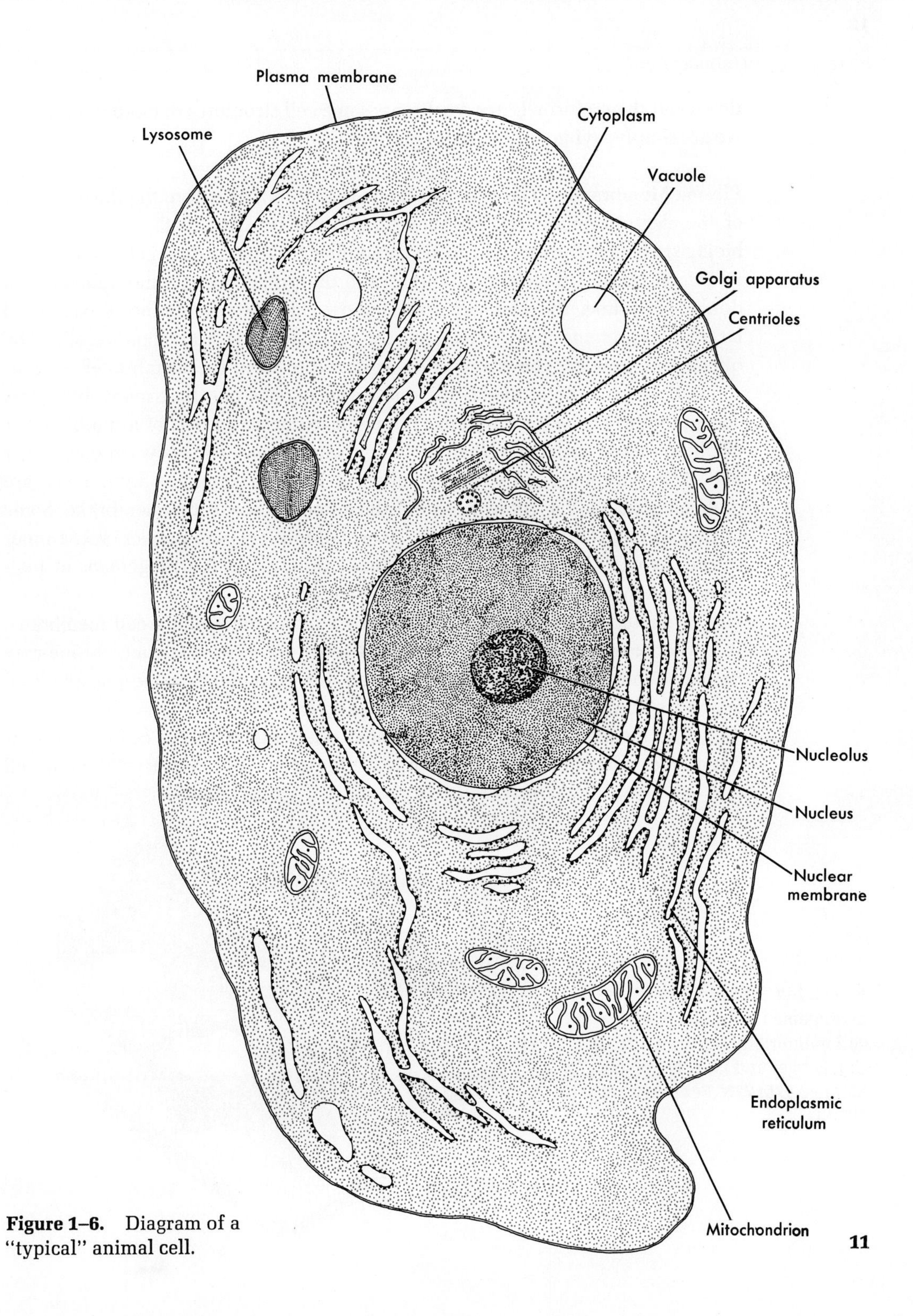

Figure 1–6. Diagram of a "typical" animal cell.

down and then reform. Later we discuss some cell structures in more detail; here we are simply laying out the basic pattern.

Plasma Membrane. The cell membrane is so thin that before the development of the electron microscope it usually could not be seen at all, although most biologists believed there had to be something to separate the cell from its environment. Under the electron microscope the plasma membrane appears as a double dark line. Biochemical studies show that the membrane is composed largely of proteins and lipids (mostly phospholipids and cholesterol). The question is, how are these molecules arranged to form the membrane? Various suggestions have been made; the one given here is one that many biologists accept today. The cell membrane is thought to consist in part of a double layer of phospholipid molecules, their fatty acid chains projecting toward each other and their phosphate ends facing outward. Embedded in the lipid layer are globular proteins that project beyond the outer surface of the membrane. Some of these proteins extend through the phospholipid layers to project on the inner surface as well. Figure 1–7 is a diagram of the proposed arrangement of molecules in a membrane.

It should be stressed that this is not an actual picture of a cell membrane. It is what scientists call a *model,* a mental construct of what the membrane may be like. Other models have been proposed, for example, that the phospholipid layers are sandwiched between continuous layers of protein.

The membrane does more than hold the contents of the cell together. It determines what substances enter or leave the cell. This may vary from one cell

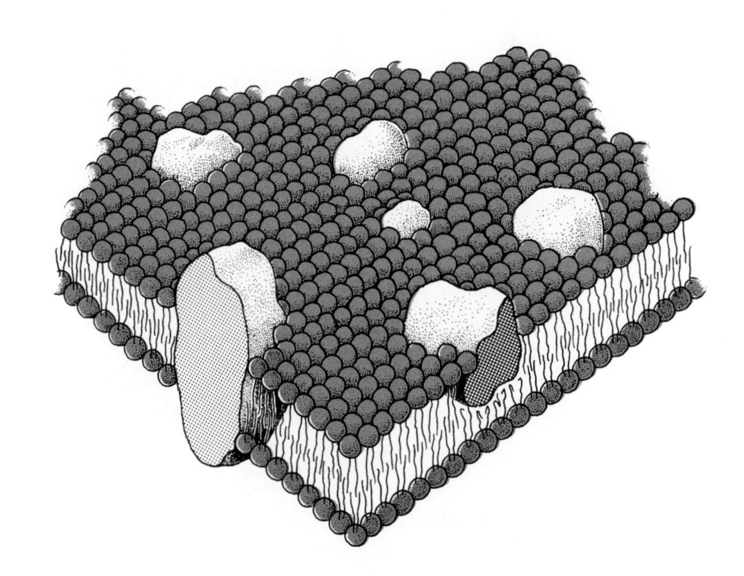

Figure 1–7. Model of the arrangement of molecules in a cell membrane. (Redrawn from S. J. Singer and G. L. Nicolson, *Science* **175**: 723, February 18, 1972.)

type to another, from time to time in the same cell, and even between different parts of the same cell. The most acceptable model of a membrane will be the one that accounts most successfully for all membrane activities. It may be that no single model will suffice.

Cytoplasm. The cytoplasm is no mere amorphous ground substance. It contains a number of *microfilaments* of protein that are so fine they can be seen only under the electron microscope. *Microtubules,* apparently formed of aggregates of ten to fourteen microfilaments, are also present. These microfilaments and microtubules apparently play a diversity of roles in the cell. The cytoplasm can flow like a liquid or it can "set" like gelatin into a more solid state. It can pass from one state to another, and within a single cell one part of the cytoplasm can be stiffer than another part. Microtubules seem to be more numerous in the stiffer regions and may be responsible for changes in solidity. They play a part in the movements of which many cells are capable. Both microfilaments and microtubules are involved in the processes of cell division.

Nucleus. The nucleus is a relatively large structure, usually rounded and frequently found near the center of the cell. It is surrounded by a double membrane and contains a type of protoplasm called *nucleoplasm.* In it lie long, tangled threads composed of nucleic acids and proteins, the *chromatin network.* At the time of cell division, the threads forming the chromatin network appear as discrete, rod-shaped bodies, the *chromosomes.* Within the nucleus are one or more smaller bodies, the *nucleoli.* We discuss the chromosomes and nucleoli further in the sections on cellular reproduction and genetics.

Endoplasmic Reticulum. When a cell is examined with an electron microscope, the endoplasmic reticulum (ER) frequently appears as one of the most conspicuous elements. It is a labyrinthine system of membrane-lined channels that meander through the cytoplasm. Apparently, the ER is derived from invaginations of the plasma membrane and gives rise to the nuclear membrane. The ER is in constant flux. New parts form, fuse with other parts, and break down again or bud off as separate vesicles. The ER may be either rough or smooth. Rough ER is encrusted on the outer surface with minute, dense particles called ribosomes, the sites of protein synthesis. Rough ER is best developed in cells that actively secrete proteins, such as the cells of the pancreas that produce digestive enzymes. On the other hand, lipids are synthesized in smooth ER, which is well developed, for example, in the cells that line the intestine. The ER in liver cells is able to detoxify poisons such as phenobarbital by breaking them down. The channels in the ER serve as pathways for the movement of molecules from one part of the cell to another.

Golgi Apparatus. The Golgi apparatus is a membranous structure associated with the endoplasmic reticulum. It is usually located close to the nucleus and looks like a stack of flattened vesicles and small sacs. Some of the proteins produced by the rough ER pass to the Golgi apparatus, where they are packaged in small membrane-bound vesicles. These bud off and either move to the plasma membrane, fuse with it, and discharge their content to the outside, or form lysosomes. The Golgi apparatus also apparently makes polysaccharides from sugars and may join them to proteins.

Lysosomes. The lysosomes are digestive organelles distributed through the cytoplasm of many cells. The proteins they contain are enzymes involved in the breakdown of complex molecules into simpler ones. They may fuse with other vesicles containing substances to be digested or take in molecules through their membranes. They do not digest the protoplasm of the cell because they are walled off from it by their membranes. But if these membranes rupture, the enzymes are released and digest the cell protoplasm. This apparently accounts for the dissolution of old and dead cells.

Peroxisomes. Peroxisomes are like little lysosomes, membrane-bound vesicles containing enzymes. They seem to bud off from the endoplasmic reticulum. Some of them contain three enzymes that catalyze reactions producing hydrogen peroxide (H_2O_2), and a fourth enzyme that destroys this very toxic substance. Other peroxisomes change fat to carbohydrate.

Mitochondria. Practically all organisms contain rod-shaped or globular structures called *mitochondria.* An individual mitochondrion consists of a matrix surrounded by two membranes, an outer one and an inner one from which extensive folds and convolutions known as *cristae* extend into the matrix. The cristae increase the amount of inner surface (see Figure 1–8). Mitochondria are rich in enzymes, most of which are lined up on the cristae with some in solution in the matrix. These organelles are known as the powerhouses of the cell, because they are the sites where most of the energy of food is transferred to ATP.

Plastids. Plant cells and many one-celled organisms contain structures known as *plastids.* A plastid is bounded by a membrane and its internal structure consists of stacks of membranes. Some plastids are colorless; they serve for the formation and storage of complex molecules, such as starch. Other plastids contain the pigments that give color to leaves, flowers, and fruit. Especially important are the chloroplasts, which house the green pigment chlorophyll. Here *photosynthesis,* the fundamental process of converting the energy of sunlight into the energy of biochemical compounds, takes place. Like the mitochon-

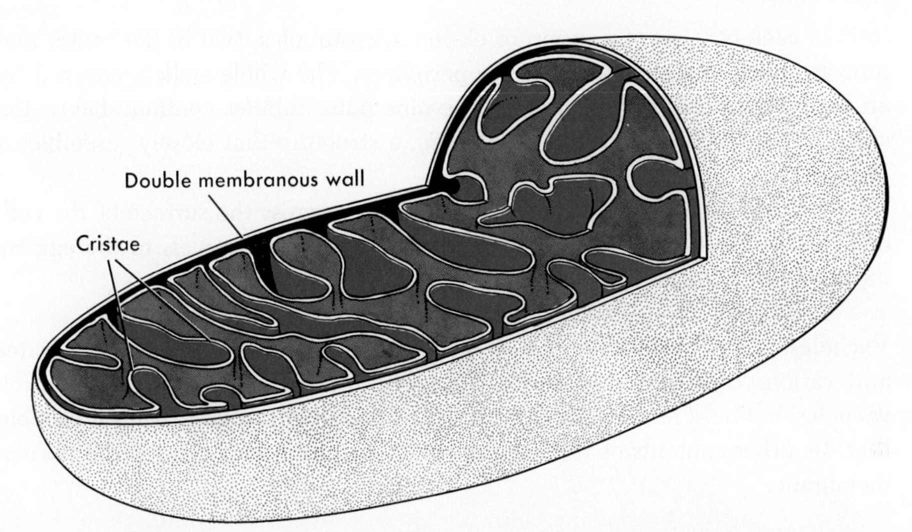

Figure 1–8. Diagram of a mitochondrion with part of the outer wall cut away to show the inner structure.

dria, chloroplasts can be thought of as cellular powerhouses. Both plastids and mitochondria contain DNA and ribosomes and are able to manufacture some of their own proteins.

Centrioles. The centrioles are a pair of small, cylindrical bodies that lie in the cytoplasm outside the nuclear membrane but adjacent to the nucleus and associated with it functionally. Each centriole is made up of a cluster of nine groups of microtubules. The elements of one centriole invariably lie at right angles to those of the other. Centrioles are typical of most animal cells but are usually not found in the cells of higher plants. They are involved in cell division.

Cilia and Flagella. Many cells have projecting, movable, hairlike structures. If these hairlike structures are short and numerous and beat with a simple bending movement like grass before the wind, they are called *cilia.* If they are longer and fewer in number and beat with an undulating motion, they are called *flagella* (see Figure 1–9). Structurally, cilia and flagella are very similar. The

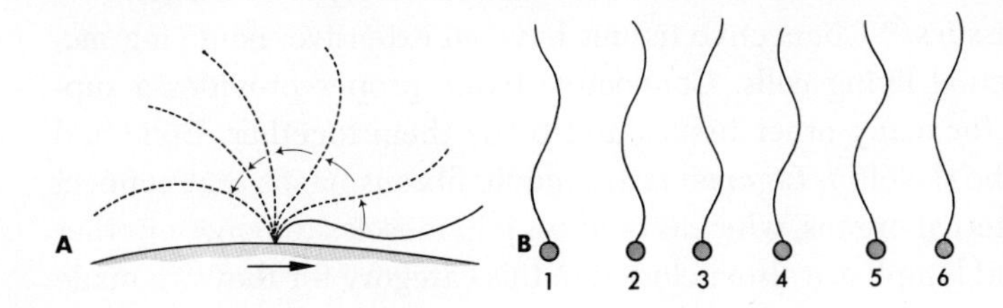

Figure 1–9. A: The beating movement of a cilium. B: The undulatory movement of a flagellum.

stalk in each consists of a group of eleven microtubules, two in the center and nine arranged in a circle around the periphery. The whole stalk is covered by an extension of the cell membrane. The nine outer tubules continue below the surface of the cell to form the *basal body,* a structure that closely resembles a centriole.

By their beating, cilia can move substances across the surface of the cell, or they can cause the cell itself to move. The sperm cell travels to the egg by means of a waving flagellum.

Vacuoles. Many cells have one or more membranous sacs that contain water and various dissolved substances. The cells of higher animals usually lack vacuoles, but most of the cells of higher plants have such a large central vacuole that the other contents of the cell appear to be pushed up against the plasma membrane.

From Cell to Organism

Thus far we have been speaking in very general terms, and most of what we have said applies to all, or nearly all, organisms. They are all similar in their chemical composition and they are all organized to the cellular level. Beyond this point we must be more specific. Some organisms exist as single cells or loosely knit colonies of cells. But most show patterns of organization beyond the cellular level; they are *multicellular,* composed of many cells. Here the differences become striking. Plants are not organized like animals, a worm is not organized like a man. For the rest of this section we are concerned with the organization of animals, and specifically of the higher animals, those that are most like man.

Cells into Tissues. From the discussion of the generalized "typical" cell it might seem that all the cells of the body are more or less alike. But this is not so. The various cells are specialized to perform particular functions. Some cells are able to contract, others manufacture and secrete various substances, and still others conduct stimuli from one region to another. Groups of cells that are similar in structure and function are known as *tissues,* which can be classed into four major types. Figure 1–10 shows a number of different types of tissues.

CONNECTIVE TISSUES. Connective tissues have an extensive, nonliving *matrix* between the actual living cells. Connective tissue proper provides a supporting framework for many other tissues and binds them together. Bone and cartilage make up the skeleton. *Ligaments* are tough, fibrous bands that connect bones or support internal organs, whereas *tendons* join muscles to bones or other structures. Blood and lymph are also included in this category for they are made

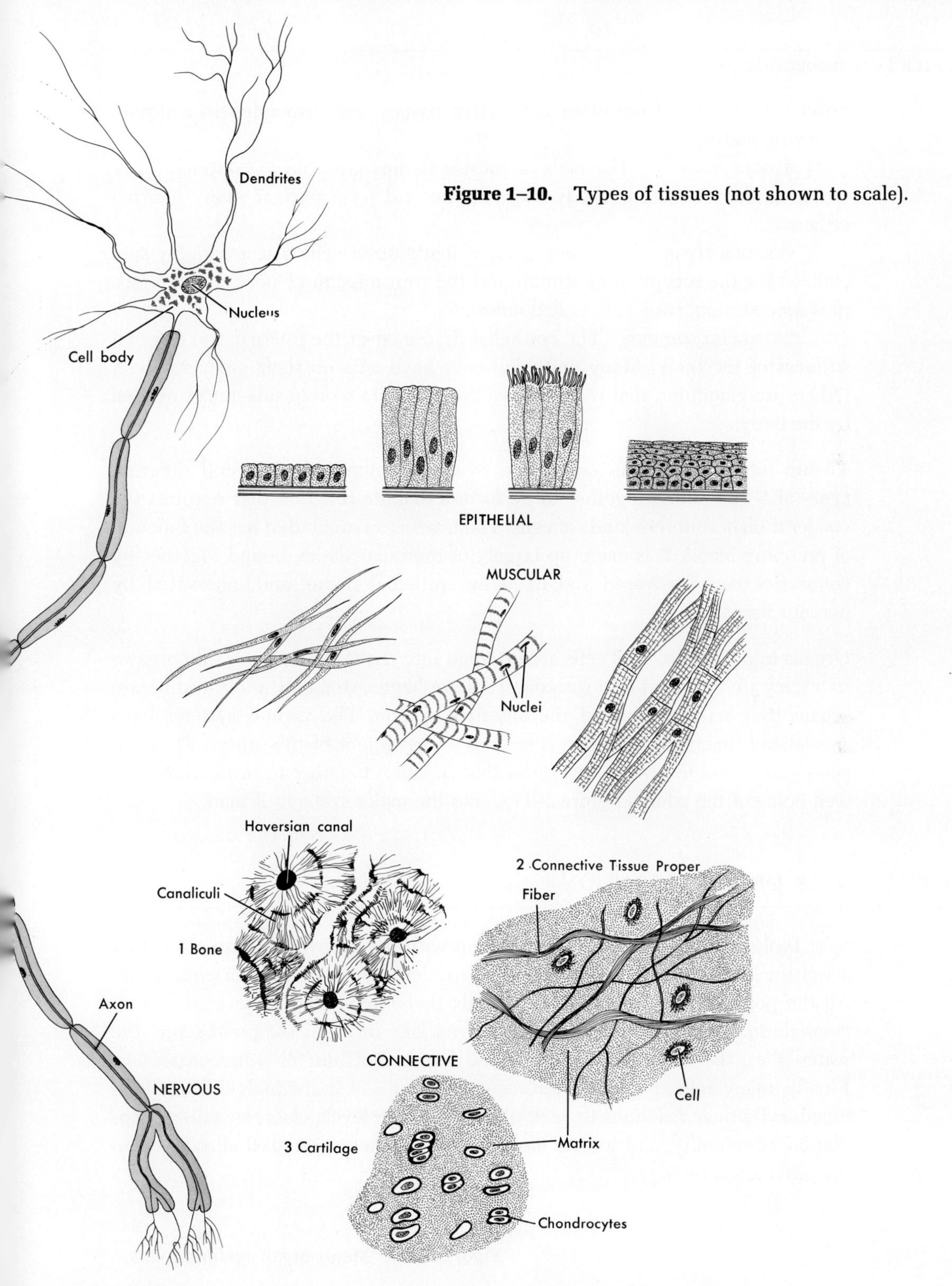

Figure 1–10. Types of tissues (not shown to scale).

up of cells derived from other connective tissues and suspended in a liquid, nonliving matrix.

MUSCLE TISSUES. The cells of muscle tissues are able to contract. They make up the bulk of the body musculature and form part of many internal organs.

NERVOUS TISSUES. Nervous tissue is made up of cells that are highly specialized for the reception of stimuli and the transmission of nervous impulses, plus special supporting cells called *neuroglia.*

EPITHELIAL TISSUES. The epithelial tissues cover the internal and external surfaces of the body. Many epithelial cells have cilia on their outer surfaces. Others are glandular, that is, they have the ability to secrete substances needed by the body.

Tissues into Organs. An *organ* is a structure composed of several different types of tissue joined together to perform a definite function that requires the cooperation of different kinds of cells. The heart is an organ that has the function of pumping blood. It is made up largely of muscular tissue, bound together by connective tissue, covered and lined by epithelial tissue, and innervated by nervous tissue.

Organs into Systems. As cells are grouped into tissues and tissues into organs, so organs are arranged into *systems.* The esophagus, stomach, and intestine are organs that make up part of the digestive system. The various systems have specialized functions, yet none is entirely independent of the others. They are coordinated and interrelated systems that function together to bring about the well-being of the whole. Figure 1–11 shows the major systems of man.

HIGHER LEVELS OF ORGANIZATION

Biological organization does not stop with the individual. Organisms that are of the same kind (belong to the same species) are members of a *population.* All the populations of plants and animals that inhabit the same area form a *community.* Thus the community in a given lake includes the pond scum, the water lilies, the mosquito wrigglers, the fish, and a host of other organisms. Finally, many animal species form *societies,* groups of individuals that are held together by their reactions to each other. All these levels of organization, population, community, and society have definite characteristics and all fall within the province of biology.

Figure 1–11. Major organ systems in man.

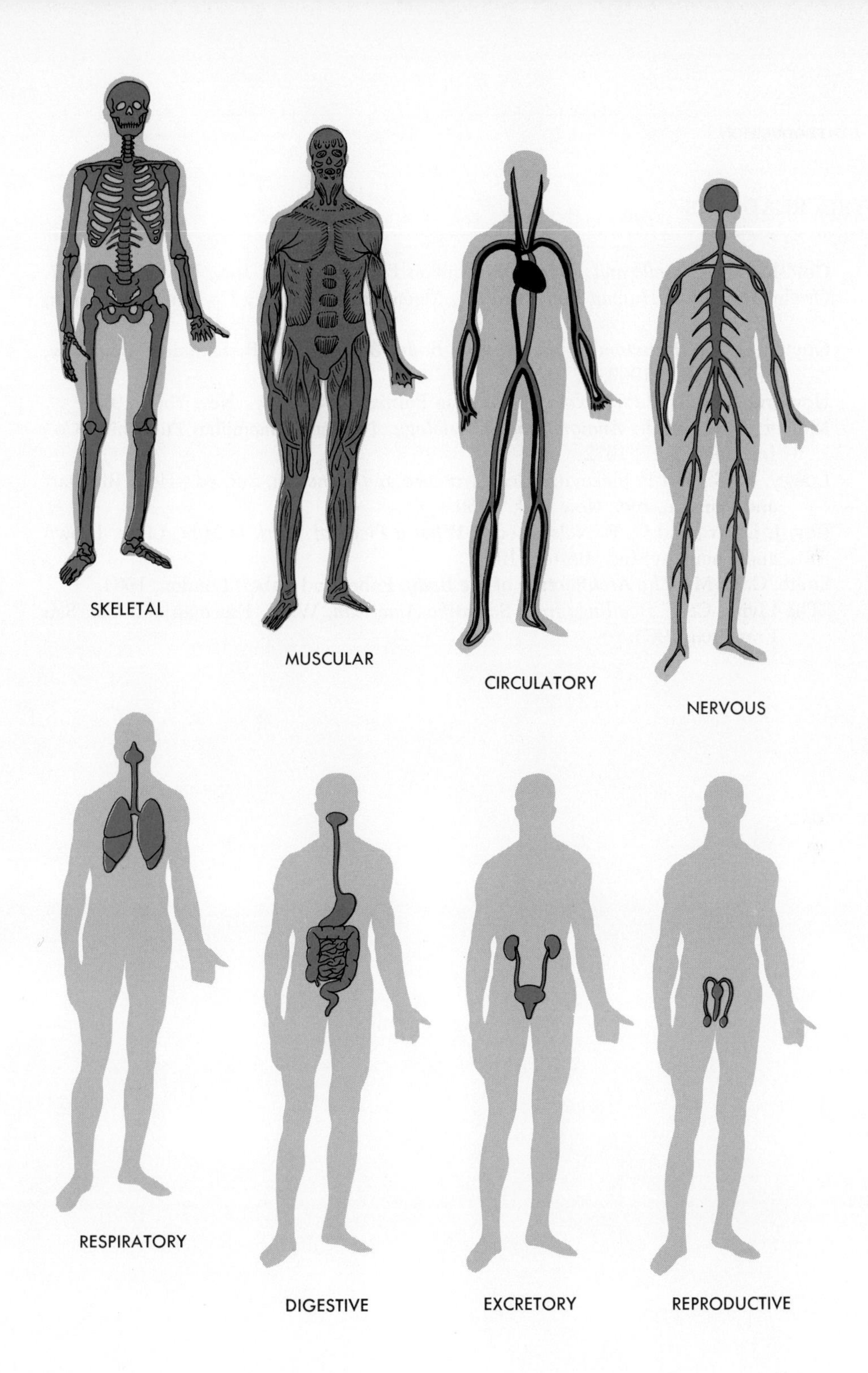
SKELETAL
MUSCULAR
CIRCULATORY
NERVOUS
RESPIRATORY
DIGESTIVE
EXCRETORY
REPRODUCTIVE

SUGGESTED READINGS

Goldsby, A. A.: *Cells and Energy,* Macmillan Publishing Co., Inc., New York, 1967.

Grollman, S.: *The Human Body,* 3rd ed., Macmillan Publishing Co., Inc., New York, 1974.

Guyton, A. C.: *Function of the Human Body,* 3rd ed., W. B. Saunders Company, Philadelphia, 1969.

Howland, J. L.: *Cell Physiology,* Macmillan Publishing Co., Inc., New York, 1973.

Kimber, D. C., et al.: *Anatomy and Physiology,* 16th ed., Macmillan Publishing Co., Inc., New York, 1972.

Loewy, A. G., and P. Siekevitz: *Cell Structure and Function,* 2nd ed., Holt, Rinehart and Winston, Inc., New York, 1969.

Ray, J. D., Jr., and G. E. Nelson, eds.: *What a Piece of Work Is Man,* Little, Brown and Company, Inc., Boston, 1971.

Smith, C. U. M.: *The Architecture of the Body,* Faber and Faber, London, 1964.

"The Living Cell," *Readings from Scientific American,* W. H. Freeman and Co., San Francisco, 1965.

The Human Body

I am fearfully and wonderfully made.
PSALMS *139:14*

CHAPTER 2

The Integument

The integument, the skin together with its various outgrowths, is the part of the body in contact with the external environment. It is not a simple, inert covering, like a coat of paint on the wall of a house. It is a complex and very active system with many functions, the source of much of man's information about his environment and the means by which he adjusts to many factors in that environment.

The skin covers all the outside of the body, folding inward to meet the tissues that line the various internal passages at the nose, the lips, the anus, and the openings of the urinary and reproductive tracts. The skin of a man nearly two meters (six feet) tall has an area of about two square meters (about three thousand square inches). The skin ranges in thickness from about 1.6 to 8.4 millimeters (1/16 to 1/3 inch). It is thinnest over the eyelids and thickest at the back of the neck, the upper back, the palm of the hand, and the sole of the foot. It has a tendency to thicken and toughen in places where it is pressed or rubbed against; it is a truism that the sole of the foot gets thicker as the sole of the shoe gets thinner.

The skin everywhere is patterned by ridges, furrows, and pores. Each part of the body has its own characteristic pattern—the skin of the elbow does not look like the skin of the back. The pattern on the hands and feet is also different for every individual and can thus be used as a basis for identification. Many hospitals take a footprint of a baby at birth. If later there is a question of whether babies have been mixed up (as has happened) the footprints provide a definite identification.

All humans have pigments in the skin unless they have an inherited defect that prevents the formation of pigment and thus causes the condition known as *albinism.* Two pigments are prominent, *melanin* and *carotene;* melanin gives a darkened color and carotene a yellowish color to the skin. Their relative amounts and distribution in any individual are determined by his inheritance. Melanin is a yellow to black pigment formed in special cells called *melanocytes.* It is able to absorb harmful ultraviolet rays and so protect the underlying tissues. Exposure to sunlight stimulates the cells to produce more melanin (you get a suntan). Human races that evolved in the tropics, where the rays of the sun are most direct, have more active melanocytes and hence darker skins than people of more temperate climes. Albinos lack an enzyme that converts the amino acid, *tyrosine,* to melanin.

Some of the color of the skin comes from the blood in the tiny vessels that course through it. If the blood is carrying a good supply of oxygen, the skin has a reddish tint, but if the oxygen content is low the skin appears bluish or purplish. A pink color is particularly evident in the skin of the face, neck, palms, soles, and nipples, regions that are richly supplied with vessels that carry oxygenated blood.

STRUCTURE OF THE SKIN

The skin is made up of two distinct layers, an outer epidermis and an inner dermis (see Figure 2–1).

Epidermis

The epidermis is composed of layers of epithelial cells; on most parts of the body this layer averages about 1/2 millimeter (1/50 inch) thick. The lower layer of the epidermis is known as the *stratum germinativum.* Here the epithelial cells form a compact sheet in close and inseparable contact with the dermis below. In contrast to many kinds of cells in the adult, these cells constantly undergo divisions to form new cells. This pushes the overlying cells outward. There are no blood vessels in the epidermis. As the cells are pushed away from the little blood vessels of the dermis from which they receive their nourishment, the supply of nutrients reaching them becomes inadequate, and the cells die. But before they die, these cells produce large stores of a fibrous protein called *keratin,* which accumulates in them and makes them tough and horny. These outer layers of dead cells that have become keratinized are known collectively as the *stratum corneum.* It is ordinarily eroded away from the body in little flakes that are scarcely noticeable, but if you get a severe sunburn it peels off in sheets. Thus the epidermis is constantly worn away and constantly renewed. It has been estimated that if a man lives his three score years and ten, he will in

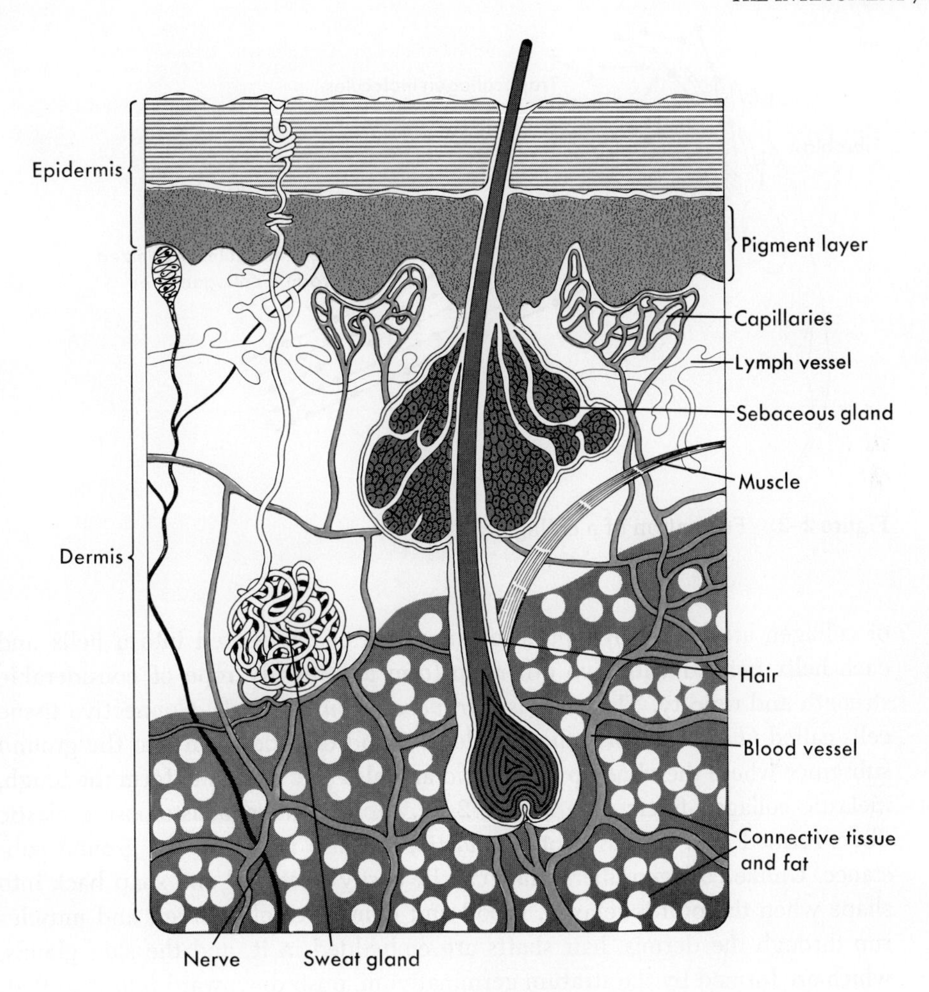

Figure 2–1. Section through the skin.

his lifetime shed twenty to twenty-five kilograms (forty to fifty pounds) of stratum corneum.

Dermis

The dermis, or lower layer of the skin, is made up of connective tissue. The nonliving ground substance, which is secreted by the cells, is composed of *mucopolysaccharide molecules,* long chain polysaccharides with groups containing nitrogen and sulfur attached to some of the subunits. Coursing through the ground substance in all directions are fibers of *collagen.* The building blocks

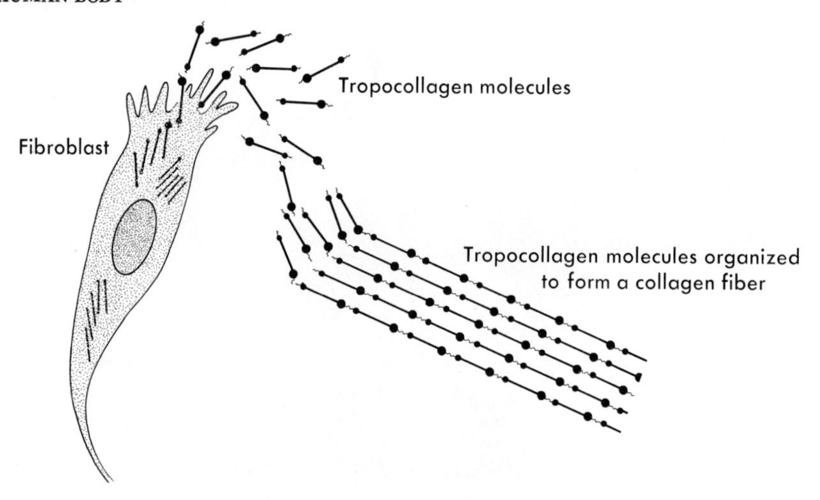

Figure 2–2. Formation of a collagen fiber.

of collagen are long polypeptide chains, each chain twisted into a helix and each helix twisted with two others to form a three-ply rope of considerable strength and rigidity. This rope is a *tropocollagen molecule.* Connective tissue cells called *fibroblasts* form these molecules and extrude them into the ground substance where they line up end to end and also side to side to form the tough, inelastic collagen fibers (see Figure 2–2). More slender, anastomosing elastic fibers formed of a fibrous protein, *elastin,* are also found in the ground substance. Unlike collagen fibers, they can be easily stretched and snap back into shape when the pull is relaxed. Blood and lymph vessels, nerves, and muscles run through the dermis, hair shafts are embedded in it, and the skin glands, which are formed by the stratum germinativum, push downward into it so that, except for their ducts, they are completely surrounded by it.

Separating the dermis from the underlying tissues is a subcutaneous layer of more loosely organized connective tissue in which the fibers are not so thick and numerous nor so closely interwoven. In many parts of the body, cells in this layer are specialized for the storage of fat.

THE GLANDS

A skin gland is composed of clusters of cells that produce a substance that is passed out through a tube, the duct. There are three main types of glands in the human skin: *sweat glands, sebaceous glands,* and *mammary glands.*

Sweat Glands

Sweat glands are of two different types, eccrine and apocrine (see Figure 2–3). They differ in their function, products, and distribution on the body.

Eccrine Glands. Eccrine glands are present all over the body except on the lips and some of the external genital organs. They are simple, coiled tubes with the coil buried deep in the dermis and the tube opening on the surface of the epidermis. The product of the eccrine glands is primarily water, with, in addition, about 0.5 per cent salt and 0.5 per cent organic compounds. The eccrine glands on the palms and soles are under psychic control; they are active at times of emotional stress, which is why your hands may get sweaty if you have to make a speech. Other eccrine glands are under thermal control; an increase

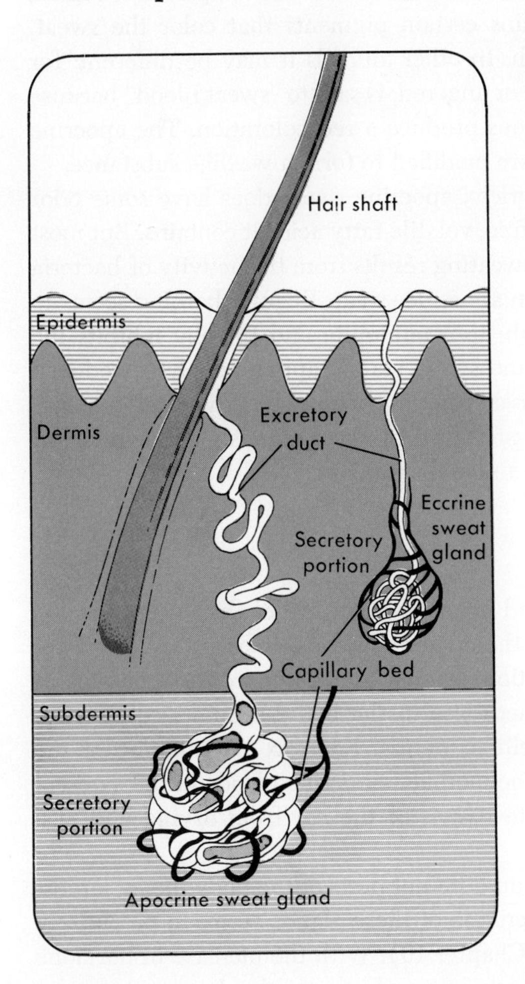

Figure 2–3. An apocrine and an eccrine sweat gland in the skin.

in temperature brings about prolific production by these glands, particularly those situated on the forehead, neck, and back.

Apocrine Sweat Glands. Apocrine sweat glands are not evenly distributed over the body but are found only in the armpits, the areas around the nipples and the navel, the region of the anus and external genital organs, in the external ear canal, and in the nasal passages. They are larger than the eccrine glands, their secretory portions are buried more deeply in the dermis, and their ducts, rather than opening on the surface of the epidermis, usually open into the canals from which the hairs grow. Apocrine sweat glands do not respond to changes in temperature, but instead seem to be entirely controlled by psychic factors. The material produced by these glands is quite different from that produced by the eccrine glands. It is a milky, sticky substance very rich in iron, proteins, sugars, lipids, and ammonia. It also contains certain pigments that color the sweat. Human sweat is generally yellowish. In other animals it may be different; for example, a hippopotamus, if excited or angered, is said to "sweat blood" because the pigments in its apocrine secretions produce a red coloration. The apocrine glands in the external ear passages are modified to form a waxlike substance.

Although eccrine sweat is odorless, apocrine sweat does have some odor of its own, primarily because of the free, volatile fatty acids it contains. But most of the body odor brought about by sweating results from the activity of bacteria that cause the decay of substances in apocrine sweat. Regular bathing seems to reduce this bacterial flora of the skin to a minimum and thereby inhibits the development of body odors. Apocrine sweat glands apparently have reduced function in modern man. In primitive societies, before the advent of bathing, perfumes, and deodorants, the odor produced by these glands may have served as a means of recognition and/or as a sexual stimulant.

Sebaceous Glands

Sebaceous glands are composed of many lobes rather than simple coiled tubes as are the sweat glands. Like the apocrine glands, they are usually associated with hairs. Their oily secretion, known as *sebum,* consists largely of waxes, which are synthesized nowhere else in the body; sebum also contains fatty acids, fats, cholesterol, and cellular debris. Ear wax is a mixture of the secretions of sebaceous glands and the modified apocrine glands in the external ear passages. It protects the delicate tissues of the ear from drying and also guards against invasion by insects.

Sebaceous glands differ in size on different parts of the body, being largest on the face, neck, and trunk. The secretion of these glands seems to be entirely under the control of hormones (see Chapter 10). With the increase in hormone production that takes place at puberty, the sebaceous glands may be over-activated. The ducts may become plugged, leading to blackhead formation. The glands may enlarge and be engorged with accumulated secretions, and may sometimes become infected, causing acne. Testosterone, a male hormone, has been implicated as one of the major causes of acne in adolescent boys, and it is believed that progesterone, a female hormone, is involved in bringing on acne in girls.

Sebum anoints the hairs and keeps them from growing brittle, it forms a protective film over the surface of the skin, and it may have bactericidal properties.

Mammary Glands

The mammary glands are modified apocrine glands that produce the milk with which babies are fed. In infancy and childhood the glands are similar in form and size in both sexes. At puberty, the female mammary glands enlarge and become firm and rounded, with the nipples centrally located. The adult female breast consists primarily of many small orange glands embedded in connective tissue that is filled with masses of lemon-yellow fat. Ducts from these glands join to form collecting ducts that open to the outside at the nipple (see Figure 2–4). The function of the mammary glands is under the control of the

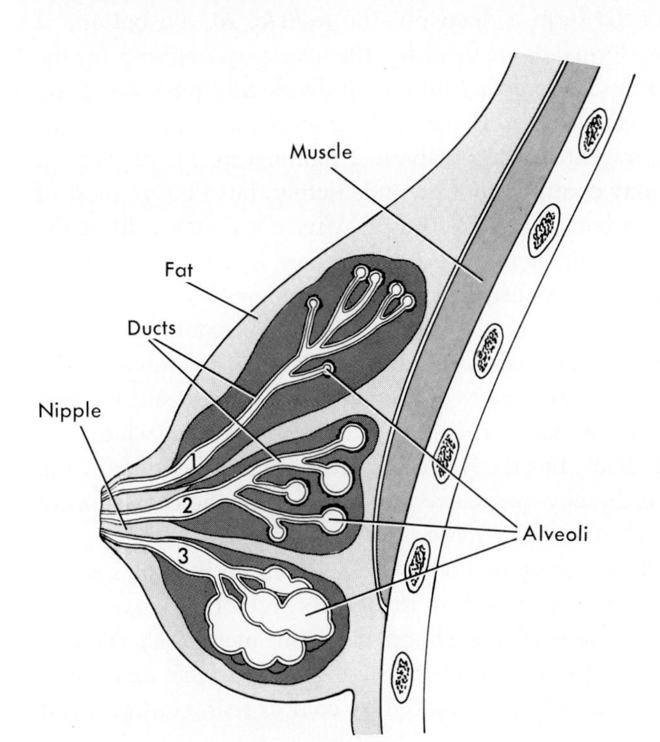

Figure 2–4. Structure of the adult female breast showing the glandular sacs (alveoli) and ducts as they appear (1) in a nonpregnant woman, (2) in mid-pregnancy, and (3) during lactation.

female hormones. After the onset of pregnancy, these hormones cause the breasts to become firmer and increase in size as a result both of an increase of the glandular tissue and of the accumulation of fat cells. After the birth of the baby, other hormones bring about the secretion and expulsion of milk (lactation). Sucking by the baby provides a nervous stimulation that increases the production of these hormones and thus increases the flow of milk. At menopause, the production of sex hormones decreases and the breasts tend to shrink, largely because of the contraction and atrophy of the fatty connective tissue; consequently the breasts sag and lie closer to the chest wall.

OUTGROWTHS OF THE SKIN

The outgrowths of the human skin are hair and nails.

Hair

A hair forms from a thickening of the stratum germinativum, which pushes down at an oblique angle into the dermis. The central cells separate from the outer ones, become keratinized, and die. They form the shaft of the hair, whereas the outer cells form a deep pit, the *follicle.* At the bottom of the follicle, a little mass of dermal tissue furnishes the necessary nutrients for the cells at the base of the shaft. These cells continue to divide and push the dead, keratinized shaft out through the skin. Figure 2–5 shows the structure of a hair and hair follicle. Hair grows at a rate of about 2 millimeters (1/12 inch) a week. Hair on the head may continue to grow indefinitely, but that on most of the body grows to a certain length and then stops. After a time the cells at the base of the shaft begin to divide again to produce a new hair that pushes the old one out. Thus hair is constantly shed and constantly renewed.

Straight hair is round in cross section, wavy hair is alternately round and oval, kinky hair is flat. The color of hair is produced by pigment granules in the cells of the shaft. With age, the production of pigment falls off, the cells become filled with air bubbles, and the hair turns gray. Hair does not turn white in a single night as a result of shock, but the shock may cause some shedding of the finer, pigmented hairs and thereby may make more prominent the coarser, gray hairs already present, which seem to be more resistant to stress.

Hair is widely distributed over the body, but is absent on the palm of the hand, the sole of the foot, and the last joints of the fingers and toes. Man actually has as many hair follicles as the gorilla or chimpanzee, but most of his follicles are so small that the hairs they produce are not visible on the surface. Associated with each hair is a muscle that, when stimulated by cold or fear, contracts and

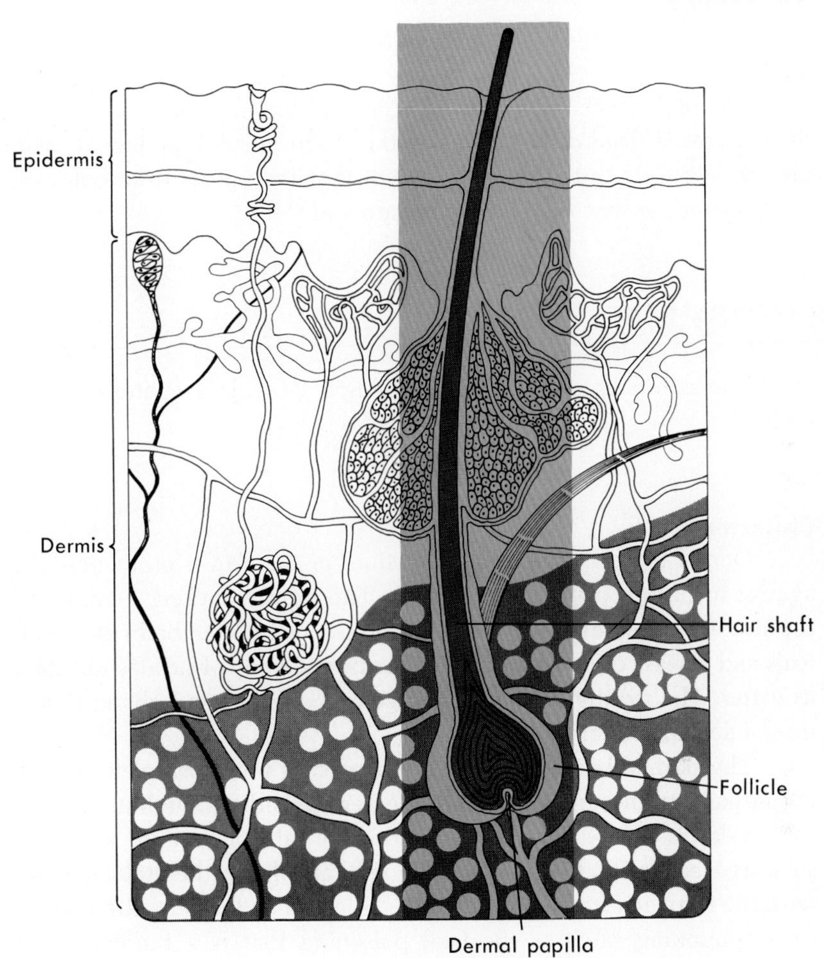

Figure 2–5. The structure of a human hair.

causes the hair to stand more erect. This movement produces a little wrinkle of skin at the base of the hair, causing the development of gooseflesh.

The eyelashes and the hairs at the entrance to the nostrils and ear canals protect the eyes, nose, and ears from invasion by foreign objects, but otherwise, human hair has little function except, perhaps, as a sexual attractant. In an evolutionary sense, mankind seems to be losing its hair. Baldness, which is genetically determined and incurable, is apparently becoming more prevalent. Perversely, many men persist in shaving the face, where hair growth is natural, and at the same time rubbing "hair restorer" on the balding scalp, where hair loss is equally natural.

Nails

The nails on the tips of the fingers and toes are made up of clear, horny, thick layers of dead cells joined together to form solid, protective plates. Their growth is similar to that of hair, except that they grow from below as well as from behind, so that they both elongate and thicken.

FUNCTIONS OF THE SKIN

The skin does much more than simply provide an outer covering for the body.

Protection

One of the most important functions of the skin is protection of the body against intrusions of the environment. The tightly packed, keratinized cells of the stratum corneum prevent both the entrance into the body of harmful chemicals and disease organisms and the loss by the body of needed substances, such as water. The stratum corneum also guards delicate underlying tissues against mechanical abrasion.

The skin, although an effective barrier, is not a complete one. Some materials pass out through it, and others enter the body through it. Most absorption by the skin is of fat-soluble compounds. The salts of some of the heavy metals, such as mercury and lead, although not fat-soluble themselves, will combine with the fatty acids of sebum to make compounds that are absorbed. They can cause poisoning, such as the lead poisoning that is a familiar complaint of painters. Some commercial insecticides are readily absorbed and can build up to dangerous levels unless the skin is carefully protected when these chemicals are handled.

Temperature Regulation

Another major function of the skin is the regulation of the flow of heat between the body and the external environment. The flow itself is under two types of control, *vasomotor* and *evaporative*.

Vasomotor control is achieved by control of the amount of blood in the numerous blood vessels of the skin. When the environmental temperature drops below about 19°C (66°F), muscles in the walls of the many small blood vessels in the dermis contract, constricting the vessels and decreasing the flow of blood to the skin. Above this temperature, these muscles relax and the vessels enlarge, bringing about an increased flow of blood, which carries heat from the internal parts of the body to the surface, where it can be dispersed to the surrounding environment. When the air temperature rises above 31°C (88°F), vasomotor control is not sufficient to disperse the excess body heat and evaporative control comes into play. The eccrine glands are stimulated to pour increased amounts of water over the surface of the body. In an environment where the humidity is low, the heat of the body is used to convert the sweat into water vapor. When the humidity is high, evaporation is slowed because the air is unable to take up much more water vapor and the sweat collects in water droplets on the skin. This is why you feel more uncomfortable in a hot and humid climate than in a hot and dry one. The skin accounts for about 95 per cent of the heat eliminated from the body. The rest is lost from the lungs in breathing and in the elimination of fecal material and urine.

Other Functions

The skin has other functions in addition to protection and heat regulation. It serves as a storage place for food in the form of fat. It excretes some waste products along with sweat. It houses many of the sense organs through which humans are made aware of the external environment (see Chapter 9). Underlying the dermis are subcutaneous sheets of tough connective tissue, called *fascia.* Fastened to them, or to the bones of the skull, are a group of muscles, the *mimetic muscles,* whose other ends are fastened to the skin of the face. By their contraction you can pull your facial skin into a frown or a smile, look puzzled, angry, grieved, or glad. This is a universal language, by which you can communicate without words to your fellows.

SUGGESTED READINGS

Crouch, J. E., and J. R. McClintic: *Human Anatomy and Physiology,* John Wiley & Sons, Inc., New York, 1971.

Grollman, S.: *The Human Body,* 3rd ed., Macmillan Publishing Co., Inc., New York, 1974.

Guyton, A. C.: *Function of the Human Body,* 3rd ed., W. B. Saunders Company, Philadelphia, 1969.

Montagna, W.: "The Skin," *Sci. Amer.,* Vol. 212, No. 2, 1965.

Weichert, C. K.: *Anatomy of the Chordates,* 4th ed., McGraw-Hill Book Company, New York, 1970.

CHAPTER 3

The Skeleton

When a man speaks of the bare bones of a matter he implies that although the framework is there, all the interesting bits that give life and color to the subject have been left out. But the human skeleton is much more than an inert framework around which the body is draped. It is composed of living and metabolically active tissues, and like the skin has many functions. It is also one of the most revealing systems. An anthropologist studying a skeleton can accurately determine the sex, age, and race of its former owner; can say whether he led an active or sedentary life; whether he was well nourished as a child; and whether he suffered from certain diseases.

COMPOSITION OF THE SKELETON

The skeleton is composed of two kinds of connective tissue, cartilage and bone. In both, the living cells lie in small spaces scattered through the nonliving matrix secreted by them. The ground substance is like that of connective tissue, collagen fibers are present, and elastic fibers are also found in some cartilage.

Cartilage

In cartilage, the cells are more or less randomly arranged in groups throughout the ground substance (see Figure 1–10). The collagen fibers may be scattered at random or they may be arranged in bundles. Cartilage lacks blood vessels, and food and oxygen reach the cells by diffusing through the matrix. This is a rather slow and inefficient process, but it is sufficient to keep the cells alive. If they die, the surrounding matrix disintegrates.

Cartilage, being less rigid than bone, gives flexibility and provides a cushioning effect. Masses of cartilage occur between the separate bones that make up the spinal column and cap the ends of the long bones that articulate with each other in the limbs. Cartilage also gives form and support to such structures as the nose and ears.

Bone

In bone, the collagen fibers are arranged in rows. Embedded between these fibers are rows of crystals of a mineral called *hydroxyapatite,* which contains both calcium and phosphorus. The presence of this mineral makes bone hard. In contrast to cartilage, bone is well supplied with small blood vessels. Bone tissue may be either spongy or compact.

Spongy Bone. Spongy bone consists of many small bony spicules and plates arranged in an intricate latticework. The bone cells are lined up more or less in rows, and the spaces (*lacunae*) in which they lie are connected with each other by little canals, *canaliculi,* which run through the surrounding matrix. The canaliculi also connect with passages in which tiny blood vessels course through the bony tissue. Thus, food and oxygen from the blood are able to reach the bone cells. The spaces between the spicules are filled with a kind of connective tissue called *marrow.* Spongy bone is found inside flat bones, such as those of the skull, and in the expanded ends of the long bones.

Compact Bone. Compact bone is somewhat more complicated in arrangement. A thin cross section examined with a microscope shows a series of rounded openings, the *Haversian canals,* through which run small blood vessels and nerves. Each Haversian canal is surrounded by more or less concentric layers (*lamellae*) of the intercellular matrix and by rings of cells that lie in lacunae between the lamellae. The cells are connected with each other and with the Haversian canal by canaliculi (see Figure 1–10).

Compact bone makes up the outer surfaces of the flat bones. In long bones it forms a hollow shaft surrounding an inner marrow cavity and extends as a thin covering over the spongy bone at the ends.

BONE GROWTH

In spite of its inert appearance, bone has a most remarkable ability to grow, to heal itself after fracture, and to change in shape in response to stress.

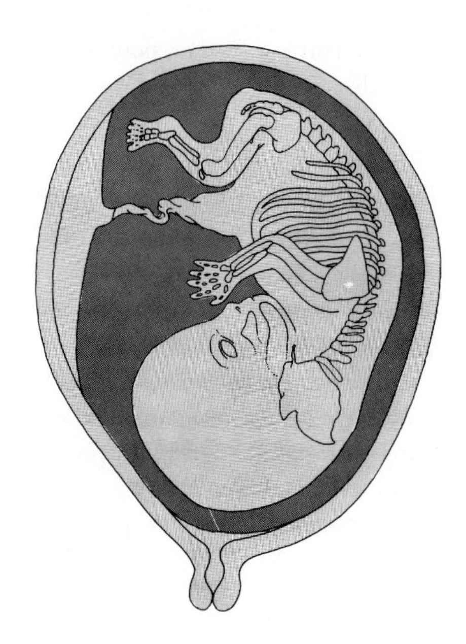

Figure 3–1. A 7-week-old human embryo. Much of the skeleton is preformed in cartilage but true bone has not yet appeared.

Embryonic Origin of Bone

Two different kinds of bone are recognized, depending on the way they develop. *Dermal bones* are formed in association with thin sheets of dense connective tissue (*membranes*) in the dermis. Most dermal bones of the human body are large, platelike structures concerned with the protection of internal organs. The bones forming the roof of the cranium are examples. *Cartilage bones* are preformed as cartilage, which later is invaded and replaced by bone. Figure 3–1 shows a human embryo about seven weeks old in which the entire skeleton is composed only of cartilage and soft membranes. The first indication of the ossification of cartilage and membranes, that is, of the formation of true bone, occurs about the eighth week in human development. In a child ready to be born the bones of the skull have not yet grown together. The spaces between them, the *fontanels,* make the skull pliable and allow for the easier passage of the head through the birth canal. The "soft spot" in the head of a young child is a fontanel that has not yet closed. The development of true bone continues gradually for the next couple of decades in humans and is generally complete at about twenty-one years of age (Figure 3–2).

Bone Growth

Lying in the membranes or surrounding the cartilages that are to be replaced by bone are some rather unspecialized cells called *osteoprogenitor* cells. These cells divide and give rise to daughter cells, some of which become

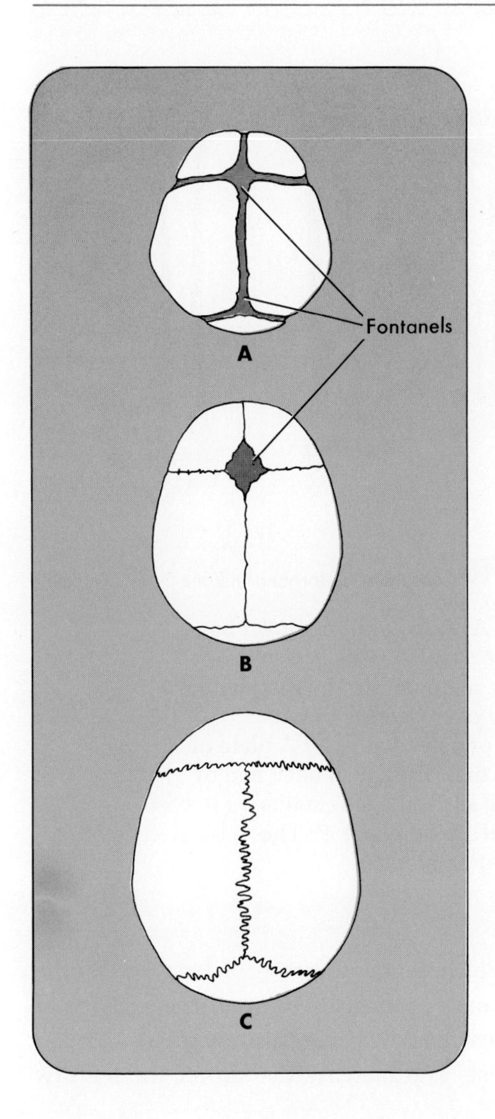

Figure 3–2. Top view of the human skull at different ages showing the gradual closure of the fontanels. A: At birth. B: At one year. C: In the adult.

specialized as *osteoblasts.* The osteoblasts begin to lay down the intercellular matrix around themselves until eventually each cell is surrounded by matrix. Osteoblasts have protoplasmic processes by which they are in contact with each other; the canaliculi are the channels surrounding the protoplasmic processes. As buds of developing bone, along with small blood vessels, grow into the cartilage, the cartilage cells die and the ground substance disintegrates. When a cell stops secreting matrix, it is called an *osteocyte.* Osteocytes are the cells that are present in the lacunae surrounding the Haversian canals.

Meanwhile, other daughter cells of the osteoprogenitor cells have become specialized as *osteoclasts.* They have the ability to dissolve bone. Figure 3–3

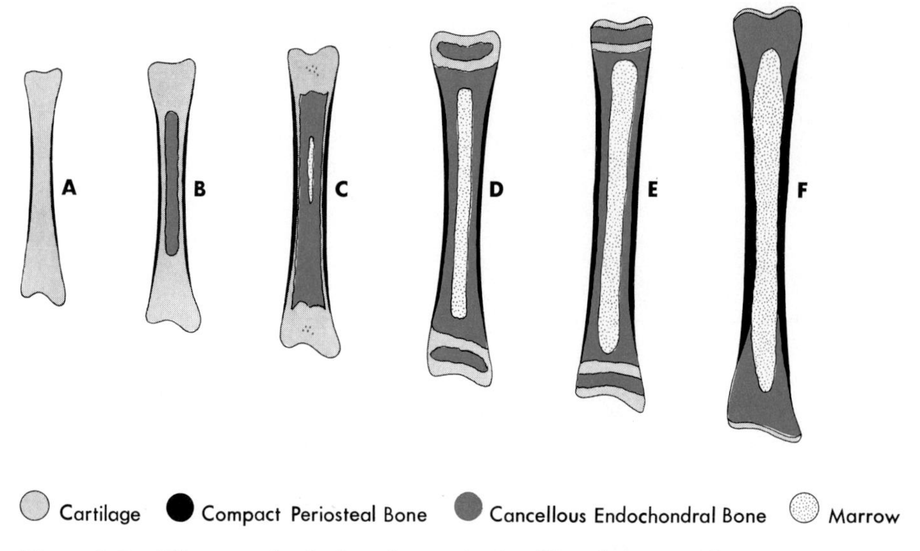

Figure 3–3. The growth of a long bone. A: A collar of compact bone (periosteal bone) is laid down around the cartilaginous precursor by osteoblasts. B: Spongy bone begins to develop within the cartilaginous shaft (endochondral bone). C: A marrow cavity forms in the endochondral bone through the activity of osteoclasts. D: A plate of endochondral bone (epiphysis) forms in the cartilage at each end of the shaft. E: Elongation continues as long as cartilage remains between the epiphyses and the bony shaft (diaphysis). F: The diaphysis and epiphyses fuse and elongation stops.

shows sections through successive developmental stages of a long bone (the femur of the thigh). As the outer diameter increases through the activity of the osteoblasts, the diameter of the inner marrow cavity also increases through the activity of the osteoclasts. Osteoclasts also increase the size of the marrow cavities in flat bones as they grow.

Reshaping of Bone

The reworking of bone through the combined activity of the osteoblasts and osteoclasts involves more than increase in size and continues throughout life. The various cell types involved in bone formation are interconvertible. Osteoblasts and osteocytes can change into osteoclasts, and osteoclasts can become osteoblasts (see Figure 3–4). The spicules of spongy bone are laid down along the lines of stress to which the bone is subjected by the pull of the muscles (see Figure 3–5). This varies from individual to individual and from

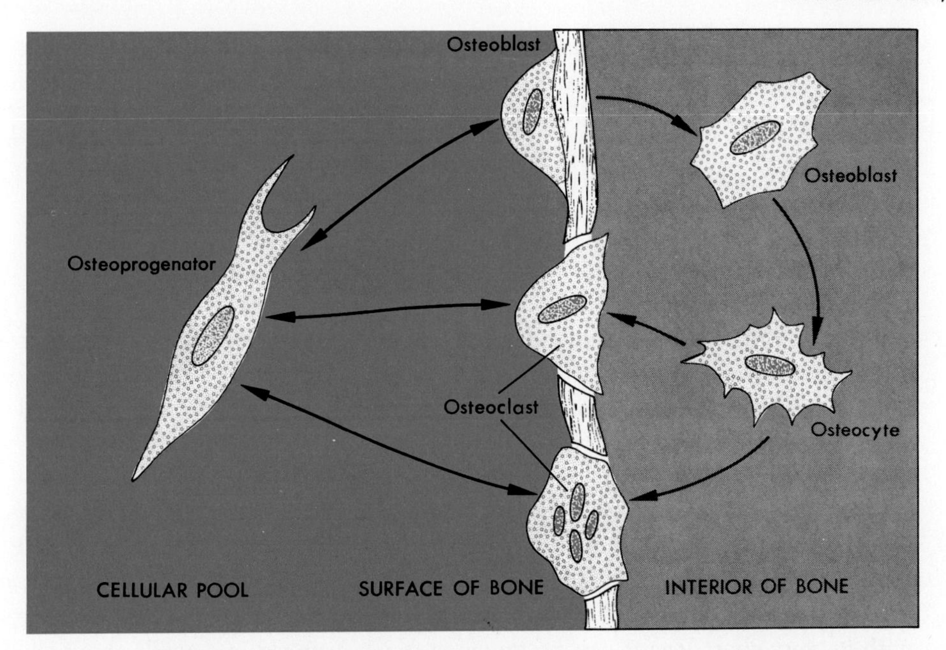

Figure 3–4. The cells involved in bone formation. Osteoprogenitor cells give rise to both osteoblasts and osteoclasts. Osteoblasts become osteocytes. Osteocytes may change into osteoclasts. Recent evidence suggests that both osteoblasts and osteoclasts can change back into osteoprogenitor cells.

time to time during the life of the individual. The pattern of the adult differs from that of the child. The pull of muscles on bone also causes the development of ridges and knobs at the sites where the muscles attach.

Stress is necessary for the maintenance of bone. In bedridden patients whose bones are deprived of the normal pull of the muscles, there is an actual shrinking of the skeleton as a result of the activity of osteoclasts. This is one of

Figure 3–5. The spicules of spongy bone at the ends of a long bone form along the lines of stress.

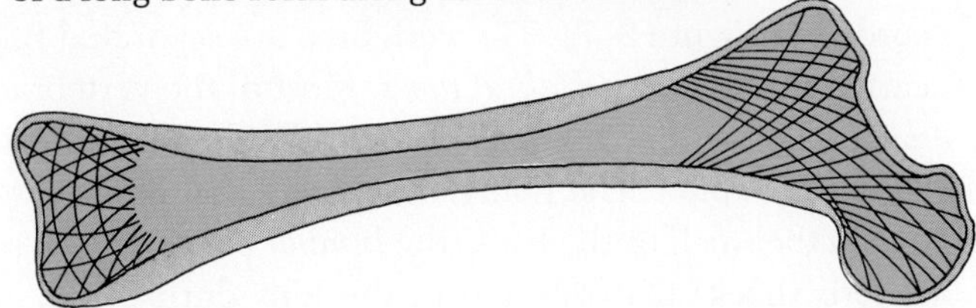

the problems facing astronauts on long space flights. In addition to muscular inactivity, astronauts suffer loss of the stress imparted to the skeleton by the pull of gravity. This is one reason why Skylabs are equipped with exercise machines.

Abnormal Bone Growth

Normal bone growth may be prevented by a number of factors. Malfunctioning of the endocrine glands may result in giants or dwarfs (see Chapter 10). A hereditary condition is known in which elongation of the long bones ceases long before the thickening of the bones stops. The result is an individual with abnormally short and thick limbs, a chondrodystrophic dwarf. Chondrodystrophic dwarfs were once favored as court jesters and are frequently shown in paintings of court scenes; nowadays many of them earn a living as wrestlers. In parts of Mexico and Africa there are local populations composed entirely of this type of dwarf. The same condition is known in other animals. The dachshund is chondrodystrophic; it is a favorite with city dwellers because with its very short legs it can be given as much exercise in a walk about the block as a rangy type dog might get in loping a country mile.

DIVISIONS OF THE SKELETON

The human skeleton consists of two major divisions, the axial and the appendicular. Figure 3–6 shows these divisions.

Axial Skeleton

The axial skeleton is made up of the skull, the vertebral column, the ribs, and the sternum or breastbone. The internal organs of the chest cavity, the heart and lungs, are supported and protected by the ribs and sternum. In addition, the movement of the ribs and sternum plays an important role in breathing (see Chapter 6). The cranium of the skull houses the brain, and the vertebrae form a series of bony blocks, the backbone or spinal column, surrounding the spinal cord (see Figure 3–7). The vertebrae are separated from each other by pads of cartilage, the *intervertebral discs.* Most of the vertebrae are slightly movable on each other so that the backbone is pliable, like an armored cable, rather than rigid like a pipe. The human backbone also has several curves in it, especially one in the small of the back, the *lumbar curve.* It acts something like a spring to absorb shocks that develop in the hips during walking, running, or jumping. This is necessary because with man's upright posture the shocks would other-

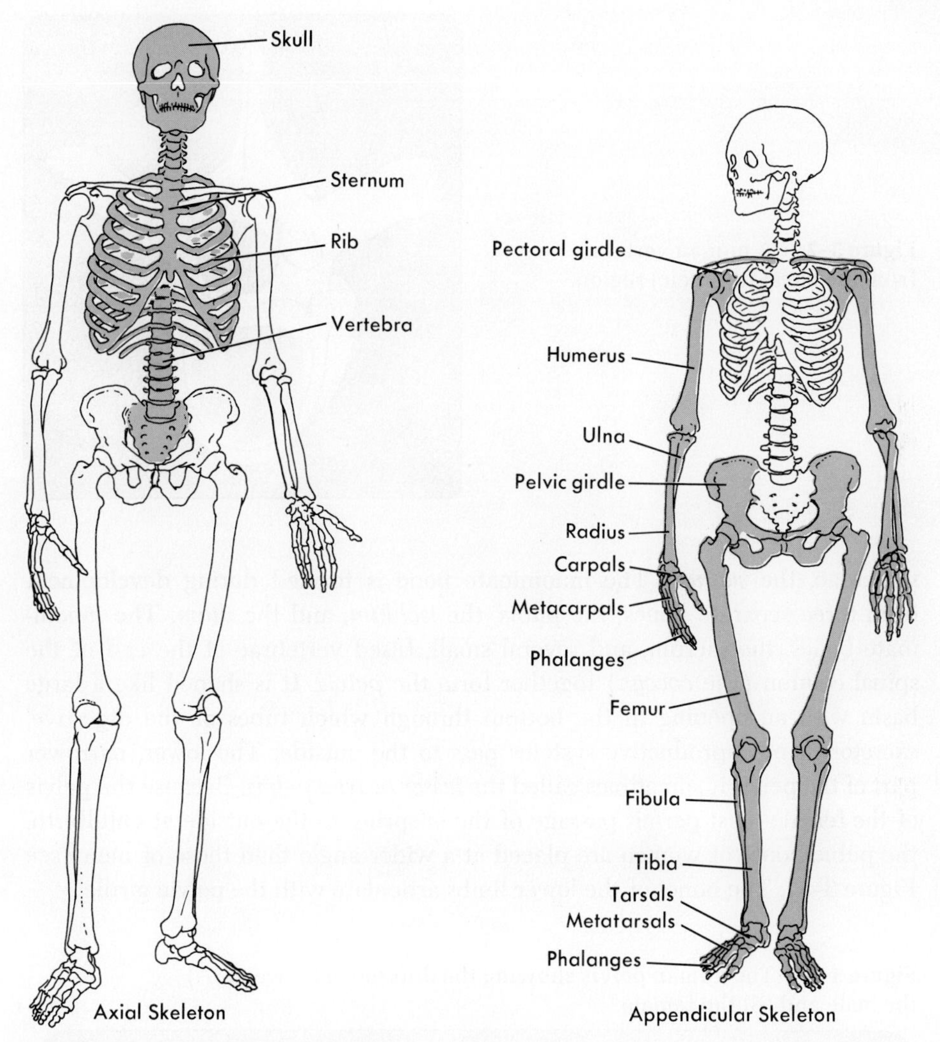

Figure 3–6. The human skeleton.

wise be transmitted to the skull and brain. The lumbar curve is not found in animals that walk on all fours.

Appendicular Skeleton

The appendicular skeleton is made up of two girdles and the limbs. The upper, or *pectoral,* girdle consists of paired shoulder blades (*scapulae*) and paired collar bones (*clavicles*), which form the support and point of attachment for the bones of the arms. The lower, or *pelvic*, girdle, is made up of a single bone on each side, the *innominate,* which is fastened behind to five fused

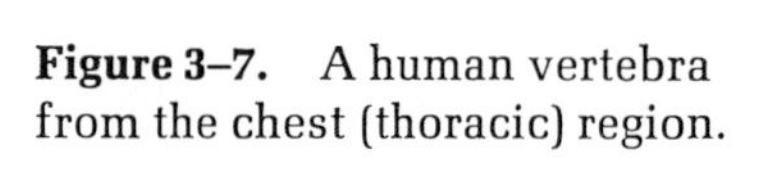
Figure 3–7. A human vertebra from the chest (thoracic) region.

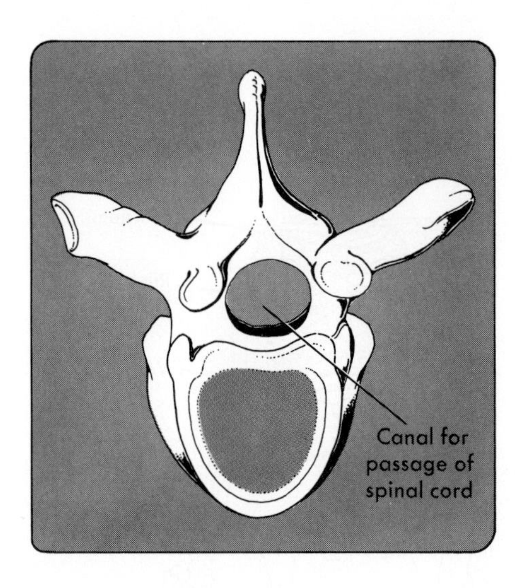

vertebrae, the *sacrum*. The innominate bone is formed during development from three separate bones, the *pubis,* the *ischium,* and the *ilium.* The innominate bones, the sacrum, and several small, fused vertebrae at the end of the spinal column (the *coccyx*) together form the *pelvis.* It is shaped like a large basin with an opening in the bottom through which tubes of the digestive, excretory, and reproductive systems pass to the outside. The lower, narrower part of the pelvis is sometimes called the *lesser* or *true pelvis.* Because the pelvis of the female must permit passage of the offspring to the outside at childbirth, the pubic bones of women are placed at a wider angle than those of men (see Figure 3–8). The bones of the lower limbs articulate with the pelvic girdle.

Figure 3–8. The human pelvis showing the difference between (A) the male and (B) the female.

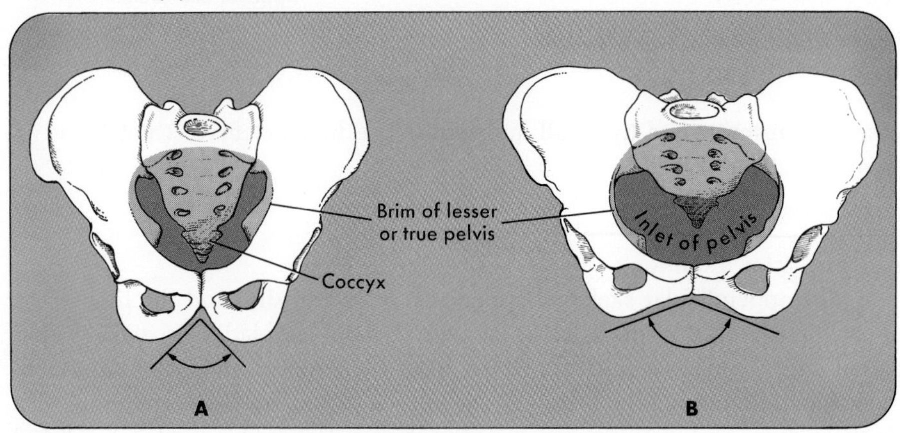

Joints

The places where bones come together are called *joints.* When the bones are held firmly in place by fibrous connective tissue or cartilage so that they cannot move, the joints are said to be *immovable.* The bones of the cranium are held together by such joints. Most joints are movable. The space between the bones is surrounded by a capsule of tough, fibrous connective tissue and filled with a sticky *synovial fluid.* Like a film of oil between the moving parts of a motor, the synovial fluid prevents friction between the cartilaginous caps of the bones.

Teeth

Although the teeth develop from the integumentary system and function as part of the digestive system, they are in man so firmly implanted in the jawbones for strength and support that they are usually considered part of the skeleton. Furthermore, they show many resemblances to bone in their composition. Figure 3–9 shows the structure of a human tooth. The root, the part that is embedded in the jawbone, is surrounded by a layer of a hard material called *cement.* The central *pulp cavity* is filled with a kind of soft connective tissue containing small nerves and blood vessels. A layer of odontoblast cells lies between the pulp cavity and the surrounding *dentin.* These cells secrete the matrix of the dentin, which is like the matrix of bone but is harder. The exposed part of the tooth, the *crown,* is covered by a layer of *enamel.* Like bone and dentin it contains crystals of hydroxyapatite, but the crystals are considerably longer and the fibers in the matrix are keratin rather than collagen. Enamel is the hardest substance in the human body—it has to be to stand up to the constant grinding when food is chewed. It is still subject to decay, as most of us know to our sorrow. When fluorine is present in the diet, atoms of fluorine replace hydroxyl groups in the hydroxyapatite crystals, making them even harder and the tooth more resistant to decay. This is why fluoride is now routinely added to the water supply in many regions where the amount of natural fluorine is low. Too much fluorine in the diet causes brown and white mottling of the enamel. Fluorine in large amounts is a violent poison.

In the front of the mouth there are, on either side of the midline, two teeth for cutting off bits of food, the *incisors.* These are followed by a single tooth, the *canine,* so named because the comparable tooth in dogs is much enlarged. Behind the canine are two flat-crowned grinding teeth, the *premolars.* Finally, the last teeth in the jaw are three large grinding teeth, the *molars.*

In man, as in most of the higher animals, the teeth are both formed and replaced in successive waves proceeding from the front of the mouth to the rear. The first set (the *deciduous* or *milk teeth*) includes incisors, canines and

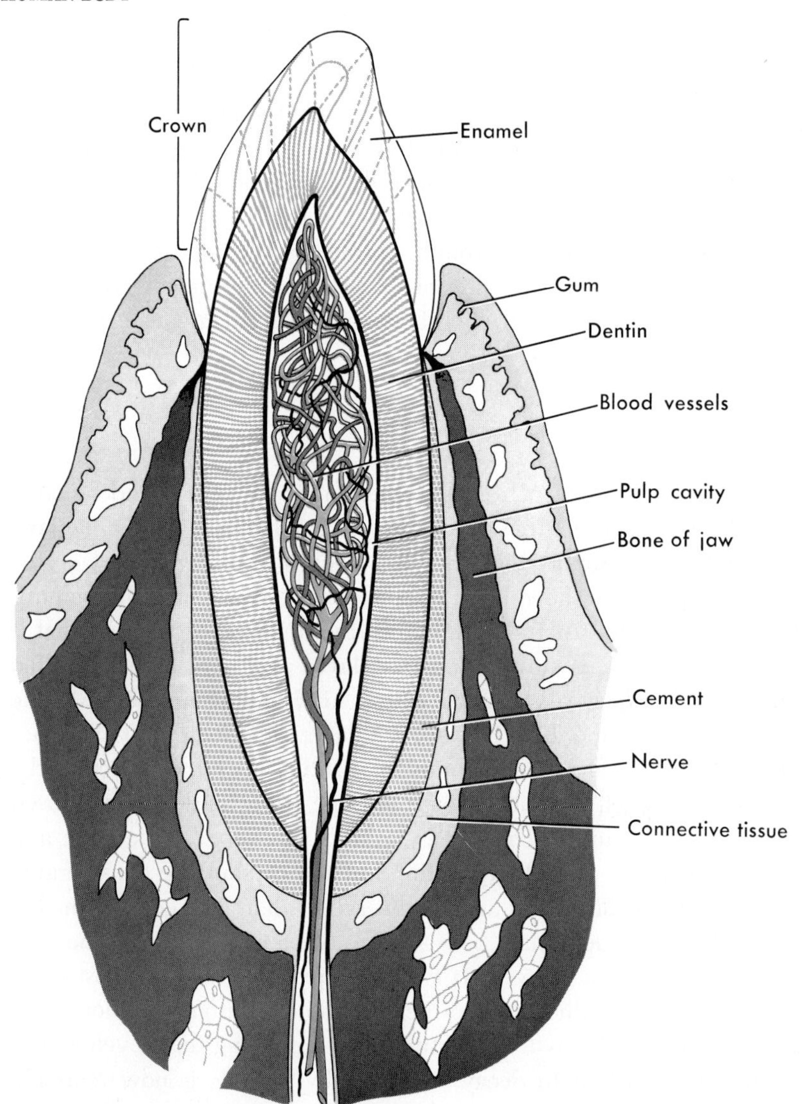

Figure 3–9. Structure of a human tooth.

premolars. By the time a child is two, all of his so-called baby teeth have usually erupted. When he is about six or seven, the roots of these teeth begin to be eroded away by osteoclasts; the teeth fall out and are replaced by the permanent teeth, which have developed below them. The molars, the last to erupt, are never replaced and are in reality remnants of the first set of teeth. The first molars appear when the child is about six. The last usually do not erupt until he is seventeen to twenty-five years old and hence are called the *wisdom teeth.* Sometimes they fail to erupt and are then said to be *impacted.*

FUNCTIONS OF THE SKELETON

The skeleton has many functions. It provides support and protection for the internal organs. It forms a system of levers that, with the muscles, make movement possible. It is a storage depot for minerals that are essential for the functioning of the body, and it is the site for the formation of blood cells.

Support and Protection

The parts of the delicate central nervous system, the brain and spinal cord, are surrounded and shielded by the skull and backbone; the heart and lungs are protected by the ribs and breastbone.

Hemopoietic Activity

In the human adult, bone is the site for the manufacture of many of the so-called *formed elements* found in the blood, including the red cells, some of the white cells, and the platelets (see Chapter 8). The formation of these structures, referred to as *hemopoietic activity,* takes places in the marrow. The rate of cell production is normally slow in the long bones, such as the femur, and the marrow in them is likely to have an accumulation of fat cells that give it a yellow color. In short or flat bones, such as those of the skull, ribs, sternum, clavicles, vertebrae, and pelvis, the rate of production of blood cells is sufficient to prevent the development of fat cells and the marrow is red in color. In time of need, though, yellow marrow can shift rapidly into the active production of blood cells. It thus serves as an immense emergency reserve for blood cell formation.

Mineral Storage

Two minerals that have to be available for the body to carry on its work are calcium and phosphorus. Calcium is needed for the clotting of blood and for controlling the excitability of nerves and muscles. Phosphorus is an essential component of every cell. It is a constituent of the nucleic acids and is necessary for the molecular energy exchanges by which cells perform their work. Bone acts as a storehouse for these two minerals. They can be released to the body as needed by osteoclastic activity and reinvested in bone by the action of osteoblasts.

Other minerals can also be stored in the skeleton. Some of these, such as sodium and magnesium, are stored until such time as they are needed by the body. Others, including radium, fluorine, and lead, are foreign to the body; when they are present in bone in excessive amounts they interfere with the bone's normal metabolism and hence lead to pathological conditions.

One of the products found in radioactive fallout is strontium 90. Strontium resembles calcium in atomic structure and may be incorporated into bone mineral in its place. When strontium 90 accumulates in bone, it may lead to the development of leukemia, a cancer of the tissues forming the white blood cells.

Movement

The fact that the skeleton is made up of many individual bones rather than a single one makes movement possible. But bones are not capable of moving themselves; they are moved by the pull of the muscles discussed in the next chapter.

SUGGESTED READINGS

Budy, A. M., ed.: *Biology of Hard Tissue,* NASA, Washington, 1968.

Grollman, S.: *The Human Body,* 3rd ed., Macmillan Publishing Co., Inc., New York, 1974.

Kimber, D. C., et al.: *Anatomy and Physiology,* 16th ed., Macmillan Publishing Co., Inc., New York, 1972.

McLean, F. C., and M. R. Urist: *Bone,* University of Chicago Press, Chicago, 1955.

Smith, C. U. M.: *The Architecture of the Body,* Faber and Faber, London, 1964.

CHAPTER 4

Muscles—Their Form and Function

The marvelously controlled and coordinated movements of a troupe of high-wire artists, the feats of a weight lifter, even the simple, but very precise act of threading a needle, are made possible by structures that have a highly developed ability to contract. These are muscles, which make up about one half of the body weight and which are able to convert chemical energy into the energy of mechanical work and heat.

TYPES OF MUSCULAR TISSUE

On the basis of form and function, there are three types of muscular tissue in the body: skeletal, visceral, and cardiac (see Figure 4–1).

Skeletal Muscle

The great bulk of the muscles of the body are made of skeletal muscle tissue. They are called skeletal muscles because they are usually associated with some part of the skeleton. They are also often referred to as striated muscles because the cells appear cross-banded, and as voluntary muscles because one is usually able to control their contractions. Skeletal muscles make possible the movements of the body by changing the relative position of the parts of the skeleton. Figures 4–2 and 4–3 show the dorsal (back) and ventral (front or belly) views of the major skeletal muscles of the human body.

Humans are born with their full complement of skeletal muscle cells and

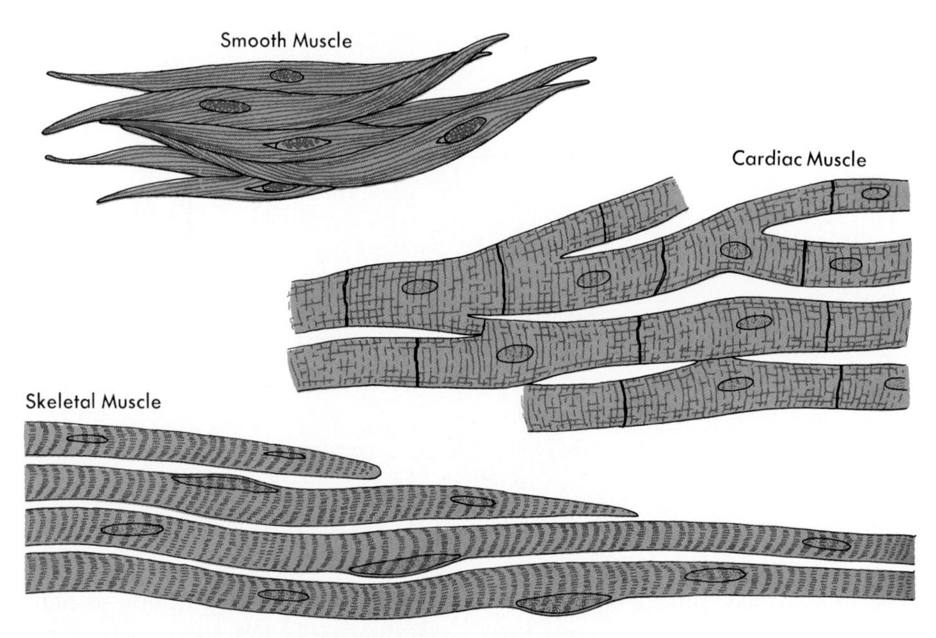

Figure 4–1. Types of muscular tissue.

the increase in size of muscles during growth or as a result of exercise comes from an increase in the size of the individual cells. Because skeletal muscle cells are not capable of dividing to produce new cells, the regeneration of muscle tissue after an injury is very slight. If the cells are destroyed they are replaced by connective tissue.

Skeletal muscle cells are usually called *muscle fibers.* When viewed under the light microscope, they are seen as elongated cylinders that appear striated, that is, they have alternating light and dark bands across them. Each fiber is formed by the fusion of a number of embryonic muscle cells and so has several nuclei instead of only one. These nuclei are located at the periphery rather than near the center of the cell. A single nervous stimulus can make a skeletal muscle fiber contract, but in the absence of continued stimulation it relaxes automatically. Skeletal muscle fibers contract quickly, in from $\frac{1}{100}$ to $\frac{1}{10}$ second.

Visceral Muscle

The second kind of muscle is usually found in the walls of tubular structures, such as the digestive tract, the blood vessels, and various ducts. It is called visceral muscle because it is associated with the viscera (the soft internal organs of the body). Visceral muscles are made up of sheets or bundles of

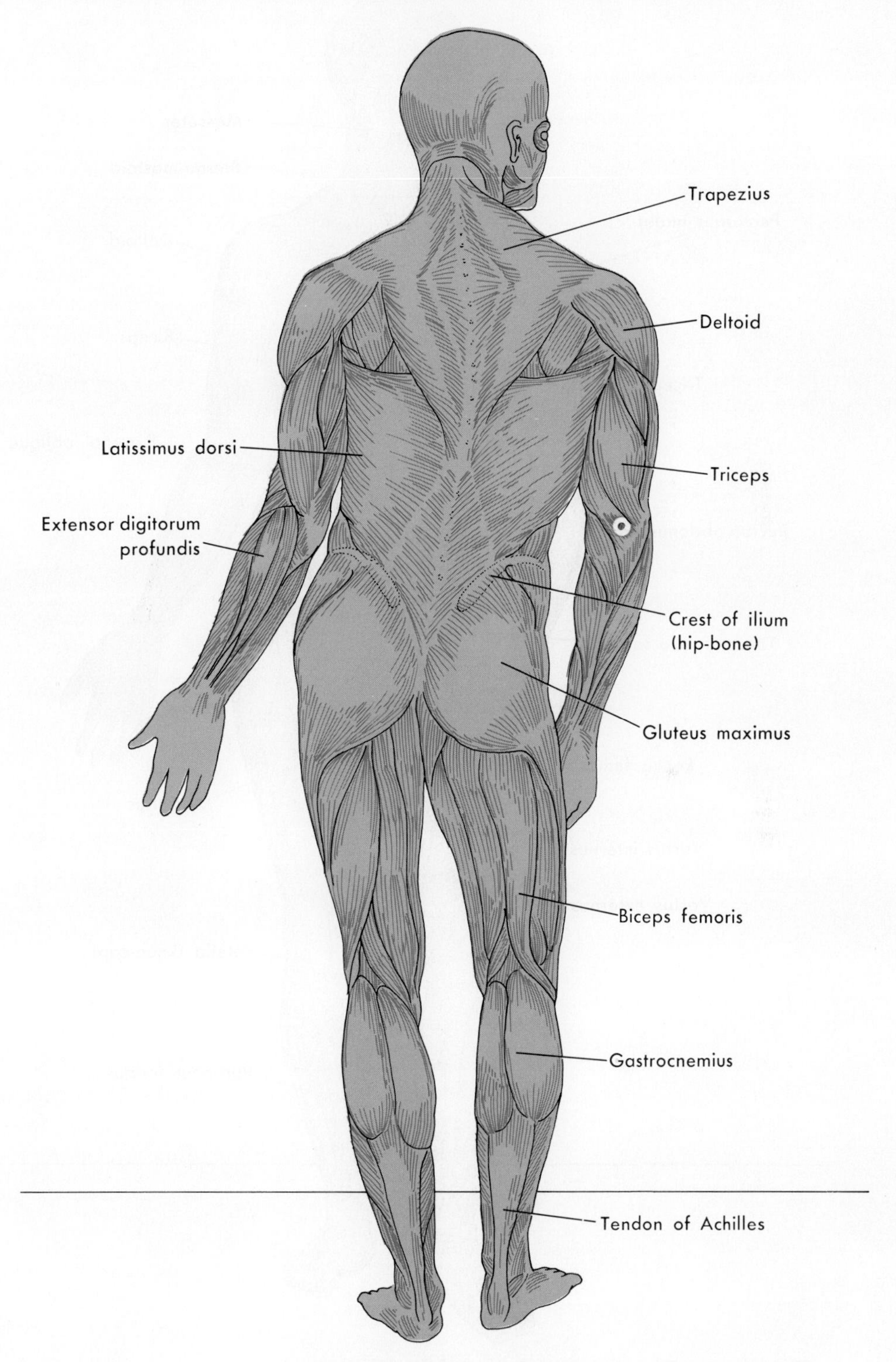

Figure 4–2. Back or dorsal view of the superficial muscles of the human body.

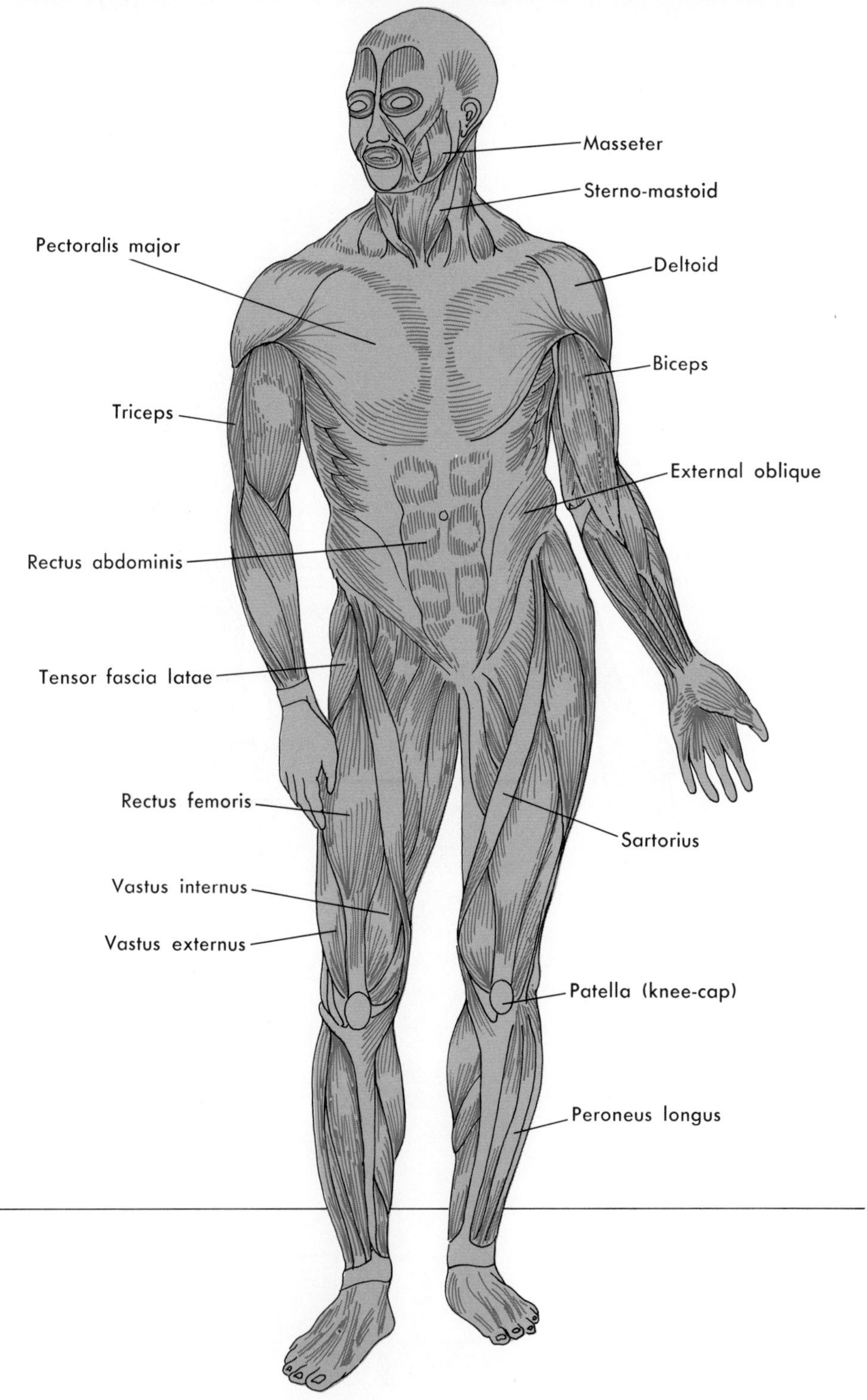

Figure 4–3. Front or ventral view of the superficial muscles of the human body.

spindle-shaped cells that lack the striations that are characteristic of skeletal muscle fibers and so are sometimes called smooth muscles. Each cell has just a single centrally located nucleus. Smooth muscle cells are slow in action, taking several seconds to contract. The individual cells are often not directly innervated but receive the stimulus to contract by transmission from nearby cells, from the stretch exerted by the contraction of other muscle cells, or from hormones. Contraction is usually not under voluntary control. Sheets of visceral muscle cells are often arranged in two layers, one with the cells running around the tube and the other with the cells running lengthwise. When the cells of the first layer contract, those of the other layer are stretched, the diameter of the tube decreases, and the tube lengthens. When the cells of the second layer contract, the first layer is stretched, the diameter increases, and the tube shortens.

Cardiac Muscle

The third type of muscle tissue is cardiac, so named because it is found only in the heart. Structurally it seems to be intermediate between skeletal and visceral muscle. The individual cardiac muscle cells are striated but each has only a single nucleus that lies deep within it, not next to the cell membrane. Cardiac tissue differs from both skeletal and visceral muscle tissue in that the individual cells branch. The cells have more or less parallel sides but the ends are uneven, with short, stubby, finger-like processes that fit very closely against the ends of other cells to form a tightly knit network. In speed of contraction cardiac muscle cells are intermediate between skeletal and visceral muscles. The heart is able to contract without direct nervous stimulation. The impulse to contract is carried from one part to another by specially modified muscle fibers. But the rate of contraction is regulated by nervous control (see Chapter 8).

Table 4–1 shows the characteristics of the different kinds of muscle cells.

Table 4–1. Characteristics of Muscle Types

Skeletal	Visceral	Cardiac
Skeleton	Internal organs	Heart
Striated	Smooth	Striated
Multinucleate fibers	Uninucleate cells	Uninucleate cells
Fast acting	Slow acting	Intermediate
Voluntary	Involuntary	Involuntary

FORM AND FUNCTION OF A SKELETAL MUSCLE

Because much more is known about the fine details of the structure and functioning of the skeletal muscles than of either visceral or cardiac muscles, the rest of this chapter is concerned with skeletal muscles. It is probable that much that we say about skeletal muscles also applies to the other kinds, but undoubtedly there are also differences although biologists do not yet know what these differences are.

Structure of a Muscle

Skeletal muscles are made up of bundles of muscle fibers bound together by connective tissue. An example of a typical skeletal muscle is the biceps, one of the largest muscles of the upper arm (see Figure 4–4).

Gross Structure of the Biceps. This muscle is called *biceps* because it has a double origin, or two heads, which are fastened to the shoulder blade by tendons on their proximal ends. Tendons are composed of a type of connective tissue in which there are many collagen fibers arranged in parallel rows. They are continuous with the connective tissue surrounding the muscle bundles and also with a layer of connective tissue covering the bone. As the muscle extends distally from these tendinous heads, the number of muscle fibers in relation to the connective tissue increases appreciably and thus the muscle enlarges in the

Figure 4–4. A skeletal muscle, the biceps.

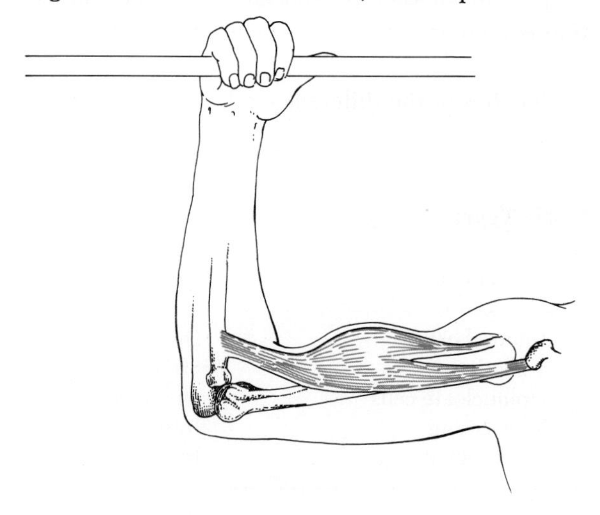

middle to form the belly. Distal to the belly, the muscle again narrows and attaches to a single tendon, which forms the insertion, by which the muscle is attached to one of the bones of the lower arm. Thus, skeletal muscles are typically comprised of an origin, a belly, and an insertion. The origin of the muscle is attached to the place of less movement and the insertion is attached to the place of greater movement.

Figure 4–5 shows a section through a muscle such as the biceps. It is divided into a number of compartments, or *fasciculi,* by sheets of connective tissue. Each fasciculus is in turn made up of a number of fibers, or muscle cells.

Structure of a Muscle Fiber. Each individual fiber is surrounded by a thin, transparent membrane known as the *sarcolemma.* The outer layer of the sarcolemma is a nonliving, elastic substance; the inner layer probably represents the plasma membrane of the fiber. The cytoplasm of a muscle fiber is called the *sarcoplasm.* It contains numerous bundles of elongate, parallel structures, the *myofibrils.* There is also much smooth endoplasmic reticulum, here called *sarcoplasmic reticulum,* and in addition an extensive system of tubes, the *T system.* These tubes form as invaginations of the sarcolemma, pass among the myofibrils, and connect with the reticulum. Because the myofibrils show alternating light and dark bands and because the bands of one myofibril are

Figure 4–5. Structure of a muscle. The individual muscle cell is the fiber. It contains columns of striated fibrils, the myofibrils. The fibers are gathered into a bundle, the fasciculus, and the whole muscle comprises a number of fasciculi with their surrounding connective tissue.

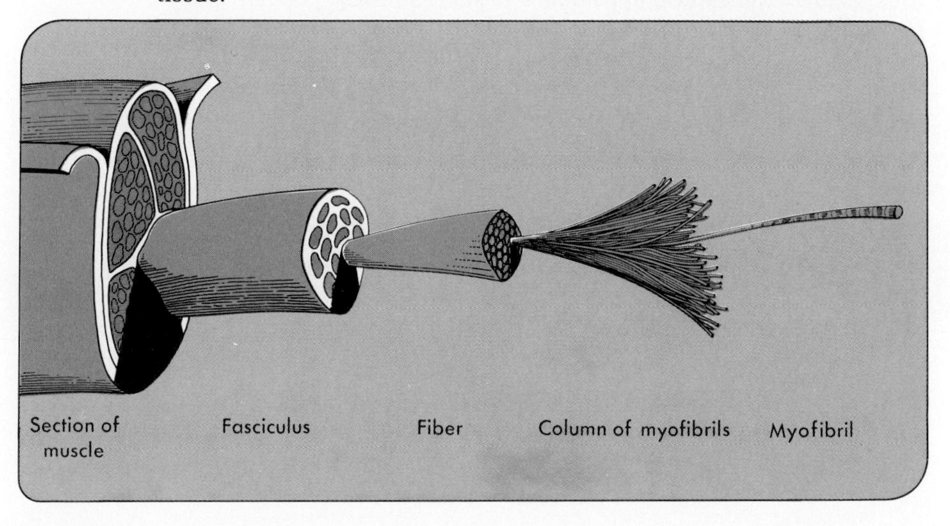

arranged to coincide with the bands of adjacent myofibrils, the fiber as a whole appears striated.

The myofibril itself is divided crosswise into definite sections, each section being separated from the adjacent one by a dark line known as the *Z-line*. Passing from one Z-line to the next, there is first a light area known as the *I-band*, then a central dark area known as the *A-band*, then another light I-band. A narrow, lighter band through the center of the A-band in a relaxed fiber is called the *H-band*. The section from one Z-line to the next is known as a *sarcomere*. (See Figures 4–6 and 4–7.)

Muscular Contraction

When the biceps contracts, it pulls the origin and insertion closer together and bends the elbow. When the arm is straightened out again, the biceps relaxes and other muscles, on the other side of the arm, contract and pull the elbow straight. Skeletal muscles are thus typically arranged in antagonistic sets.

The nerve that stimulates a muscle to contract is composed of a number of nerve fibers, each an extension of a single nerve cell. The individual nerve fibers are much branched at their endings, and each terminal branch makes contact with a muscle fiber at a specialized area on the sarcolemma called the *motor end plate*. An individual nerve fiber may send terminal branches to from a few to several hundred muscle fibers. The muscle fibers that are innervated by a single nerve fiber all contract together. They are called a *motor unit*. When

Figure 4–6. Schematic drawing of two sarcomeres of a muscle fiber. The banded appearance results from the parallel banding of the individual myofibrils.

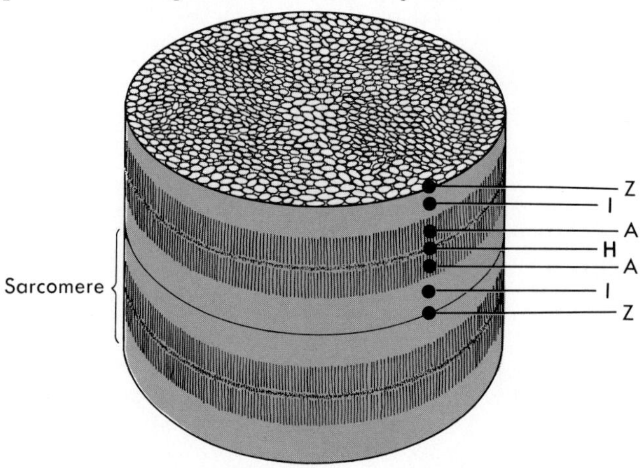

Figure 4–7. Photograph through the electron microscope of several myofibrils. The section from one Z line to the next is a sarcomere. (Courtesy of H. E. Huxley, Cambridge University.)

a single muscle fiber contracts, it contracts as much as it can, about one third of its resting length. How much the whole muscle contracts depends on the number of motor units that are stimulated to contract.

When a muscle fiber contracts, all parts of it contract simultaneously. This means that there must be a rapid transmission throughout the fiber of a stimulus received at one point on the surface. The T system is apparently the structure by which this is achieved, but just how it works is not known.

The details of the contraction of an individual muscle fiber are not yet thoroughly understood, but it is known that the shortening of the fiber is brought about by a reduction in the width of the I-band. Two lines of evidence are being studied to try to determine what actually happens when a myofibril contracts. One is based on an analysis of the fine structure of the myofibrils as revealed by the electron microscope, the other on the nature of the chemical constituents and their reactions. These two lines of evidence are converging, but they have not yet coalesced to form a single, unified picture.

Chemical Constituents of Muscle Fiber. About 75 per cent of the muscle fiber is water. Of the remaining 25 per cent, about four fifths, or 20 per cent of the total fiber, is protein. The remaining 5 per cent of the cell is made up of inorganic compounds, certain biocompounds containing nitrogen, such as ATP and creatine, and carbohydrates. Calcium and magnesium ions are present. One of the proteins is a globular protein called *myoglobin,* which is joined to a nonprotein group known as a *heme group.* This group is a flattish assemblage of atoms of carbon, hydrogen, nitrogen, and oxygen with a single iron atom in the center. The heme group is able to combine reversibly with oxygen, holding it in reserve or releasing it as needed. The actual contractile structures of the cell, the myofibrils, consist almost entirely of proteins. Three of them appear to be especially important.

ACTIN. Actin comprises about 25 per cent of the muscle protein. The individual molecules are globular and aggregate to form long, thin filaments. Each molecule has a binding site for ATP and one for calcium.

MYOSIN. Myosin is the most abundant protein found in muscle, making up about 54 per cent of the total muscle protein. Myosin molecules are long and thin with a globular portion at one end; like actin molecules they aggregate to form filaments. The globular portion has a binding site for actin and is also able to act as an enzyme to split ATP and release energy.

TROPOMYOSIN. Tropomyosin is also present in the myofibrils in association with actin, but the role it plays in contraction has not been determined.

Structure of a Myofibril. The individual myofibrils are apparently made up of many alternating filaments of myosin and actin (see Figure 4–8). As seen under the electron microscope, the thinner actin filaments seem to arise from the Z-line at the end of the sarcomere. The light I-band is a region where only actin filaments are present, the dark part of the A-band is a region of overlap between the actin and myosin filaments, and the H-band in the center of the A-band is the region where only myosin filaments occur. In the region of the A-band each myosin filament is surrounded by six actin filaments. At regular intervals, cross bridges, which correspond to the globular portions of the myosin molecules, extend between the myosin and actin filaments.

A Theory of Myofibril Contraction. During contraction, the A-bands remain the same length while the I- and H-bands grow shorter as a result of the actin filaments sliding along the myosin filaments. Biologists are still not sure just what takes place at the actual moment of contraction. It involves a complicated series of interactions between actin and myosin, magnesium and calcium ions, ATP, and sarcoplasmic reticulum and mitochondria. When the fiber is relaxed the calcium ions are sequestered in the reticulum and the actin and myosin molecules are not linked by cross bridges. Stimulation of the fiber causes release of the calcium into the sarcoplasm, bridges form to link the actin and myosin into an actomyosin complex, ATP is released from the mitochon-

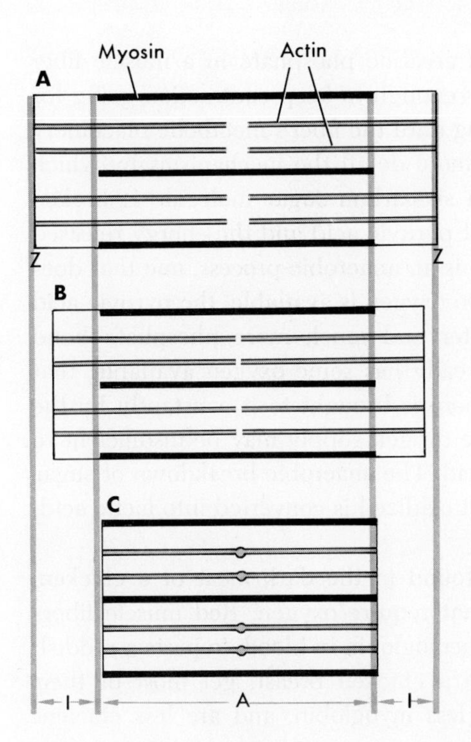

Figure 4–8. Diagram of contraction of a sarcomere. A: Relaxed state. B: Partially contracted. C: Fully contracted (the ends of the actin filaments have met or slid past each other).

dria, binds to the complex, and breaks the bridges. The actin filaments slide along the myosin and new bridges form in the new position. This is contraction. It exerts a pull on the elastic connective tissue surrounding the fiber. When the fiber relaxes, the bridges are broken, and the calcium ions are taken up into the reticulum again. The elastic connective tissue pulls the filaments back into their original position.

Both actomyosin and myosin can act as enzymes to split ATP and provide energy for the process. Magnesium ions inhibit the formation of bridges, but when calcium is added the bridges develop spontaneously, i.e. their formation does not require energy from ATP. But the addition of ATP to the actomyosin breaks the bridges. This explains the development of *rigor mortis* after death. When ATP production stops, actin and myosin remain locked together by the spontaneously formed bridges.

Energy of Muscular Contraction. Because not much ATP is stored in a fibril, muscle activity would be very limited unless there were a quick way to reconstruct ATP. A compound found in the muscle cell, *creatine phosphate,* transfers its phosphate group to ADP in a very rapid reaction:

$$\text{creatine phosphate} + \text{ADP} \rightarrow \text{creatine} + \text{ATP}$$

There is about five times as much creatine phosphate in a muscle fiber as there is ATP, but even so, this is only enough to keep contraction going for a few seconds. It carries the process along until the fiber's metabolic machinery can produce more ATP. We discuss in more detail the mechanisms by which ATP is formed in Chapter 6. Briefly, a six-carbon sugar molecule is broken down into two three-carbon molecules of pyruvic acid and the energy released is used to bind phosphate to ADP. This is an anaerobic process, one that does not require the presence of oxygen. When oxygen is available, the pyruvic acid is oxidized to carbon dioxide and water and much more phosphate bond energy is produced. The muscle cell already has some oxygen available, that which is stored in its myoglobin, and more is brought to it constantly by the blood. But during strenuous exercise, the oxygen supply may be insufficient to oxidize all or even most of the pyruvic acid. The anaerobic breakdown of sugar continues and the pyruvic acid that is not oxidized is converted into lactic acid, which accumulates.

Red muscle fibers, such as those found in the dark meat of a chicken, form most of their ATP by processes that require oxygen. Red muscle fibers have a lot of myoglobin, which, like the hemoglobin in blood, imparts a reddish color. White fibers, like those found in a chicken breast, get most of their energy from anaerobic processes, have less myoglobin, and are less efficient than red fibers.

SOME ASPECTS OF MUSCULAR ACTIVITY

Muscles are like bones—they must be used or they will atrophy. If the nerve to a muscle is cut so that the muscle no longer receives stimuli causing it to contract, the muscle shrinks. If a joint is immobilized so that the muscles that move it are not doing their normal work, they also shrink and lose their ability to contract. They can be restored by exercise after the joint is freed.

Exercise

When a muscle receives a single stimulus, it contracts once and then relaxes. This is known as a *muscle twitch.* Usually the muscle receives a series of stimuli that follow each other so rapidly that the muscle does not have time to relax in between. The twitches fuse into a single sustained contraction. This is tetanus, the ordinary state of a muscle when it is doing work.

Warming Up. When a muscle is first stimulated by an impulse sufficient to produce a reaction, a contraction of a certain strength results. For a while, the strength of the contraction increases with repeated stimulation. This increase is thought to be brought about partly by changes in the concentration of ions within the muscle and partly by a simple increase in temperature. Both of these changes make the muscle fibers more sensitive to stimulation and hence increase the number of fibers that are responding. Thus, a muscle is capable of greater contraction after it has been used for a short time than it is after a period of rest. The more favorable condition for additional work has been brought about by a warming up both chemically and physically. Athletes know well the importance of this warming up period.

Fatigue. Once warmed up, the muscles continue to show maximal contractions for a period of time. Sooner or later, though, continued stimulation brings about a decrease in contraction. The impulse to contract is transmitted from the nerve fibers to the muscle cells by a chemical, *acetylcholine*, which is released by the terminal branches of the nerve fibers. As their supply of acetylcholine is depleted, the individual nerve fibers stop transmitting, fewer and fewer motor units are stimulated, and contraction of the muscle grows weaker and eventually stops.

However, the weariness and pain you feel in your arm when, for example, you have to carry a heavy suitcase for some distance is not so much muscle fatigue as it is the result of continued strain on the ligaments and joint capsules.

Recovery. After strenuous exercise, you continue to pant for a while because you need an extra supply of oxygen to restore the energy sources of the muscle cells. The lactic acid that accumulated in your muscles is carried by the blood to your liver, reconverted to pyruvic acid, to glucose, and then built back up into glycogen. Glucose is passed from the liver to the blood, taken up by the muscles, and built into muscle glycogen. Phosphate is transferred from ATP to creatine to form creatine phosphate again. Aerobic oxidation provides the energy for these processes.

Conditioning. Repeated exercise not only increases the size of the muscle fibers and hence the amount of work they are able to do but it also stimulates and strengthens the respiratory and circulatory systems. More oxygen can be brought to the muscles and carbon dioxide and other waste products can be carried off more readily. The athlete in training can run a mile with less fatigue than some people show after a hundred yards. In addition, one of the side effects of conditioning is the accumulation of buffers in the blood that permit higher acid concentrations in the tissues than can normally be tolerated. Thus, the average adult not highly conditioned cannot tolerate more than about 0.2 per cent lactic acid in his tissues. On the other hand, one trained athlete has been reported to be able to tolerate a concentration of 0.3 per cent lactic acid in his muscles.

Muscle Tone

Even relaxed muscles exert some pull. This comes partly from the elasticity of both muscle tissue and the associated connective tissues and partly from periodic contractions of one group or another of muscle fibers within the largely relaxed muscle. This mild state of contraction is known as *tonus* or *muscle tone.* It is muscle tone that holds your body up against the pull of gravity and keeps your jaw shut when you are not talking or chewing. The contractions of tonus are not tiring because only a few fibers in any muscle are contracted at any one time. Then they relax and other fibers contract.

Isotonic and Isometric Contractions

In ordinary muscular contractions, the muscles shorten and a weight is lifted. It may not be an actual weight that is lifted from the floor—it may be an arm or leg or finger, but something moves. Such contractions are called *isotonic.* Sometimes, though, muscles contract and no movement results. Perhaps the weight to be lifted is too heavy. If the muscles on both sides or the arm contract simultaneously so that the pull is equal, the arm does not move. Such contractions are called *isometric.* Isometric exercises are ones in which antagonistic sets of muscles are contracted at the same time. They may be very important on long space flights. Astronauts face not only prolonged periods of inactivity but also the effect of lack of gravity on the tone of the postural muscles. If the astronauts are taught to do isometric exercises in their space suits, they may be able to minimize muscular atrophy and the concomitant bone atrophy.

Heat Production

The chemical energy involved in muscular contraction is not all converted into the mechanical energy of work. Most of this energy appears as heat. Even under the best of conditions, the muscles are only about 40 per cent efficient in converting the energy stored in glucose into work and the normal level of efficiency is around 30 per cent—about the same as that of an automobile engine.

The heat produced by muscular contraction does not represent simply an unfortunate, though inevitable, loss of energy by the body. Much of the heat by which the body maintains its internal temperature at the high and constant level at which it functions most efficiently is produced by the muscles. When the body temperature drops below this level, the heat produced by rhythmic contractions of antagonistic sets of skeletal muscles (shivering) returns the body temperature to the proper level. If you are coming down sick, you may have a chill. The toxins produced by disease organisms stimulate a heat control center in your brain. This stimulation sends messages to the blood vessels in your skin, causing them to constrict and shunt the blood to the internal parts of your body, thus reducing heat loss through the skin. Your skin feels cold to the touch, and the cold receptors in it are stimulated by the drop in skin temperature. Meanwhile, the heat control center has also stimulated the muscular contractions that result in shivering. During such a chill, although you feel cold, the internal temperature of your body is above normal and is rising rapidly. This is a defense mechanism of the body, an attempt to use the heat of muscular contraction to raise the internal temperature to a level that will be detrimental to the invading disease organism.

The reactions taking place during muscular contraction that involve ATP and creatine phosphate are cyclic, that is, the materials in the cell are used over and over again. Those involving glycogen and oxygen are not cyclic. During contraction, the oxygen and the sugar stored in the glycogen are converted into carbon dioxide and water and are lost to the cell. The supply of oxygen and sugar must be constantly renewed. The next chapter takes up the problem of how sugar, and other nutrients, are made available to the cells of the body.

SUGGESTED READINGS

Gordon, M. S., et al.: *Animal Physiology: Principles and Adaptations,* 2nd ed., Macmillan Publishing Co., Inc., New York, 1972.

Grollman, S.: *The Human Body,* 3rd ed., Macmillan Publishing Co., Inc., New York, 1974.

Guyton, A. C.: *Function of the Human Body,* 3rd ed., W. B. Saunders Company, Philadelphia, 1969.

Huxley, H. E.: "The Mechanism of Muscular Contraction," *Sci. Amer.,* Vol. 213, No. 6, 1965.

Kimber, D. C., et al.: *Anatomy and Physiology,* 16th ed., Macmillan Publishing Co., Inc., New York, 1972.

Weichert, C. K.: *Anatomy of the Chordates,* 4th ed., McGraw-Hill Book Company, New York, 1970.

CHAPTER 5

Digestion—The Intake of Food

The digestive system is the gateway into the body. Nearly everything that enters the body does so by passing through the delicate tissues lining the digestive tract. There is one major exception, oxygen, which comes in by way of the lungs. In a sense, material inside the digestive tract is still outside the body proper. It has not been incorporated into the body as water flowing through a hose has not been incorporated into the substance of the hose. The gateway is a very selective one, moreover, because the digestive tract shares with the skin the function of keeping out unwanted substances. The cells of the tissues lining the tract are closely packed, and substances that are taken in pass mostly through the cells, not between them. This imposes strict limitations on the size of the particles that can enter. Chewing is not enough; the very molecules of which food is composed must be broken down into smaller molecules. This chemical digestion is brought about by enzymes that are able to split molecules. The digestive tract, then, breaks up food mechanically into small particles, secretes enzymes and mixes them with the food to break it down chemically, absorbs the substances necessary for the body, and passes the unusable remainder to the outside.

STRUCTURE OF THE DIGESTIVE SYSTEM

The digestive system is essentially a tube, the gut or *alimentary canal,* which passes through the body. Associated with the canal are a number of glands. Figure 5–1 shows the relationship of the various parts of the digestive system.

Mouth

The mouth is made up of a *buccal pouch* (the space between the lips and the teeth) and an *oral cavity* (the space between the teeth and the opening into the next part of the tube, the pharynx). The oral cavity is separated from the nasal chambers above by the *palate*. In front is the hard palate, hard because the tissue lining the mouth here covers a bony shelf. Behind is the soft palate, a movable flap. The buccal pouch is bounded by the cheeks, which prevent food from spilling out of the mouth while it is being ground up by the teeth. Like every tube and body cavity opening to the outside, the mouth is lined by a mucous membrane. As used in this sense, *membrane* means a thin, pliable sheet of tissues. Epithelial cells on the surface of the mucous membrane secrete mucus, a lubricating fluid.

During chewing, the food is mixed with saliva secreted by *salivary glands* opening into the mouth. The largest of these glands, the *parotids,* are the ones that swell up in mumps. Saliva is a watery secretion containing *amylase,* the first of the digestive enzymes encountered by the food, which begins the digestion of starches. More than a liter and a half (a quart and a half) of saliva may be produced by the salivary glands during a day. Salivation is stimulated not only by chewing but also by the sight and smell of food, or even the thought of it. Your mouth truly does water in the presence of appetizing food.

Food is manipulated about by the tongue and chewed until it forms a soft mass; then the tongue forces it back between two muscular columns at the rear of the mouth. The opening between these columns, the *fauces,* leads into the pharynx.

Pharynx

The pharynx is a large chamber, some 125 millimeters (five inches) long, shared by the respiratory and digestive systems. The upper part of the pharynx, into which the nasal chambers open, is the *nasopharynx.* The part into which the mouth leads is the *oral pharynx.* At its lower end the pharynx divides into two tubes, one leading to the lungs and the other, the esophagus, leading to the stomach. Food in the pharynx must pass only into the esophagus. Swallowing, the movement of food from the mouth into the esophagus, involves the closing of the other openings of the pharynx. The back part of the tongue presses against the columns of the fauces, preventing the return of food to the mouth. The soft palate moves up and shuts off the nasopharynx. A special valve, the *epiglottis,* closes the opening to the passage leading to the lungs. If you cough while swallowing, your epiglottis is forced open and a bit of food may enter your respiratory

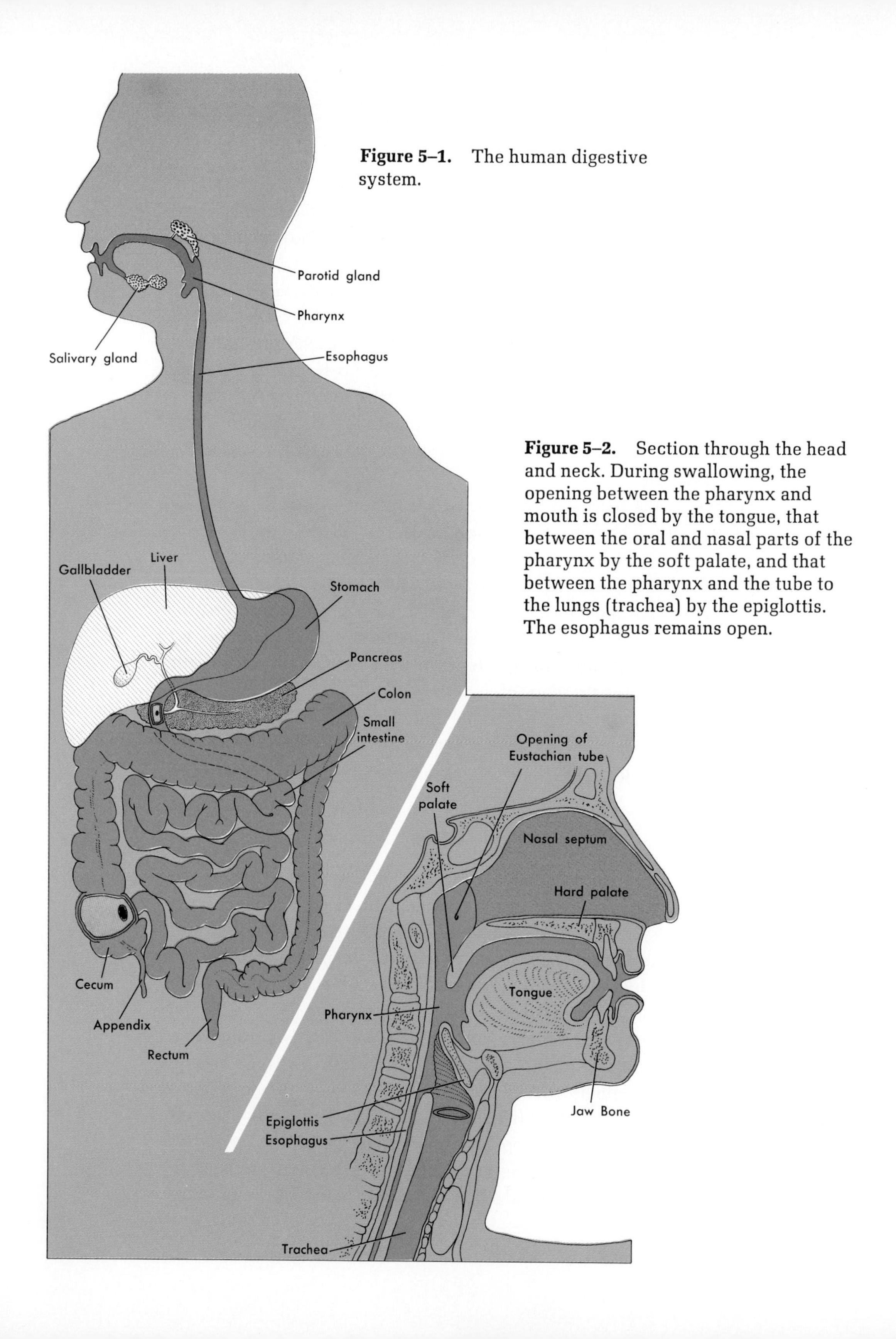

Figure 5–1. The human digestive system.

Figure 5–2. Section through the head and neck. During swallowing, the opening between the pharynx and mouth is closed by the tongue, that between the oral and nasal parts of the pharynx by the soft palate, and that between the pharynx and the tube to the lungs (trachea) by the epiglottis. The esophagus remains open.

tract. You choke and gasp in an effort to dislodge the offending particle, and you may say, "I swallowed something the wrong way." This is literally true. Figure 5–2 shows the relationship of the various parts involved in swallowing.

Esophagus

The esophagus is a tube about 250 millimeters (ten inches) long leading from the pharynx to the stomach. Its walls are extensively folded vertically and the central cavity, the *lumen,* is usually practically closed. Because of the folding, the walls are expansible so that you can swallow rather large mouthfuls.

Underlying the mucous membrane of the esophagus is a layer of connective tissue, the *submucosa,* and surrounding this are two layers of muscles. In the upper part of the esophagus, the muscles are striated but in the lower part they are smooth. The fibers of the inner layer are arranged in a circular manner around the tube; those of the outer layer are longitudinal. Food is moved down the esophagus by *peristalsis,* that is, by waves of alternating contraction and relaxation of these two muscle layers (see Figure 5–3).

The esophagus passes down through the chest behind the heart and through an opening in the *diaphragm,* the dome-shaped sheet of muscle that separates the chest region from the abdominal cavity. No digestive enzymes are secreted by the esophagus, and food passes through it in just a few seconds to enter the first large digestive part of the tube, the stomach.

Stomach

The stomach is an expanded part of the alimentary canal that lies in the upper part of the abdominal cavity. This cavity is lined by a smooth, shiny

Figure 5–3. Diagram of the movement of food through the digestive tract by peristalsis.

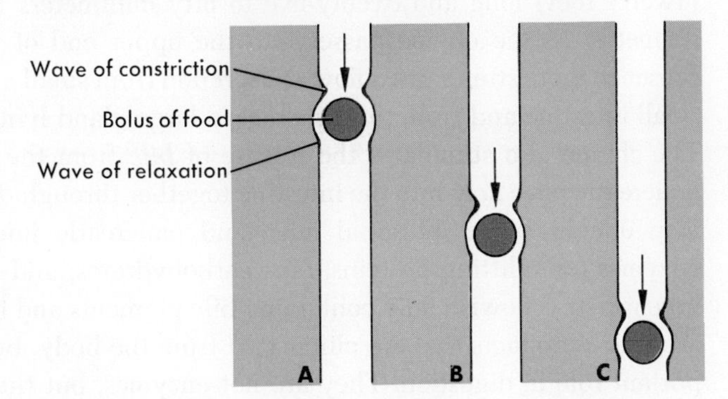

membrane, the *peritoneum.* From the midline of the back part of the cavity, this membrane extends inward as a double sheet, the *dorsal mesentery,* to enclose the alimentary canal. Thus, the wall of the part of the gut that lies within the abdominal cavity consists of the mucous membrane (mucosa), the submucosa, the muscle layers, and the investing peritoneum (the serosa). The dorsal mesentery holds the canal in place, yet, because it is flexible, allows the movements by which food is mixed and pushed along.

In the stomach food is churned and further broken up by the muscular activity of the stomach wall and is mixed with *gastric juice,* a fluid secreted by cells in the wall of the stomach. Gastric juice contains hydrochloric acid and the enzyme *pepsin.* The amylase secreted in the saliva works in a more or less neutral situation; as food is mixed with the gastric juice it becomes more and more acid and the activity of the amylase is inhibited. On the other hand, pepsin is secreted in an inactive form, *pepsinogen,* and is only able to function after it is mixed with hydrochloric acid. Pepsin breaks down large proteins into shorter polypeptide chains. The mucus formed by the cells of the mucous membrane protects the wall of the stomach from damage by the pepsin and hydrochloric acid. Sometimes, for some reason, this barrier breaks down, a part of the stomach wall is digested, and a peptic ulcer develops. The strongly acid environment in the stomach is destructive to many bacteria that may be swallowed with the food.

The activities of the stomach reduce the food to a thin, soupy liquid called *chyme.* After several hours in the stomach, the chyme passes into the intestine through a small opening, the *pyloric orifice.* This opening is surrounded by the *pyloric sphincter,* a ring-shaped muscle formed by a thickening of the circular muscle layer.

Small Intestine

The small intestine is called small because it is smaller in diameter than either the stomach or the large intestine. In life it is about three-and-a-half meters (twelve feet) long and twenty-five to fifty millimeters (one to two inches) in diameter. As the chyme passes into the upper end of the small intestine, its presence there stimulates a flow of secretion from small glands in the wall of the small intestine and from the *pancreas,* a large gland lying outside the intestine. The chyme also stimulates the release of *bile* from the *gall bladder.* Bile and pancreatic juice flow into the intestine together through ducts that join at a common opening. The intestinal juice and pancreatic juice both contain many enzymes for splitting proteins, fats, carbohydrates, and nucleic acids. Bile is a greenish or yellowish fluid containing bile pigments and bile salts. The pigments are waste products that are eliminated from the body, but the salts play an important role in digestion. They are not enzymes, but they are able to emulsify fats, that is, to break them up into small droplets. These droplets present a larger surface area to the fat-splitting enzymes and thus bile salts speed up the digestion of fats. They also make fats more soluble in water. Bile, pancreatic juice, and intestinal juice are all alkaline and change the acid chyme into an alkaline substance. Some of the intestinal enzymes are released by disintegrating cells that are shed continuously from the mucosa.

Food is moved through the small intestine by peristalsis in about four to six hours. The lining of this part of the gut contains innumerable minute, fingerlike processes called *villi,* which enormously increase the surface area of the intestine (see Figure 5–4). The villi are the structures where the digested foods, now in the form of simple sugars, amino acids, fatty acids, glycerol, and nucleotide components, are actually absorbed. The sugars, amino acids, and some of the fatty acids pass through the cells lining the intestine to enter blood vessels in the intestinal wall. These substances are then transported to the liver. Most of the products of fat digestion, instead of passing to the liver, are recombined into fat in the mucosal cells and are absorbed by tiny lymphatic vessels that ultimately transport them back into the main circulatory system (see Chapter 8). In the lymphatic vessels the fats form small globules surrounded by sheaths of absorbed protein molecules. The protein sheaths prevent the droplets from coalescing or sticking to the vessel walls and so keep them suspended in the fluid lymph. Because the concentration of these droplets makes the content of the lymphatic vessels of the intestine look milky, these vessels are known as *lacteals.* Other substances in solution in the chyme, such as salts and vitamins, are also absorbed into the body in the small intestine. The remainder of the chyme together with epithelial debris from the walls of the gut, the bile pigments, and bacteria, are passed into the large intestine.

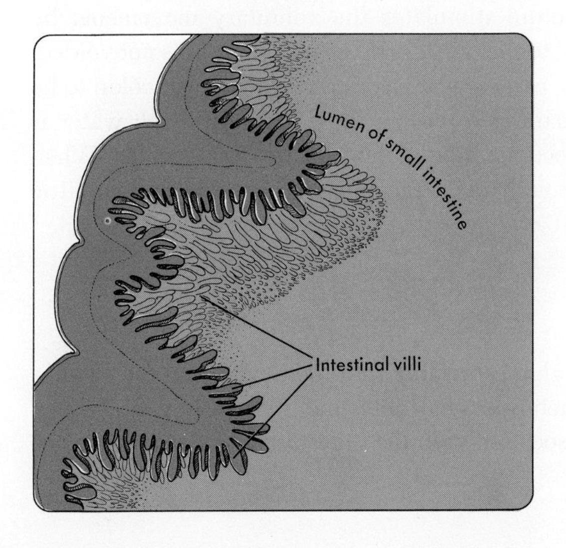

Figure 5–4. Section through the wall of the small intestine.

Large Intestine

The large intestine is about one-and-a-half meters (five feet) long; it is wider than the small intestine and its walls are more distensible. The small intestine enters the large intestine a few inches back from the end, leaving a blind pouch, the *cecum.* In man, the end of the cecum is shrunken, forming the appendix. Just before it enters the large intestine, the small intestine is surrounded by a sphincter muscle, a ring of smooth muscle tissue that acts as a valve. When peristaltic waves from the small intestine reach the region of the sphincter, the muscle relaxes, allowing the chyme to pass into the large intestine. The valve prevents backflow into the small intestine. The main part of the large intestine is called the *colon;* passage of food through it may take as long as twelve hours. Many bacteria are present as normal and useful inhabitants of the large intestine. These bacteria complete the digestion of foods, and, in return for the nutrients they absorb, produce vitamins, which are in turn absorbed by their human host. Much of the water, both that taken in with the food and drink and that in which the enzymes are dissolved or suspended, is removed from the chyme through the wall of the large intestine.

The remains of the chyme, which is no longer liquid, is molded into masses of fecal material. These masses are then passed on to the lower part of the large intestine, the *rectum,* which by muscular activity can force them to the outside through the anus. The anal opening is guarded by two sphincter muscles, an upper involuntary one and a lower voluntary one. Voiding of bowels generally takes place in the morning for the simple reason that food is not eaten overnight, persistaltic activity wanes, and the fecal material tends to accumulate in the part of the colon just above the rectum. Eating breakfast in the morning stimulates peristaltic activity, which moves the fecal material from the lower colon to the rectum, and pressure in the rectum stimulates the voluntary movements by which the fecal material is moved to the outside. If the fecal matter is not voided, a reverse peristalsis in the rectum can move it back up into the lower colon to be held until a later time. If it remains too long in the colon, too much water is absorbed, the fecal material becomes hard, and constipation results. When diarrhea occurs, food passes through the large intestine so quickly that little water is removed.

Liver and Pancreas

In addition to the salivary glands opening into the mouth and the smaller glands in the walls of the stomach and small intestine, two large glands lying outside the digestive tract are associated with the digestive processes. These are the liver and the pancreas.

Liver. The liver, the largest gland in the body, has many functions. It secretes bile, which helps convert the contents of the alimentary canal from the acid state in which they leave the stomach into an alkaline state so that the enzymes of the pancreas and intestine can function in a more suitable environment. The bile salts bring about emulsification of fats, making them more susceptible to digestion. Bile is also the vehicle by which bile pigments, resulting from the breakdown of red blood cells, are eliminated from the body. Bile is produced constantly by the liver and stored in the *gall bladder,* a pear-shaped sac embedded in the liver substance. The gall bladder is stimulated to release bile by a hormone secreted by the small intestine in the presence of fat.

The simple sugars absorbed through the intestinal wall are transported to the liver and the excess sugar not needed immediately by the body is converted into glycogen. The liver is capable of splitting this glycogen back to simple sugars and releasing them to the circulatory system so that they can be transported to the cells of the body. Excess lactic acid from muscular activity is also converted to glycogen in the liver.

The liver is able to rebuild fatty acids taken in as food into forms appropriate for human needs, to oxidize them with the release of energy, or to build them into other substances needed by the body. Many of the amino acids carried to the liver from the small intestine simply pass through on their way to other cells of the body. Some of the amino acids are *deaminated,* that is, the amino group (NH_2) is removed. The nonnitrogenous part that remains may be oxidized for energy or built into sugars or fat. The amino group may be used to form other amino acids, other nitrogenous products needed by the body, or it may be converted into a waste product, *urea,* to be excreted from the body. The liver is also able to build amino acids into important proteins found in the blood. With all this intense chemical activity, the liver is the greatest source of internal heat in the body.

Pancreas. The pancreas is a gland with a dual function. It produces hormones and is thus part of the endocrine system (see Chapter 10). It also produces enzymes, which flow into the small intestine through a duct. Figure 5–5 shows the relationship of the liver, pancreas, esophagus, stomach, and upper part of the small intestine.

CHEMICAL DIGESTION OF FOODS

Not everything taken in through the mouth is considered a food. The body has no enzymes to digest the carbohydrate cellulose that makes up a large part of the plant material eaten. Nor can waxes or phospholipids be digested. These

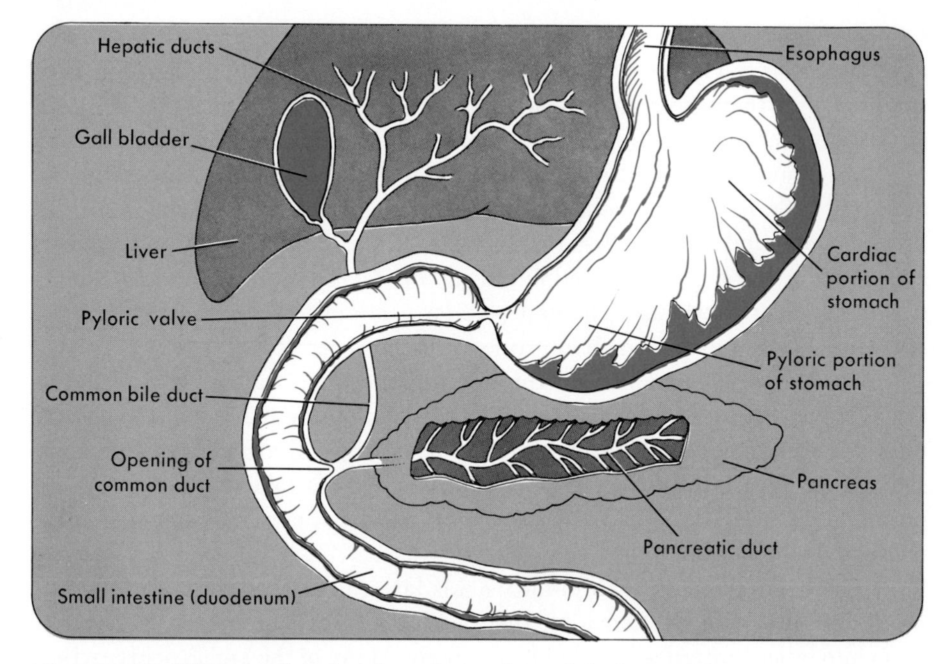

Figure 5–5. Diagram showing the relationship of the liver and pancreas to the small intestine.

substances simply pass through the digestive tract. (They may be broken down by the bacteria of the colon but the body gets no nourishment from them.) In addition, spices such as pepper may be eaten simply because they stimulate the taste buds. You may inadvertently ingest organisms that cause disease. These are not considered foods.

What Are Foods?

Foods are substances absorbed through the digestive tract that are used by the body to yield energy, to build tissue, and to regulate body processes. Water, vitamins, and mineral salts are absorbed without further change. Molecules of carbohydrate, fat, protein, and nucleic acid are too large to be absorbed and must first be broken down chemically.

Digestion

Chemical digestion is the process that changes complex molecules into smaller ones that can be absorbed. Usually this is done by hydrolysis, in which water is added to a larger molecule causing it to split into smaller molecules.

Carbohydrate Digestion. Some simple sugars are taken in as food; they can be absorbed without further digestion. Most carbohydrates, though, are ingested in the form of starches. They are broken down into double sugars and finally into simple sugars in the digestive tract. The stages in starch digestion might be diagrammed thus:

$$\underset{\text{starch molecule}}{C_{1200}H_{2000}O_{1000}} + \underset{\text{water}}{100\ H_2O} \longrightarrow \underset{\text{double sugar}}{100\ C_{12}H_{22}O_{11}}$$

$$\underset{\text{double sugar}}{C_{12}H_{22}O_{11}} + \underset{\text{water}}{H_2O} \longrightarrow \underset{\text{simple sugar}}{2\ C_6H_{12}O_6}$$

Protein Digestion. The digestion of proteins follows a similar path, but the intermediate stages are short chains known as *peptides* and the final products are amino acids. Some protein-digesting enzymes, though, break off the end amino acid of a chain rather than breaking the chain in the middle. Some of the amino acids needed for protein formation can be synthesized by the body cells but eight cannot and two are not formed in adequate amounts. These, called the *essential amino acids,* must be supplied in the food. Animal proteins, found in meat, milk, cheese, and eggs, are richer in essential amino acids than are most plant proteins. For this reason, a diet very poor in animal proteins is likely to be deficient in the essential amino acids and result in a deficiency disease.

Fat Digestion. In fat digestion, the fat is broken down into its component fatty acids and glycerol. It seems probable that some of the unsaturated fatty acids needed by humans cannot be formed by the liver from other fatty acids but, like the essential amino acids, must be present in the diet.

Nucleic Acid Digestion. Enzymes secreted by the small intestine are able to split the nucleic acids into nucleotides. Other enzymes break the nucleotides into phosphoric acid, sugar, and nitrogenous bases. Little is known about the dietary role of these essential ingredients of every cell.

A summary of the main stages of digestion in man is shown in Figure 5–6.

A Balanced Diet

To be adequate, a diet must include protein to provide nitrogen and the essential amino acids, and fat to provide the essential fatty acids. But because both protein and fat can be oxidized for energy, and because sugars can be formed in the liver from either, it might seem that one could get along without carbohydrates. But when excess amounts of fat are oxidized for energy, certain toxic substances called *ketone bodies* accumulate in the blood. These are acid

GLAND	ENZYME	FOOD AFFECTED	PRODUCT
SALIVARY	Amylase	Starch	Double sugars
GASTRIC	Pepsin (peptidase)	Protein	Peptides
	Rennin	Milk protein	Coagulates milk
PANCREAS	Amylase	Starch	Double sugars
	Lipase	Fat	Free fatty acids and glycerol
	Trypsin and Chymotrypsin (peptidase)	Proteins and Polypeptides	Smaller polypeptides
	Nuclease (?)	Nucleic acids	Nucleotides
INTESTINAL	Peptidases	Polypeptides	Amino acids
	Maltase	Maltose	Simple sugars
	Lactase	Lactose	Simple sugars
	Sucrase	Sucrose	Simple sugars
	Nuclease	Nucleic acids	Nucleotides
	Nucleotidase	Nucleotides	Phosphoric acid, sugars, and nitrogenous bases

Figure 5–6. Summary of human digestion showing the sources of the digestive enzymes, the food each digests, and the products.

and can upset the acid-base balance of the body tissue, leading to acidosis. A high protein diet can also produce ketone bodies. In addition, when too many amino acids are deaminated before being oxidized, an excess of ammonia is produced. This ammonia must be converted to urea and excreted. The kidneys become overworked and kidney trouble may develop. You thus need all three of the basic foods in your diet.

ENZYMES

The breakdown of food molecules is brought about by enzymes. The enzymes of digestion were the first to be recognized and studied, but it is now known that practically all of the chemical reactions of the body, those that take place within the cells as well as those of the digestive tract, are mediated by enzymes. Thus, the various reactions carried on in the liver and those discussed in the section on muscular contraction (Chapter 4) require the presence of enzymes.

The Function of Enzymes

A *catalyst* is a substance that increases the rate of a chemical reaction without being itself changed by the reaction. An enzyme is a biochemical catalyst. It binds to the reactant(s) and releases the product(s) after the reaction has taken place. Different enzymes can catalyze all the various types of reactions described in Appendix A. Also, in order for a reaction to take place, even one that releases energy, some energy must be fed into the system. This is called the *energy of activation*. In muscles, the sugar glucose is oxidized with the release of energy. A lump of table sugar can also be oxidized (burned) with the release of energy if one applies energy of activation in the shape of a lighted match. An enzyme is able to reduce the energy of activation required to initiate a chemical reaction.

The Structure of Enzymes

An enzyme is a protein molecule with a complex folded structure that gives it a characteristic shape. On its surface is an *active site* that fits (chemically) the shape of a particular substance, the *substrate*, that is acted on by the enzyme. A crude analogy is the way one piece of a jigsaw puzzle fits into one other piece, but does not fit the other pieces of the puzzle. Figure 5–7 shows in highly diagrammatic fashion how this occurs. Once the reaction has taken place, the substrate separates from the enzyme, which is then free to act again. An enzyme can thus be used many times.

Many enzymes require the presence of certain ions in order to function. An enzyme may have a *prosthetic group* attached to it. Such a group is a nonprotein molecule tightly bound to a protein and necessary for its activity. The heme part of the myoglobin molecule is a prosthetic group. An enzyme may be inactive unless it is more or less closely linked to another nonprotein compound called a *coenzyme*. A coenzyme is less tightly bound to the protein than a prosthetic group is. Many coenzymes seem to be relatively nonspecific, that is, they are able to function with a number of different enzymes. The hydrogen acceptors NAD and NADP mentioned in Chapter 1 are such coenzymes, whereas FAD is a prosthetic group.

Figure 5–7 suggests that an enzyme should be able to work in either direction, to break as well as make. If A and B fit the surface of the enzyme, then AB should also fit. In a test tube, many enzymes can cause a reaction to go in either direction, but in the living cell they may not. The products of enzymatic reactions do not accumulate at the site of the enzyme. They enter into other reactions or are transported to other parts of the cell. In the organism, then, many enzymatic reactions go only one way.

Sometimes, though, when enough of the end product has accumulated, it

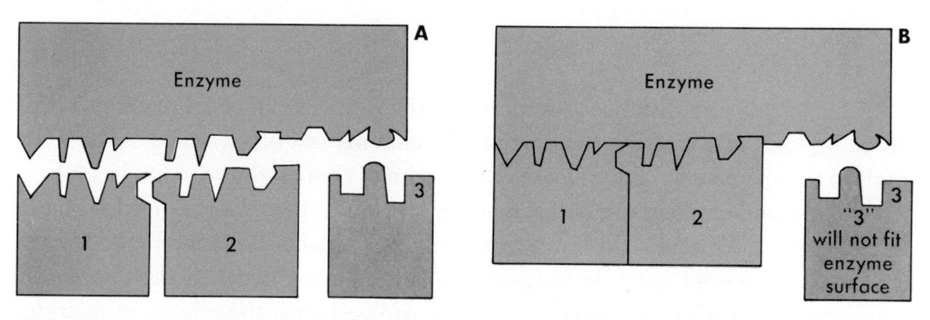

Figure 5–7. Model of enzymatic activity. A: The enzyme and other molecules (1, 2, and 3). B: Molecules 1 and 2 are bound to the surface of the enzyme and can react. They are substrates of the enzyme, while 3, which does not "fit," is not.

competes with the substrate for the site and may block further production. This is one way that enzymatic activities of the cell can be regulated.

Some enzymes are said to be *allosteric.* In addition to the active site, these enzymes have another site that can bind to a substance other than the substrate. This binding changes the shape of the active site so that it is no longer able to bind to the substrate. Frequently the enzymatic activities of the cell consist of chain reactions. Enzyme A changes substrate A to B, enzyme B changes substrate B to C, and so on. The final end product may bind allosterically to enzyme A and thus turn off the sequence. End product control, whether by competitive or allosteric binding, prevents the cell from producing more than it needs of a given substance.

The Naming of Enzymes

In the early days of work on the digestive enzymes, many of them were given more or less arbitrary names. *Pepsin* comes from a Greek word meaning "digestion"; *ptyalin* (an early name for amylase) comes from the Greek for "spittle." Then it became customary to form the name of the enzyme by adding the suffix *-ase* to the stem of the name of the substrate. *Amylon* means "starch" in Greek, so *ptyalin* became *amylase*. Later, as more was learned about the many enzymes within the cells, this system broke down because a single substrate can be acted on in different ways by different enzymes. The current system of naming enzymes involves adding *-ase* to the particular reaction the enzyme catalyzes. The resultant names are often formidable—*glyceraldehyde-3-phosphate dehydrogenase* is an example. For the digestive enzymes we here continue to use the "common" names, *proteases* for those that split proteins and *lipases* for those that split fats.

The Characteristics of Enzymes

The knowledge that enzymes are proteins helps clarify some of their characteristics.

1. Being protein, enzymes can be destroyed by heat, which breaks the chemical bonds by which they are held in their characteristic shapes. They are then said to be *denatured.* Most human enzymes function most effectively at the normal temperature of the body.
2. Enzymes do their best work at rather specific conditions of acidity, because changes in ionic concentration can also break hydrogen bonds, and so alter the shape of the enzyme.
3. Enzymes can be inactivated by poisons. Certain substances can combine with and permanently block the active site or can interfere with the functioning of the nonprotein part of the enzyme.
4. Enzymes are specific, that is, they catalyze only a single reaction. The degree of specificity varies. Lipase is able to break the bond between any fatty acid and glycerol, but some proteases will break the bond only between two specific amino acids.

Vitamins

The body is able to synthesize some of the coenzymes it needs; others, or their precursors, it must take in from outside. The latter are called *vitamins.* (Not all vitamins have been shown to function as coenzymes, but most are known to do so.)

Many vitamins, or the coenzymes formed from them, act as acceptor-donors of hydrogen. Others are acceptor-donors of carboxyl groups (COOH) or of amino groups (NH_2), or of electrons.

Because vitamins are not used up in the reactions in which they take part, they can be used over and over again. You need to take in only enough vitamins to replace those lost from your body through natural wear and tear. Extremely large doses of at least some of the vitamins can be harmful. Polar bear liver is very rich in Vitamin A, and arctic explorers who feast on it may suffer severe headaches and vomiting. Vitamin D stimulates absorption of calcium by the intestine. Overdoses of vitamin D can result in the calcification of soft tissues and in kidney damage.

Not all vitamins are necessarily taken in with food. The bacteria that are normal inhabitants of the intestine manufacture certain B vitamins and vitamin K. Antibiotics that destroy harmful bacteria may also destroy these useful bacteria, so that it may be necessary to give the patient vitamin supplements.

Vitamin D can be synthesized in the skin from a precursor in the presence of sunlight and absorbed into the bloodstream.

Table 5–1 lists the more familiar vitamins and the ailments that result from their deficiency.

Most animals spend a good bit of time and energy getting their food. The kind of food an animal eats is largely determined by the structure of its digestive tract and the kinds of digestive enzymes it has. But the process of getting food involves much more than the animal's digestive tract. For example, a horse is adapted for grazing on grass but would make a poor show of stalking a rabbit.

Table 5–1. Some Common Vitamins

Vitamin	Major Sources	Effect of Deficiency
Vitamin A	Green and yellow vegetables, dairy products	Night blindness, keratinization of epithelial tissue
Vitamin B_1 (thiamin)	Cereal, yeast, pork	Fatigue, changes in heart and circulation, neuritis; extreme cases are called beriberi
Vitamin B_2 (riboflavin)	Milk, liver, wheat germ	Sores on the lips, blurred vision
Nicotinic acid (niacin)	Lean meat, eggs, yeast	Headache, stomach ache; extreme cases are called pellagra
Vitamin B_6 (pyridoxine)	Liver, fish, beans, cabbage, molasses	Itching, scaly skin, sleeplessness, irritability
Pantothenic acid	Green vegetables, liver, intestinal bacteria	Deficiency not seen in man; essential for oxidation of pyruvic acid and synthesis of lipids
Folic acid	Present in many foods	Certain types of anemia
Vitamin B_{12} (cyanocobalamine)	Liver, lean meat, kidney	Pernicious anemia
Vitamin C (ascorbic acid)	Citrus fruit, tomatoes	Fragility of capillaries, bleeding gums, tenderness of legs; extreme cases are called scurvy
Vitamin D	Cod liver oil, sunshine on skin	Soft, fragile bones, bowlegs (rickets in children)
Vitamin K (naphthoquinone)	Green plants, intestinal bacteria	Inhibition of blood clotting

The evolution of the whole animal is a reflection of the evolution of its digestive enzymes.

SUGGESTED READINGS

Crouch, J. E., and J. R. McClintic: *Human Anatomy and Physiology,* John Wiley & Sons, Inc., New York, 1971.

Green, D. E.: "Enzymes in Teams," *Sci. Amer.,* Vol. 181, No. 3, 1949.

Grollman, S.: *The Human Body,* 3rd ed., Macmillan Publishing Co., Inc., New York, 1974.

Guyton, A. C.: *Function of the Human Body,* W. B. Saunders Company, Philadelphia, 1964.

Kimber, D. C., et al.: *Anatomy and Physiology,* 16th ed., Macmillan Publishing Co., Inc., New York, 1972.

Robinson, C. H.: *Fundamentals of Normal Nutrition,* 2nd ed., Macmillan Publishing Co., Inc., New York, 1973.

CHAPTER 6

Respiration—Exchange and Use of Gases

Man does not live by bread alone. If he is to release the energy locked in the food molecules, he must also have oxygen. And he must have it constantly. He cannot store much oxygen in his body as he stores glycogen and fat. If a man is deprived of oxygen for even a few minutes, his cells soon exhaust their stores of energy, their multitudinous activities cease, and irreparable damage is soon done.

Respiration in man involves three processes: breathing, by which fresh air is brought into contact with the lining of the lungs; physical respiration, by which oxygen is transferred from the lining of the lungs to the cells where the oxygen is used, and carbon dioxide is transferred from the cells where it is formed to the lungs to be expelled; and cellular respiration, by which the energy of food is converted to the energy of ATP. In this last process, oxygen is used and carbon dioxide and water are formed.

BREATHING

Gas exchange between organisms and the environment takes place across cell membranes. These membranes must be kept moist to prevent desiccation; the gases are in solution in the fluid with which the cells are bathed. This poses problems for large animals that live on land. Remember that the outer layer of dead, keratinized cells of the skin has an important protective function. If large areas of unprotected cells were exposed to the air, not only would they be very susceptible to mechanical damage but also intolerable amounts of water would be lost through evaporation. In man, as in most land-living vertebrates, the surface through which the exchange of gases takes place is the lining of the lungs. These are saccular organs that lie inside the body although they are connected to the outside by narrow passages. The location of the respiratory surfaces inside the body means there must be some mechanism for bringing air to them. And there must be a regular exchange between the air in the lungs and outside air; if there were not, the air in the lungs would soon be depleted of the necessary oxygen. This exchange of air is accomplished by breathing.

Structure of the Respiratory System

The respiratory system is made up of a pair of nasal chambers, a series of passages (the pharynx, larynx, trachea, and bronchi), and the lungs. Figure 6–1 shows the upper part of the respiratory tract.

Nasal Chambers. The nasal chambers are large cavities separated by a medium septum and opening to the outside through the nostrils. Hairs at the opening of the nostril prevent the entrance of foreign particles. On the walls of the chambers are projecting, scroll-shaped bones called *turbinates.* The nasal chambers, like the rest of the respiratory tract, are lined with a mucous membrane. Air drawn in through the nose swirls around the projecting turbinate bones and is warmed and moistened. Specks of dust and detritus settle on the film of mucus on the membranes. Beating cilia on cells lining the chambers

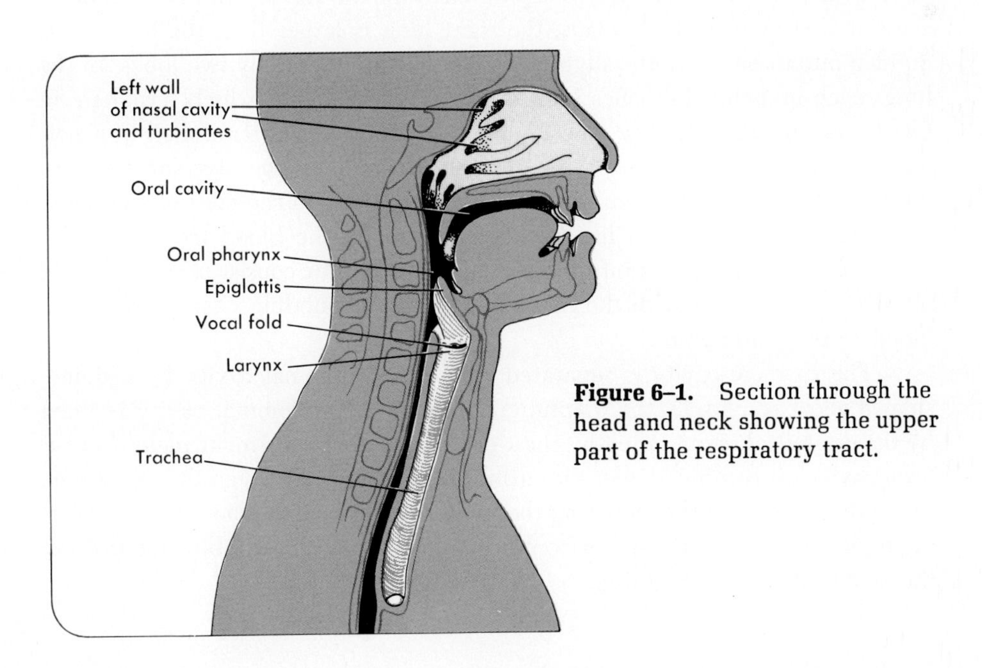

Figure 6–1. Section through the head and neck showing the upper part of the respiratory tract.

pass the mucus, along with the entrapped particles, backward to the pharynx where it is either swallowed or expectorated. These are all protective devices, ensuring that the air reaching the delicate lining of the lungs is warm, moist, and relatively free of foreign particles.

Respiratory Passages. From the nasal chambers the air passes into the pharynx. Except during the act of swallowing, the epiglottis, the flap at the entrance to the *larynx,* remains folded back so that air can pass from the pharynx into the larynx, or voice box. This is the expanded upper end of the tube leading from the pharynx to the lungs. It is less prominent in women than in men, in whom it forms the Adam's apple. The walls of the larynx are supported by cartilages and across it from front to back run two fibrous bands, the vocal cords. In speaking, air expelled from the lungs sets these cords vibrating. From the larynx, the air passes into the *trachea,* a flexible, permanently open tube whose walls are supported by sixteen to twenty horseshoe-shaped cartilages (see Figure 6–2). These incomplete rings are open toward the back so that the esophagus, which lies behind the trachea, can make a temporary expansion at the expense of the trachea when food is swallowed. The trachea divides at its lower end into two *bronchi,* which pass into the lungs, one into the right lung, the other into the left.

Lungs. The lungs are large, cone-shaped organs occupying the thoracic (chest) cavity; the right lung is separated from the left by the heart and its associated structures. In humans, the right lung is larger than the left and is divided into three lobes; the slightly smaller left lung has only two lobes. In the lungs each bronchus branches into a series of *bronchioles,* which terminate in blind sacs, the *alveolar sacs* (see Figure 6–3). Each sac is enmeshed in a network of capillaries. Oxygen passes through the walls of the alveolar sacs into these minute blood vessels to be carried by the blood to the cells of the body, and carbon dioxide, which has been picked up by the blood from the cells, passes from the capillaries into the sacs. The lungs thus consist of a number of alveolar sacs together with the bronchioles and surrounding connective tissue, blood vessels, and nerves.

The thoracic cavity is separated from the abdominal cavity by a dome-shaped sheet of muscle, the diaphragm. Just as the abdominal cavity is lined by the peritoneal membrane, the chest cavity is lined by a similar *pleural membrane,* which is continued over the surface of the lungs. Between the membrane lining the cavity and that covering the lungs is a very small amount of lubricating liquid. This keeps the lungs from becoming irritated by rubbing against the thoracic wall during breathing.

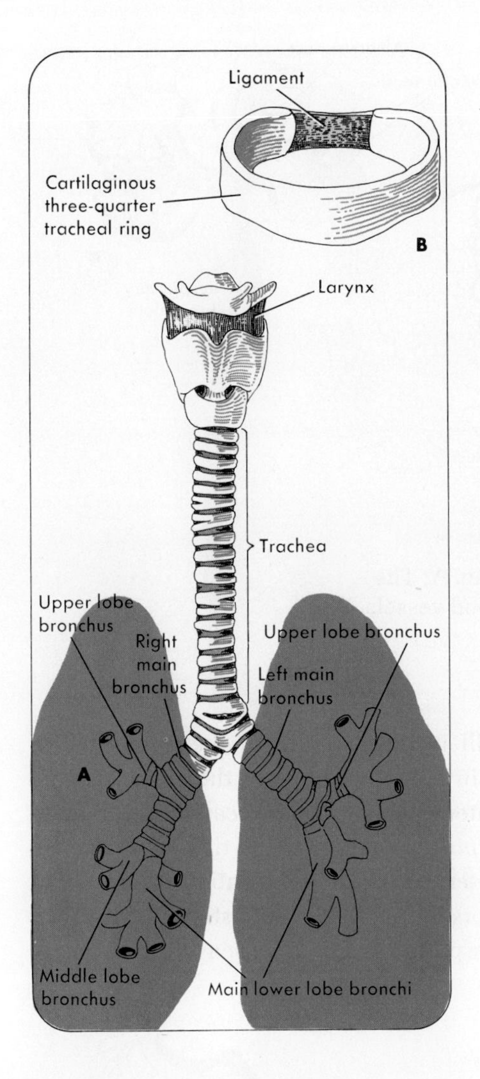

Figure 6–2. A: The human larynx and trachea. B: One of the tracheal cartilages.

Function of the Respiratory System

A rather complicated nervous and chemical coordinating system makes rhythmic breathing possible even during sleep. Not all parts of this respiratory control system are completely understood, but the basic outline seems clear.

Mechanics of Breathing. An *inspiratory center* in the brain sends nervous stimuli to the diaphragm and the muscles running between the ribs (*intercostals*). As a result of the contraction of these muscles, the diaphragm becomes flatter and lower and the ribs are raised and rotated (see Figure 6–4). This brings about an expansion of the chest cavity. Lung tissue is elastic; as the chest

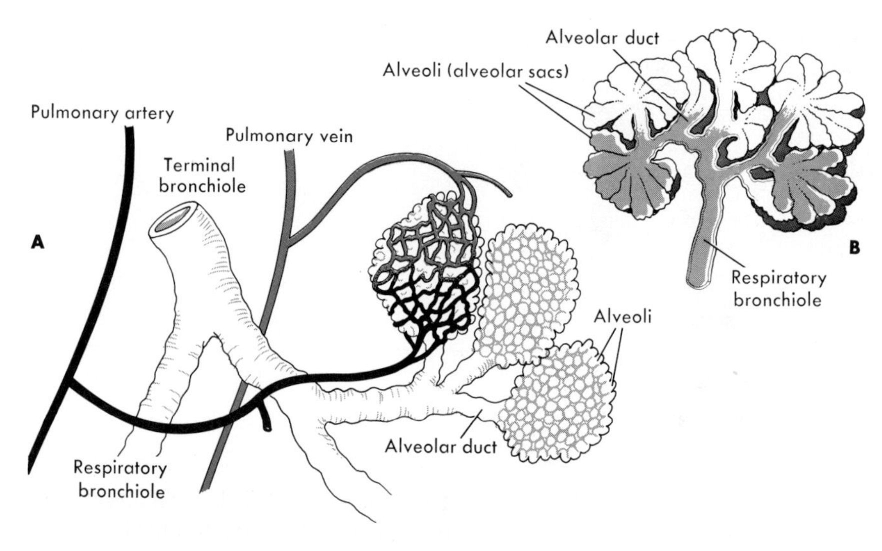

Figure 6–3. The fine structure of the lung. A: The relationship of the ducts, alveoli, and blood vessels. B: Enlarged view of alveoli.

cavity enlarges the lungs expand to fill it, air pressure in them falls below atmospheric pressure, and air rushes in from outside. At the same time, a stimulus passes from the inspiratory center to a *pneumotaxic center* in the brain that in turn sends a stimulus to the *expiratory center*. This then sends an inhibitory stimulus to the inspiratory center. Also, when inhalation brings about the inflation of the lungs, stretch receptors in their walls send stimuli that excite the expiratory center and inhibit the inspiratory center. When the diaphragm

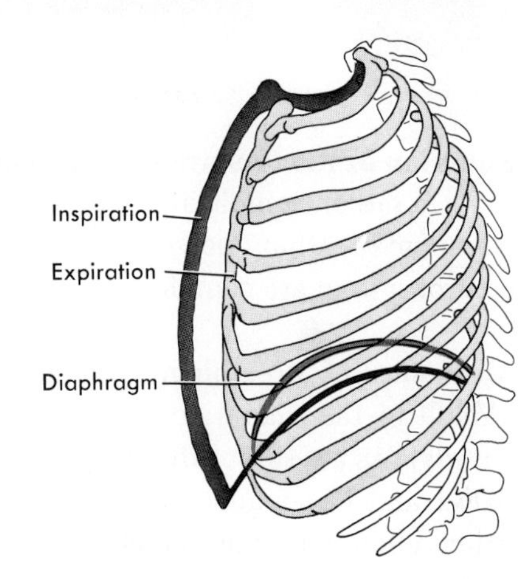

Figure 6–4. Changes in shape of the chest cavity during breathing. During inspiration the diaphragm is lowered and the ribs are raised and rotated outward.

and intercostal muscles no longer receive impulses from the inspiratory center, they relax. At the same time, the expiratory center sends impulses to certain abdominal muscles; these pull down on the ribs and push the abdominal contents against the diaphragm. Lowering the ribs and raising the diaphragm decreases the size of the chest cavity and air is pushed out of the lungs by the elastic recoil of the lung tissue.

The inspiratory center is believed to function continuously unless it is being actively inhibited by stimuli from the expiratory center and the stretch receptors in the lungs. The rhythmic pattern of normal quiet breathing results from the alternating inhibition and release from inhibition of this center.

Chemical Control of Breathing. The amount of oxygen needed by the body and the amount of carbon dioxide that must be eliminated vary from time to time. When you are exercising, your need for oxygen is greater than when you are sitting quietly. Although other chemical factors are involved, the most important stimulus for an increase in the rate and depth of breathing is an increase in the carbon dioxide content of the blood. When the carbon dioxide reaches a certain level, receptors in the brain are stimulated to send impulses to the inspiratory center and the rate and depth of breathing increase.

Another type of chemical control is possible but is not usually involved in breathing. In certain blood vessels of the neck and chest are sensory receptors, the *aortic* and *carotid bodies,* that respond to changes in the level of oxygen in the blood. When the oxygen level drops too low, these receptors send stimulatory impulses to the inspiratory center. Remember that the cells of the body take oxygen from the blood and dump carbon dioxide into it. Hence, as the level of oxygen is lowered in the blood, the level of carbon dioxide is raised. Ordinarily, the carbon dioxide increases enough to stimulate the centers in the brain before the oxygen decreases enough to stimulate the centers in the blood vessels. Among people who live at very high altitudes, such as the Indians of the Peruvian Andes, this is not so. Where the air is very thin, carbon dioxide passes out of the blood more easily than oxygen enters. The oxygen level drops low enough to stimulate the inspiratory center before the carbon dioxide level increases. These interacting factors are diagrammed in Figure 6–5.

The diaphragm and intercostal muscles are skeletal muscles and are thus voluntary. You can exercise some control over your rate of breathing—you can even hold your breath temporarily. But you have no control over the chemical factors in your blood. You cannot, by an act of will, stop breathing entirely.

Lung Volume. The total volume of air in the lungs varies from individual to individual and is usually more in males than in females. The following figures, which are probably close to the human average, give some idea of the amount

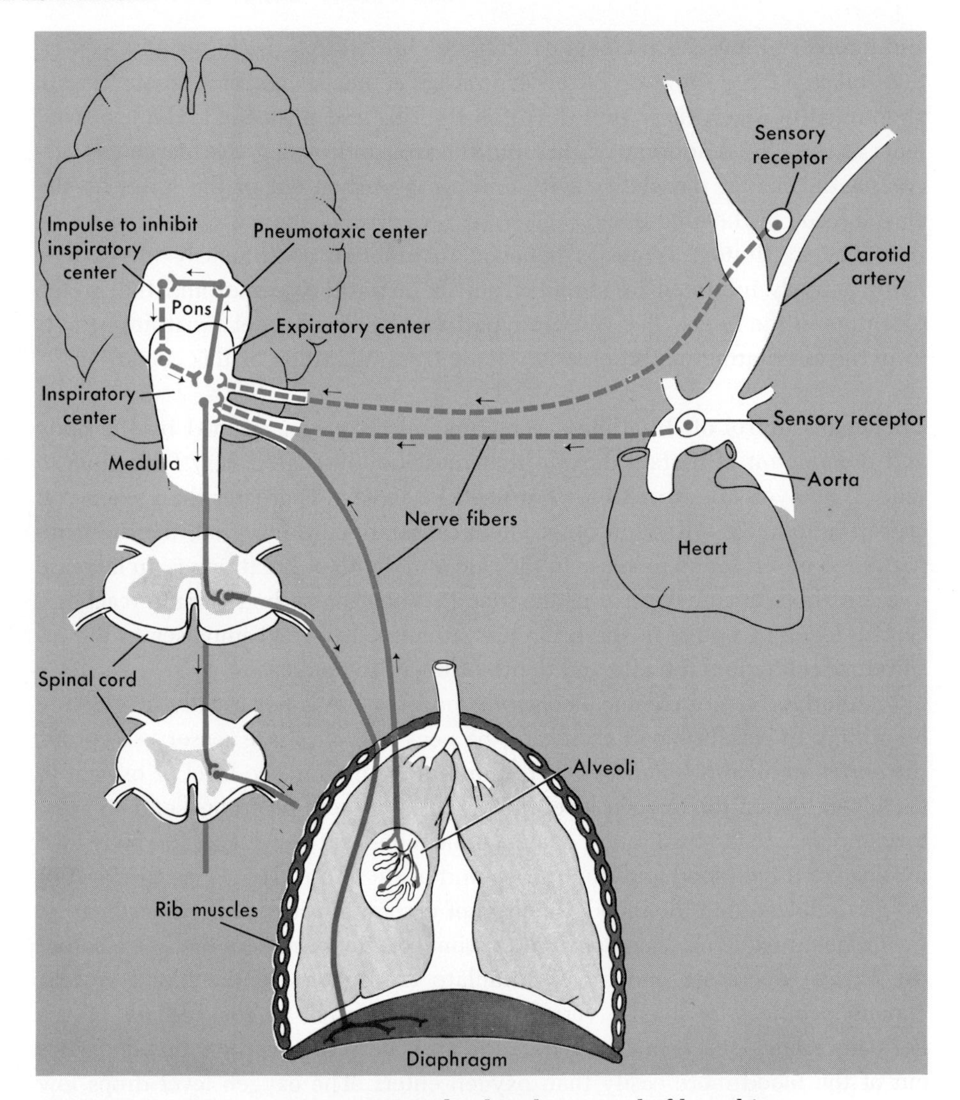

Figure 6–5. Nervous structures involved in the control of breathing include stretch receptors in the lungs, chemical receptors in the brain and blood vessels, nerve fibers passing through the brain and spinal cord, and several separate centers in the brain.

of air in the lungs and the amount moved in and out. In normal, quiet breathing about 500 cubic centimeters (thirty cubic inches) of air are taken in and expelled at each breath. The air that moves in and out in quiet breathing during rest is known as *tidal air.* In addition, you can, by a strong exertion, take in another 2,500 cubic centimeters (150 cubic inches) of air, known as *reserve inspiratory air.* On the other hand, you can, by violent expiration, deflate your lungs by 1,000 cubic centimeters (sixty cubic inches) more than you normally expel in quiet breathing. This is called *reserve expiratory air.* The amount of reserve expiratory air, tidal air, and reserve inspiratory air together make up what is known as the *vital capacity* of the lungs, about 4,000 cubic centimeters (240 cubic inches). In a grown person, even after the most violent expiration, about 1,500 cubic centimeters (ninety cubic inches) of air remains in the lungs. It cannot be evacuated under any circumstances by normal breathing mechanisms. This is known as the *residual air.* When a child is born there is no air at all in the lungs. After he starts breathing his lungs accumulate residual air, which then can no longer be evacuated except by removing the lung or by letting air in between the lung and chest cavity. The pressure of this air causes the lung to collapse. This is sometimes done surgically in cases of tuberculosis to give the diseased lung a chance to rest and heal.

Modified Respiratory Movements. Because breathing is under voluntary control as well as involuntary control, respiratory movements can be modified for talking, singing, and whistling. In addition to normal breathing and these voluntary modified respiratory movements, there are a number of involuntary respiratory reflexes.

In *sighing,* a deep, slow breath is drawn in, followed by a shorter but correspondingly larger expiration. *Yawning* is similar to sighing, but instead of being drawn in through the nose, the air is taken in through the mouth while the chin is held in a characteristic manner. A *hiccup* is brought about by a very sudden contraction of the diaphragm while the larynx closes; the entering air drawn down through the rapidly narrowing opening causes the characteristic sound. *Coughing* is a reflex to clear the bronchial tubes, trachea, and larynx. It consists of a full inspiration followed by a very violent and rapid expiration with the laryngeal opening held closed during the first part of the expiration; this builds up a tremendous pressure and the increased air flow through the trachea tends to blow out any entrapped particles. The rate of air flow in the trachea in a violent cough may be as much as 112 kilometers (seventy miles) an hour. *Sneezing* is similar to coughing but results from an irritation in the wall of the nasal chamber. The columns of the fauces at the back of the mouth are held together and the soft palate is lowered so that the opening from the pharynx to the mouth is closed and the blast is driven through the nostrils. A sneeze is one of the most difficult of the modified respiratory movements for an actor to portray. *Laughing* consists of a series of short expirations following a long, single inspiration. In laughing, the larynx is kept open all the time and the vocal chords are set in vibration. *Crying,* physiologically, is strikingly similar to laughing, and at times one may pass into the other. The main difference between the two lies in the emotions involved.

PHYSICAL RESPIRATION

As used here, the term *physical respiration* involves both the movement of gases between the cavity of the lungs and the blood (external respiration), and movement between the blood and the cells (internal respiration). Physical respiration is brought about by gradients in the concentration of the gases between the cavity of the lungs, the fluid in the lining of the lungs, the blood in the circulatory system, the tissue fluid in the tissues (see Chapter 8), and the cells, which are surrounded by the tissue fluid. Gases are able to diffuse through the membranes lining the lungs and surrounding the capillaries. In diffusion, molecules of a gas move from a region of higher concentration to a region of lower concentration. If oxygen is being used in a cell and is being brought into the lung by breathing, then the concentration of oxygen will be highest in the lung and lowest in the cell, and the gradient will cause the oxygen to move from the lining of the lung into the capillaries around the alveoli, through the circulatory system to the tissues, and then through the tissue fluid into the immediate cell where oxygen is being used. Conversely, carbon dioxide passes from the cell in which it is produced into the tissue fluid, thence into the capillaries, through the circulatory system to the lung, and from the lung capillaries through the lining of the lung into the alveolar sacs (see Figure 6–6).

Figure 6–6. The paths of oxygen and carbon dioxide in physical respiration. Oxygen enters the bloodstream in the lungs and is carried through the heart to the cells of the body. Carbon dioxide enters the bloodstream from the body cells and is carried through the heart to the lungs.

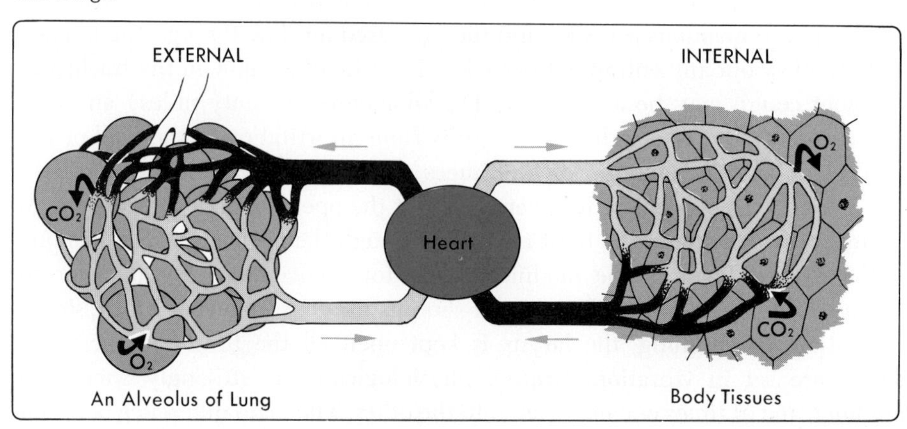

CELLULAR RESPIRATION

The ultimate function of cellular respiration is to convert the energy of a food molecule, such as glucose, $C_6H_{12}O_6$, into the energy of ATP molecules. Respiration can be either aerobic, in which oxygen is used, or anaerobic, in which oxygen is not used. Much more ATP energy is produced by the aerobic respiration of a molecule of glucose than by its anaerobic respiration. The energy needs of most organisms, including man, are so great that they can be met only by aerobic respiration.

Aerobic Respiration

The classical empirical presentation of the breakdown of glucose is given in the equation:

$$C_6H_{12}O_6 + 6\ O_2 \rightarrow 6\ H_2O + 6\ CO_2 + \text{energy}$$

It is a gross understatement to say that this is an oversimplification. As we pointed out in Chapter 5, a similar release of energy takes place when a lump of table sugar is burned. When energy is produced in such a single, large packet, though, most of it is lost as heat. If you were to oxidize glucose in this way in your cells, you could retrieve and store little of the energy for later use. And the heat would destroy the cell proteins. The transfer of energy from the glucose molecule to the ATP molecules within the cell is achieved by a complicated series of smaller reactions that are mediated by enzymes. This process can be divided into four separate stages.

Stage I—Glycolysis. Glycolysis, which means the splitting of glycogen, is really a misnomer for the processes of Stage I. It is glucose, not glycogen, that is broken down, and the glucose need not have come from glycogen. Also, glycolysis does not require oxygen and so is really an anaerobic process. But glucose cannot enter directly into the respiratory processes that do require oxygen. Glycolysis includes preparatory steps by which glucose is made ready for aerobic respiration and so must be considered in any discussion of aerobic respiration. During glycolysis, the six-carbon glucose molecule is broken down into two three-carbon molecules with energy provided by two molecules of ATP. Each of the three-carbon molecules (phosphoglyceraldehyde or PGAL) has two phosphate groups joined to it. It passes two hydrogens to the hydrogen acceptor NAD and two phosphate groups to ADP to form ATP. The three-carbon molecule that remains is pyruvic acid. Thus, from the glycolysis of a single glucose molecule, four ATP molecules, two molecules of NAD·H_2, and two molecules of pyruvic acid are formed. Because two ATP molecules were used to start the process and four are present at the end, there is a net gain of two ATP's. This process is summarized in Figure 6–7.

Stage II—Formation of Acetyl Coenzyme A. In Stage II, each of the pyruvic acid molecules releases two hydrogens to NAD and loses a molecule of carbon dioxide. It is thereby converted into a molecule of acetic acid. Each acetic acid molecule combines with coenzyme A, which is formed from pantothenic acid, one of the B vitamins. Thus, in stage II, the two pyruvic acid molecules yield two molecules of acetyl coenzymes A, two molecules of NAD·H_2, and two molecules of CO_2. This stage is outlined in Figure 6–8.

Stage III—Krebs or Citric Acid Cycle. The citric acid cycle is a series of chemical reactions in which the two-carbon compound in acetyl coenzyme A, produced in stage II, combines with a four-carbon molecule to build a six-carbon molecule (citric acid). Through a series of steps, the citric acid breaks down into a four-carbon molecule again with the release of H_2, CO_2, and the formation of ATP. The hydrogen produced is attached to the acceptor NAD or

Figure 6–7. The first stage in the cellular respiration of glucose.

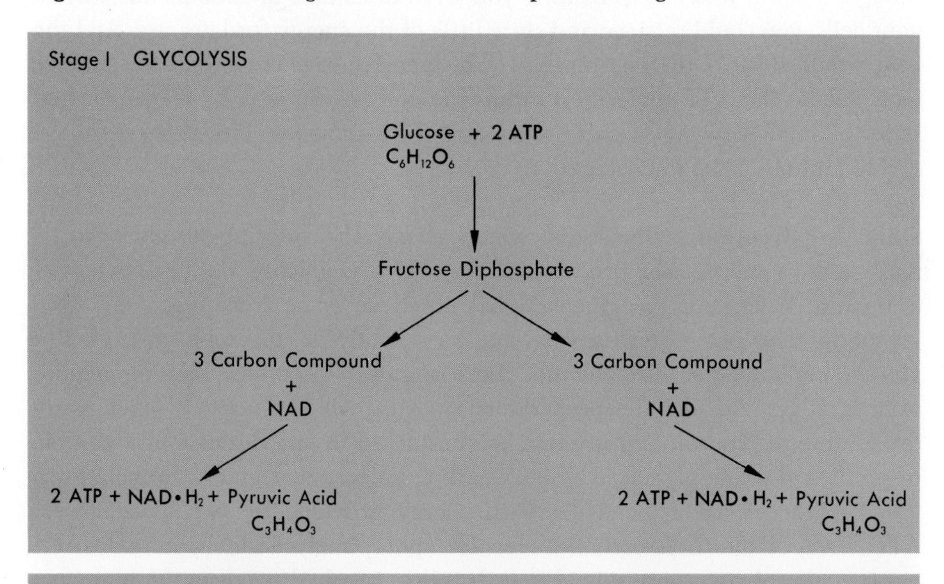

Stage II FORMATION OF ACETYL CO-A

2 Pyruvic Acid + 2 Co•A ⟶ 2 Acetyl Co•A + 2 CO_2 + 2 NAD•H_2
(2 Carbon Compound)

Figure 6–8. The second stage in the cellular respiration of glucose.

another acceptor. As the result of two molecules of acetyl coenzyme A going through the cycle, four molecules of NAD•H_2, four of CO_2, and two of ATP are produced. The coenzyme A and two four-carbon molecules are left over and can be recycled. This series of steps is summarized in Figure 6–9.

Stage IV—Oxidative Phosphorylations. Stage IV is a sequence of chemical reactions whereby the hydrogen that has become attached to the hydrogen acceptors in the three previous stages is passed on to successive acceptors, with a corresponding release of energy, to the final acceptor, oxygen. Actually, it is the hydrogen electrons that are transferred; the protons are released to the surrounding medium as hydrogen ions. The oxygen atom accepts two electrons

Figure 6–9. The third stage in cellular respiration.

Stage III KREBS OR CITRIC ACID CYCLE

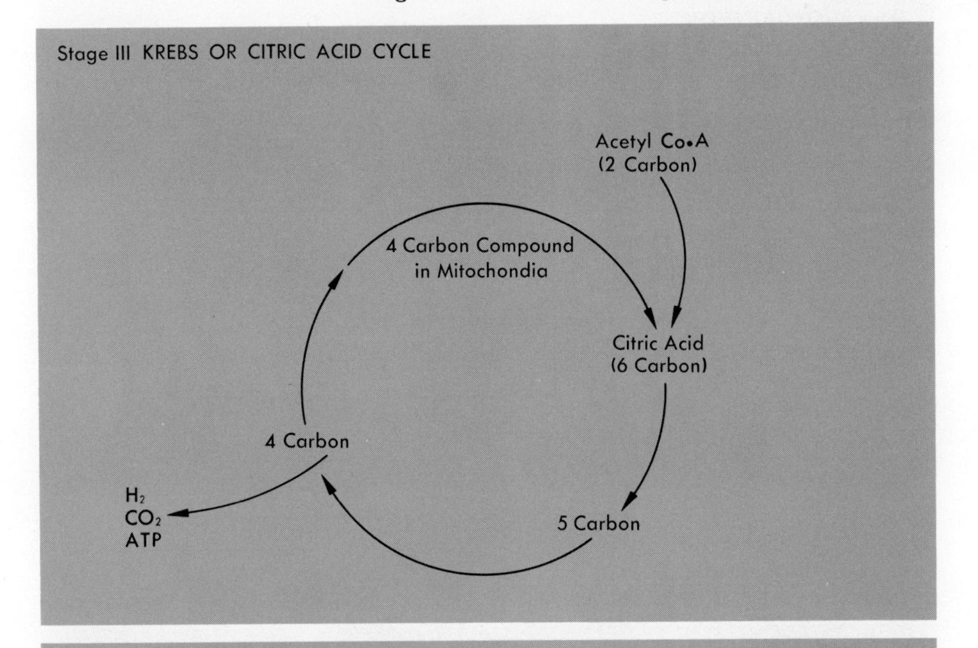

SUMMARY:

2 Acetyl Co•A + 6 H_2O + 2 ADP + 2 P ⟶ 4 CO_2 + 2 ATP + 8 NAD•H_2 + 2 Co•A

and thus has two extra negative charges. It combines with two positively charged hydrogen ions in the medium to form a molecule of water, H_2O. Thus, the actual function of the oxygen taken in during breathing is to serve as the ultimate acceptor of the hydrogen released in the preceding rather complicated reactions. In the transfer of the electrons from NAD through the various cytochromes down to oxygen, there is a large release of energy and most of the ATP is now formed—thirty-four (some authorities say thirty-five) molecules of ATP. This stage is summarized in Figure 6–10.

The cytochromes are proteins carrying prosthetic groups similar to the heme group found in myoglobin. The atom of iron in the center of the ring combines with the electrons. The poison cyanide binds the electrons to the iron so that they cannot be passed on down and thus prevents cellular respiration from proceeding. Almost instantly, the cells run out of energy and stop funtioning and death ensues.

Because thirty-four molecules of ATP are produced in stage IV, two in stage III, and two in stage I, it follows that the oxidation of one glucose molecule

Figure 6–10. The fourth stage in cellular respiration. Transfer of hydrogen through the cytochrome series to the final acceptor, oxygen (not all the acceptors are shown here).

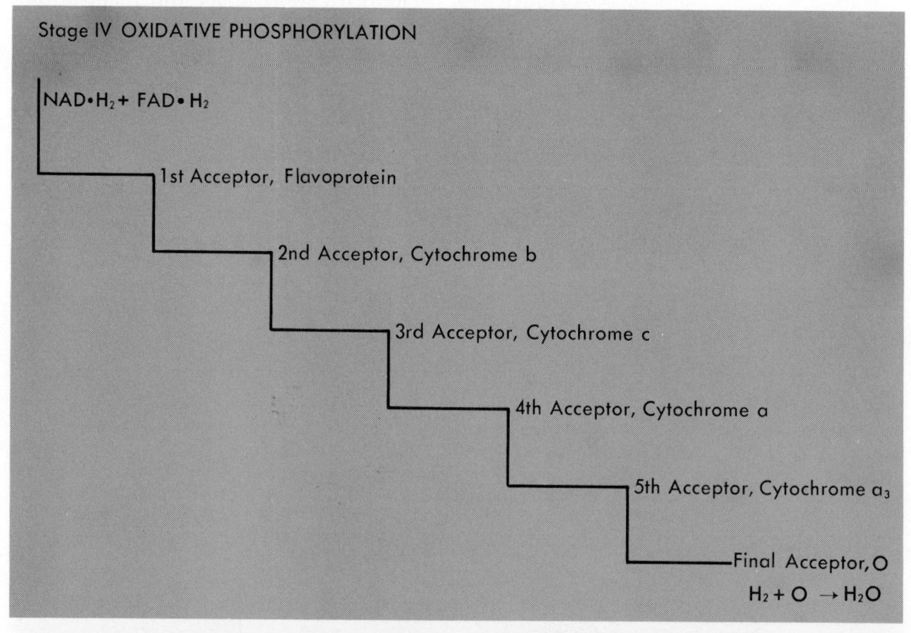

yields thirty-eight molecules of ATP. We can now rewrite the original equation for cellular respiration a little more precisely:

$$C_6H_{12}O_6 + 6\ O_2 + 38\ ADP + 38\ \textcircled{P} \rightarrow 6\ CO_2 + 6\ H_2O + 38\ ATP$$

These ATP molecules provide the immediate energy by which the cell carries on its activities.

These processes involve a great number of separate steps and require a great many enzymes. The enzymes of glycolysis are apparently free in the cytoplasm and not bound to specific organelles. The pyruvic acid formed during glycolysis enters the mitochondria. The enzymes involved in the Krebs cycle are believed to be located in the matrix of the mitochondrion, and it is suggested that the molecules of the electron transport mechanism are arranged in an orderly fashion on the cristae. It is in the mitochondria, then, that most of the ATP required for cellular activity is produced.

By a number of different pathways, fats and proteins can be broken into fragments, each containing a small number of carbon atoms, and these can be converted into some of the same compounds (e.g., pyruvic acid) that appear during the respiration of glucose. They can be fed into the system and oxidized in the same way.

Anaerobic Respiration

Although aerobic respiration is the typical process in most animals and plants, it is possible for some respiration to take place in the absence of free oxygen. This is known as *anaerobic respiration* or *fermentation.* Anaerobic respiration proceeds through the first stage, glycolysis, only. It is also somewhat different in plants than in animals. In animals, the pyruvic acid produced by glycolysis accepts hydrogen back from the molecules of $NAD \cdot H_2$ to form molecules of lactic acid, $C_3H_6O_3$, and NAD. As we pointed out in the chapter on muscles, lactic acid is produced during muscular contraction. If the muscular activity is not too great, and sufficient oxygen reaches the cells, part of this lactic acid is converted to pryruvic acid and is oxidized, the rest built back to glucose. If you carry on strenuous physical activity for a period of time, the oxygen supply to your muscles is not sufficient and lactic acid tends to accumulate in your body temporarily.

In many microscopic plants and some higher ones as well, in the absence of oxygen, the pyruvic acid produced by glycolysis combines with the hydrogen from the $NAD \cdot H_2$ to yield alcohol (C_2H_6O), carbon dioxide, and NAD, as shown in the following equation:

$$C_3H_4O_3 + NAD \cdot H_2 \rightarrow C_2H_6O + CO_2 + NAD$$

If juice from berries is combined with yeast plants and placed in a bottle where ample oxygen is not present, alcohol and carbon dioxide are produced by the anaerobic respiration of the sugar in the juice by the yeast. The result is wine. The same thing happens when yeast is combined with wheat flour in making bread; the carbon dioxide accumulates as a gas and makes the dough rise, and the alcohol evaporates as the bread is baked.

SUMMARY OF RESPIRATION

The relationship between the various aspects of respiration should now be clear. The oxygen taken in during breathing is the ultimate acceptor of the hydrogen released when food is broken down. The water thus formed, called *water of metabolism,* helps replace the water lost through the lungs in the form of water vapor. You can see your breath on a frosty morning because this water vapor condenses when it hits the cold air. Some of the carbon dioxide formed by cellular respiration combines with water in the blood to form carbonic acid, H_2CO_3. An excess of carbon dioxide leads to acidity of the body fluids. This is why carbon dioxide must be constantly eliminated. The energy released from food is used to bind phosphate groups to ADP to form ATP. Both proteins and fats can be converted into carbohydrates and oxidized in the same way. When this happens, however, certain waste products are left over that are not in the form of gases and cannot be eliminated through the lungs. Their removal is part of the function of the excretory system, which we discuss in the next chapter.

SUGGESTED READINGS

Baker, J. J. W., and G. E. Allen: *Matter, Energy and Life,* Addison-Wesley, Reading, Mass., 1970.

Chapman, C. B., and J. H. Mitchell: "The Physiology of Exercise," *Sci. Amer.,* Vol. 212, No. 5, 1965.

Comroe, J. H., Jr.: "The Lung," *Sci. Amer.,* Vol. 214, No. 2, 1966.

Gordon, M. S., et al.: *Animal Physiology: Principles and Adaptations,* 2nd ed., Macmillan Publishing Co., Inc., New York, 1972.

Grollman, S.: *The Human Body,* 3rd ed., Macmillan Publishing Co., Inc., New York, 1974.

Guyton, A. C.: *Function of the Human Body,* 3rd ed., W. B. Saunders Company, Philadelphia, 1969.

CHAPTER 7

The Excretory System—Homeostatic Control

Although any real understanding of the excretory system and its functioning had to await the development of modern chemistry, man has long been aware of a connection between the appearance of the urine and health. Sending a specimen to be analyzed is an ancient practice. Shakespeare knew and apparently distrusted it.

> FALSTAFF: Sirrah, you giant, what sayeth the doctor to my water?
> PAGE: He said, Sir, the water itself was a good, healthy water, but for the party that owned it, he might have more diseases than he knew for."

The excretory system is also called the *urinary system,* its product is *urine,* and one of the constituents of urine is *urea,* a waste product formed from the nitrogen-containing parts of amino acids. This nomenclature is unfortunate. It tends to focus attention on a single aspect of the system's far broader function of maintaining a proper environment for the cells of the body. These cells are bathed in a fluid, the *interstitial* or *tissue fluid,* which seeps out of and into the vessels of the circulatory system. It is mostly water with many substances dissolved in it—ions of various salts, glucose, amino acids, oxygen, vitamins, and hormones. This fluid is the internal environment, and the cells are very sensitive to changes in it. For the cells to survive and function properly, the right substances must be present in the right amounts, within narrow limits. *Homeostasis,* the maintenance of the stability of the internal environment, is the true function of the excretory system. Removal of waste products is an important part of this process, but it is only a part.

STRUCTURE OF THE EXCRETORY SYSTEM

The excretory system in humans is made up of a pair of kidneys, each drained by a tube, the ureter, which leads to a median, unpaired storage sac, the bladder. The bladder in turn drains to the outside through a single tube, the urethra. The arrangement of these structures is shown in Figure 7–1.

The Kidney

The kidney is a bean-shaped structure, about 115 millimeters (4½ inches) long, lying in the lumbar region of the back deep to the muscles of the small of the back. A lengthwise cut of the kidney shows that it is divided into two rather definite layers, an outer *cortex* and an inner *medulla.* Tubes from the medulla open into a large chamber, the *renal pelvis,* which leads directly into the ureter (see Figure 7–2).

The cortex is made up of about 3 million microscopic units known as *nephrons.* Each nephron is composed of a renal tubule and associated blood vessels. A diminutive branch of the renal artery (the afferent arteriole) divides to form a mass of capillaries, the *glomerulus.* The capillaries of the glomerulus join again to form a small *efferent arteriole.* Surrounding the glomerulus is a double-layered, tulip-shaped cup, *Bowman's capsule.* The glomerulus and Bowman's capsule together are called the *Malpighian body.* The outer layer of

Figure 7–1. The human excretory system.

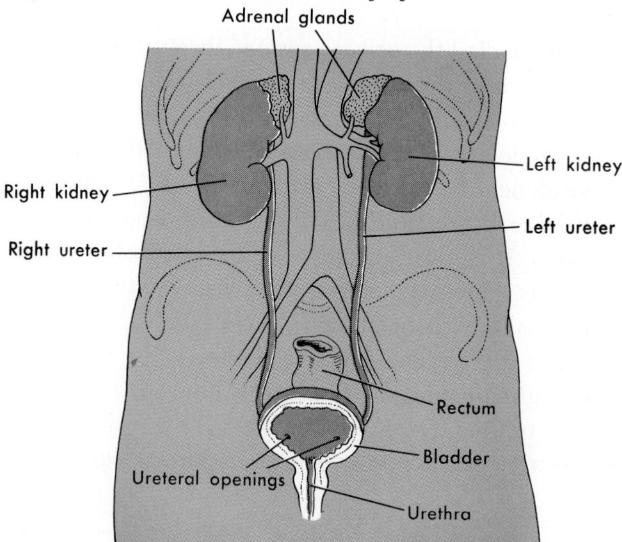

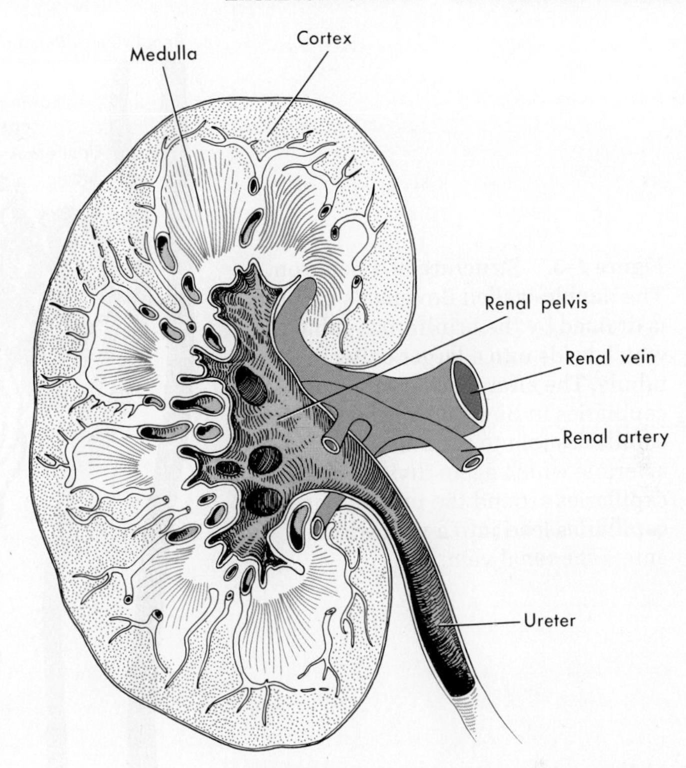

Figure 7–2. Section through a kidney.

Bowman's capsule leads into a small tubule that at first is convoluted (proximal convolutions) and then becomes straight. The straight part passes downward into the medulla of the kidney and then turns back into the cortex, forming a loop known as *Henle's loop*. As it approaches the region of the proximal convolutions the tubule again becomes convoluted (distal convolutions). After this convoluted portion, the tubule joins with the tubules from other nephrons to form a larger collecting tubule. This passes down through the medulla and ultimately opens into the renal pelvis. After the efferent arteriole leaves the glomerulus it again breaks down into capillaries that encircle and enmesh the proximal and distal convolutions of the tubule as well as parts of Henle's loop. The structure of an individual nephron is shown in Figure 7–3. The pelvis of the kidney leads into the ureter.

Ureter

The ureter is a muscular tube capable of peristaltic activity. It has an outer, longitudinal layer and an inner, circular layer of smooth muscles in its

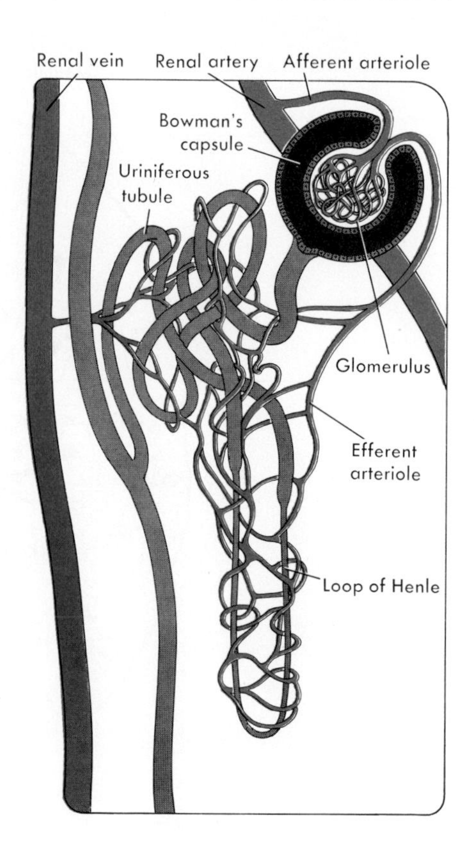

Figure 7–3. Structure of a nephron. The double-walled Bowman's capsule is drained by the uriniferous tubule which leads into a larger collecting tubule. The afferent arteriole leads into capillaries in Bowman's capsule. The capillaries join to form the efferent arteriole which again divides into capillaries around the tubule. These capillaries lead into a vein which enters the renal vein.

wall. Near the bladder a less distinct layer of longitudinal fibers underlies the circular layer (see Figure 7–4). The ureters lead into the median, unpaired bladder.

Bladder

The bladder is a hollow, muscular bag, like a collapsed balloon when empty, rather pear-shaped when full. The ureters enter it at an angle, and their openings are covered by folds of tissue that act like valves to prevent urine from backing up into the ureters from the bladder.

Urethra

A median tube, the urethra, leaves the ventral surface of the bladder and passes to the outside. In the female, it is independent of the reproductive system, but in the male it passes through the tissue of the prostate gland and it transports the seminal fluid as well as urine. The urethra of the female opens

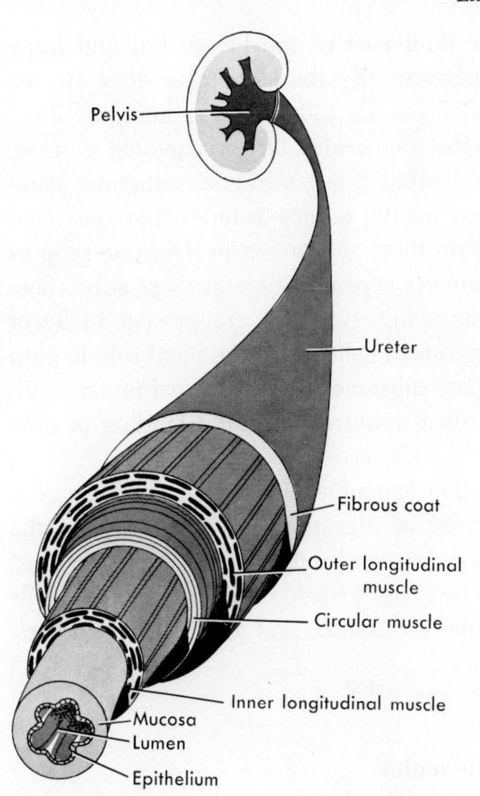

Figure 7–4. Diagram of the structure of the ureter.

directly to the outside; in the male it passes through the copulatory organ, the penis, and thence to the outside. Bundles of muscle fibers from the bladder extend down the urethra, and an accumulation of elastic tissue is present in the region where the two join. A constant urine flow is prevented by an elongation of the urethra, which closes off its internal cavity (lumen). Below its junction with the bladder, the urethra is surrounded by a band of skeletal muscle, the external sphincter, which can be contracted at will to close the urethra even though reflex action may cause the tube to shorten so that the lumen is opened.

FORMATION OF URINE

The formation of urine begins with a passive filtration of fluid through the walls of the capillaries of the glomerulus and of Bowman's capsule and so into the tubule. Blood in the glomerular capillaries is under pressure. The walls of the capillaries and of Bowmen's capsule are permeable, that is, they permit the passage of fluid through them. This fluid, the *glomerular filtrate,* is essentially nothing more than blood plasma, the fluid part of the blood. Fat and large protein molecules are almost lacking because, like the blood cells, they are too large to pass readily through the capillary walls and the inner lining of Bowman's capsule. In addition to water, the glomerular filtrate contains glucose, ions of various salts, amino acids, and other small nitrogen-containing compounds such as urea. The fluid passed by the kidney tubules into the renal pelvis, now called *urine,* does not contain these substances in the same proportion as does the filtrate that enter Bowman's capsule. Some of these substances are present in urine in larger amounts, whereas others are present in lesser amounts. It is thus apparent that the epithelial cells lining the renal tubule both absorb substances from the fluid and pass substances into it. A real understanding of the functioning of the kidney, then, requires an understanding of how substances move into and out of cells.

This, of course, is not a process that is confined to the kidney. The cells lining the intestine take up the products of digestion and pass them to the blood, oxygen and carbon dioxide move into and out of cells, and lactic acid leaves the muscle cells and is taken up by cells of the liver. In short, all the cells of the body are constantly taking in some substances and getting rid of others. How do they do it?

Membranes and the Movement of Molecules

The cells of the body are made up mostly of water, with many other substances dissolved or suspended in it, and they are surrounded by tissue fluid, which is also mostly water with various substances dissolved or suspended in it. Separating the intracellular fluid from the tissue fluid is the plasma membrane of the cell. Remember that the plasma membrane is made up in part of lipid molecules, which are apparently arranged in two layers, each one molecule in thickness. Protein molecules are associated with the lipid layers (see Figure 1–7. If we disregard the protein molecules for the time being, we can consider the plasma membrane as a sheet of lipid separating two bodies of water. It is through this lipid sheet that the movement of molecules between the extracellular and intracellular fluid must take place.

Diffusion. The molecules in a fluid are in constant motion. Each molecule continues on a straight path until it collides with another molecule and is sent off in a new direction. In a solution this is true both for the molecules of the dissolved substance (the *solute*) and the molecules of the substance in which they are dissolved (the *solvent*). If you dissolve a spoonful of sugar in a cup of coffee, at first the sugar molecules will be concentrated near the bottom of the cup. Because there are more sugar molecules near the bottom, there is more chance that one of them will be knocked toward the upper part of the cup than that one in the upper part of the cup will be knocked back down. There will be a net movement of sugar molecules upward and this will continue until they are evenly distributed throughout the coffee. There is then as much chance that a sugar molecule will move down as that one will move up, and the even distribution of the molecules is thereby maintained. The movement of molecules from a region of high concentration to a region of low concentration is called *diffusion.*

If a substance is readily lipid soluble, the lipid layer of the plasma membrane will present little resistance to the movement of the molecules. If there are more molecules in the tissue fluid than in the intracellular fluid, they will readily diffuse into the cell. If the concentration gradient of the molecules is the other way, they will diffuse out. For example, oxygen and carbon dioxide are lipid soluble and move into and out of cells by simple diffusion. This is a passive process that requires no energy expenditure by the cells.

Facilitated Diffusion. Many of the substances that are important in the economy of the cell, such as glucose, do not dissolve readily in lipids. They would diffuse very slowly into or out of a cell, too slowly to meet the cell's needs. Yet they do, in fact, move readily between the cells and the tissue fluid. It is obvious that some mechanism other than simple diffusion is at work. One suggestion is that a molecule of low lipid solubility is joined to a molecule that is lipid soluble and is "carried" by it through the lipid layer. Some of the protein molecules on the outer surface of the membrane may act as enzymes to join the molecules to be transported to the carriers or may act as carriers themselves. Once through the lipid layer, the transported molecule is released by another enzyme and the carrier is free to diffuse back through the membrane.

Active Transport. Active transport resembles facilitated diffusion in that substances are enzymatically coupled to carrier molecules. The processes differ in two important respects. In diffusion, either simple or facilitated, the substance moves from a region of higher concentration to one of lower concentration. This does not require the expenditure of energy by the cell. In active transport, movement of the molecules can be against the concentration gradient, that is, from a region of lower to a region of higher concentration. Energy for this movement is provided by ATP. Although glucose moves from the tissue fluid into the body cells by facilitated diffusion, during digestion it apparently moves from the intestine into the mucosal cells by active transport.

Some ions are also quite insoluble in lipids and they also move into and out of cells against the concentration gradient. The concentration of potassium ions is much greater inside the cell than in the tissue fluid, and the concentration of sodium ions is much less. As with glucose, some mechanism of active transport seems to be involved.

Osmosis. The processes we have discussed so far involve the movement of molecules dissolved in the water of the intracellular and tissue fluids. What about the water molecules themselves? They too are in constant motion and they too tend to move from a region of higher concentration to a region of lower concentration. Now because two bodies cannot occupy the same place at the same time, the thing that determines the number of water molecules in the fluid on either side of the plasma membrane is the number of particles dissolved in it. If the fluid inside the cell has relatively more dissolved particles than the fluid outside the cell, it has relatively fewer water molecules and water will move into the cell. If the difference is too great, water will continue to move in until the cell bursts. If there are relatively more dissolved particles outside the cell, water will move out of the cell. Again, if the difference is too great, the results are disastrous. The cell shrivels, loses its ability to function, and soon dies. The movement of water into and out of cells through the plasma membrane is called *osmosis*.

It is easy to see why it is important to maintain the osmotic balance between the cells and the extracellular fluids of the body within narrow limits. If the number of dissolved particles is the same inside the cell as it is outside, water molecules move in at the same rate as they move out. The two solutions are in osmotic balance and are said to be *isosmotic*. If the two solutions do not have the same number of dissolved particles, the one with more particles and fewer water molecules is said to be *hyperosmotic*, the one with fewer particles and more water molecules is *hypo-osmotic*.

The discovery that the plasma membrane contains layers of lipid molecules poses difficulties to this somewhat simple picture of osmosis. Water is very slightly soluble in lipids. Yet it moves into and out of cells very rapidly, far more rapidly than can be accounted for by movement through the membrane. One suggestion is that the plasma membrane may not be continuous, that there are in it pores, perhaps lined with protein molecules, through which water can flow into or out of the cell (see Figure 7–5). Whether these pores are permanent or temporary openings is not known. Other models have been suggested to explain the movement of water across the plasma membrane, and much active research is still in progress on the problem.

Small, negatively charged ions seem able to move through the pores in the membrane just as water does, but positively charged ions are excluded. It is suggested that the substances lining the pores carry positive charges that repel the positively charged ions.

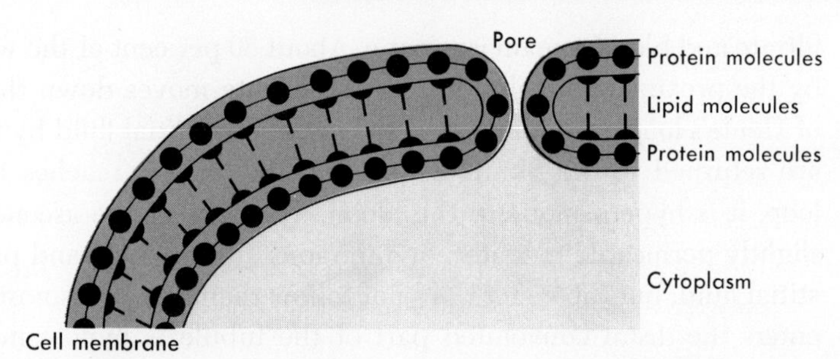

Figure 7–5. Model of a cell membrane showing a protein-lined pore.

Pinocytosis. Some large molecules or aggregates of molecules that are too large to pass into the cells by the processes already discussed are still taken in by the cells. Folds from the plasma membrane extend out to enclose part of the surrounding medium. The folds fuse and pinch off to form little vesicles inside the cell. If the molecules thus engulfed are in solution, the process is called *pinocytosis* (drinking by cells); if solid particles are taken in, it is *phagocytosis* (eating by cells).

Function of the Kidney

By the various mechanisms for moving molecules across membranes, the cells of the kidney tubules convert the glomerular filtrate into urine.

Absorption of Solutes. Many of the substances contained in the glomerular filtrates are useful to the body. Such things as glucose, amino acids, and ions of various salts are actively reabsorbed from the filtrate by the cells of the proximal convolutions and passed back into the blood in the capillaries surrounding the tubule. Sodium ions are also reabsorbed by the cells of the ascending arm of Henle's loop and are passed to the interstitial fluid in the medulla of the kidney. Sodium is also removed from the filtrate by the cells of the distal convolution.

Absorption of Water. One of the things the body needs to recover from the filtrate is water. About 170 liters (180 quarts) of filtrate, most of it water, pass through the glomerular filters in a day. A man could not possibly drink enough water to replace this loss. As particles are returned to the blood by the proximal convolutions, water also passes back into the capillaries by osmosis so that the filtrate and blood remain isosmotic. About 80 per cent of the water is reabsorbed by the proximal convolutions. As the filtrate moves down the descending arm of Henle's loop, sodium ions passed to the interstitial fluid by the ascending arm are returned to the filtrate so that by the time it reaches the bottom of the loop, it is hyperosmotic to the blood. The cells of the ascending arm are only slightly permeable to water. Sodium ions are removed and passed to the interstitial fluid, but the water does not follow them out by osmosis. The filtrate that enters the distal convoluted part of the tubule is hypo-osmotic to the blood. The cells of the distal convolutions are more permeable to water, which is removed from the filtrate by osmosis.

The collecting tubule passes down through the kidney medulla parallel to the loops of Henle. The interstitial fluid here is concentrated by the sodium that is being constantly pumped out of the ascending arms of the loops. The filtrate is hypo-osmotic to the interstitial fluid. More water leaves the filtrate by osmosis, and the final product that the collecting tubules dump into the renal pelvis is somewhat more concentrated than the blood. Of the original 170 liters (180 quarts) of filtrate, only between 1 and 2 liters (about 1 to 2 quarts) are passed as urine during a day.

The cells of the distal convolutions and the collecting tubules vary in their permeability. They are under the control of an *antidiuretic hormone* secreted by the pituitary gland (see Chapter 10). This hormone acts on the membranes of these cells to increase their permeability. If you are short on drinking water or are losing a great deal of water through sweat, your pituitary gland increases its production of the antidiuretic hormone, the cells of the proximal convolutions and collecting tubules become even more permeable, more water is reabsorbed, and the urine becomes even more concentrated. On the other hand, if you drink a great deal of water, the secretion of the hormone drops, the tubule cells become less permeable to water, and your urine becomes more dilute. Some substances, such as alcohol, suppress the secretion of the antidiuretic hormone. The copious flow of dilute urine that results dehydrates the body. This is why a person with a hangover has an overpowering thirst.

Secretion. Because potassium and hydrogen ions, urea, and creatinine (a nitrogenous waste product produced by the breakdown of creatine) occur in urine in amounts larger than can be accounted for by simple filtration, it follows that there is an active movement of these materials from the blood in the capillaries surrounding the tubule through the tubule cells into the cavity of the tubule. Creatinine and probably also urea are apparently passed into the tubule in Henle's loop. The distal convolutions move potassium and hydrogen ions into the filtrate in exchange for the sodium ions they remove.

Summary of Kidney Functions. Figure 7–6 summarizes the formation of urine by the kidney tubules. The kidney performs its homeostatic function by differential absorption and secretion by different parts of the tubules. It removes waste products of metabolism from the blood; it recovers useful substances, such as glucose; by its ability to absorb and excrete ions it ensures that the various kinds of ions are present in the body in the right proportions; and by its control of water excretion it regulates the osmotic balance of the body. If the concentration of some substance in the blood exceeds the optimum level, the excess is passed out through the urine. If the concentration is less than optimum, none passes out. This is why it is possible, by examining the urine, to make fairly accurate estimates of the level of many substances in the blood. For example, in diabetes the uptake of glucose by the body cells is interfered with so that glucose reaches abnormally high levels in the blood. The tubules are unable to reabsorb this excess and the sugar is excreted in the urine. In the seventeenth century, Thomas Willis discovered sugar in the urine of diabetics by the simplest of chemical tests. He tasted the urine.

PASSAGE OF URINE

The urine formed in the tubules passes into the pelvis of the kidney and starts on its way from the pelvis down the ureter. This movement is partly brought about by gravity, but it is also under control of peristaltic waves in the muscular walls of the ureter (see Figure 7–4). The urine that constantly passes down the ureter from the kidney enters the bladder. When empty, the bladder is shrunken because the smooth muscles in its walls are normally contracted. As the ureters pour their urine in the bladder, it becomes more and more distended, but without an appreciable increase in pressure simply because the muscles in its wall tend to relax as the bladder expands.

The voiding of the bladder, or urination, is a reflex act, but in all except young children, it is also normally under voluntary control. The centers for the control of urination are in the lower part of the brain and the lower part of the spinal cord. If the cord is cut so that the brain control is separated from the control of the lower cord, the bladder will empty itself at regular intervals without any voluntary control. Under such conditions, the bladder empties when it contains about 250 cc's (half a pint) of urine. With both the brain and spinal cord connections intact, urination is initiated when sensory receptors in the bladder wall are stimulated by the stretching of the wall. This stimulation brings about a reflex contraction of the bladder muscles themselves to force the urine out, and a relaxation of the urethra. As the urine passes from the bladder into the urethra, a second impulse is initiated that relaxes the ex-

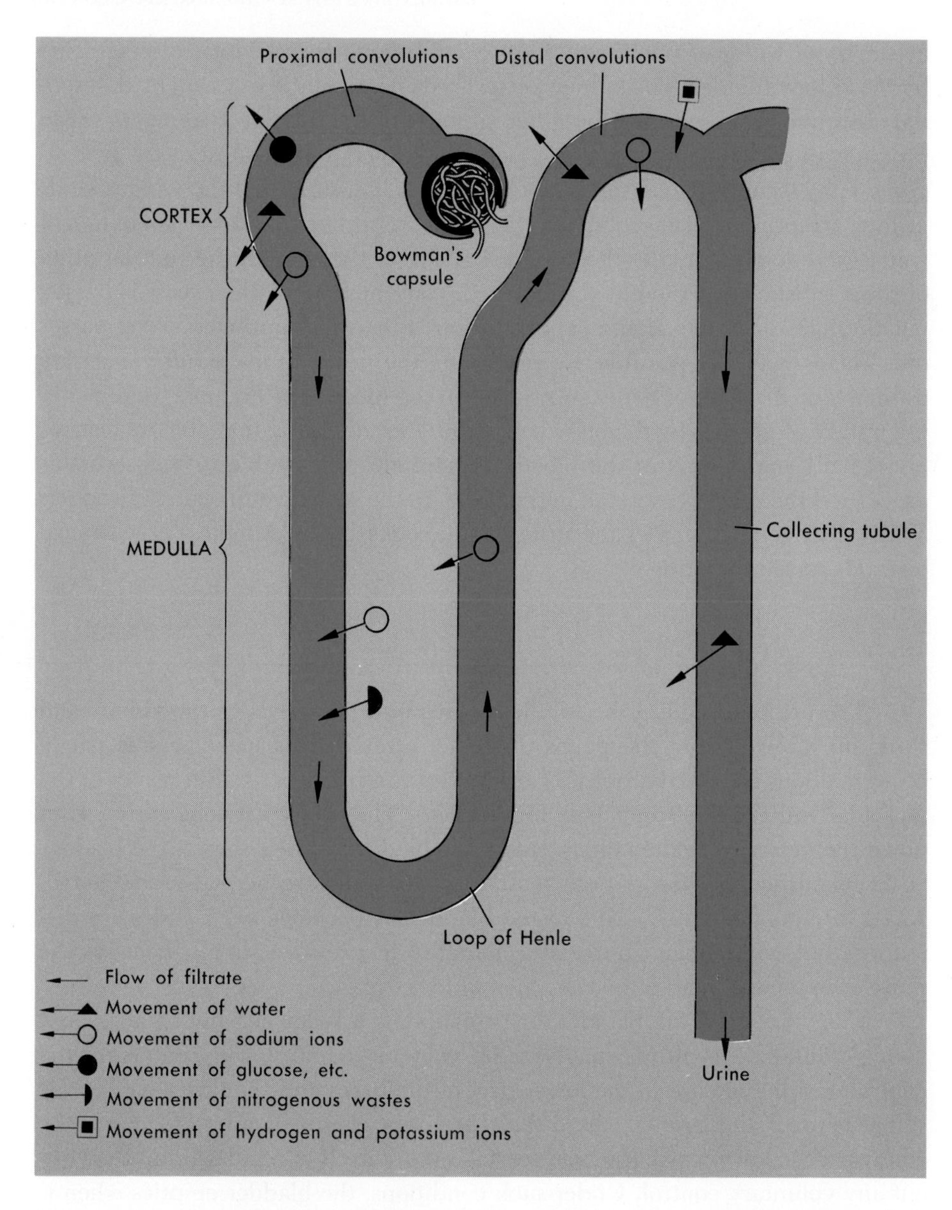

Figure 7–6. Formation of urine by the kidney tubules. Glucose and other needed substances are removed from the filtrate in the proximal convolutions. Water moves out by osmosis from the proximal convolutions, distal convolutions, and collecting tubule. Sodium ions are removed from the filtrate in both proximal and distal convolutions and in the ascending arm of Henle's loop but move into the filtrate in the descending arm. Nitrogenous wastes are secreted into the filtrate in Henle's loop. Hydrogen and potassium ions are exchanged for sodium ions in the distal convolutions.

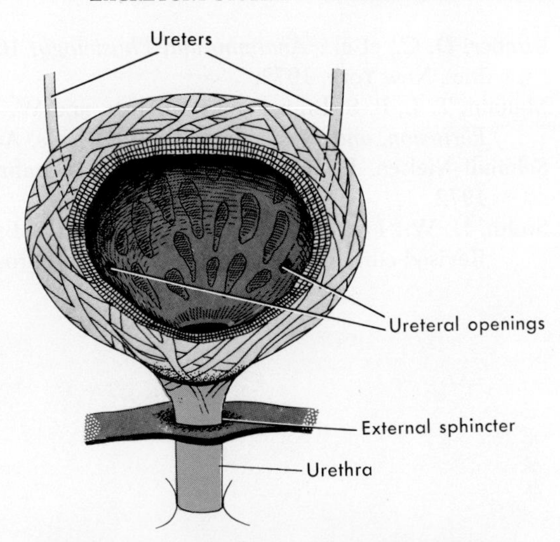

Figure 7–7. The bladder and urethra. A section of the bladder wall has been cut away to show the openings of the ureters.

ternal sphincter. This permits the discharge of the urine to the outside. The relationship of these muscles to the bladder and urethra is shown in Figure 7–7. The external sphincter cannot be made to relax voluntarily, but it can be made to contract; thus urination can be consciously halted or deferred. When you are expected to urinate by an effort of the will, as when you are asked to give the doctor a urine sample, impulses arising in the conscious part of your brain pass down the cord and initiate the original reflex action in the urination center of the spinal cord. Thus, although you do not have voluntary control of the muscles in the wall of the bladder itself, you can exert a conscious control over the reflex center that controls these muscles.

In the last three chapters we have assumed that there is a system by which materials are transported in the body—oxygen from the lungs to the cells, digested food from the intestine to the liver, and waste products from the cells to the kidneys. The next chapter deals with the transport system of the body.

SUGGESTED READINGS

Crouch, J. E., and J. R. McClintic: *Human Anatomy and Physiology,* John Wiley & Sons, Inc., New York, 1971.

Grollman, S.: *The Human Body,* 3rd ed., Macmillan Publishing Co., Inc., New York, 1974.

Kimber, D. C., et al.: *Anatomy and Physiology,* 16th ed., Macmillan Publishing Co., Inc., New York, 1972.

Malinin, T. I., B. S. Linn, A. B. Callahan, and W. D. Warren, eds.: *Microcirculation, Perfusion, and Transplantation of Organs,* Academic Press, New York, 1970.

Schmidt-Nielsen, K.: *How Animals Work,* Cambridge University Press, New York, 1972.

Smith, H. W.: *From Fish to Philosopher,* Little, Brown and Company, Boston, 1953. Revised edition by CIBA Pharmaceutical Products, Inc., 1959.

CHAPTER **8**

Circulation— The System of Transport

In the last few chapters we have discussed how the human body meets two of its basic needs: to take in from the environment those substances it requires in order to maintain itself and grow; and to return to the environment waste products formed by its own activity. These needs man shares with all living things. Very small organisms are able to pass material directly from cell to cell at a rate sufficient to fill the requirements of the individual cells. The human body is composed of many billions of cells. Their needs could never be met by the simple passage of material from cell to cell. If a cell in the little toe had to wait for oxygen to diffuse down to it from the lung, it would long since have died of oxygen starvation. Large organisms need a means by which materials can be transported rapidly from one part of the body to another. In animals this is the function of the circulatory system. The human circulatory system consists of a pump, the heart; a series of vessels, the arteries, veins, capillaries, and lymphatics; and their included fluids, blood and lymph. The transport of nutrients, oxygen, waste products, and other substances is not the only function of the circulatory system. Like other systems, it has several duties. It is concerned in the regulation of temperature, an important factor in homeostasis. And it is a main line of defense against disease organisms that invade the body.

BLOOD

Blood is a red or purplish liquid made up of a fluid, the *plasma*, in which the formed elements, the red and white blood cells and the platelets, are suspended. An adult has five to six liters (about six quarts) of blood in his body.

Plasma

Plasma, the fluid part of blood, constitutes about 55 per cent of the blood's total volume. Plasma is largely made up of water (about 92 per cent). In it are dissolved blood proteins, which form about 7 per cent of the plasma, together with smaller amounts of nonprotein nitrogen compounds, inorganic salts, blood sugars, and lipids. Dissolved gases, vitamins, and hormones are also carried by the blood.

Blood Proteins. The proteins in solution in the blood have many important functions. They play a chief role in maintaining the osmotic balance between the blood and the cells. They act as *buffers;* that is, if the concentration of hydrogen ions in the blood increases, the proteins combine with the ions; if the concentration drops, the proteins release the ions. The blood proteins thus maintain the acid-base ratio of the internal environment within the narrow limits required for the survival of the cells. Some of the proteins serve as carriers, combining with molecules such as lipids, which are practically insoluble in water, thereby transporting these molecules to the cells. Fibrinogen and other blood proteins are concerned with the clotting of the blood. In cases of starvation, the blood proteins can act as a reserve amino acid source.

A large group of proteins known as *immunoglobulins* defend the body against invasion by disease organisms or other foreign substances. If bacteria break through the barrier of the skin or mucous membrane and begin to multiply in the body, there is an increased production of immunoglobulin. Part of the globulin molecule is unstable and can be modified in such a way that it is able to attach to the invaders (see Figure 8–1). Attachment may cause the invading organisms to clump together, to be more sensitive to the attack of macrophages (see the section in this chapter about leukocytes), or to break apart. The invader is called an *antigen,* the globulin an *antibody.* Antibodies are usually specific; that is, each kind of antibody reacts only with the specific antigen that caused its appearance. Moreover, once the body has learned the trick of producing a certain antibody, it often retains that ability. If you ever had whooping cough it took some days for you to produce antibodies to overcome the attack. In the meantime you were sick. The next time your body was invaded by the bacterium that causes whooping cough, antibodies appeared so quickly that you probably were not even aware that anything was wrong. You have become immune to whooping cough.

The ability of the blood proteins and white blood cells to attack foreign substances is the greatest stumbling block to successful organ transplants. The cells of the body have specific antigenic proteins on their plasma membranes. At least thirty-one different proteins of this kind have been identified, with

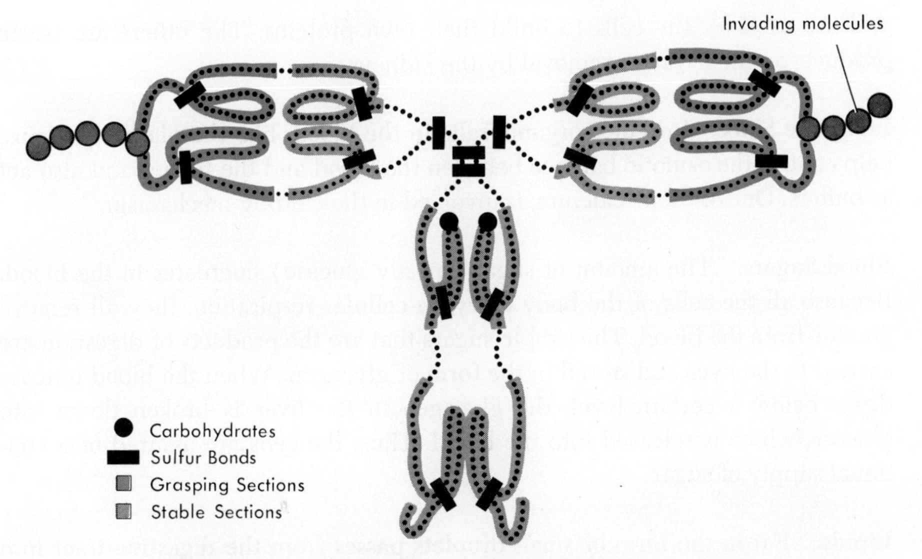

Figure 8–1. Diagram of a molecule of gamma globulin, one of the immunoglobulins.

each individual having his own particular combination. The immune system "recognizes" these proteins and refrains from attack. This is why the cells of your body are safe from your own immune system (except in certain autoimmune diseases such as arthritis, in which, for some reason, the immune system attacks tissues of its own body). If the donor of a transplanted organ does not have the same, or nearly the same, combination of antigenic proteins as the recipient, the organ is attacked and destroyed (rejected). Antigenic proteins are inherited; identical twins have the same combination, and close relatives are more likely to have similar ones than unrelated people.

It is possible to suppress the immune reaction by X-ray treatment or by certain drugs, with the concomitant disadvantage of lowering the body's resistance to disease. LSD is apparently able to substitute for one of the amino acids, *tryptophan,* when a protein chain is being built. But it cannot form a peptide bond with the amino acid that should follow it in the chain, so formation of the protein is broken off at that point. The immunoglobulins contain tryptophan. Persons taking LSD have a reduced production of immunoglobulin and consequently have a reduced ability to combat infections.

Nonprotein Nitrogen Compounds. The nonprotein nitrogen-containing compounds, familiarly known as NPN, are such things as urea, amino acids, creatinine, and uric acid. The amino acids are products of the digestion of proteins and are used by the cells to build their own proteins. The others are waste products on their way to removal by the kidneys.

Inorganic Salts. Ions of inorganic salts in the blood, like the plasma proteins, help control the osmotic balance between the blood and the tissues and also act as buffers. One of them, calcium, is involved in the clotting mechanism.

Blood Sugars. The amount of sugar (mostly glucose) fluctuates in the blood. Because all the cells of the body carry on cellular respiration, they all remove glucose from the blood. The simple sugars that are the products of digestion are carried to the liver and stored in the form of glycogen. When the blood glucose drops below a certain level, the glycogen in the liver is broken down into glucose, which is released into the blood. Thus, the cells are assured of a continual supply of sugar.

Lipids. Fat in the form of small droplets passes from the digestive tract into the lymphatic system and thence into the blood to be carried to the cells for storage or use. Phospholipids formed in the liver are also found in the blood. Cholesterol, another component of the plasma, enters the bloodstream from the digestive tract. If enough of it is not taken into the body by way of the digestive system, cholesterol can be synthesized in the liver and some other organs of the body. Cholesterol is a component of cell membranes and is a precursor of some of the hormones. Because cholesterol is readily excreted by both the liver and the small intestine, under ordinary circumstances it does not tend to accumulate in the body. However, in some people whose diets are very rich in animal fat or who have uncontrolled diabetes, cholesterol may build up in the walls of the arteries, a condition known as atherosclerosis.

Formed Elements

The solid fraction of the blood is made up of three types of formed elements: *erythrocytes*, commonly called *red blood cells; leukocytes*, commonly called *white blood cells;* and *thrombocytes*, commonly called *platelets.*

Erythrocytes. Red blood cells, as they are seen in whole blood, are small structures, concave in shape, and without nuclei (see Figure 8–2). The interior mass is known as the *stroma* and consists mostly of protein and lipid material. *Hemoglobin,* a red substance bound up in the stroma, makes up approximately one third of the cell's volume and 95 per cent or more of the protein in the cell. Hemoglobin is composed of two pairs of polypeptide strands; each strand is folded around a heme group that has an atom of iron in its center. The heme

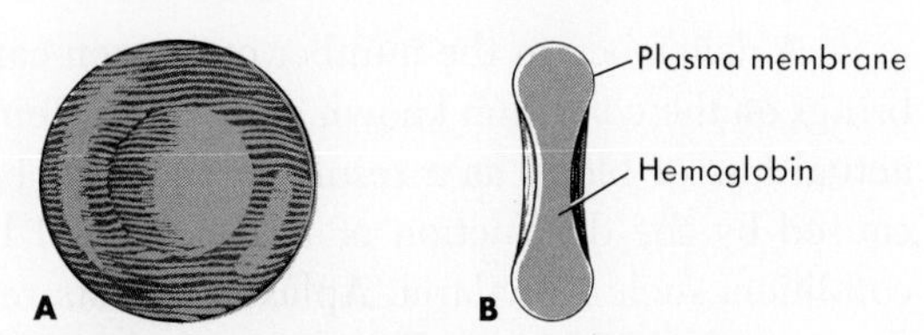

Figure 8–2. Red blood cell. A: Surface view. B: End view of a cell cut in half.

group is very similar to the prosthetic groups of the cytochromes involved in cellular respiration. The muscle protein, myoglobin, is like a single strand of hemoglobin.

One of the most important functions of the blood is the transport of oxygen to the cells of the body. Oxygen is not very soluble in water, not enough for the blood to carry it in solution in sufficient supply for the needs of a large and active animal. But oxygen does combine very readily, although loosely, with the iron of the heme group. As it diffuses into the blood, the oxygen is picked up by the hemoglobin of the red blood cells. About 95 per cent of the oxygen enters the red blood cells; the rest remains dissolved in the plasma. In this way the blood is able to carry very much more oxygen than that which is dissolved in the plasma. When the blood reaches the tissues of the body, the oxygen separates from the hemoglobin and diffuses into the cells. At the same time carbon dioxide diffuses out of the cells into the blood. Hemoglobin that is combined with oxygen is bright red in color so that blood that is rich in oxygen is also bright red. This is why the color of the blood is an index of how much oxygen it is carrying. When much of the hemoglobin is unoxygenated, the blood is purplish. Carbon monoxide combines with the heme iron much more firmly than oxygen does. In carbon monoxide poisoning, so much of the hemoglobin is thus blocked that little oxygen reaches the cells.

Erythrocytes are formed in the red bone marrow from nucleated precursor cells called *erythroblasts* and are released into the bloodstream as enucleate (without nuclei) erythrocytes. In men there are approximately 5 million erythrocytes per cubic millimeter, in women the number is somewhat less. The number of erythrocytes in the blood is not constant; it is determined largely by the need of the individual for oxygen. An athlete has more erythrocytes than a sedentary office worker. Erythrocytes tend to increase in number during warm weather and decrease in cold weather. The familiar expression, "Your blood thins when you become acclimated to warm weather" is the reverse of the real situation. An increase in the red blood cell count also occurs after a full meal. Persons who live at high altitudes, where there is less oxygen available, have more red blood cells than those who live at sea level.

A deficiency in the number or oxygen-carrying capacity of red blood cells brings on the condition known as *anemia.* Hemorrhagic anemia results from the actual loss of blood as a result of injury or hemorrhage. Hemolytic anemia is caused by the destruction of the individual blood cells by some pathological condition, such as malaria. Aplastic anemia results from a decrease in the production of red blood cells; it may be caused by exposure to X-rays, atomic radiation, or drugs. Pernicious anemia indicates a defect in the formation of the red blood cells as a result of a dietary deficiency of vitamin B_{12}.

Erythrocytes, which lack nuclei, cannot sustain themselves indefinitely and eventually disintegrate. It is estimated that the average life span of a red blood cell in the circulatory system is about four months. As they break up, the cell fragments are destroyed by special cells called *phagocytes* in the spleen and liver, which literally engulf and digest them. In the liver, the iron is removed from the heme group and saved for reuse; the heme group is excreted in the bile.

Leukocytes. Leukocytes, called white blood cells because they lack hemoglobin, are much rarer in the blood than are the red blood cells; there are 6,000 to 10,000 leukocytes in a cubic millimeter of blood in a normal, healthy person. Actually there are a number of different kinds of white blood cells. Some are phagocytes, which are capable of independent movement. They can creep through minute pores in the walls of the capillaries and migrate to areas where tissue damage has occurred either through mechanical injury or through the activity of invading disease organisms. There they engulf by phagocytosis both the invaders and the debris from the cells that have been destroyed. After it has ingested a certain number of invaders, the phagocytic cell itself dies. The pus that forms at the site of an acute infection is largely an accumulation of dead phagocytes, along with living leukocytes, bacteria, and cellular debris.

Other leukocytes are called *lymphocytes.* Their precursors are formed in bone marrow like the erythrocytes, but they do not lose their nuclei. Some of the leukocytes migrate to the thymus gland, a large, two-lobed structure that lies in the upper chest cavity above the heart. It is a very prominent organ at birth and reaches its greatest relative size at puberty, when each lobe may be nearly a third as long as a lung. From then on the thymus undergoes a gradual regression during which some of the thymal tissue is replaced by fat and connective tissue. In the thymus, the leukocyte cells proliferate and are somehow modified into special *T-cells.* The T-cells move to various lymph organs in the body, the tonsils, adenoids, appendix, spleen, and lymph nodes (see the section on lymphatics in this chapter), and take up residence there.

Other lymphocyte precursors leave the bone marrow but do not migrate to the thymus. They become *B-cells,* which circulate in the blood. They have receptor sites on their surfaces that bind to antigens. As a result, the B-cells are stimulated to produce and secrete antibodies.

B-cells and T-cells divide responsibility. B-cells form antibodies against certain bacteria, such as those that cause pneumonia. T-cells attack other bacteria and viruses, and seem to be the cells involved in graft rejection. (How they do it is not known). The enormous versatility of the immune system is suggested by the estimate that the body is able to produce about a million different kinds of antibodies.

Sometimes children are born who lack either the B-system or the T-system, or rarely both. Such children usually die before the age of two because they are unable to combat infections. Attempts have been made to correct the lack of an immune system by grafts of either bone marrow or thymus. Such grafts are particularly risky because, should the cells produced by the grafted tissue fail to recognize the host's cells, they attack and destroy the tissues of the patient.

The efficiency of the immune system declines in old age. Some doctors believe that potentially cancerous cells appear frequently in the body and are destroyed by the T-cells. The fact that cancer develops most frequently in old people is thought to result from the reduced ability of the T-system to combat the disease. Individuals who develop cancer when young usually show other deficiencies in the immune system.

Thrombocytes. The thrombocytes, or blood platelets, are fragments of cytoplasm pinched off from large cells called *megakaryocytes* in the bone marrow (see Figure 8–3). They exist in the blood in numbers ranging from 200,000 to 300,000 per cubic millimeter. When a blood vessel is broken or injured, the thrombocytes adhere to the broken parts and disintegrate. This initiates a chain of reactions that results in the formation of a clot, a structure that seals the broken or injured blood vessel and prevents the loss of blood by the body.

Blood clotting is a complex process. When the platelets break down, they release a substance that enters into a series of reactions with other substances in the blood. These reactions result in the formation of a substance called *thromboplastin.* The thromboplastin in turn reacts with a number of other factors, including calcium ions and the proteins *prothrombin* and *proconvertin,* to form *thrombin.* Thrombin is an enzyme that splits several small peptide chains from the blood protein *fibrinogen.* The modified fibrinogen molecules are then able to aggregate end to end and also side to side to form long chains of *fibrin.* Fibrin is insoluble and precipitates as fibers. The mass of fibers forms a network that traps the solid elements in the blood, and the fibers contract to squeeze out the remaining fluid. This thin, yellowish fluid, called *serum,* differs from blood plasma in that it lacks fibrinogen.

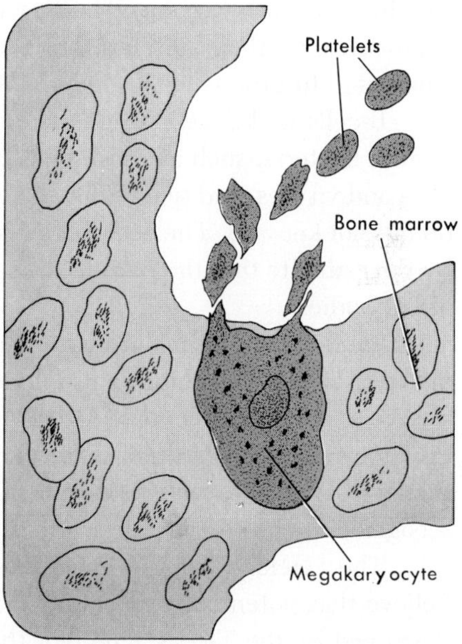

Figure 8–3. The formation of platelets from a megakaryocyte in bone marrow.

Vitamin K is necessary for the formation of prothrombin and proconvertin by the liver. This is why a deficiency of this vitamin reduces the clotting ability of the blood.

Because platelets are themselves broken down from time to time in the blood vessels, a natural anticoagulant reaction must be constantly going on within the blood to keep it from forming clots in normal blood vessels. A substance called *heparin* is secreted into the blood continually by certain cells in the connective tissue around the capillaries. Heparin blocks the reactions by which thrombin is formed and also probably destroys thrombin once it has been formed. Sometimes, of course, there is internal damage to the walls of the blood vessel and a clot does form in the vessel. If it forms in, or is carried to, an artery in the brain, blocking it and shutting off the blood supply, a stroke results.

MECHANICS OF CIRCULATION

The blood would be of little use to the body if it were not kept in constant circulation. This is achieved by a system of pumps and associated vessels.

Heart

The heart, a cone-shaped, muscular structure about the size of a man's fist, is really two pumps housed in a single organ. It lies between the two lungs in a separate compartment of the chest cavity called the *mediastinum.* The heart is made up of four chambers, two *atria* and two *ventricles.* In adults, the atrium and ventricle of the right side are completely separated from the left atrium and ventricle by partitions. The blood from most of the body flows into the right atrium through two large veins, the *superior* and *inferior vena cava.* The right atrium leads into the right ventricle through the *atrioventricular opening.* The *pulmonary artery* carries blood from the right ventricle to the lungs. Blood is returned from the lungs to the left side of the heart through the *pulmonary veins* and enters the left atrium. This leads into the left ventricle by way of the left *atrioventricular opening.* Blood from the left ventricle passes out to the body through a large artery called the *aorta.* Valves are present on the ventricular sides of the atrioventricular openings and also at the beginning of the pulmonary artery and of the aorta. The walls of the atria, which simply move blood into the ventricles, are thinner and less muscular than the walls of the ventricles, which pump the blood out to the lungs and to the body. The structure of the heart is shown in Figure 8–4.

Like any active muscle, the heart needs a good supply of blood, but it cannot exchange materials with the blood passing through its chambers. It is provided with two special arteries that branch from the aorta, the *coronary arteries.* If a clot blocks one of the branches of these arteries, the part of the heart muscle beyond the clot is deprived of blood and the cells die. This is a *coronary occlusion.* Like skeletal muscle, heart muscle uses creatine as a temporary reservoir for phosphate. The enzyme that mediates the transfer of the phosphate to ADP is called *creatine phosphokinase,* CPK for short. When part of the heart muscle dies, CPK is liberated into the blood. The earliest clinical sign that a pain in the chest is caused by a coronary occlusion may be the presence of large amounts of CPK in the plasma.

Heartbeat. It is useful in discussing heartbeat to employ the words *systole* and *diastole. Systole* refers to the phase of contraction, *diastole* to the phase of relaxation. A heartbeat begins with the contraction of the atria. Atrial systole takes about 0.1 second. The atria then relax as the ventricles contract; ventricular systole takes about 0.3 second. Then there is a quiescent period of about 0.4 second when both the atria and ventricles are relaxed. Thus, the diastole and rest period of the atria lasts 0.7 second and that of the ventricles 0.5 second. Because the complete beat cycle takes about 0.8 second, contrary to popular opinion, the heart, rather than working constantly, is at rest over half the time.

When a doctor listens to a normal heart with a stethoscope, he hears a sound: lub, dupp—lub, dupp—lub, dupp—lub, dupp. These sounds seem to come largely from the snapping shut of the heart valves as pressure changes occur. When the ventricles begin their systole, the increasing pressure snaps the atrioventricular valves shut and causes the initial *lub* sound. With the valves

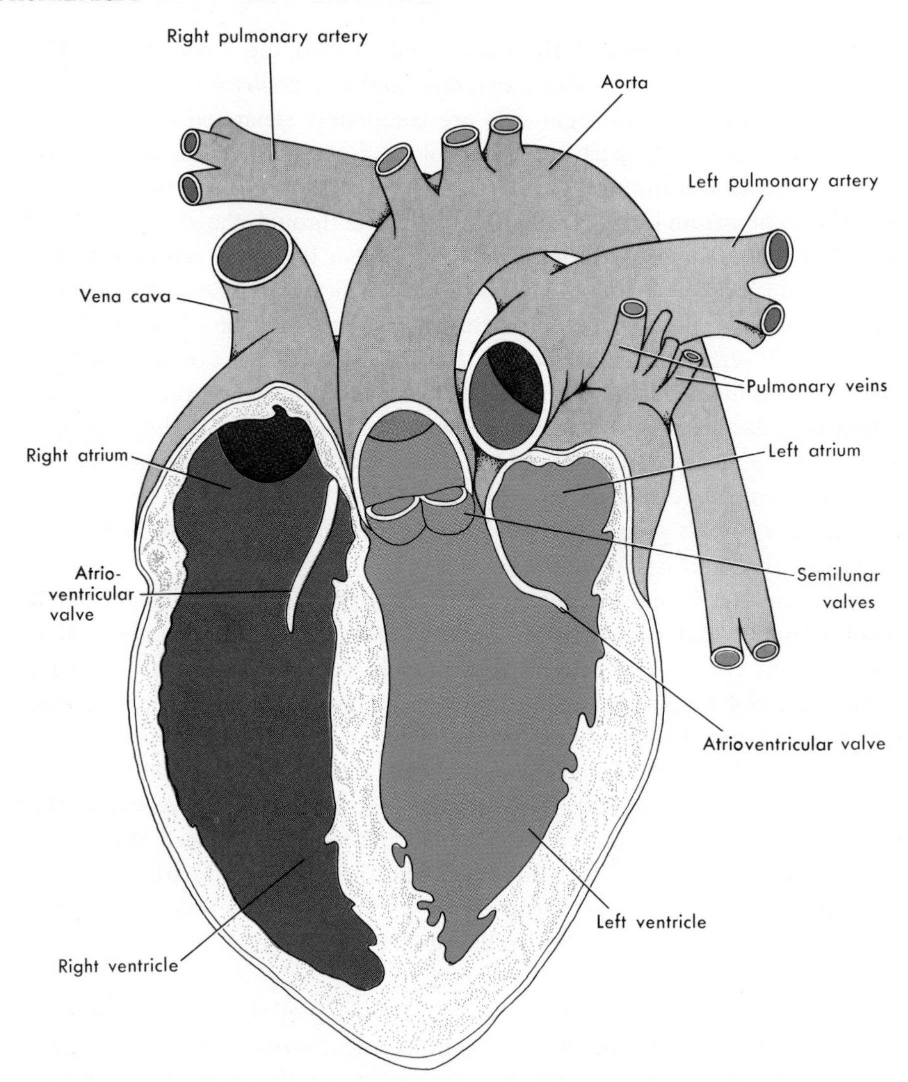

Figure 8–4. The human heart from in front. The wall has been cut away to show the chambers.

shut the blood cannot be pushed back into the atria, and the systole of the ventricles forces the blood into the elastic aorta and pulmonary artery. When the ventricles begin their diastole, pressures in the expanded and taut aorta and pulmonary close the valves at their entrances, thus preventing a backflow of blood into the ventricles. The snapping shut of the valves of the aorta and pulmonary artery forms the second noise, *dupp*. During the quiescent period, the atria and ventricles are relaxed and blood flows into the atria from the superior and inferior vena cava and the pulmonary vein.

Blood is kept flowing through the system by a pressure gradient. The pressure is greatest in the ventricles at maximum systole, where it approaches 140 millimeters of mercury above atmospheric pressure. From here the pressure greatly diminishes in the arteries, capillaries, and veins, and in the large vena cava and the heart itself during diastole it may actually drop a few millimeters below atmospheric pressure. Figure 8–5 shows the pressures and cycles of the heart as plotted against time.

Control of Heartbeat. Cardiac muscular tissue is able to contract and relax alternately so long as it is furnished with the necessities of life. The contractions of the various parts, though, must be coordinated. In the region where the two vena cava enter the right atrium there is a mass of specialized conductive muscle fibers known as the *sinoatrial (SA) node,* and in the band of dense, fibrous tissue that separates the muscular walls of the atria from the ventricles there is another node, known as the *atrioventricular (AV) node*. A bundle of fibers extends from the AV node into the wall that separates the ventricles and divides into two branches, one going to the right ventricle, the other to the left. The SA node is known as the pacemaker of the heart. Impulses from it radiate to the walls of the atria and to the AV node and from thence to the walls of the ventricle. It is by this means that the contractions of the various chambers of the heart are coordinated. If the conduction system from the pacemaker is damaged, the part of the heart beyond the damaged area will continue to contract and relax but more slowly so that it is no longer coordinated with the rest of the heart. This

Figure 8–5. Pressure during the heart cycle. The bottom two lines represent the systole (solid line) and diastole (broken line) of the atria and ventricles.

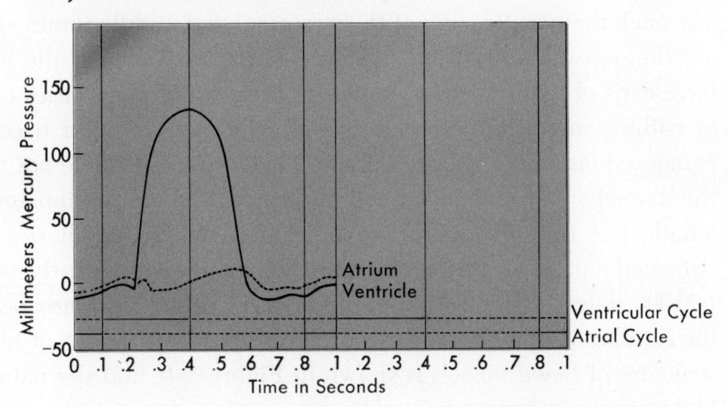

condition greatly impairs the pumping efficiency of the heart; it is known as *heart block*. It is now possible to install an artificial pacemaker, a tiny battery with electrodes implanted in the region of the heart from which the impulses from the true pacemaker are blocked. Little electric shocks, timed to the rhythm of the SA node, restore coordination to the beat.

The heart is also innervated by two sets of nerves—these do not make it beat but simply control its rate. The complete story of the control of heartbeat is long and complex; what follows is merely a review of the major points. Sensory receptors in the large veins entering the heart, in the aorta, and in the wall of the heart itself transmit impulses to a region of the brain known as the *cardiac center*. These impulses carry information to the cardiac center about the degree of tautness or stretching in these structures. Passing to the SA node and the AV node are two sets of nerve fibers, one from nerves coming from the spinal cord in the neck region and another by way of a long nerve (the *vagus*) that originates at the base of the brain. The vagus is in inhibitory nerve; stimuli reaching the heart from it slow the heart by acting on the SA node, decreasing both the rate and the force of the contraction. Impulses from the nerves in the neck region bring about an increase in the heartbeat by shortening both the systolic and diastolic periods and increasing the force of atrial and ventricular contraction.

Blood Vessels

The blood vessels of the body are arteries, capillaries, and veins. An artery is made up of three layers, called *tunics*. The external tunic is a tough connective tissue, the middle tunic is made up primarily of smooth muscles, and the internal tunic consists of a basement membrane composed mostly of elastic fibers, a thin, delicate layer of connective tissue, and a sheet of epithelial cells that line the artery. The epithelium lining blood vessels is called *endothelium*. The same tunics are present in veins but the external and middle tunics are much thinner. Because of their external and middle tunics are reduced, veins are thinner-walled and not so rigid as arteries. Veins of the limbs and viscera have flaps of tissue forming valves in their walls. Capillaries consist essentially of nothing more than the endothelial cells with an outer basement membrane composed largely of collagen fibers. They are small, some not much wider than the diameter of a red blood cell nor more than one millimeter (1/25 inch) in length, but they are so numerous that if all the capillaries in a man were stretched out in a single line they would measure nearly 96,000 kilometers (60,000 miles). They form dense networks called *capillary beds*. It is through the capillaries that substances in solution pass into or out of blood vessels. The structure of blood vessels is shown in Figure 8–6, and the nature of a capillary bed is shown in Figure 8–7.

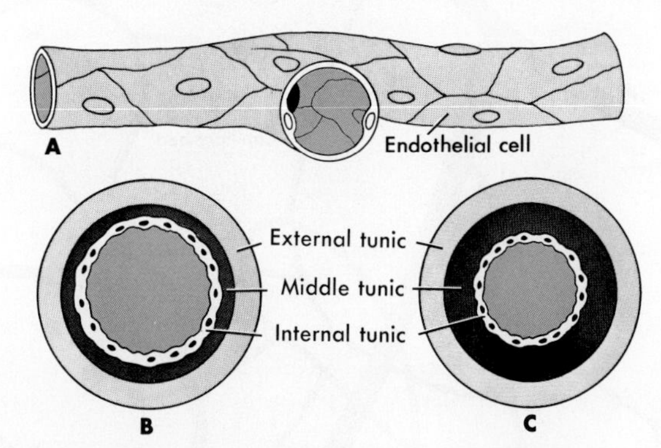

Figure 8–6. Structure of blood vessels. A: A capillary. B: Cut end of a vein showing the width of the tunics. C: Cut end of an artery.

General Circulation

The basic pattern of blood circulation is from the heart to the arteries, to the capillaries, to the veins and back to the heart again. For example, blood circulating to the leg leaves the heart from the left ventricle and passes into the aorta and down into the leg through successively smaller arterial branches. It flows into capillary beds in the muscles and other tissues of the leg. From these capillary beds it moves back into small veins, from these into larger veins, and ultimately into the inferior vena cava, which empties into the right atrium of the heart. The valves of the walls of the veins prevent the backflow of the blood. From the right atrium the blood passes to the right ventricle, from the right ventricle into the pulmonary artery, out to the lungs, through the capillaries of the lungs into the pulmonary veins, back into the left atrium, then into the left ventricle and once again it is ready to leave the heart and go through another circulation. The advantage of this double circulation through the heart is that all of the blood going to the tissues at each circuit through the body is richly supplied with oxygen. For the general course of circulation, see Figure 8–8.

The typical pattern of blood flow from artery to capillary to vein does not always hold. The veins coming from the intestine do not pass directly into larger veins and then into the heart, but instead they come together to form a large vein, the portal vein, which enters the liver and there breaks down into capillaries. The blood from the liver is then collected by a series of small veins and flows through the larger hepatic veins into the vena cava and thence into the heart. In the kidney the small artery entering a glomerulus divides to form capillaries, which then join to form the efferent arteriole. Another exception to the artery-capillary-vein sequence is found in the skin of the nose, eyelids, ears, and digits, where small arteries connect directly with small veins. Nothing in

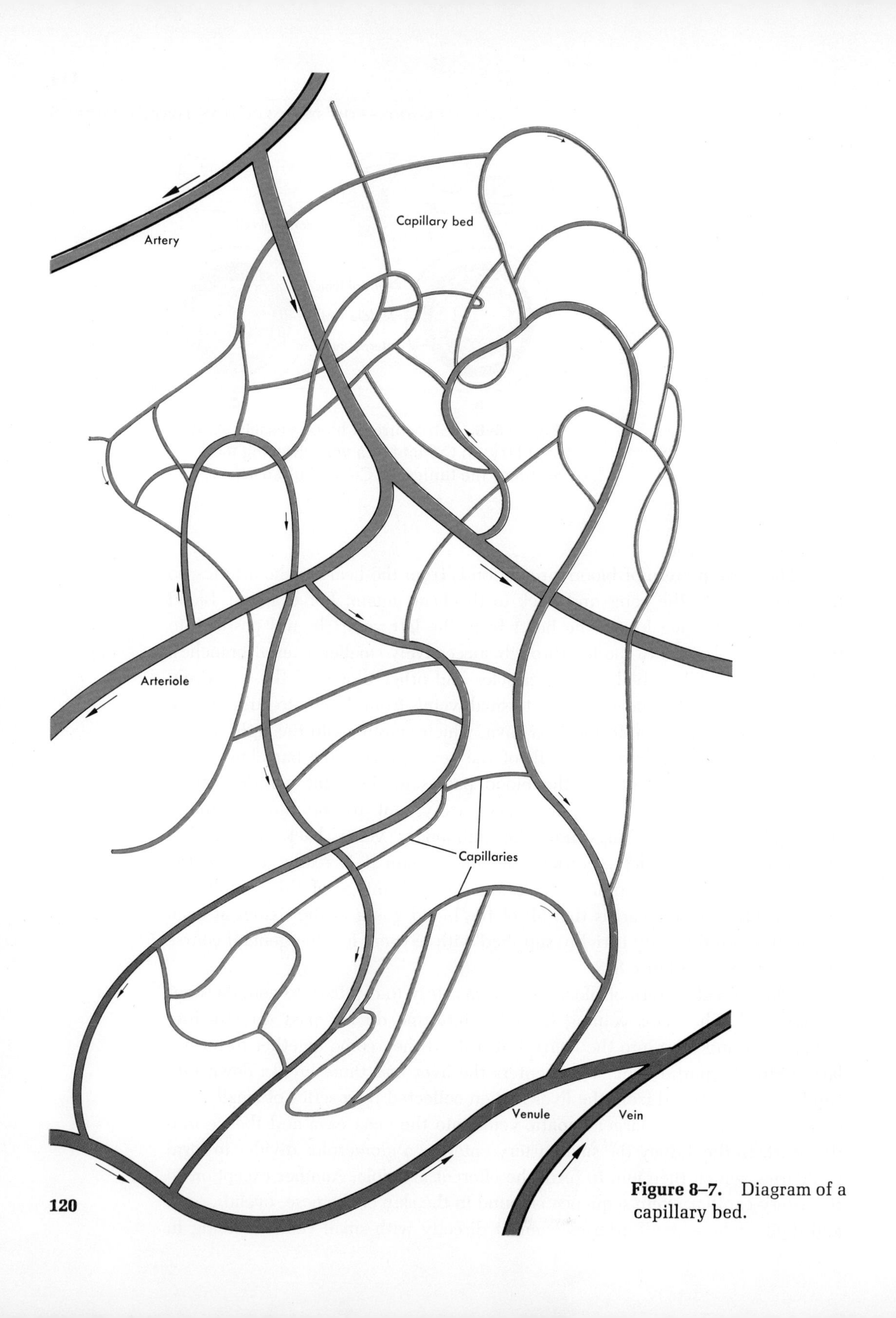

Figure 8–7. Diagram of a capillary bed.

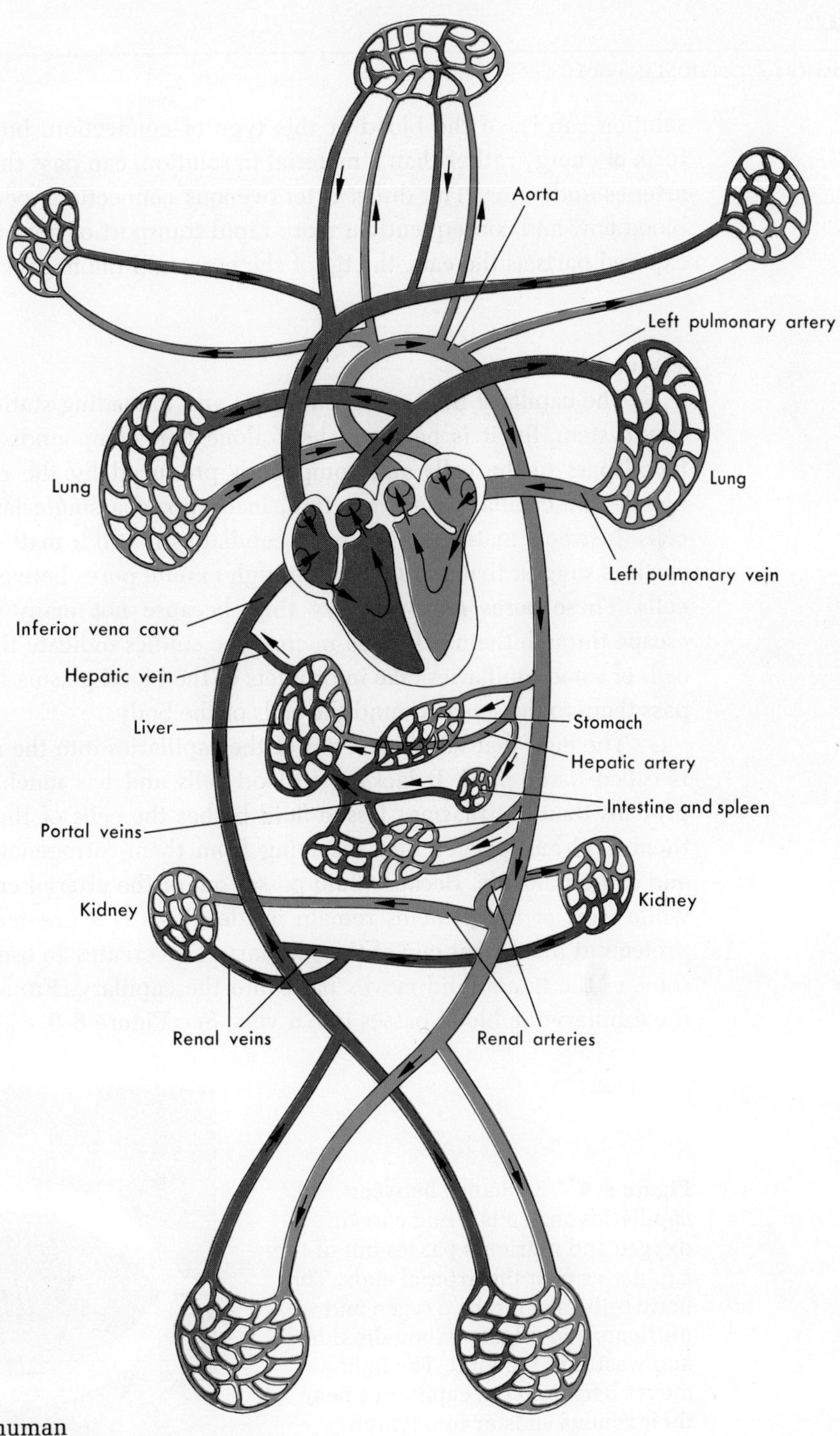

Figure 8–8. The human circulatory system. Arrows indicate the direction of blood flow through the vessels.

solution can leave the blood at this type of connection, but heat, which is a form of energy rather than a material in solution, can pass through the walls of arteries and veins. The direct arteriovenous connections permit a more rapid blood flow and consequently a more rapid transport of heat to the skin of such exposed parts as the ears, the tip of the nose, and the fingers.

Tissue Fluid

The capillary beds are the loading and unloading stations of the circulatory system, for it is here and here alone that compounds in solution in the blood pass to the cells and compounds produced by the cells pass into the blood. The capillary walls are thin, made up of a single layer of endothelial cells. Just how materials get out of capillaries is still a matter of debate. Some workers suggest that it is largely through minute pores between the endothelial cells. These pores must be very tiny, because not many protein molecules escape through them. Electron microscope studies indicate that the endothelial cells of some capillaries take in droplets of the blood plasma by pinocytosis and pass them to the space around the cells of the body.

The fluid that has passed out of the capillaries into the surrounding tissue is called *tissue fluid*. It lacks red blood cells and has much less of the blood proteins than has plasma. Tissue fluid bathes the cells of the body, passing to them food and oxygen and receiving from them nitrogenous waste products and carbon dioxide. Because fluid passes out of the arterial end of the capillary while most of the proteins remain inside, there is a greater concentration of proteins at the venous end of the capillary. This creates an osmotic gradient and some of the tissue fluid moves back into the capillary. From the distal end of the capillary the blood passes into a vein. See Figure 8–9.

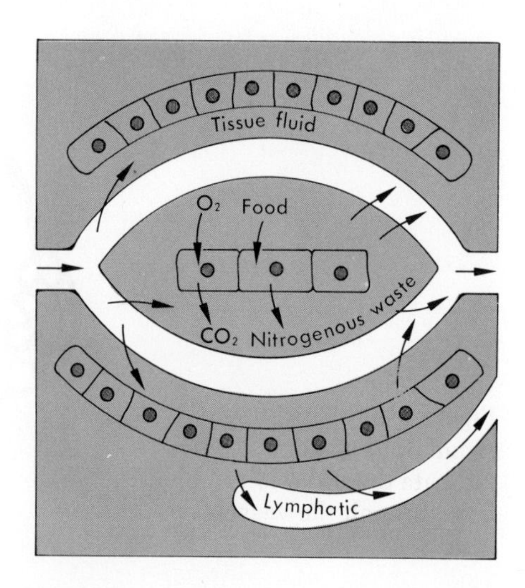

Figure 8–9. Exchange between capillaries and cells. Fluid carrying oxygen and nutrients passes out of the capillaries near the arterial ends. The body cells take up the oxygen and nutrients and return carbon dioxide and wastes to the fluid. The fluid moves back into the capillaries near their venous ends or into lymph vessels.

LYMPHATIC SYSTEM

Not all of the tissue fluid makes its way back into the capillaries. Some of it returns to the bloodstream by another route. In addition to arteries, capillaries, and veins, there is another set of vessels in the body associated with fluid movement. These are the lymphatics. Rather than forming a continuous circuit, the lymphatic vessels terminate in closed ends. At the closed end, the wall of the lymphatic is thin like that of a capillary and is even more permeable. As tissue fluid accumulates between the cells, the part of it that does not return to the capillaries flows into the lymphatic vessels, where it is known as *lymph*. Figure 8–10 shows the relationship between lymphatics and a capillary bed.

The lymphatics, like the veins, have valves at intervals that permit the enclosed fluid to flow in only one direction. The pressure of the tissue fluid and movements of the body that squeeze the lymphatic vessels cause the lymph to flow through the vessels. These vessels fuse with one another and ultimately lead back into the venous system in the shoulder region. The blood proteins that escape from the capillaries are returned to the blood by this route. If, for some pathological reason, tissue fluid does not pass back into the capillaries and lymphatics at a rapid enough rate, it tends to accumulate in the interstices be-

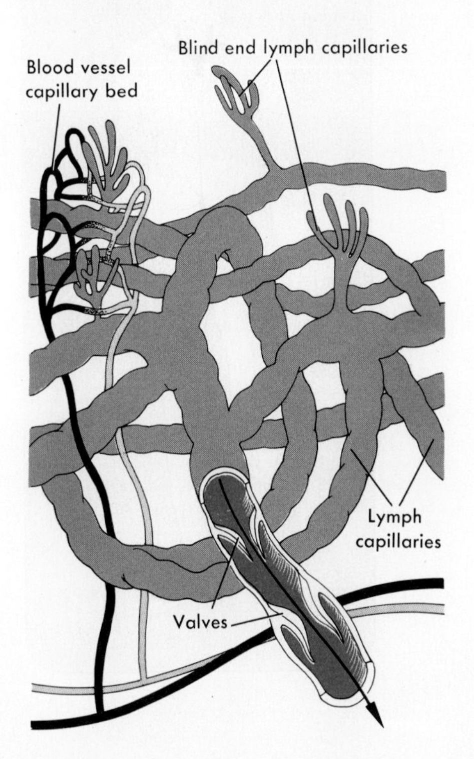

Figure 8–10. The relationship between lymph vessels and blood vessels. A part of the wall of a lymph vessel has been removed to show the valves that prevent the back flow of lymph.

tween the cells and the condition known as *edema* results. Figure 8–11 shows the lymphatic circulation.

Scattered along the lymphatics are small swellings, the lymph nodes. On the walls of the lymph channels through the nodes are large cells, the *fixed macrophages*. They phagocytize disease organisms and proteins foreign to the body and so play an important role in cleansing the body fluids. T-cells are present and B-cells may be processed in the lymph glands.

SPLEEN

The spleen is an oblong, flattened, dark purplish organ, about 120 millimeters (4¾ inches) long, which lies on the left side between the stomach and the diaphragm. Its functions include phagocytosis, especially of worn-out red blood cells, and the production of antibodies by the T-cells that have migrated

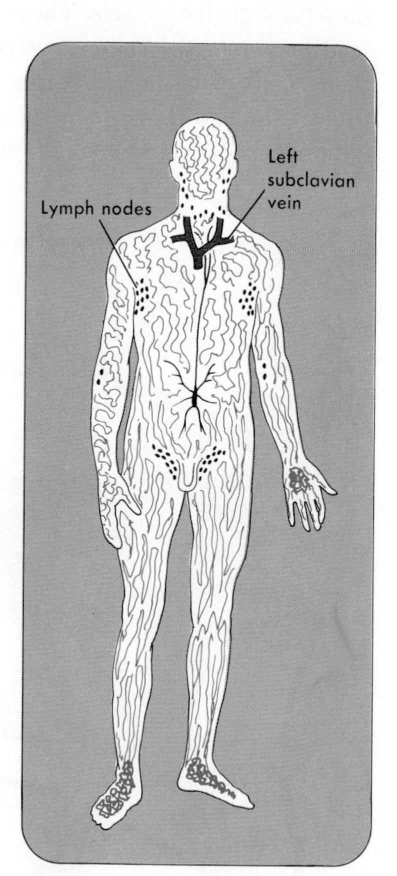

Figure 8–11. The lymphatic system.

to the spleen from the thymus. These antibodies are then released to the bloodstream and travel to the site of the antigenic infection.

By now you should have some idea of the complexity of the interactions between the various parts of the body. These parts must act in coordination with one another and the entire organism must adjust to changes in the outside environment. Coordination and adjustment are the functions of the systems we discuss in the next two chapters.

SUGGESTED READINGS

Crouch, J. E., and J. R. McClintic: *Human Anatomy and Physiology*, John Wiley & Sons, Inc., New York, 1971.

Grollman, S.: *The Human Body*, 3rd ed., Macmillan Publishing Co., Inc., New York, 1974.

Malinin, T. I., B. S. Linn, A. B. Callahan, and W. D. Warren, eds.: *Microcirculation, Perfusion, and Transplantation of Organs*, Academic Press, New York, 1970.

Mayerson, H. S.: "The Lymphatic System," *Sci. Amer.*, Vol. 208, No. 6, 1963.

Porter, R. R.: "The Structure of Antibodies," *Sci. Amer.*, Vol. 217, No. 4, 1967.

Zweifach, B. W.: "The Microcirculation of the Blood," *Sci. Amer.*, Vol. 200, No. 1, 1959.

CHAPTER 9

The Nervous System—Central Coordination

Pick up a pencil from your desk. If you did so, your eyes were stimulated by a series of black marks on a white page, nervous signals were transmitted from your eyes to your brain and there interpreted, and your brain decided to act on the suggestion. Your eyes were instructed to search for the pencil, information about its location was conveyed to your brain, and signals were sent to the muscles in your arm. These muscles contracted to bring your hand into position, other muscles contracted so that your thumb and first finger closed around the pencil, and still other muscles contracted to lift your hand. This trivial example shows the complexity of an apparently simple act. It involved the reception of information from the environment, the processing of that information in the brain, and the coordination of a series of muscular contractions. Response to the environment and coordination of the activity of the parts of the body are functions of the nervous and endocrine systems. The distinction between the two is not really clearcut, and there is an overlap of function, but, in general, the nervous system is concerned with rapid responses to environmental changes, both external and internal, whereas the endocrine deals with slower, but longer lasting, internal adjustment.

THE FUNCTIONAL UNIT

The structure and functioning of the human nervous system reach a degree of complexity that is nearly unimaginable. Nevertheless, its fundamental working unit is a single-type cell, the neuron.

Neuron

Neurons come in a variety of shapes, but all have certain common features. A typical neuron is illustrated in Figure 9–1. It is made up of a cell body containing a nucleus and a number of more or less branching processes. The processes that carry signals toward the cell body are called *dendrites,* the relatively unbranched process that carries a signal away from the cell body is called an *axon.* Dendrites and axons are known as *nerve fibers.* All neurons have a cell body, an axon, and one or more dendrites. The cell body is grayish in color; nervous tissue composed primarily of cell bodies is usually called *gray matter.* The dendrites and axons are often covered with a fatty insulating material,

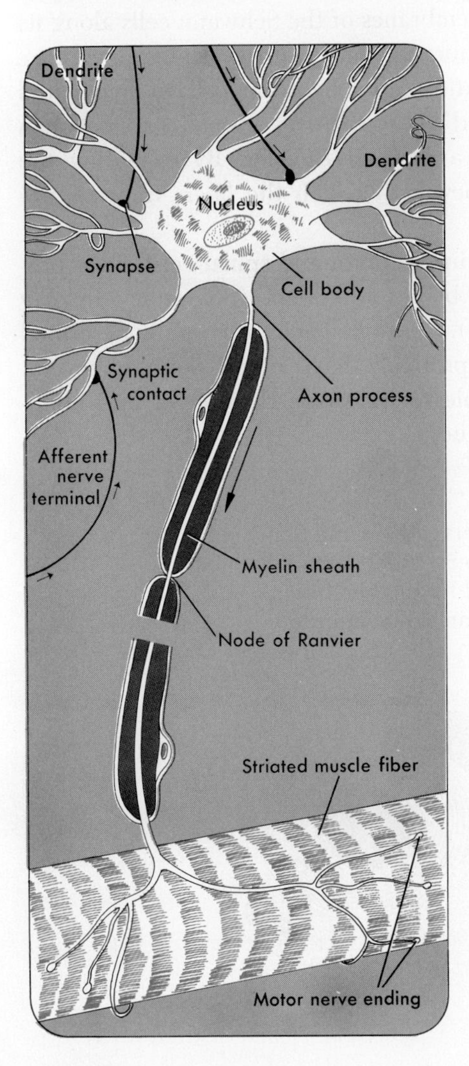

Figure 9–1. The nerve cell shown here is a motor neuron, one that carries impulses to a muscle fiber. Axons from other neurons make contact with the dendrites and cell body at the synapses. The myelin sheath is formed of Schwann cells. The node of Ranvier is the gap between two Schwann cells. Arrows indicate the direction of flow of nerve impulses.

myelin; where these fibers predominate, the tissue is yellowish in color and is called *white matter.* Nerve fibers may be short or very long, running perhaps from the tip of the toe to the center of the back. Nerves are simply bundles of myelinated dendrites and axons. They branch into smaller and smaller bundles, and finally separate into the fibers of individual neurons.

Myelin Sheath

Closely wrapped around the axons and dendrites are companion cells called *Schwann cells.* In unmyelinated tissues, several fibers are embedded in deep pits in a single Schwann cell. In myelinated tissue, a single fiber is wrapped around by many layers of the plasma membranes of the Schwann cells along its length. It is the fatty layers of the membranes that give the white appearance to these fibers. Figure 9–2 shows the relationship of Schwann cells to myelinated and unmyelinated fibers. In myelinated fibers there is a small gap between adjacent Schwann cells that is called a *node of Ranvier.* By insulating the fibers, the myelin increases the efficiency with which the fibers transmit impulses.

Neuroglial cells, which are found in the brain and spinal cord, have processes that weave around the nerve cell bodies and fibers. (Schwann cells may be considered special types of neuroglia). Neuroglial cells are usually described as supporting the neurons, but they probably have other, more important functions. Some neuroglial cells are able to transform themselves into phagocytes and remove damaged nervous tissue.

Figure 9–2. The relationship between nerve fibers and Schwann cells. A: In unmyelinated tissue several nerve fibers are embedded in a single Schwann cell. B: In myelinated tissue the fiber is surrounded by many layers of the Schwann cell membrane.

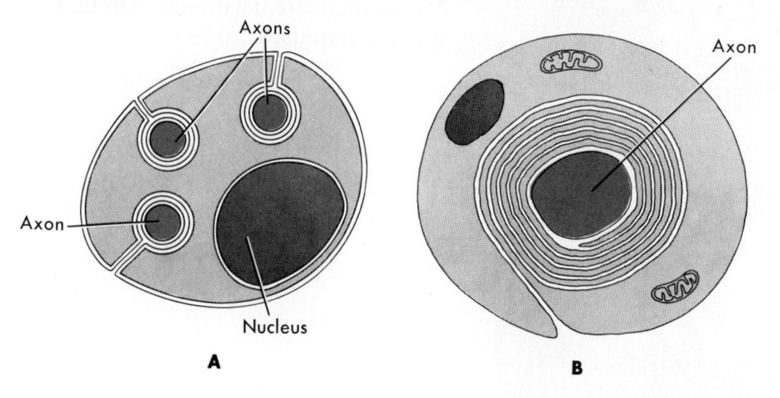

How Neurons Work

Functionally, neurons are of three types—*sensory, internuncial,* and *motor.* Sensory neurons carry signals to the spinal cord and brain from sensory receptors. Internuncial neurons transmit signals from the axon of one neuron to the cell body or dendrites of another. Motor neurons relay the signals to a gland or muscle cell and thus bring about some reaction.

Nature of the Stimulus. A neuron can be stimulated electrically, thermally, chemically, or mechanically. The stimulus sets up an impulse that travels from one part of the neuron to another. Neurons, like skeletal muscle fibers, respond in an all-or-none manner. A stimulus that is just capable of producing an impulse is said to be of *threshold value,* and one too weak to produce an impulse is said to be *subthreshold.* A subthreshold stimulus may still affect the neuron in such a way that repeated subthreshold stimuli build up the excited state of the neuron to the point where an impulse will be initiated. Increasing the intensity of a stimulus beyond the threshold value does not increase the strength of the impulse. But it does increase the frequency with which impulses travel along the neuron. Also, the different neuronal fibers that make up a nerve vary in their thresholds. A stimulus usually affects more than one fiber and the more intense the stimulus, the greater is the number of nerve fibers carrying impulses. You are able to distinguish between a weak and a strong stimulus by the frequency of the impulses and the number of fibers carrying the impulses.

The Impulse. The initiation and transmission of an impulse along a fiber is a chemically and physically complicated process that is not yet completely understood. The concentration of ions is different on either side of the cell membrane of the neuron. There are more positively charged sodium ions outside the membrane, and more positively charged potassium ions inside. The positive charges inside the cell are more than counterbalanced by the presence in the cell of negatively charged proteins that are too large to pass through the cell membrane. Thus, the inside of the membrane is electrically negative relative to the outside. The membrane is said to be polarized. A stimulus has the ability to bring about a change in the permeability of the membrane. Sodium ions rush in through the membrane from the region of higher concentration outside the cell to the region of lower concentration inside. For a moment, at the particular point where the fiber was stimulated, the charges are reversed. The inside of the membrane becomes positive relative to the outside. This sets up a current on the surface of the membrane, which flows from the negatively charged surface of the depolarized zone to the adjacent positively charged polarized region and alters the permeability there. Thus, the excited region of the fiber shifts away from the point of stimulation.

Meanwhile, at the point of initial stimulation, potassium ions flow out from their region of high concentration inside the membrane to the region of low concentration outside. Thus, the positive charge outside the membrane is restored and the fiber is ready to be stimulated again. The number of sodium and potassium ions exchanged with each impulse is relatively very small compared to their total number on either side of the membrane. Eventually, though, the sodium ions would accumulate inside the membrane to the point where the cell could no longer respond to stimulation if there were not some means for returning the ions to the outside. An active transport mechanism, nicknamed the sodium pump, is constantly at work, exchanging sodium ions that have entered through the membrane with potassium ions in the surrounding medium, thereby restoring the original concentrations of ions on either side of the membrane. The flow of an impulse along a fiber is illustrated diagrammatically in Figure 9–3.

A myelinated fiber is heavily insulated by its sheath, which is closely applied to the plasma membrane of the fiber and prevents the exchange of ions between the fiber and the external medium. The current flows inside the membrane from one node of Ranvier to the next, where ionic exchange is again possible. This speeds up transmission of the impulse, which is faster in myelinated than in unmyelinated fibers.

Figure 9–3. Diagram of the movement of an impulse along a nerve fiber through changes in the concentration of ions on either side of the membrane.

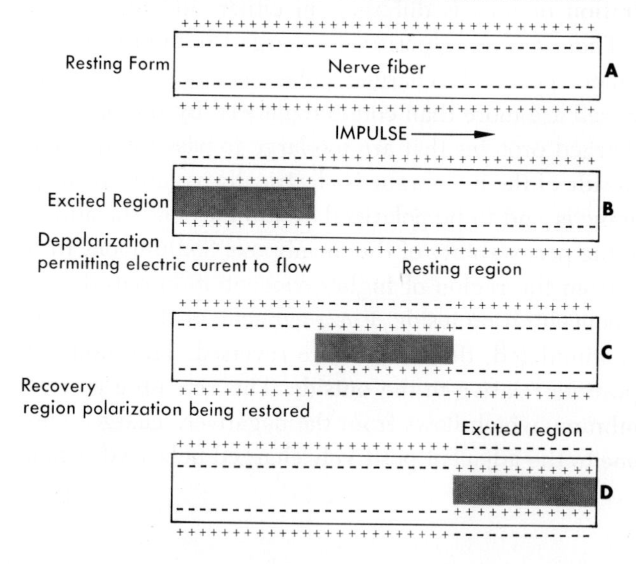

The Synapse. There is a small gap between the tip of a terminal branch of an axon and the cell membrane of the next cell, be it muscle, gland, or neuron, to which the axon is applied. This gap is called a *synapse*. The electrical impulse traveling along the axon does not jump this gap. The terminal branches of the axon end in small swellings, which house many little vesicles each containing a transmitting substance. When an impulse reaches the terminal swellings of the axon, the vesicles discharge their content into the synaptic space between adjacent cells. The transmitter substance diffuses across the space and depolarizes the membrane of the next cell and thereby initiates an impulse in it. In producing and secreting these substances, neurons act like gland cells. Impulse transmission is a one-way affair because only the tips of the axons have this secretory ability. An impulse can travel both ways in a single neuron, but only the axon can pass the impulse along to another cell. In synapses between neurons and voluntary muscles and between many neurons in the brain and spinal cord the transmitter is a chemical called *acetylcholine*. Other transmitting substances, including *epinephrine* and *norepinephrine*, are also known. The effect of a transmitter is sometimes inhibitory; that is, it blocks the postsynaptic cell by increasing its threshold of response. Whether a transmitter stimulates or inhibits may depend on the postsynaptic cell. Acetylcholine stimulates voluntary muscle but inhibits heart muscle.

The Reflex Arc. A reflex is an involuntary response that is always the same for a given stimulus. When the doctor taps the ligament in your knee, you respond by jerking your leg.

A reflex arc consists of a *receptor* (sense organ), two or more neurons, and an *effector* (muscle or gland). The simplest type of reflex involves only a sensory and a motor neuron. A stimulus received from a sense organ somewhere in the body sets up an impulse in the sensory neuron. The impulse travels along the fiber to its synaptic contact with a motor neuron. The motor neuron is stimulated and in turn transmits the impulse to an effector. Because the sensory neuron is in direct contact with the motor neuron, no alternative paths are open and only a single response is possible.

Postural reflexes are of the two-neuron type. As the body begins to sag, sensory dendrites in the body muscles are stimulated. The impulses are carried to motor neurons in the cord and impluses from these pass to the muscles, causing them to contract enough to maintain body posture (see Figure 9–4).

Many reflexes involve three types of neurons—sensory, internuncial, and motor. Figure 9–5 shows a reflex of this type. Because a reflex response may involve a number of different sets of muscles in different parts of the body, the internuncial neurons are necessary to pass the impulse to the different appropriate motor neurons.

Although a healthy nerve fiber with an adequate blood supply will con-

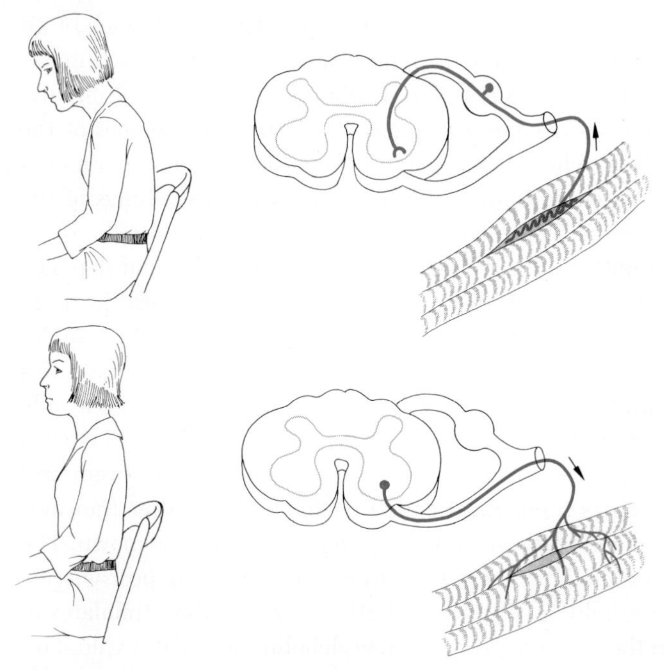

Figure 9–4. A two-neuron reflex (postural).

tinue to work indefinitely, a reflex arc is very susceptible to fatigue. This probably results from the exhaustion of the synaptic mechanism. Under such conditions, the transmitter chemical is used up faster than it is produced by the cell.

THE SYSTEM AS A WHOLE

The nervous system can be divided into two main parts, the central nervous system (CNS), which includes the brain and spinal cord, and the peripheral nervous system, the nerves.

Nerves are bundles of fibers passing from the brain or spinal cord to some structural unit of the body. The *cranial nerves* leave the brain and pass through holes in the cranium. The *spinal nerves* pass from the spinal cord between the vertebrae. The spinal nerves are all built on more or less the same pattern, but

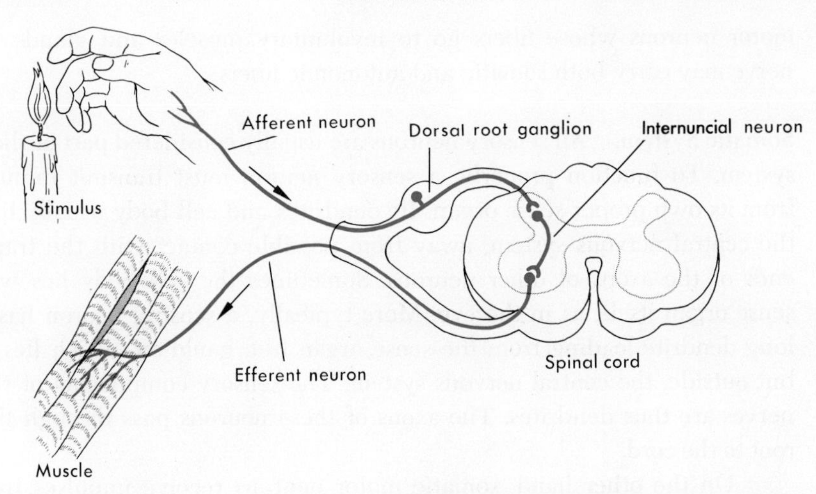

Figure 9–5. A three-neuron reflex.

the cranial nerves are not. Some of the cranial nerves have only motor fibers, such as the nerve controlling the muscles that rotate the eyeball; some have only sensory fibers, such as the nerve carrying the impulses from the inner ear to the brain; and others carry both types of fibers, such as the nerve that goes to the face.

In man there are thirty-one pairs of spinal nerves. A typical spinal nerve is formed by the fusion of a *dorsal* and a *ventral root* that extend from the side of the spinal cord. The dorsal root contains only sensory fibers carrying impulses into the cord. The nerve cell bodies of the sensory neurons are packed together in a swelling on the dorsal root called the *dorsal root ganglion.* (A ganglion is a collection of nerve cell bodies lying outside the central nervous system.) The nerve cell bodies of motor neurons usually lie within the cord so that the ventral root contains the axons of motor neurons carrying impulses away from the cord. Shortly after the dorsal and ventral roots join to form the spinal nerve, the nerve divides into three branches: a dorsal branch going to the skin and muscles of the back, a ventral branch going to the skin and muscles in front, and a visceral branch going to the internal organs. These branches contain both motor and sensory fibers. The three main branches divide into smaller and smaller bundles of fibers until, at last, the terminal branches of a single fiber pass to individual cells. The terminal branches of a single motor axon may innervate several hundred cells. See Figures 9–4 and 9–5.

The peripheral nervous system can be divided into the somatic system, which contains both the sensory neurons and the motor fibers going to the voluntary muscles and skin, and the autonomic system, which contains only motor neurons whose fibers go to involuntary muscles and glands. A single nerve may carry both somatic and autonomic fibers.

Somatic System. All sensory neurons are usually considered part of the somatic system. To function properly, a sensory neuron must transmit impulses only from its own proper sense organ. Its dendrites and cell body usually lie outside the central nervous system, away from possible contact with the transmitting ends of the axons of other neurons. Sometimes the cell body lies within the sense organ itself, as in the eye. More typically, a sensory neuron has a single long dendrite leading from the sense organ to a ganglion, which lies close to, but outside, the central nervous system. The sensory components of the spinal nerves are thus dendrites. The axons of these neurons pass through the dorsal root to the cord.

On the other hand, somatic motor neurons receive impulses from more than one source. You kick in response to a wide variety of stimuli, not just when a doctor taps the tendon of your knee to elicit the knee jerk reflex. The dendrites and cell bodies of somatic motor neurons lie within the central nervous system, in close contact with internuncial neurons. The axons of these motor neurons leave the CNS as components of the nerves.

Autonomic System. The autonomic system differs structurally as well as functionally from the somatic system. In some of the cranial nerves and some of the spinal nerves, an internuncial fiber leaves the central nervous system and makes a synapse with a motor neuron in a ganglion outside of the brain or spinal cord. This second neuron then carries the impulse to a visceral muscle or gland. Activities regulated by these two neuron series include such functions as the peristaltic contractions of the digestive system and the release of sweat by the sweat glands. These functions are usually beyond the control of the will, and hence the nerve fibers and ganglia involved are spoken of as the *autonomic system.* This system has two subdivisions, the *sympathetic* and the *parasympathetic.* Fibers of the sympathetic system are present in spinal nerves of the chest and lower back region; fibers of the parasympathetic system occur in some of the cranial nerves and the spinal nerves of the sacral region. The main pathways of the autonomic system are shown in Figure 9–6 and Figure 9–7.

Organs innervated by the autonomic system are under a dual control provided by the sympathetic and parasympathetic systems. Axons of the parasympathetic system release acetylcholine at the neuromuscular synapse, just as do the motor neurons going to the voluntary muscles. Most axons of the sympathetic nervous system, on the other hand, secrete norepinephrine, but sympathetic fibers going to the sweat glands and the uterus secrete acetylcholine. These two systems have antagonistic effects on their target organs. Stimulation

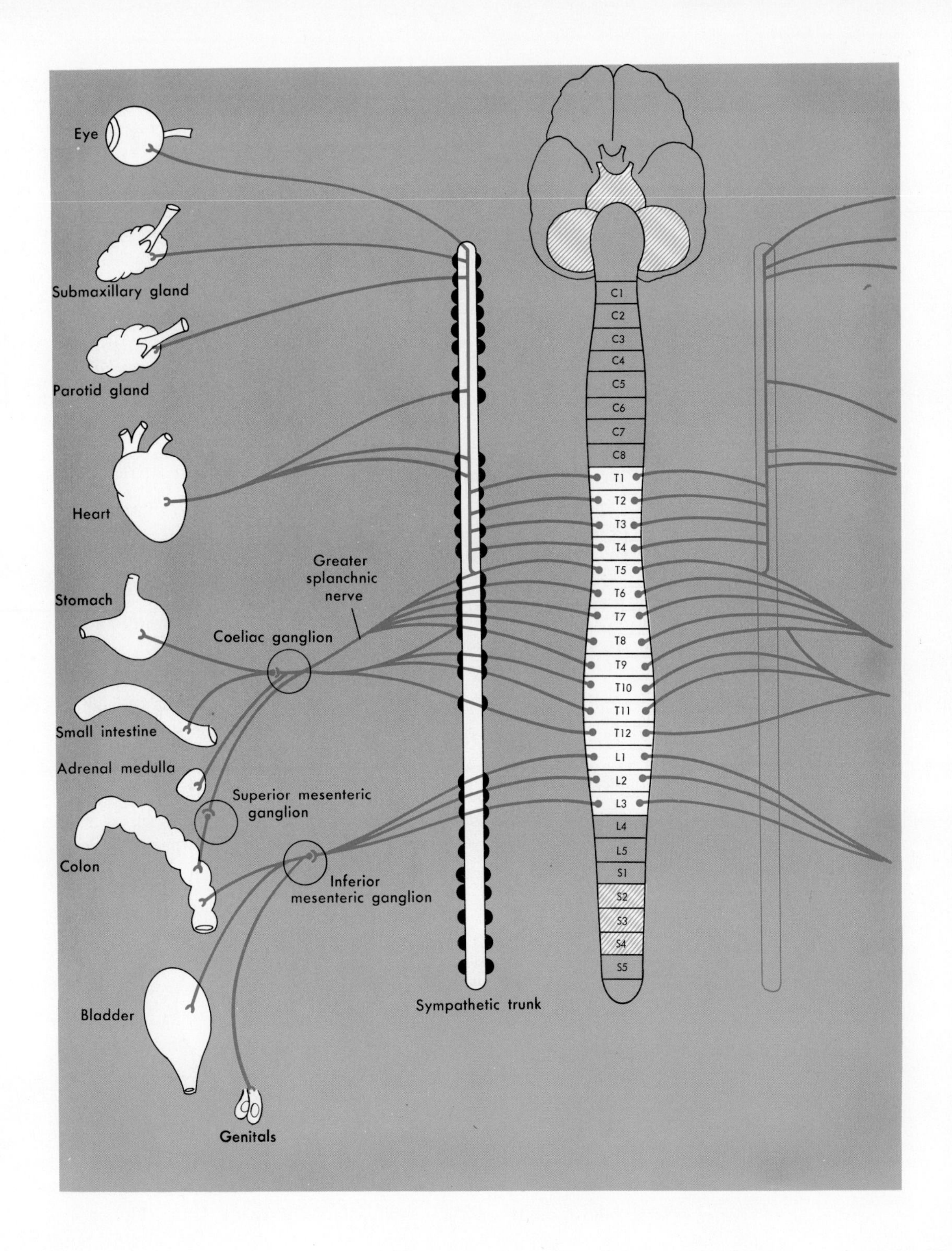

Figure 9–6. The sympathetic nervous system.

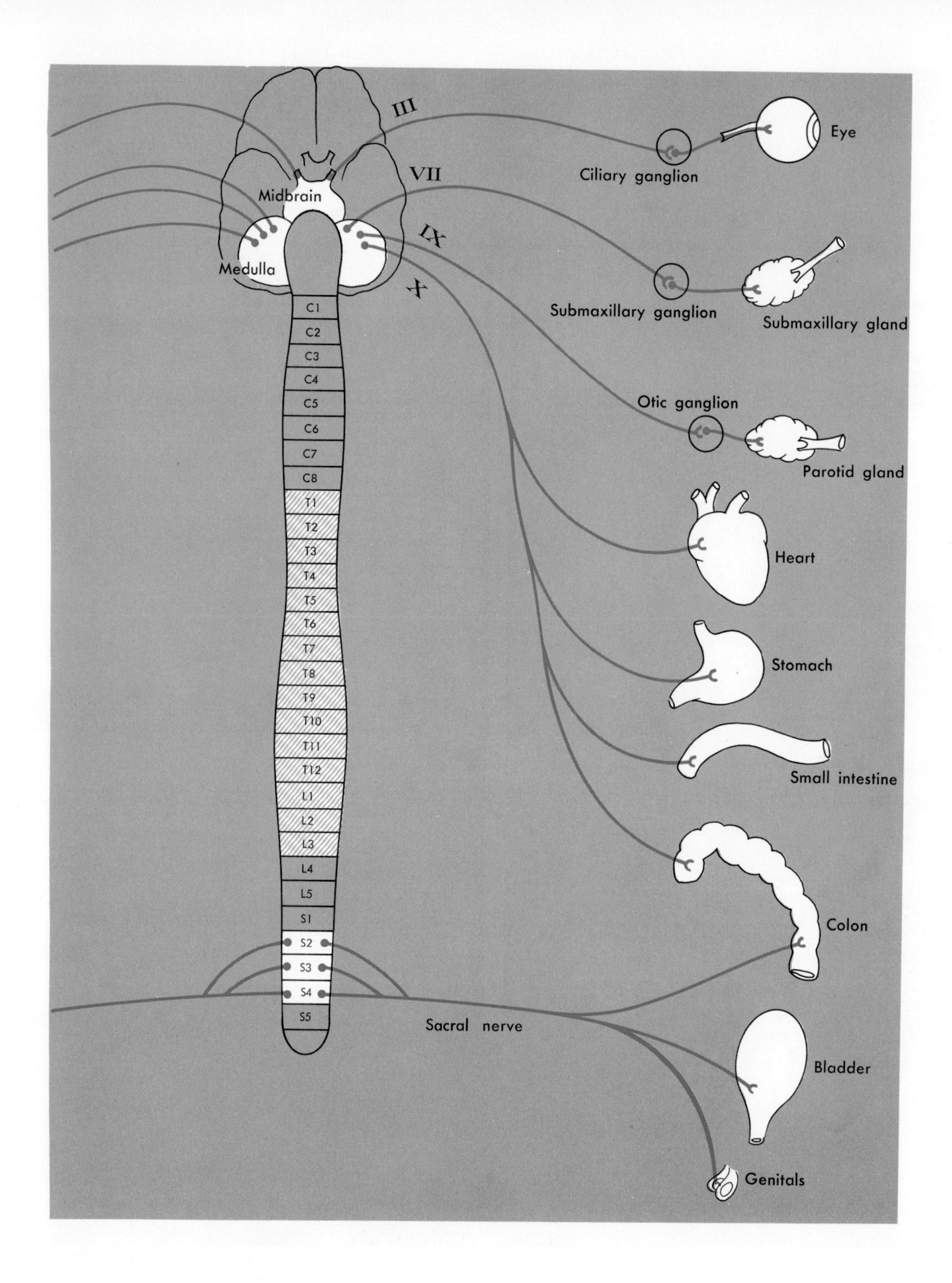

Figure 9–7. The parasympathetic nervous system.

by the sympathetic nervous system prepares the body for action—the heartbeat quickens, the blood pressure increases, and the activities of the digestive tract slow down so that blood can be shunted to the muscles. This is the "fight or flight" syndrome. When you are angry or frightened, your sympathetic system is active. On the other hand, the parasympathetic system slows down the heartbeat, lowers the blood pressure, and stimulates the digestive organs. The action of the two parts of the autonomic system on some of the organs is summarized in Table 9–1.

Central Nervous System

The central nervous system consists of the brain and spinal cord. They are both made up of internuncial neurons; the dendrites, cell bodies, and proximal parts of the axons of motor neurons; and the terminal parts of the axons of sensory neurons. There is a small, central cavity in the spinal cord that connects with a series of cavities, called *ventricles,* in the brain. Both the brain and the spinal cord are surrounded by three membranes, the *meninges,* which form a protective pad around the delicate central nervous system. Between the meninges and the nervous tissue is a cushioning layer of *cerebrospinal fluid.* Inflammation of the meninges is known as *meningitis.*

The CNS is composed of literally billions of cells, each in synaptic contact with other cells. Some hint of the complexity of these contacts is suggested by Figure 9–8. It has been estimated that the dendrites and cell body of an average neuron in the central nervous system may have a thousand synaptic contacts. The terminal branches of the axon of the neuron may in turn be in contact with many other neurons. An impulse reaching the central nervous system spreads out like water sprayed from a hose. The number of potential pathways is inconceivably large, and biologists are still far from understanding what deter-

Table 9–1

Organ	Sympathetic	Parasympathetic
Heart	Stimulates	Inhibits
Blood vessels	Mostly constricts	Mostly dilates
Iris of eye	Dilates	Constricts
Sweat glands	Stimulates	No action
Bronchi	Dilates	Constricts
Salivary glands	Inhibits	Stimulates
Intestinal tract	Inhibits	Stimulates
Bladder	Inhibits	Stimulates

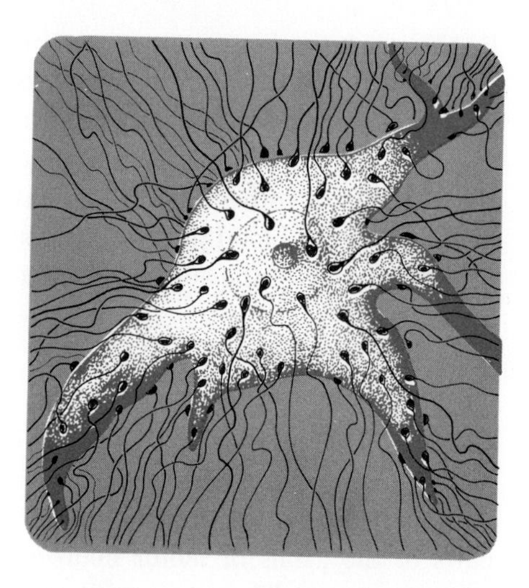

Figure 9–8. Diagram showing some of the synaptic contacts on the dendrites and cell body of a single neuron in the central nervous system.

mines the path a given impulse will follow. Some pathways are apparently fixed and innate (e.g., those involved in the control of breathing); some may become fixed through repetition (habit formation); some vary according to circumstances.

The central nervous system is not a *tabula rasa,* a mere transmitter of impulses from the sense organs to the muscles and glands. Even during deep sleep the brain is in a state of activity. There are constant, apparently spontaneous "firings" across synapses. Input from the sense organs is imposed upon this ongoing activity, both modifying it and being modified by it.

The Spinal Cord. The spinal cord extends from the brain down the neural canal of the vertebral column. It is perhaps twelve millimeters (half an inch) in diameter. In cross section, the spinal cord is made up of an outer portion of myelinated fibers, yellow in color, and a central, H-shaped portion, where the cell bodies of internuncial and motor neurons are located and that is therefore, gray in color. This can be seen clearly in Figure 9–4.

When a child is first born, the spinal cord extends well down into the region of the lower back, but as he grows the vertebral column lengthens at a rate greater than the spinal cord does so that in an adult the cord is much shorter than the neural canal. This differential growth is compensated for by an elongation of the branches of the spinal nerves before they pass out between the vertebrae. In an adult, the lower end of the canal thus contains a bundle of nerves rather than a single spinal cord (see Figure 9–9).

The Brain. The most conspicuous part of the human nervous system is the brain. It includes clusters of nerve cell bodies, called, confusingly, *nuclei* or

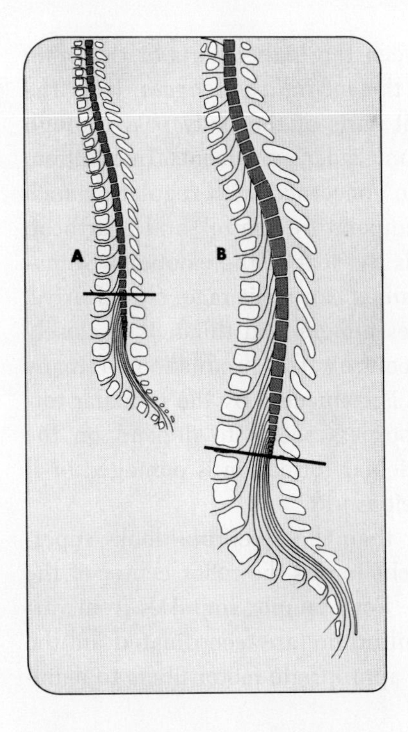

Figure 9–9. Diagram illustrating the difference in the extent of the spinal cord in A: a newborn child and B: an adult. The line is drawn between the first and second lumbar vertebrae. In the adult the cord terminates at this point. In the child it continues to the level of the fourth lumbar vertebra.

nerve centers. The brain also has *fiber tracts,* which are bundles of myelinated fibers, and bundles of mixed myelinated and unmyelinated fibers called the *reticular formation.* The gray matter of the brain comprises nerve cell bodies and unmyelinated fibers, and the white matter comprises fiber tracts. The parts of the brain are the *brain stem,* the *cerebellum,* and the *cerebrum.* (See Figure 9–10.)

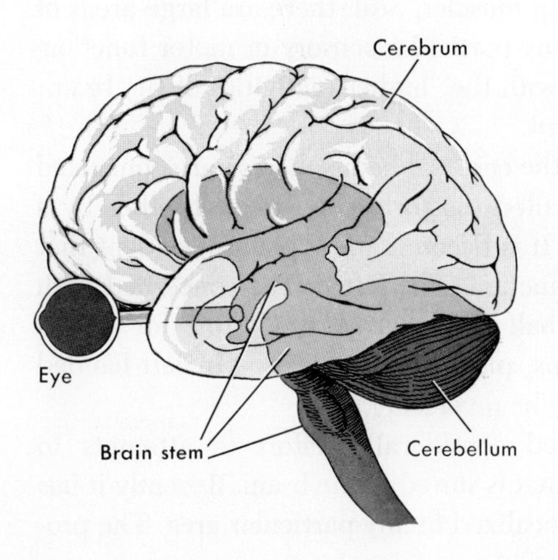

Figure 9–10. The human brain.

The brain stem is the passageway between the higher parts of the brain and the rest of the body. Its posterior part, the *medulla oblongata,* joins the brain and spinal cord. Sensory fibers from all parts of the body pass through the brain stem, as well as fibers controlling all muscular movements. In addition, the brain stem contains nerve centers that are concerned with regulating such essential, automatic functions as breathing, temperature control, and heartbeat. The *hypothalamus,* a part of the brain stem, is the seat of the emotions. Stimulation of one part or another of the hypothalamus can elicit rage, sexual drive, pleasure, or pain. The hypothalamus regulates hunger and thirst. It is closely connected with the endocrine system and controls the production of many hormones (see next chapter). The brain stem is connected by the reticular formation to other parts of the brain. Consciousness seems to depend on the integrity of the reticular formation. If the reticular formation is damaged or if its activity is reduced by drugs, one loses consciousness.

The cerebellum, which is much smaller than the cerebrum, looks superficially like a miniature version of it. The cerebellum is the reflex center of the brain; it is here that impulses are coordinated. For example, impulses from various sensory receptors concerned with equilibrium are coordinated in the cerebellum and the impulses passed on to the appropriate motor fibers to maintain body equilibrium.

The largest, most complex, most fascinating, and least understood part of the human brain is the cerebrum, a heavily convoluted, bilobed structure that overlies the rest of the brain and fills most of the cavity of the skull. It is the center of control of voluntary motor actions; it is the place where sensory impulses entering the brain are interpreted; and it is the site of memory, reasoning ability, and judgment. It has been possible to localize the sensory and motor areas of the cerebrum. Thus, one area receives stimuli from the thumb, whereas another controls the movement of the lip muscles. Still, there are large areas of the cerebrum that are not assigned to any particular sensory or motor function. These unassigned areas are concerned with the "higher" activities of the brain: memory storage, reasoning, and judgment.

The two halves of the cerebrum (the cerebral hemispheres) are connected to the hypothalamus by a fiber tract called the *fornix* and to each other by a similar fiber tract, the *corpus callosum.* It is becoming increasingly evident that the two halves of the brain differ in function. The left half is concerned with speech and logical thought; the right half is primarily responsible for spatial concepts and the recognition of persons, places, and objects. (In left-handed people, the localizations of function may be reversed.)

Neurophysiologists have expended considerable effort in attempts to identify the *engram,* the memory trace that is stored in the brain. Recently it has been suggested that the engram is not localized in any particular area. The process of learning even a simple task seems to affect enormous numbers of cells in many different parts of the brain. One can monitor activity in the brain by recording the electrical discharges (brain waves) that indicate neuronal activity. With microelectrodes it is possible to record simultaneously from many different areas of the brain. It has been shown that when a cat performs a specific learned task, a specific pattern of brain waves appears in many different parts of the *cortex* (the gray matter of the cerebrum). With a different task, a different pattern appears. Any given engram seems to involve most of the neurons of the cortex. Conversely, any given neuron must participate in the storage of an enormous number of engrams. The diffuse nature of the engram results in a considerable built-in safety factor in the brain. A man has many more neurons in his cerebrum than he needs for normal functioning. Many of these neurons can be destroyed without significantly impairing the ability to learn, to remember, and to reason. It has been estimated that several thousand neurons in the adult human brain die every day. They are not replaced, yet the brain continues to function adequately for many years.

The cerebrum, cerebellum, and brain stem are all interconnected. Thus, the cerebrum interprets the sensory impulses received by the brain stem. On the other hand, impulses to motor areas from the brain stem can be modified by activities of both the cerebellum and the cerebrum. A person has some control (from the cerebrum) over his rate of breathing.

SENSORY RECEPTORS

Like all organisms, a man faces the continual necessity of adjusting to changes in his environment. He can do this only if information about these changes is available to him. Awareness of these changes is made possible by specialized structures, the sensory receptors. Changes that take place in the environment excite receptors, either specialized endings of sensory dendrites or specialized cells that transmit stimuli to the sensory neurons. Each receptor normally responds to only one type of stimulus. Thus, a salt taste receptor responds only to the presence of salt. It is not stimulated by sugar or heat or sound. This specificity is advantageous, for it permits you to respond to specific stimuli. Just think of the utter confusion that would result if you could not differentiate between stimuli. Regardless of the excitatory stimulus, though, or the structure of the sense organ, the impulse that travels along a sensory neuron is always the same—a change in the electrical potential along the membrane. The interpretation of the stimulus, then, must reside in the particular part of the brain reached by the impulse. If you could switch the location of the endings of the fibers coming from your ears and from your eyes, your brain would interpret thunder as sight and lightning as sound. Moreover, it is the brain that localizes the sensation. If a man has a leg amputated, he may still feel pain in the foot that is no longer there. The cut ends of the fibers in the stump are being stimulated and the impulses are interpreted by the brain as coming from the foot.

Several methods of classifying receptors have been proposed; one that seems satisfactory groups receptors as *exteroceptors, proprioceptors,* and *interoceptors.* Exteroceptors, those that receive stimuli from the outer environment, are usually located on the surface of the body. Proprioceptors are located in the muscles and joints and convey information about the state of contraction of the muscles and the relative position of the parts of the body. Those embedded in the body muscles concerned with posture are good examples. Interoceptors are located in internal organs, such as the ones in the wall of the heart and in the arteries mentioned in the section about heartbeat control in Chapter 8, or the ones in the wall of the lungs that take part in breathing control as discussed in Chapter 6.

When one speaks of sensory receptors, he usually has the exteroceptors in mind; these are the only receptors that we will discuss further. Exteroceptors can be classified as cutaneous receptors and special receptors. See Table 9–2.

Cutaneous Receptors

Figure 9–11 shows a section through the skin with the location of various cutaneous receptors.

Tactile Receptors. The tactile receptors respond to light touch. Because they are more common in some areas of the skin than in others, the sense of touch is not uniform over the body. The tactile receptors are spaced furthest apart on the back, much closer together on the fingertips, and most closely packed on the tip of the tongue. Your interpretation of the size of something you feel is deter-

Table 9–2

Cutaneous Receptors	Special Receptors
Tactile (light touch)	Auditory—sense of hearing
Pressure	Equilibrium—sense of balance and position
Pain	Photoreceptors—sense of sight
Warmth	Gustatory—sense of taste
Coolness	Olfactory—sense of smell

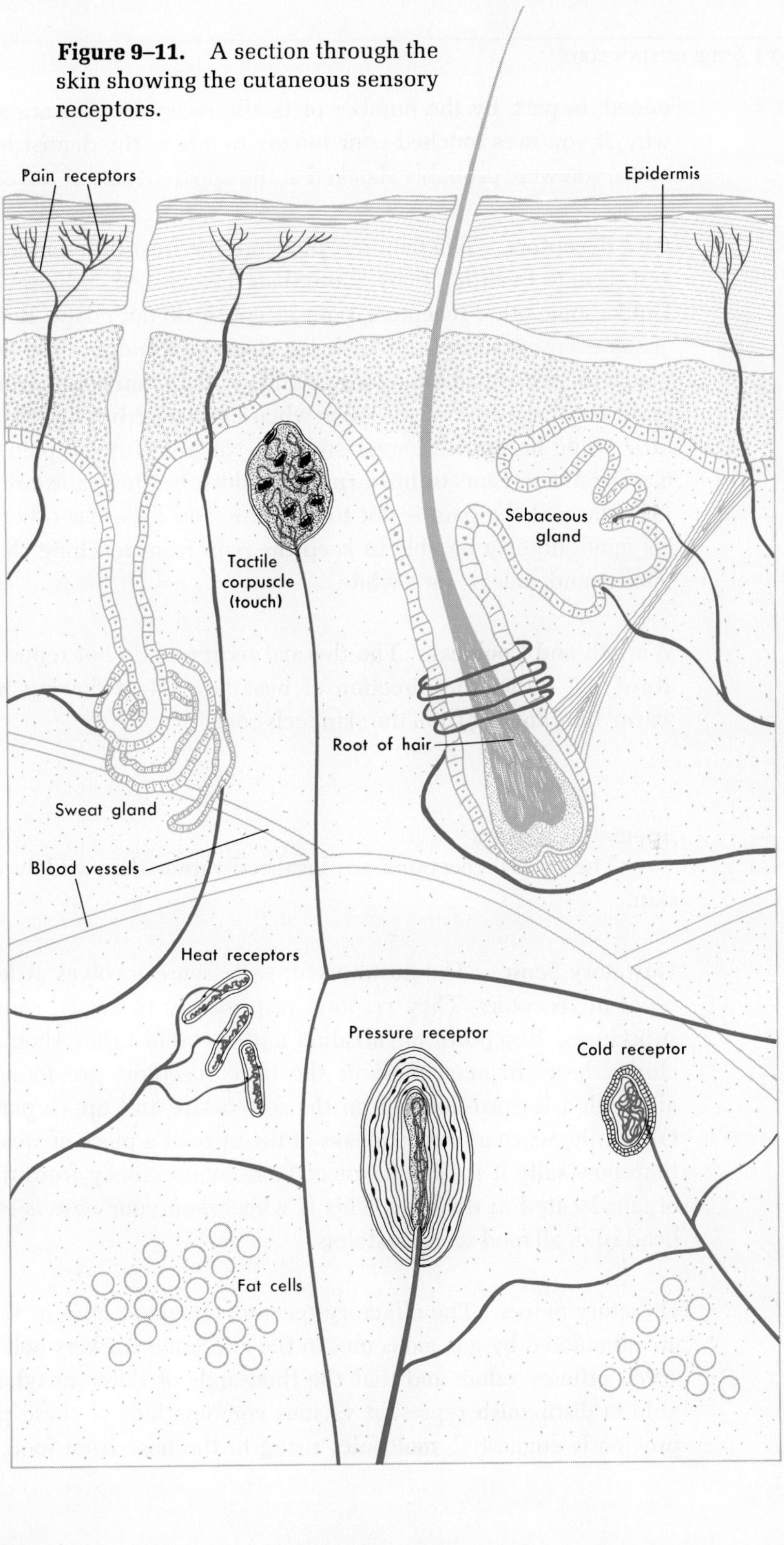

Figure 9–11. A section through the skin showing the cutaneous sensory receptors.

mined, in part, by the number of tactile receptors that are stimulated. This is why, if you ever touched your tongue to a hole the dentist had drilled in your tooth, you were probably shocked at the apparent size of the cavity.

Pain Receptors. The pain receptors are the least differentiated of all receptors and seem to be little, if any, more than the exposed endings of sensory neurons. Unlike some other receptors, pain receptors do not adapt. Repeated stimulation of some sensory receptors may bring about the condition known as *adaptation*, in which the receptor no longer responds to the stimulus. A worker in a chemical plant may be aware of the odor when he first arrives in the morning, but in a little while he can no longer smell it. Not so with the pain receptors. A man may be unconscious of his own body odors, but his pains are always with him. (However, if the pain is not too intense, and if he has other things to occupy his mind, he may be able to keep the pain from reaching the level of his conscious mind, at least for a while).

Warmth and Coolness. The thermal receptors do not report absolute temperature, but rather the direction of heat flow. Heat flowing into the skin feels warm, heat flowing from the skin feels cool.

Special Receptors

The special receptors are localized rather than widely distributed in the skin.

Gustatory Sense. The gustatory or taste sense involves at least four different types of receptors. They respond respectively to sweet, sour, salt, and bitter substances. Receptors for alkaline and metallic tastes should possibly be included here. In man, most of the taste receptors are found on the tongue, although a few are located on the soft palate and upper part of the pharynx. Obviously, when a person speaks of the taste of a piece of steak, he means more than how salty it is. The flavor of food comes chiefly from the olfactory sense organs located in the nose. This is why, when your nose is stopped up with a head cold, all food seems tasteless.

Olfactory Sense. The olfactory receptors, located high in the nasal chamber, are stimulated by gas molecules in the air. Some workers believe that there are seven primary odors and that the thousands of different odors a man may be able to distinguish represent various combinations of these primary odors. As previously suggested, molecules rising to the nose from food being chewed in the mouth produce stimuli that are interpreted by the brain as coming from the mouth.

The Eye. The eye is a mechanism that transforms light energy into the energy of a nerve impulse. Impulses from the eye are translated in the brain as visual images. Figure 9–12 is a diagram of a section through the human eye. It is shaped like a hollow ball, divided into two very unequal chambers by a biconvex *crystalline lens* and the muscles and ligaments that hold the lens in place. The outer coat of the eye, the *sclera* or white of the eye, is tough and fibrous. In front of the lens, the sclera is transparent and bulges slightly outward to form the *cornea.* The cornea and lens both serve to focus the incoming light rays on the light-sensitive back part of the eye. The middle layer of the eye is the *choroid coat.* It is absent in the corneal region. In the back part of the eye, the choroid coat is heavily pigmented and contains many blood vessels. In front, *ciliary* muscles pass from the edge of the choroid coat to the edge of the lens. These muscles can change the shape of the lens and make it possible for the eye to *accommodate,* that is, to focus rays of light from sources at various distances on the back part of the eye. Between the ciliary muscles and the cornea in front is the *iris,* a colored, disc-shaped structure with a hole in the center, the *pupil.* By means of circulatory and radially arranged smooth muscle fibers in the iris, the size of the pupil can be changed. The iris thus functions as an adjustable light diaphragm. The pupil opens wide in the dark to let in as much light as possible and contracts in bright light. The size of your pupil also indicates your

Figure 9–12. Section through the human eye.

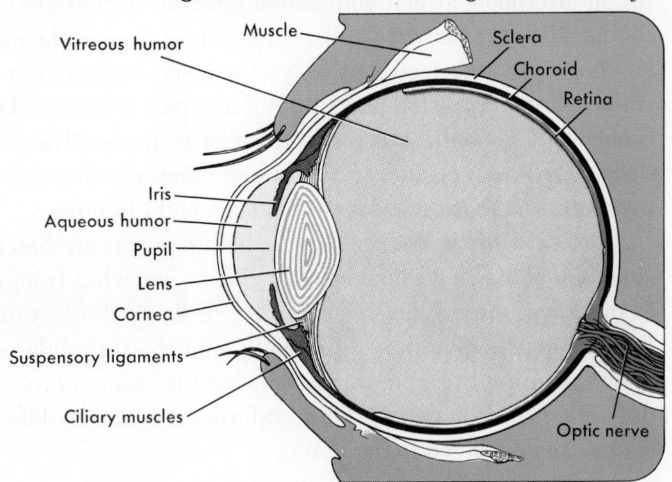

reaction to what you are looking at. If you see something that interests you, the size of your pupil increases. To say someone is "wide-eyed with interest" is quite true.

The innermost layer of the eye is the light-sensitive *optic retina.* The chamber of the eye between the cornea and the lens is filled with a watery fluid and is known as the *aqueous chamber;* that between the lens and the retina is filled with a jelly-like substance and is known as the *vitreous chamber.* The function of the structures that lie in front of the retina is to admit the incoming light rays and focus them on the retina, which contains the light-sensitive cells. In the human eye, these are of two main types, *rods* and *cones.* They differ slightly in shape, more so in function. Each consists of a cell body and a long, thin, outer segment composed primarily of a stack of membranous discs.

ROD VISION. Vision by rods is a dark-light or *scotopic* vision, that is, these cells are sensitive to differences in the intensity of light but not to color. Lined up on the membranes of the outer segment of a rod cell are many molecules of *rhodopsin* (visual purple). Rhodopsin is a large protein molecule connected with ten small pigment molecules. When rhodopsin is exposed to light, the light energy alters the arrangement of the atoms of the pigment molecules. This sets up a stimulus that is transferred by the rod cell to one of the sensory neurons lying in the retina. When the pigment molecules change, they separate from the protein. Rhodopsin thus breaks down into a protein, *opsin,* and the pigment, *retinene.*

Rhodospin is resynthesized both in the light and in the dark, but in a bright light it is broken down more rapidly than it is reconstructed, and soon there is little left in the rods. Then vision depends almost entirely on color or *photopic* vision by the cone cells. When you enter a darkened movie house on a bright afternoon, at first you cannot see your way around because there is little rhodopsin left in your rod cells. After a short time in the dark, enough rhodopsin has been resynthesized so that you can see to find your seat. In a dim light, color (photopic) vision plays little if any part and light-dark processes are used exclusively. Vitamin A is a precursor of retinene. If a person has a vitamin A deficiency he is not able to synthesize much rhodopsin and hence suffers from poor light-dark vision and is said to have night blindness.

CONE VISION. Less is known about cone vision than about rod vision. The cones contain visual pigments that differ somewhat from rhodopsin. According to one idea, three types of cones are present: those stimulated by yellow-red light, those stimulated by green light, and those stimulated by blue light. Each neuron leading to the brain is connected by synaptic pathways to at least two different types of cones. Thus, information about different combinations of colors can be relayed to the brain.

The Ear. The human ear has three functions: the reception of sound, the reception of stimuli concerning the position of the body, and the reception of stimuli concerning the motion of the body.

The ear is made up of outer, middle, and inner portions. The outer ear consists of an essentially nonfunctional flap, the *pinna,* and a tube at the bottom of which is the delicate *tympanic membrane* (eardrum). The middle ear is a chamber filled with air and connected with the pharynx by means of the *Eustachian tube.* Across the chamber a chain of three bones carries vibrations from the tympanic membrane to the *oval window* of the inner ear (see Figure 9–13).

The inner ear contains the sensory receptors. It consists of the *cochlea* and the *semicircular canal* system. Together they comprise a series of interconnected, fluid-filled, membranous tubes (the *membranous labyrinth*) closely surrounded by bony, fluid-filled tubes of the same shape (the *bony labyrinth*). See Figure 9–14.

HEARING. The cochlea is the part of the inner ear concerned with hearing. It is a coiled tube divided longitudinally into three chambers, the *scala tympani,* the *scala vestibuli,* and lying between them, the *cochlear duct.* The cochlear duct is separated from the scala vestibuli by the *vestibular membrane* and from the scala tympani by the *basilar membrane.* These membranes represent the membranous labyrinth part of the cochlea. The relationship of the chambers of the cochlea is shown in Figure 9–15. The scala vestibuli and scala tympani are continuous at their distal ends. The receptors of the vibrations that are interpreted as sound by the brain are located on the basilar membrane.

Air vibrations set the tympanic membrane in motion, and this motion is transmitted by the bones in the middle ear to the oval window at the base of the cochlea. The vibrations set up waves in the fluid of the scala vestibuli. These waves are transmitted to the fluid of the scala tympani and are damped out by the flexible *round window.* The waves put the basilar membrane in motion. Situated on the basilar membrane are specialized cells with minute projections or "hairs" on one end. These hairs project up into an overarching gelatinous structure, the *tectorial membrane.* By some mechanism as yet unknown, the up-and-down-movement of the hair cells against the tectorial membrane sets up stimuli that are transferred from these cells to the dendrites of adjacent sensory neurons. By this rather complex mechanism, you are able to "hear" air vibrations that range from about forty cycles per second to some 17,000 cycles per second. This greatly exceeds the range of the human voice, which ranges from about 42 cycles per second in the lowest bass to the note of c^4 (2,048 cycles), the highest recorded note by a soprano.

STATIC AND KINETIC SENSES. The receptors for the position (static) and

Figure 9–13. The middle ear. The small bones transmit vibrations from the tympanic membrane to the oval window of the inner ear.

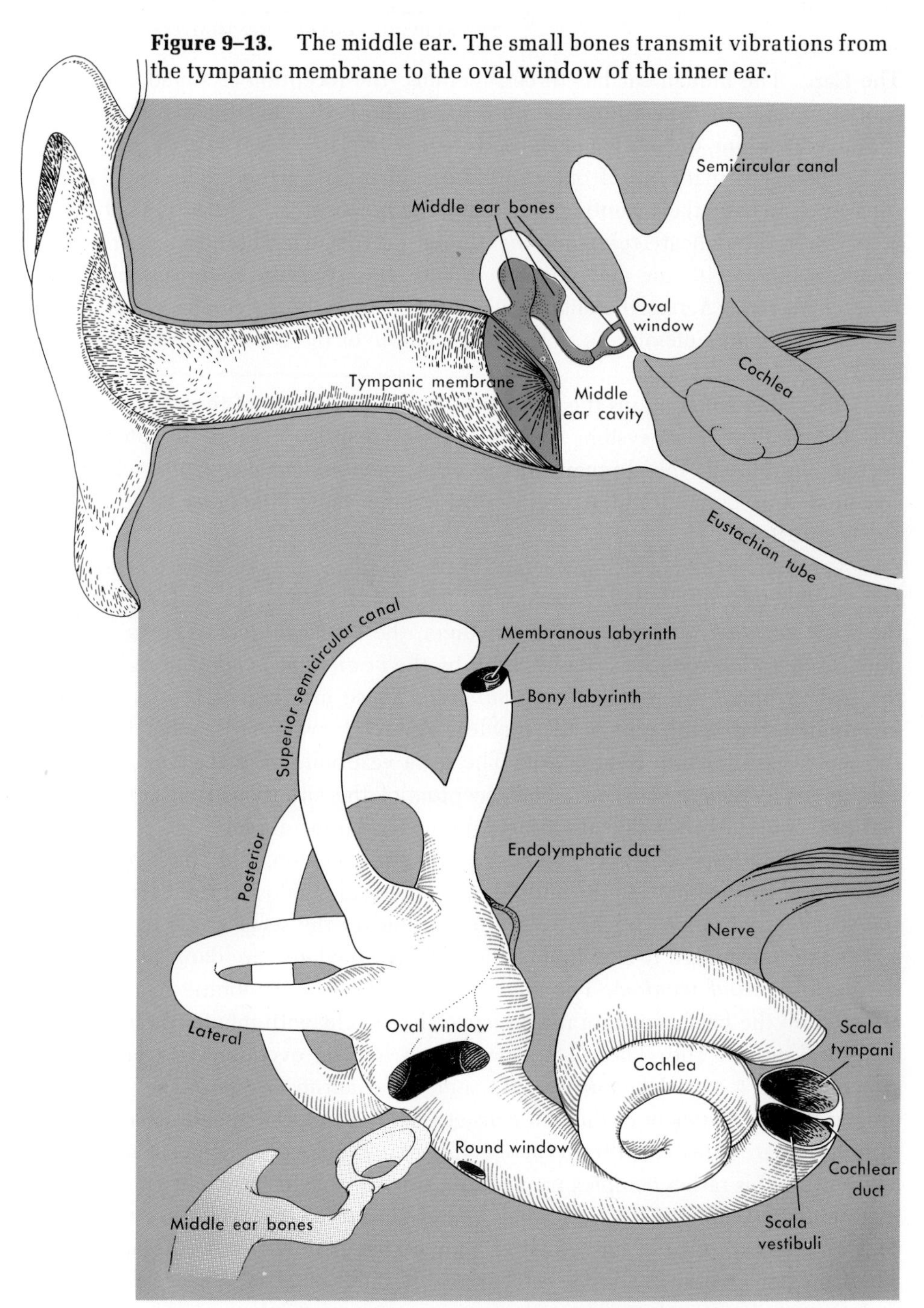

Figure 9–14. The inner ear. Parts of the bony labyrinth have been cut away from the superior semicircular canal and the cochlea to show the membranous labyrinth.

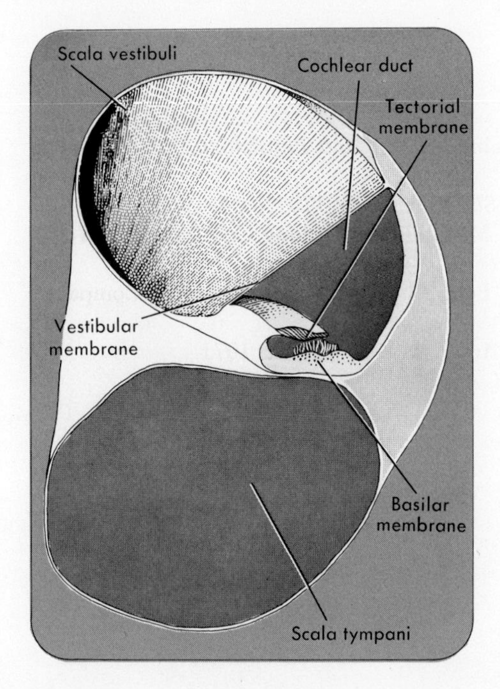

Figure 9–15. Section through the cochlea.

motion (kinetic) senses are both located in the inner ear. At the base of the semicircular canals is a chamber, the *utriculus,* within which are tufts of hair cells. The hairs of these cells are embedded in a gelatinous material containing many small, heavy, bony masses called *otoliths.* As you move your head around, the pull of gravity on the otoliths distorts the gelatinous mass and thereby moves the hairs of the sensory hair cells. You judge the position of your head in relation to the pull of gravity by the degree and angle of distortion of the tufts of hair.

The three semicircular canals are set at right angles to each other and are filled with fluid. They are so arranged that no matter at what angle you turn your head the fluid is caused to flow through the canals by its own inertia. This movement of fluid stimulates hair cell receptors located at the base of each canal. Stimulation of these receptors allows you to judge which way and how rapidly your head is rotating. Unfortunately, there seems to be a close connection in the brain between the centers for the reception of stimuli from the semicircular canals and from the digestive system. Overstimulation of the equilibrium centers in the canals tends to spill over and be misinterpreted by the brain as indicating something wrong in the stomach. Seasickness and airsickness are common and unhappy results of overactivity by the semicircular canals.

SUGGESTED READINGS

Crouch, J. E., and J. R. McClintic: *Human Anatomy and Physiology,* John Wiley & Sons, Inc., New York, 1971.

Gray, G. W.: "The Great Ravelled Knot," *Sci. Amer.,* Vol. 179, No. 4, 1948.

Grollman, S.: *The Human Body,* 3rd ed., Macmillan Publishing Co., Inc., New York, 1974.

Guyton, A. C.: *Function of the Human Body,* 3rd ed., W. B. Saunders Company, Philadelphia, 1969.

Heimer, L.: "Pathways in the Brain," *Sci. Amer.,* Vol. 225, No. 1, 1971.

Rushton, W. A. H.: "Visual Pigments," *Sci. Amer.,* Vol. 207, No. 5, 1962.

CHAPTER **10**

The Endocrine System—Internal Control

Many activities of the body are coordinated by nerves, which exercise their control by relaying information about environmental changes to the central nervous system and conveying stimuli for the appropriate responses from the CNS to the muscles and glands. But this is not enough. Not only muscles and glands but all tissues, all cells, must function together harmoniously for the good of the whole. The liver must at times be stimulated to remove glucose from the blood and build it into glycogen for storage, at other times to break down glycogen and return glucose to the blood. The rate of growth of the legs must be proportionate to the rate of growth of the rest of the body. Activities such as these are regulated by a group of chemical messengers called *hormones*, which are secreted by specialized gland cells. The glands we have considered so far, such as the sweat glands or the salivary glands, pour their secretions onto epithelial surfaces through special tubes or ducts. They are called *exocrine glands.* Glands that produce hormones, however, have no ducts. Their products make their way through the tissue fluid into the circulatory vessels, by which they are then widely and promptly distributed throughout the body. These glands are called *endocrine,* or *ductless, glands.*

Chemically, hormones fall into three groups. Some are steroids, others are polypeptides, still others are derivatives of amino acids.

The nervous and endocrine systems interact and overlap in function, but hormonal control is predominant in three major areas: growth and development of the individual, homeostasis, and reproduction. Malfunctioning of the endocrine glands can lead to serious disturbances in one or more of these main areas. These malfunctions are basically of two sorts, underactivity (hypo-) or overactivity (hyper-). Mechanical injury or disease can cause a gland to produce less than the normal amount of a hormone; a tumorous growth can stimulate it to produce more. Congenital defects are known in which a child is born with an overactive or underactive gland. In addition to obvious abnormalities, many differences in physique, and perhaps in temperament, that fall within the normal range of the population are probably the result of minor individual variations in the activity of the various endocrine glands.

THE ENDOCRINE ORGANS

Not all cells that produce hormones are gathered into discrete glandular organs. Thus certain cells of the stomach mucosa secrete a hormone called *gastrin,* which is carried by the bloodstream to the stomach glands and stimulates the production of hydrochloric acid. Generally, though, when biologists speak of hormones they have in mind the productions of special organs that have as their sole function, or as a major function, the secretion of these chemical messengers. These organs are (1) the pituitary, (2) the thyroid, (3) the parathyroids, (4) the adrenals, (5) the pancreas, and (6) the gonads (ovary and testis). To this list should be added the hypothalamus, a part of the brain, and perhaps the pineal body, a small structure buried deep in the brain. Figure 10–1 shows the location of the various endocrine glands. The endocrine function of the gonads is discussed under the topic of human reproduction in Chapter 13.

Pituitary

The pituitary gland is a lobed structure, about one centimeter (⅖ inch) in diameter, that hangs from the floor of the brain by a stalk, the *infundibulum* (Figure 10–2). The pituitary gland has two main parts, an anterior lobe and a posterior lobe. (There is a small intermediate lobe that is concerned with the distribution of pigment granules in lower animals but has no known function in man.)

Anterior Lobe. Seven hormones, all of them proteins, are known to be produced by the anterior lobe of the pituitary gland. Several of these hormones, which go under the general name of *trophins,* have as their main function the control of other endocrine glands. The anterior pituitary hormones are (1) *growth hormone* (GH), which stimulates protein synthesis and promotes growth, particularly that of bone, although soft tissues are also affected; (2) *lactogenic hormone,* the primary function of which is to stimulate the produc-

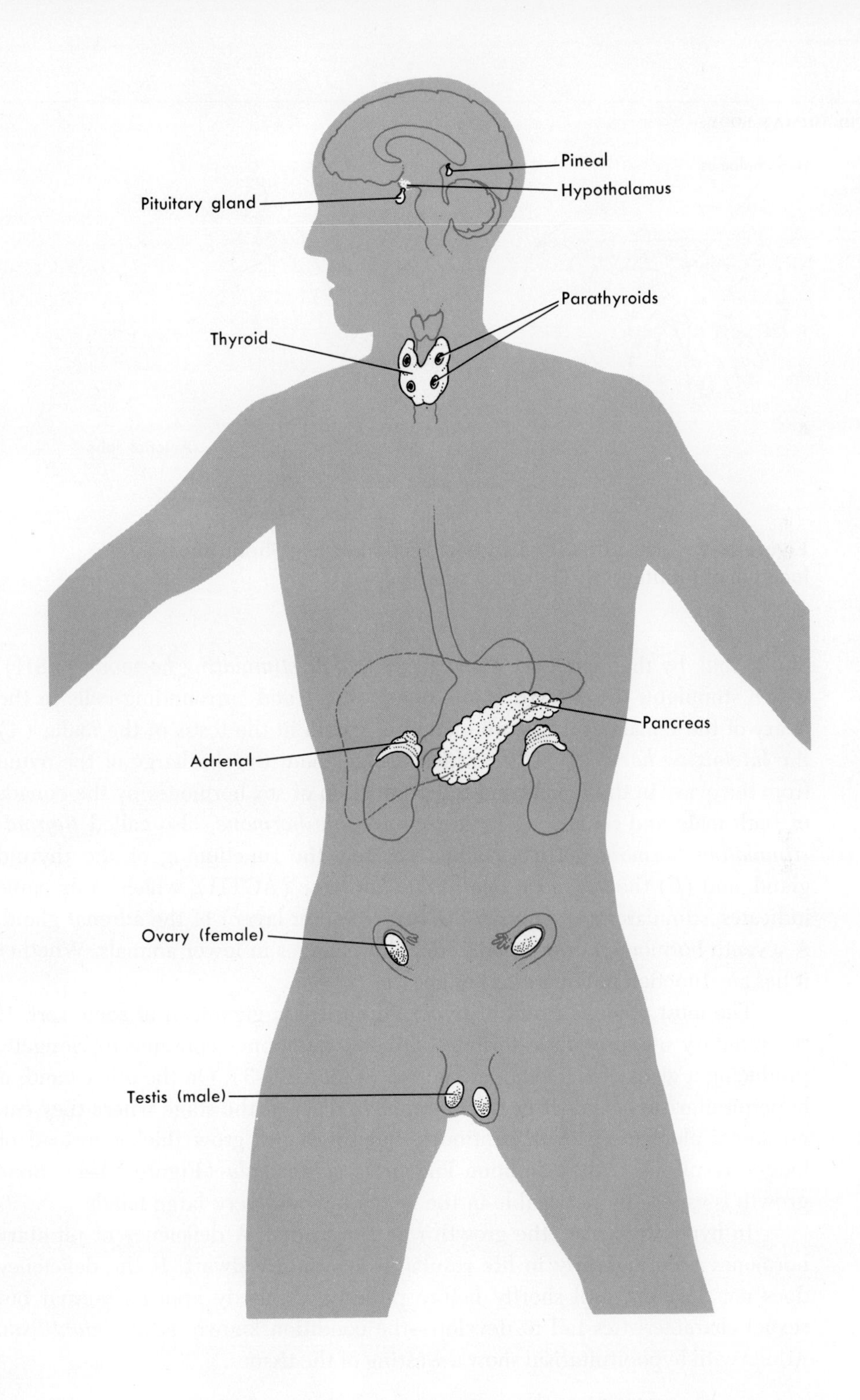

Figure 10–1. The human endocrine glands.

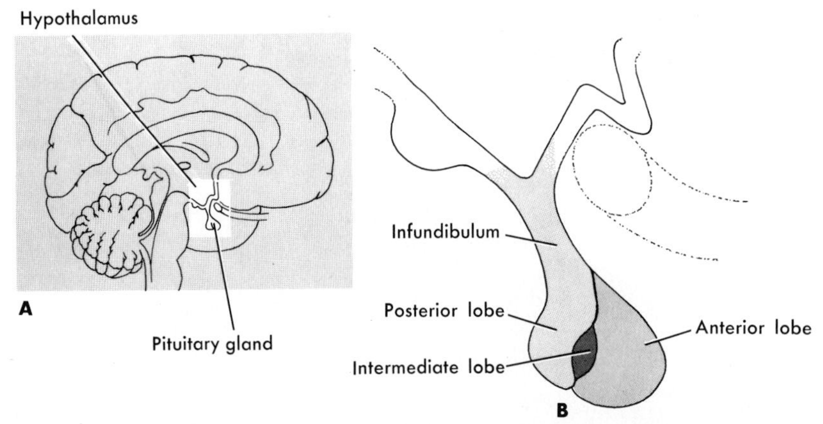

Figure 10–2. The pituitary gland. A: Section through brain showing location of pituitary. B: The gland enlarged.

tion of milk by the mammary glands; (3) *follicle-stimulating hormone* (FSH), which stimulates the growth of the ovum (egg) and surrounding cells in the ovary of the female and the production of sperm in the testis of the male; (4) the *luteinizing hormone* (LH), which brings about the discharge of the ovum from the ovary in the female and the production of sex hormones by the gonads of both male and female; (5) the *thyrotrophic hormone*, also called *thyroid-stimulating hormone* (TSH), which controls the functioning of the thyroid gland; and (6) the *adrenocorticotrophic hormone* (ACTH), which as its name indicates, stimulates secretion by the cortex (outer layer) of the adrenal gland. A seventh hormone is concerned with color changes in lower animals. Whether it has any function in humans is not known.

The most obvious result of hyperpituitarism is gigantism of some sort. If the pituitary overproduces during childhood, the bones continue to elongate, producing a giant of the long, lanky type (Figure 10–3). On the other hand, if hyperpituitarism occurs after the bones have reached the stage where they can no longer elongate, growth continues, but the bones grow thicker instead of longer, resulting in the condition known as *acromegaly* (Figure 10–4). Bone growth is especially noticeable in the heavy jaw and very large hands.

In hypopituitarism, the growth rate is retarded. A deficiency of pituitary hormones occuring early in life results in a pituitary dwarf. If the deficiency does not develop until shortly before puberty, the body appears normal but sexual characteristics fail to develop—the condition known as *sex infantilism.* Adults with hypopituitarism show a wasting of the tissues.

Posterior Lobe. The posterior lobe of the pituitary gland does not itself produce any hormones. The two hormones secreted by it are formed by

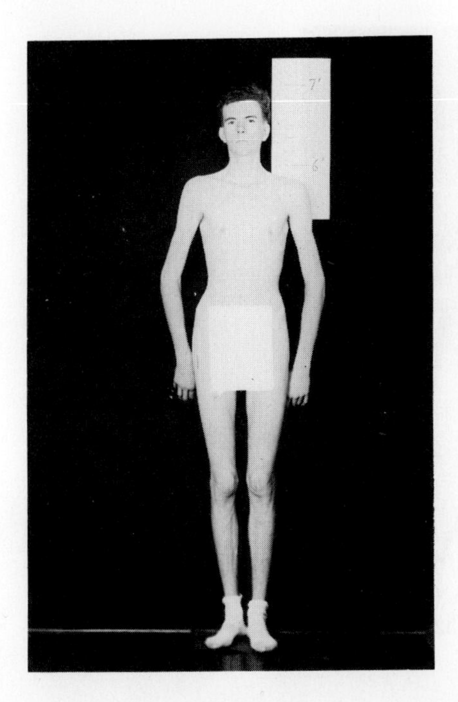

Figure 10–3. A hyperpituitary giant. (Courtesy of Armed Forces Institute of Pathology, print 0–8196–1.)

neurosecretory cells in the hypothalamus, the part of the brain from which the pituitary depends. These cells are shaped more or less like typical neurons. Each has dendrites in synaptic contact with other neurons in the brain, and each has a long axon extending down the infundibular stalk to the posterior pituitary. Two substances, *vasopressin* and *oxytocin*, are produced by those cells, passed down the axons, and released into the bloodstream in the posterior pituitary. They are very similar, each consisting of eight amino acids, six of which are the same. But they differ in effect. *Vasopressin* increases the amount of water reabsorbed by the collecting ducts of the kidneys and is thus also known as *ADH* (*antidiuretic hormone*). In amounts larger than those normally secreted, it raises the blood pressure by constricting the peripheral blood vessels and so is sometimes used to maintain blood pressure during surgery. *Oxytocin* stimulates contraction of smooth muscles, particularly those in the wall of the uterus during childbirth. Because of this it is sometimes used in obstetrics to reduce postpartum hemorrhage. It also promotes the ejection of milk by the mammary glands.

Hypothalamus

In addition to vasopressin and oxytocin, the neurosecretory cells of the hypothalamus produce at least ten other hormones. These substances enter a minute portal system, a capillary-vein-capillary chain that passes from the hypo-

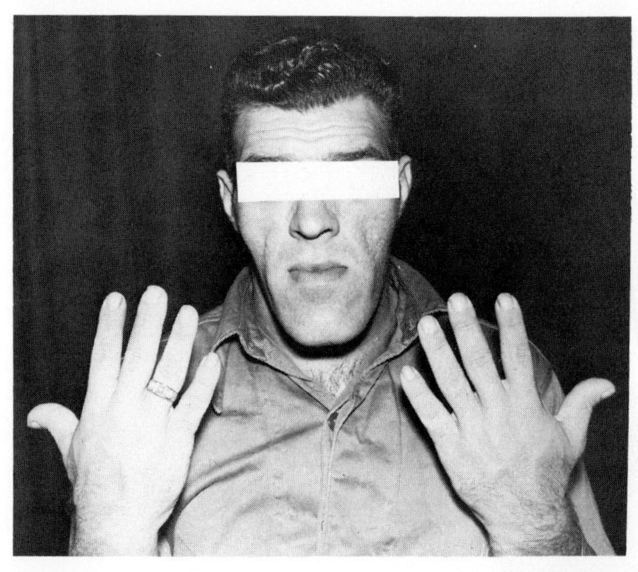

Figure 10–4. Acromegaly in a human male. (Courtesy of Armed Forces Institute of Pathology, print 321.)

thalamus to the anterior pituitary. These hormones regulate the rate of secretion of the anterior pituitary hormones.

Thyroid

The thyroid gland is a double-lobed structure lying in the neck region at about the junction of the larynx and the trachea (see Figure 10–5). It is supplied with blood by no less than five separate arteries and is said to be the organ with relatively the richest blood supply of any in the body.

The principal hormone produced by this gland, *thyroxine,* is a single amino acid to which four iodine atoms are joined. Thyroxine regulates the rate of metabolic processes in the body. A second thyroid hormone, *thyrocalcitonin,* is a polypeptide. It acts to decrease the level of calcium in the blood.

The thyroid can malfunction in three different ways: it can fail to produce sufficient thyroxine simply because of a lack of iodine in the diet; it can produce insufficient amounts of thyroxine even though all the materials needed for synthesis are present; or it can produce an excess of thyroxine. When there is a deficiency of iodine, the thyroid gland tends to compensate by enlarging, the condition known as *endemic goiter—endemic* because it is characteristic of those parts of the world where there is not a sufficient iodine supply in the food or water. This condition has practically been eliminated in modern civilization by the simple technique of adding sodium or potassium iodide to the table salt.

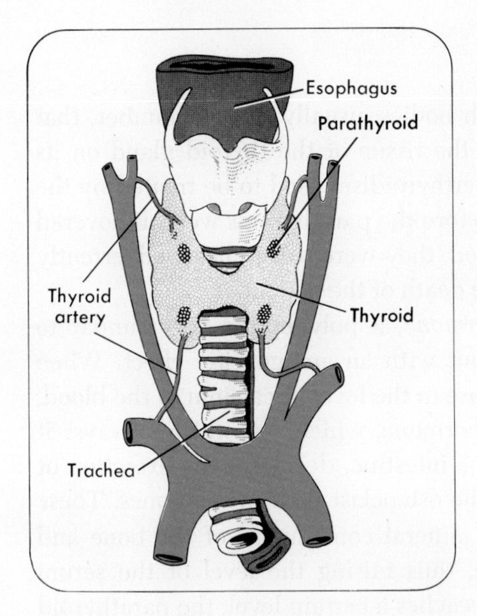

Figure 10–5. The thyroid gland lies in the neck region below the voice box and in front of the trachea. The parathyroids lie on the under surface of the thyroid.

In endemic goiter, if the iodine deficiency is not too great, the enlarged gland may be able to remove sufficient iodine from the blood to produce an adequate supply of the hormone. Then the other symptoms of thyroid deficiency, lethargy, mental dullness, obesity, and loss of hair, do not appear.

If the thyroid gland does not secrete enough hormone even though the iodine supply is sufficient, *hypothyroidism* results. If it occurs in very early childhood it produces a condition known as *cretinism*. This is characterized by a low metabolic rate, together with the early cessation of mental and physical development. If hypothyroidism occurs later in childhood, it may bring on juvenile *myxedema*. A child suffering from this condition tends to be short and squat, with a head proportionately large for his age. The child's neck is short and heavy, the skin is dry, and the face is bloated and has a dull, stupid expression. Often the mouth hangs partly open with the tongue protruding.

Hyperthyroidism produces nervousness, irritability, excessive body activity, fatigue, loss of weight, and general emotional instability. Many of these effects appear to be related to a general increase of metabolic activity in the tissues. Sometimes a person with hyperthyroidism has prominent and staring eyes with protruding eyeballs. This is *exophthalmic goiter*. There is some indication that when hyperthyroidism is linked with exophthalmos it is the result of overproduction of thyroid-stimulating hormone by the anterior pituitary. The appearance of the eyes may be caused by the TSH rather than by the thyroid hormones. This suggests that there may be two kinds of hyperthyroidism, one caused by an overactive response of the thyroid to TSH, the other by the response of a normal thyroid to too much TSH.

Parathyroids

The parathyroids are tiny, brownish bodies, usually four in number, that are generally intimately associated with the tissue of the thyroid gland on its posterior surface (see Figure 10–5). Hyperthyroidism used to be treated by the surgical removal of part of the thyroid. Before the parathyroids were discovered and something of their function understood, they were sometimes inadvertently removed at the same time, resulting in the death of the patient.

The parathyroids produce *parathormone,* a polypeptide very similar to thyrocalcitonin in amino acid content, but with an antagonistic effect. When the parathyroids are activated by a decrease in the level of calcium in the blood, they secrete increased amounts of parathormone, which acts in three ways: it increases the absorption of calcium by the intestine, decreases the excretion of calcium by the kidneys, and stimulates the osteoclast cells in the bones. These cells secrete chemicals that dissolve the mineral constituents of the bone and so release calcium into the bloodstream, thus raising the level of the serum calcium. When the calcium in the serum reaches a certain level, the parathyroid glands produce less parathormone, the thyroid produces more thyrocalcitonin, the osteoclasts become less active, and there is a net gain in the calcium deposited in the bones by the osteoblasts. Thus, the level of calcium in the blood fluctuates only within very narrow limits and homeostasis is maintained. Calcium is important in regulating the permeability of cell membranes. When the level of calcium in the tissue fluid drops too low, the permeability of the nerve and muscle fiber membranes is altered and they may "fire" in a spontaneous and irregular manner. The result is muscular weakness, irritability, and tetany (see Figure 10–6). The condition can be treated under a physician's direction by the administration of calcium. Hyperparathyroidism, on the other hand, removes too much calcium from bone and may result in the softening and bending of the long bones, together with the formation in other parts of the body of calcium deposits, such as kidney stones.

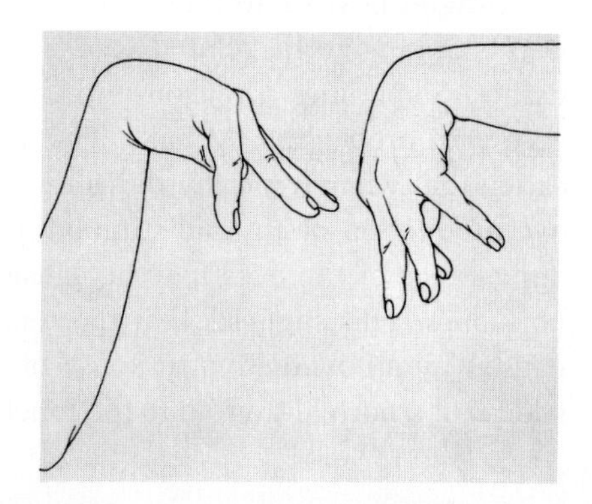

Figure 10–6. Muscular spasms of the hands shown in tetany, the result of hypoparathyroidism.

Adrenal Glands

The adrenals lie, one on each side, just above the kidneys. Each is really a dual structure, a gland within a gland. The two parts develop separately in the embryo at first. Then the cells that are to form the *medulla* of the gland migrate along the course of a vein to reach the central region of the structure that forms the *cortex*.

Adrenal Medulla. The inner part of the adrenal, the medulla, produces two very similar hormones, *epinephrine* and *norepinephrine,* also called *adrenalin* and *noradrenalin.* They are derived from the amino acid tyrosine. Epinephrine causes increased glucose secretion by the liver. It constricts blood vessels in the viscera but dilates them in the skeletal muscles. It increases the heartbeat, decreases persistalsis, accelerates breathing, and increases sweating. The effects of norepinephrine are similar, but it does not increase glucose secretion and causes a greater rise in blood pressure. The reactions caused by the medullary hormones should sound familiar—they are none other than those produced by the sympathetic nervous system. As we discussed in the last chapter, structures innervated by the autonomic system receive impulses through a chain of two fibers, one running from the CNS to a ganglion (*presynaptic fiber*) and the other from the ganglion to the organ (*postsynaptic fiber*). The adrenal medulla is the exception. It receives only presynaptic sympathetic fibers; the whole gland itself apparently acts as the postsynaptic part, secreting the same substances and playing the same role.

The adrenal medulla is seldom diseased. Overactivity may cause high blood pressure but no deficiency states are known. If the adrenal medulla is removed surgically, its function is taken over by the sympathetic nervous system and the body can remain normally active.

Adrenal Cortex. All the hormones considered so far have been formed of amino acids, but the hormones of the outer part of the adrenal, the cortex, are *steroids* generally derived from cholesterol. A bewildering variety of these steroids have been identified in extracts from the adrenals; probably most of them are not secreted as active hormones but represent intermediate steps in the synthesis of hormones. The active hormones fall into three major groups. Some are concerned with carbohydrate metabolism (*glucocorticoids*). They stimulate the conversion of amino acids into carbohydrates by the liver and at the same time interfere with the utilization of glucose by the tissues so that both the glycogen stored in the liver and the blood glucose increase in amount. Other hormones are involved in mineral metabolism (*mineralocorticoids*). They stimulate absorption of sodium by the kidney tubules and control the distribution of sodium and potassium ions between the tissue fluids and the cells. Still other adrenal cortex hormones (*androgens*) are very similar to the sex hormones secreted by the testes and ovaries and are involved in the development of secondary sex characteristics.

Although the adrenal medulla can be removed with impunity, total removal of the cortical parts is fatal within a short period of time. A reduced flow of hormones from the cortex, hypoadrenocorticalism, brings on the condition known as *Addison's disease*. The symptoms of this disease are an increased pigmentation of the skin, loss of appetite, loss of weight, weakness, low blood pressure, and some anemia. Hyperadrenocorticalism can have varied effects. If the corticoids are produced in excess, the result is Cushing's disease. The excess salt reabsorbed by the kidneys upsets the ionic balance of the body fluids and results in circulatory troubles. Fat is stored in the face, neck, and trunk. Excess production of androgens produces what is known as the *adrenogenital syndrome*. If it develops in a young boy, it causes an early development of the penis and pubic hair with early bone growth and precocious development. In a newborn girl, it may bring about the enlargement of the clitoris to such an extent that she is mistaken for a boy and raised as one. If the syndrome develops in an older girl, there is a repression of female characteristics and the assumption of male characteristics, such as excess facial hair and reduced breasts.

Pancreas

The pancreas, discussed as an exocrine gland with the digestive system, is also an endocrine gland. Within its tissue are found small clusters of endocrine cells known as the *islets of Langerhans*. Here two different types of cells produce two hormones, *insulin* and *glucagon*, both proteins. Insulin increases the ability of the body cells to remove glucose from the blood, stimulates the conversion of glucose to glycogen in the liver and muscles, and increases the utilization of glucose in the synthesis of fats and proteins. Hyperinsulinism can result in a reduction below normal of blood sugar levels. Nerve cells are especially sensitive to low levels of sugar in the blood (*hypoglycemia*), the symptoms of which are blurred vision, weakness, faintness, tremors, and even convulsions. Hypoinsulinism brings on the condition known as *diabetes mellitus*, in which the cells are unable to remove glucose from the blood, and the breakdown of glycogen and the conversion of fats and proteins into glucose in the liver are accelerated. This, of course, results in a rapid rise in blood sugar. Diabetic patients are treated by the administration of insulin, which increases the ability of the cells to remove glucose from the blood and of the liver to convert glucose into glycogen.

Unfortunately, insulin, as a protein, is broken down by the enzymes of the digestive tract and, therefore, must be given by injection. Recently some medicines for diabetes have been developed that can be taken by mouth. They either increase the production of the insulin-forming cells that are still functional, or they act like insulin itself to increase the uptake of glucose by the cells and to regulate carbohydrate metabolism in the liver.

Glucagon acts as an antagonist to insulin, stimulating the release of glucose by the liver, thereby raising the level of blood sugar.

MECHANISM OF HORMONE ACTION

The hormones are a diverse group of different substances with different target organs and different effects. It has thus been surprising to find that many of them operate in the same way. The cells of the target organ(s) have membrane-bound proteins that bind to the specific hormone involved. In some way, this binding stimulates the production of an enzyme, *adenyl cyclase,* which acts to form *cyclic AMP* (cyclic adenosine monophosphate) from ATP. The cyclic AMP then alters the rate of one or more of the processes carried on by the cell. It can change the permeability of the cell membrane, stimulate or depress the formation or activity of various proteins, or increase or decrease the secretion of a number of substances. Hormones whose activity has been shown to be mediated by cyclic AMP include epinephrine, ACTH, glucagon, vasopressin, LH, parathormone, and TSH.

REGULATION OF HORMONE SECRETION

The endocrine glands do not secrete continually or at a constant rate. They can be turned on or off, speeded up, or slowed down. This means that they are subject to a control system. There must be some type of receptor capable of being stimulated by some change in the internal or external environment and usually a transmitting mechanism that is capable of transferring the stimulus to the gland.

Patterns of Control

The simplest control system is one in which the gland itself acts as the receptor (see Figure 10–7A). The parathyroids respond directly to a decrease in blood calcium by an increased production of parathormone and the islet cells of the pancreas respond to changes in the blood sugar level by secreting insulin or glucagon as needed.

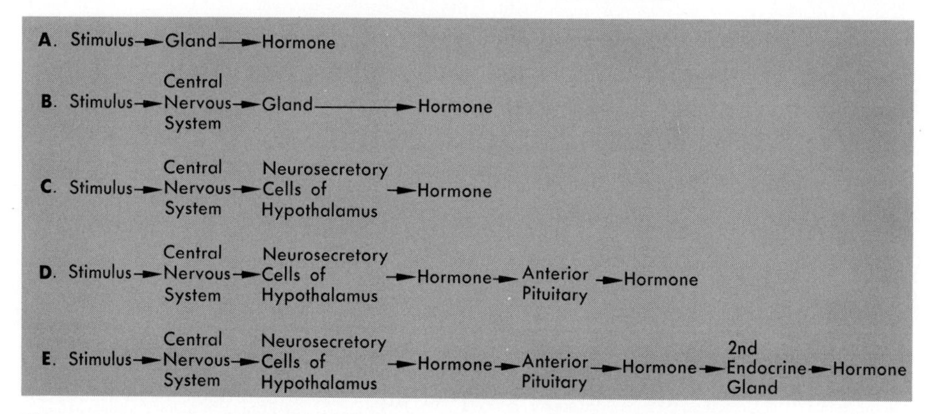

Figure 10–7. The control of hormone secretion. See discussion in text.

The secretion of most hormones is mediated by the central nervous system. The hormones of the adrenal medulla are secreted in direct response to stimulation by nerve fibers of the sympathetic system (see Figure 10–7B). The posterior pituitary hormones are also secreted in response to nervous stimuli. Osmotic receptors in the brain are stimulated by an increase in the osmotic concentration of the blood. Pressure receptors in the wall of the heart and in arteries in the neck are stimulated by a drop in blood volume. These stimuli are transmitted to the neurosecretory cells of the hypothalamus and bring about the secretion of vasopressin, which, by its antidiuretic action, returns water to the blood from the urine and so restores the osmotic balance and increases the volume of the blood (see Figure 10–7C).

More typically, hormones are not secreted in response to direct nervous stimulation. The transmitting mechanism involves both nervous and hormonal components. Stimuli received in the brain and relayed to the hypothalamic neurosecretory cells cause the release of substances that have a stimulating or inhibiting effect on the anterior pituitary. In turn, the hormones of the anterior pituitary may act directly on the body tissues, as the growth hormone does (see Figure 10–7D), or they may stimulate other endocrine glands, such as the thyroid or the adrenal cortex, to release hormones (see Figure 10–7E).

Negative Feedback

Basic to the regulation of most hormones is a pattern of control known as *negative feedback*. For example, when the level of blood sugar rises, the pancreas is stimulated to secrete insulin. This increases the rate at which the cells remove sugar from the blood, which decreases the amount of sugar circulating in the blood, which in turn suppresses the production of insulin by the pancreas. Similarly, a reduction in the amount of thyroxine in the blood stimulates the neurosecretory cells in the hypothalamus. These cells secrete a thyrotrophic releasing factor (TRF) that is transmitted to the anterior pituitary. The anterior pituitary is stimulated to release TSH, which in turn stimulates the thyroid to secrete thyroxine. As the level of thyroxine in the blood increases, it inhibits the production of TRF. This in turn inhibits the release of TSH so that the thyroid is no longer stimulated to produce thyroxine. As the thyroxine circulating in the blood is used up or excreted, the level falls and the whole cycle begins again. Thus the level of thyroxine is maintained within narrow limits.

Such negative feedback systems are very common in the body. Remember the way a drop in body temperature causes shivering, which increases heat production by the muscles, and the way in which changes in the level of carbon dioxide in the blood regulate breathing. Indeed the whole body can be considered a complex of interlocking negative feedback mechanisms.

By such finely tuned mechanisms all the cells of the body are induced to work in harmony for the good of the whole. Indeed it has been said that the function of all the complex organs and systems of the body, all the elaborate, interlocking reactions, is to ensure that the individual cells are maintained in an environment that allows them to survive. A man is his cells. If too many of them die, he dies. But life involves more than the survival of the individual, cell or man. It also involves an increase in the number of individuals. This aspect of life is considered in the next chapter.

SUGGESTED READINGS

Davidson, E. H.: "Hormones and Genes," *Sci. Amer.*, Vol. 212, No. 6, 1965.

Donovan, B. T.: *Mammalian Neuroendocrinology*, McGraw-Hill Book Company, 1970.

Frieden, E., and H. Lipner: *Biochemical Endocrinology of the Vertebrates*, Prentice-Hall, Inc., Englewood Cliffs, N.J., 1971.

Frye, B. E.: *Hormonal Control in Vertebrates*, Macmillan Publishing Co., Inc., New York, 1967.

Grollman, S.: *The Human Body*, 3rd ed., Macmillan Publishing Co., Inc., New York, 1974.

Kimber, D. C., et al.: *Anatomy and Physiology*, 16th ed., Macmillan Publishing Co., Inc., New York, 1972.

Prosser, C. L., and F. A. Brown, Jr.: *Comparative Animal Physiology*, 2nd ed., W. B. Saunders Company, Philadelphia, 1961.

II

Reproduction—Organic Perpetuation

Crescite et multiplicamini.
VULGATE, *Gen. 1:28*

CHAPTER **11**

Reproductive Processes

Different cell types differ in size, but almost all cells are microscopically small. This is largely because most molecular traffic within the cell takes place by diffusion. The larger the cell, the longer it takes a molecule to travel from the place where it enters the cell, or is formed by the cell, to the place where it is used. This means that there is an upper limit to the size at which the cell can function efficiently. No matter how well nourished a cell may be, it does not continue to grow indefinitely. When it reaches a certain size it either stops growing or, more usually, divides to give rise to two smaller daughter cells, which then go through another cycle of growth and division.

If the daughter cells remain together, the organism grows through an increase in cell number. If the daughter cells separate to give rise to new organisms, reproduction has occurred. This ability of an organism to produce copies of itself is one of the features that sets life off from the nonliving world.

If reproduction involves only the formation of daughter cells from a divided mother cell, it is said to be asexual. If it involves the formation of a specialized cell that fuses with another specialized cell to form a new organism, it is said to be sexual.

ASEXUAL REPRODUCTION

Watching a single-celled animal, such as an amoeba, in the process of dividing, gives the impression that the animal simply pinches itself in two as shown in Figure 11–1. But when more refined techniques are used to investigate

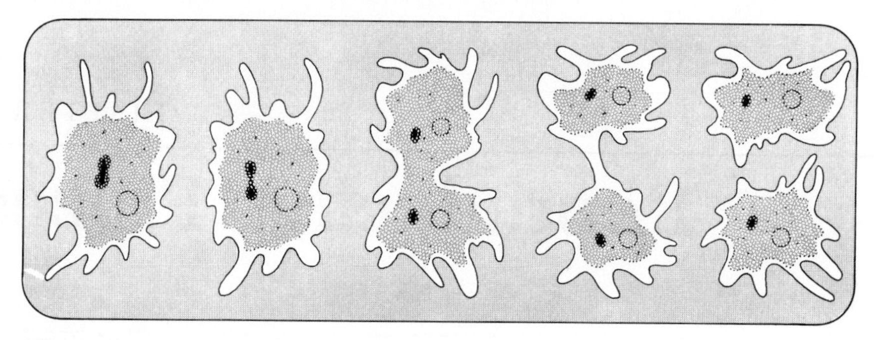

Figure 11–1. An amoeba dividing.

what goes on in the amoeba, particularly in its nucleus, the process of division is found to be much more complicated. There is first a duplication of the nucleoprotein material that makes up the chromatin network and then a precise qualitative and quantative division of this material. Each daughter cell receives the same kind and same amount of chromatin the mother cell had originally—a type of cell division called *mitosis*. This is the most usual kind of cell division. Many animals and plants reproduce by mitotic division; it is also the process by which growth and tissue repair take place in multicellular organisms.

Mitosis

Formerly a cell not in the process of actively dividing was said to be in the resting stage. But the cell is not really resting, it is diligently carrying on metabolic activities. It may never divide again, though it may live for many years. The concern here, however, is only with cells that do undergo periodic divisions. In actively dividing cells, the period between one cell division and the next is called the *mitotic cycle*. It is customarily divided into five stages, which are really arbitrary divisions of a single, continuous process. These stages are diagrammed in Figure 11–2.

Interphase. Interphase is the period between the completion of one cell division and the first appearance of the obvious changes that lead to the next. In an interphase cell the chromatin network in the nucleus does not appear to be organized into discrete bodies, the nucleolus is well developed, the nuclear membrane is distinct and well organized, and the two centrioles lie close together outside the nucleus. Each centriole has already given rise to a small daughter centriole that lies at right angles to it, but because each mother-daughter centriole complex moves as a unit during cell division, it is customary to speak of it still as a single body. During interphase the cell usually grows

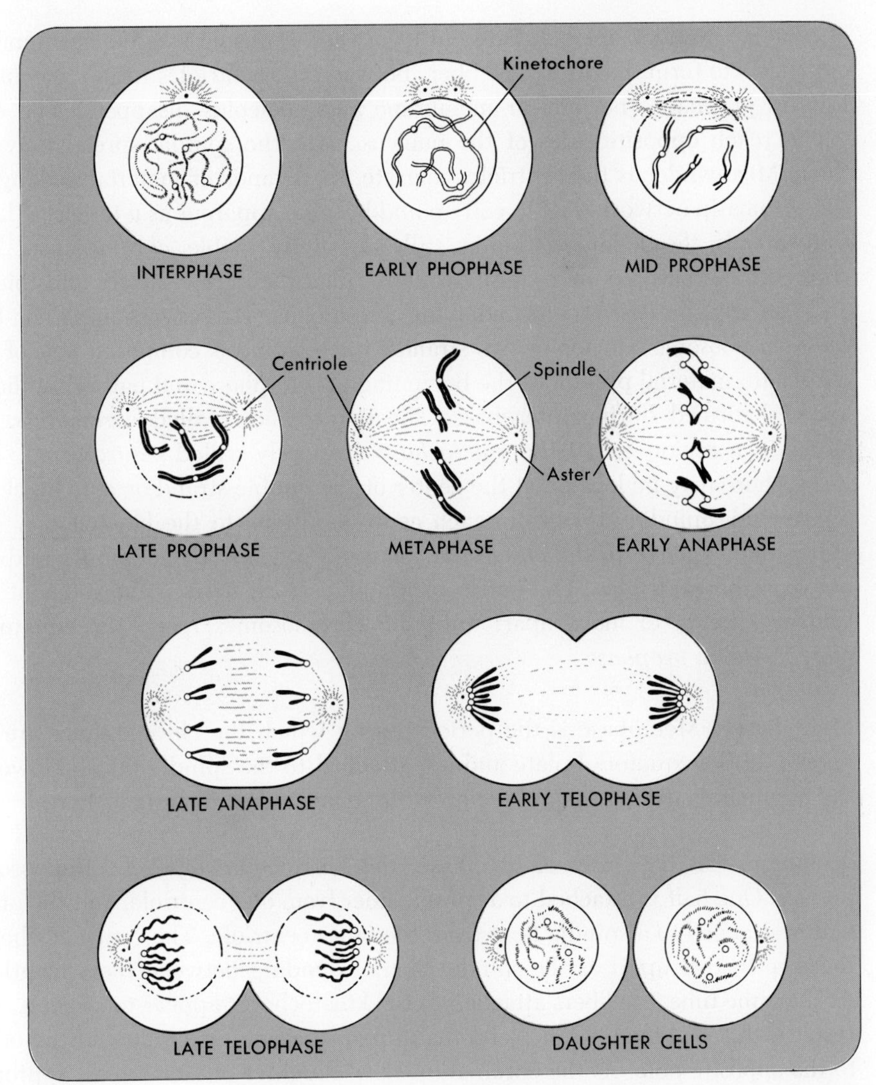

Figure 11–2. The stages of mitosis. See discussion in text.

until it reaches the size of the parent cell. There is then a precise duplication of the nucleoprotein material of the chromatin network. Energy stores are accumulated for the work the cell must do during the next stages of mitosis.

Prophase. At the beginning of prophase in an animal cell, the two centriole complexes begin to move toward opposite poles of the nucleus. As they separate, they appear to be connected by fibers called *spindle fibers* (really microtubules of protein). Similar fibers radiate out from each centriole into the surrounding cytoplasm to form the *aster*. The fibers between the centrioles form the *spindle*. Late in prophase, the nuclear membrane and nucleolus disappear. The centrioles reach opposite sides of the nucleus, with the spindle spread between them. Meanwhile, as the centrioles migrate, the strands of material making up the chromatin network start to coil up and become apparent as a tangle of long, thin threads, the definitive, though still only dimly visible, *chromosomes*. The chromosomes become more tightly coiled so that they grow shorter and thicker and can now be easily seen under the microscope. It becomes apparent that each chromosome is made of two strands. Each strand is composed half of the chromatin material present at the beginning of interphase and half of duplicate material formed during interphase. The two strands, called *chromatids,* are joined at a region called the *kinetochore* (formerly called *centromere*). The kinetochore may be located at the center of the double strand or may be closer to one end. Spindle fibers from each centriole attach to the kinetochore. The chromosomes move to the *equatorial plate,* the region of the spindle midway between the centrioles. The entire continuous process, from the moment the centrioles begin to move apart until the chromosomes reach the equatorial plate, is called *prophase*.

Metaphase. Metaphase is the period during which the chromosomes are arranged on the equatorial plate and are attached to the spindle fibers. However the chromatids during metaphase are still connected at the kinetochores.

Anaphase. At the onset of anaphase, the kinetochore of each chromosome divides. One half is attached to a spindle fiber from one centriole, and the other half is attached to a spindle fiber from the other centriole. The centrioles begin to move further apart as the spindle fibers extending between them lengthen. At the same time, the fibers attached to the kinetochores appear to shorten. The result is that one member of a chromatid pair moves to one pole and the other to the opposite pole. As the chromatids, now daughter chromosomes, approach the poles, division of the cytoplasm (*cytokinesis*) begins; in animal cells it is indicated by a pinching in of the plasma membrane to form a furrow around the cell. The pinching in seems to be caused by the contraction of a band of microfilaments that apparently forms under the influence of the asters.

Telophase. The last phase, telophase, is the reconstruction stage. Sacs and vesicles that apparently are fragments of endoplasmic reticulum fuse around each cluster of chromosomes to form the double-layered nuclear membranes. The chromosomes uncoil, become indistinct, and fade back into the chromatin network. The mother-daughter centriole complex separates, and each now gives rise to a new daughter centriole. The nucleolus reforms, and a physical division of the cytoplasm ensues. At the end of telophase there are two daughter cells in interphase where there was one to begin with.

Mitosis is thus a process involving a precise replication of the chromosomes and their equal distribution to the daughter cells. Because the chromosomes contain the DNA that determines the characteristics of the organism, mitotic cell division is a phenomenon that produces two essentially similar daughter cells that are also like the mother cell from which they were formed. Figure 11–3 is a photomicrograph of cells undergoing mitosis in a developing whitefish egg.

Types of Asexual Reproduction

Asexual reproduction by mitotic cell division occurs not only in single-celled animals but also in some multicellular ones and in many plants. There is a wide variation in asexual reproductive methods, but practically all can be classified as either budding or fission. In budding, a new organism starts to grow out of the parent, from which it eventually separates. An example of this can be seen in the small aquatic animal hydra (see Figure 11–4). Reproduction

Figure 11–3. Mitosis as seen in a developing whitefish egg. (Courtesy Ward's Natural Science Establishment, Inc., Rochester, N.Y.)

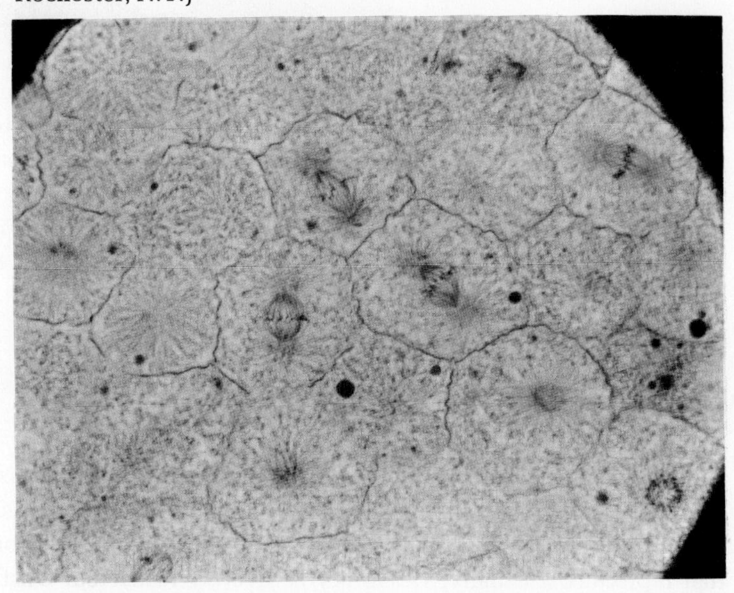

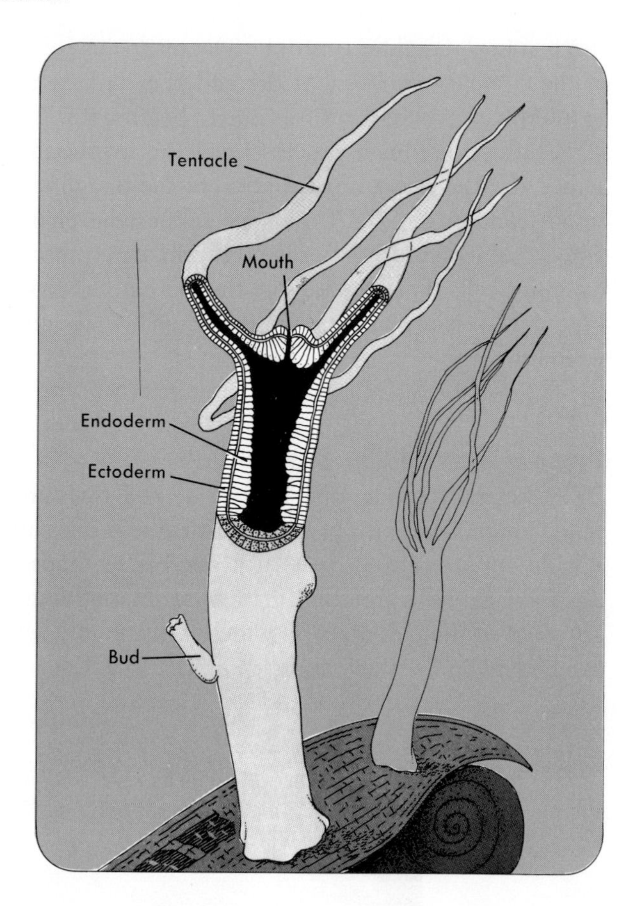

Figure 11–4. Reproduction by budding in hydra. In animals this type of reproduction is most often found in forms that show little tissue differentiation. A section of the body wall of the animal on the left has been cut away to show the simplicity of the internal structure.

in the amoeba is a type of fission, but not the only type. Sometimes the nucleus of a single-celled organism divides several times before the cytoplasm divides to form a number of daughter cells at one time. Many of the multicellular flatworms are able to break up into a number of fragments with each fragment growing a new head and/or a new tail by mitotic cell division. Fission may thus be either binary (in which two daughter forms are produced) or multiple (in which a number of offspring are produced at a single time).

ASEXUAL VERSUS SEXUAL REPRODUCTION

Asexual reproduction has its disadvantages. Unless a chance change in the nuclear material occurs, reproduction by mitotic cell division is a process that produces offspring essentially like the parent and also like each other. In species that reproduce asexually, variability is severely restricted. On the other hand, sexual reproduction usually involves the fusion of cells from two different parents. As discussed later, this makes new combinations of characters possible. The offspring differ from the parents and also from one another. Sexual reproduction thus increases variability among the members of a species. A variable population is better able to evolve, to adapt to long-range changes in the environment, or to move into new environments. Almost all organisms, even those that normally reproduce asexually, also reproduce sexually at some time during their life cycle, and for many it is the only method of reproduction.

SEXUAL REPRODUCTION

In asexual reproduction, the cells may not only contribute to new individuals by mitotic division but may also play a part in the maintenance and functioning of the organism. On the other hand, in all but the simplest organisms, there are cells that make no contribution to the functioning of the body but instead are concerned only with the production of succeeding generations. These cells are called *germ cells,* in contrast to the *somatic,* or *body cells,* which do not give rise to new individuals.

Germ and Soma

The Greek philosopher Democritus (about 400 B.C.) believed that there were representative particles, *pangens,* in all parts of the body and that these particles passed from the specialized tissues into the semen that was introduced into the female during copulation. Thus, he thought that the "seeds" from which the new individual developed were formed by the conjunction of pangens from every part of the father's body. The child that developed as a result of copulation had at the beginning representatives of all the specialized parts he would have as an adult. Aristotle pointed out that if this were so, a one-armed man should have one-armed children, but the idea was slow to perish and indeed was revised in a modified form more than two-thousand years later by no less a person than Charles Darwin.

It remained for the German biologist August Weismann (1834–1914) to suggest that the reproductive (germ) cells form an entirely separate line from the body (somatic) cells, and that although germ cells can, by division and specialization, give rise to the somatic tissues of the body, specialized body cells cannot give rise to germ cells. Instead, the germ cells develop from undifferentiated cells that have never been involved in the formation of the specialized epithelial, muscle, nervous, or connective tissues. A comparison of pangenesis and the modern germ and soma concept is shown in Figure 11–5.

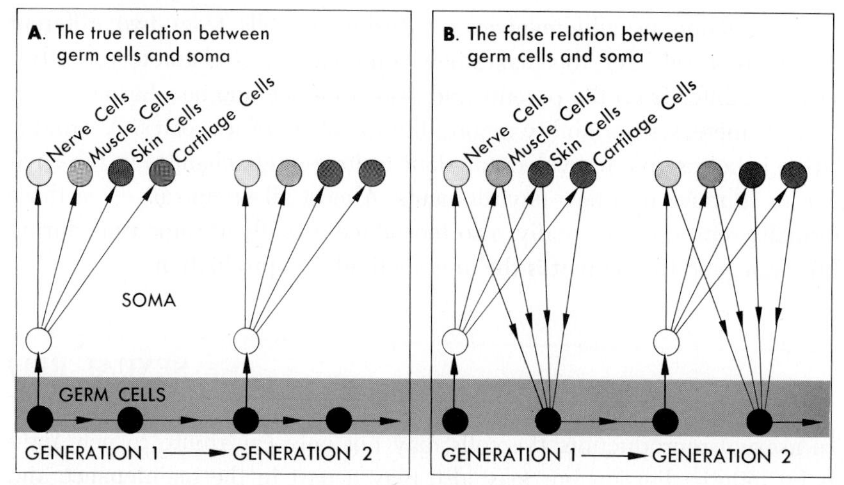

Figure 11–5. The modern idea of germ and soma A: contrasted with the idea of pangenesis B.

Thus, in most animals, the somatic cells become specialized into various types of tissues that function for the life of the individual, but when death ensues they go the way of all flesh and lead to nothing. A germ cell, on the other hand, may have given rise to a line of cells that developed into a new individual, including both the somatic and germ cells of that individual. The germ cell line may continue from generation to generation; it is potentially immortal.

Meiosis: Germ Cell Division

Mitotic cell division results in the production of daughter cells that have the same number of chromosomes as the mother cell did. Sexual reproduction involves the fusion of two cells. Although different species have different numbers of chromosomes, the number is typically constant for all individuals of the same species. Humans have forty-six chromosomes, hellbenders (a kind of salamander) have sixty-two, bullfrogs have twenty-six, and fruit flies have eight. If the cells that fused in sexual reproduction had the same number of chromosomes as the cells of the parents, the number would double with each generation and would soon be impossibly large. *Meiosis* is a type of cell division in which the daughter cells receive only half the number of chromosomes present in the parent cell. As a result, when two of these cells fuse to form the single cell from which the new individual will develop, this cell has the same chromosome number as each parent did. The specialized cell produced by meiosis, which is capable of fusing with another cell, is known as a *gamete*. In animals, the male gamete is the *sperm*, the female gamete is the egg or *ovum*. Fertilization results from the fusing of two gametes; the single cell formed at fertilization is called a *zygote*.

The chromosomes in the body cells are typically paired. The two chromosomes that form a pair are similar both in size and general appearance and in structural details that are considered in the section on genetics. They are said to be *homologous*.

Meiosis is not a single division but two. In a cell dividing by meiosis, homologous chromosomes come to lie side by side (go into *synapsis*). As in mitosis, each chromosome has doubled, so that the synapsing pair forms a four-stranded structure known as a *tetrad*. At the first division, the homologous chromosomes separate and each daughter cell receives one member of the pair. Each member consists of two chromatids; it is known as a *diad*. The second meiotic division separates the chromatids that make up the diad. Each of the resulting gametes has a single chromosome strand from one member of each of the homologous pairs of chromosomes. Because it has only one chromosome for each pair of chromosomes in the parent cell, it has half the number of chromosomes. Figure 11–6 shows this in diagrammatic form. If you have absorbed the essentials at this time, the more detailed outline of gamete formation given later and shown in Figure 11–7 should not obscure the basic simplicity of meiosis.

Sperm Formation

In man as in other animals, sperm are formed in the testes. Arranged in tubules in the testes are some rather large cells called *spermatogonia*. These cells are the direct descendants by mitotic divisions of primordial germ cells

Figure 11–6. Simplified diagram of meiosis. In the first division, one member of each pair of chromosomes goes to each daughter cell. In the second division, one chromatid of each chromosome goes to each daughter cell.

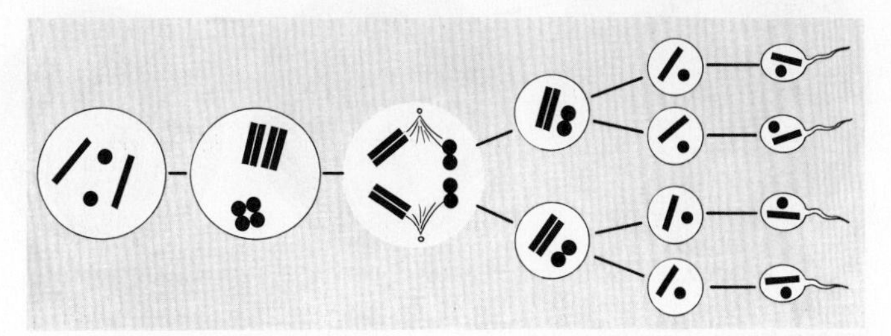

Figure 11–7. The details of meiosis.

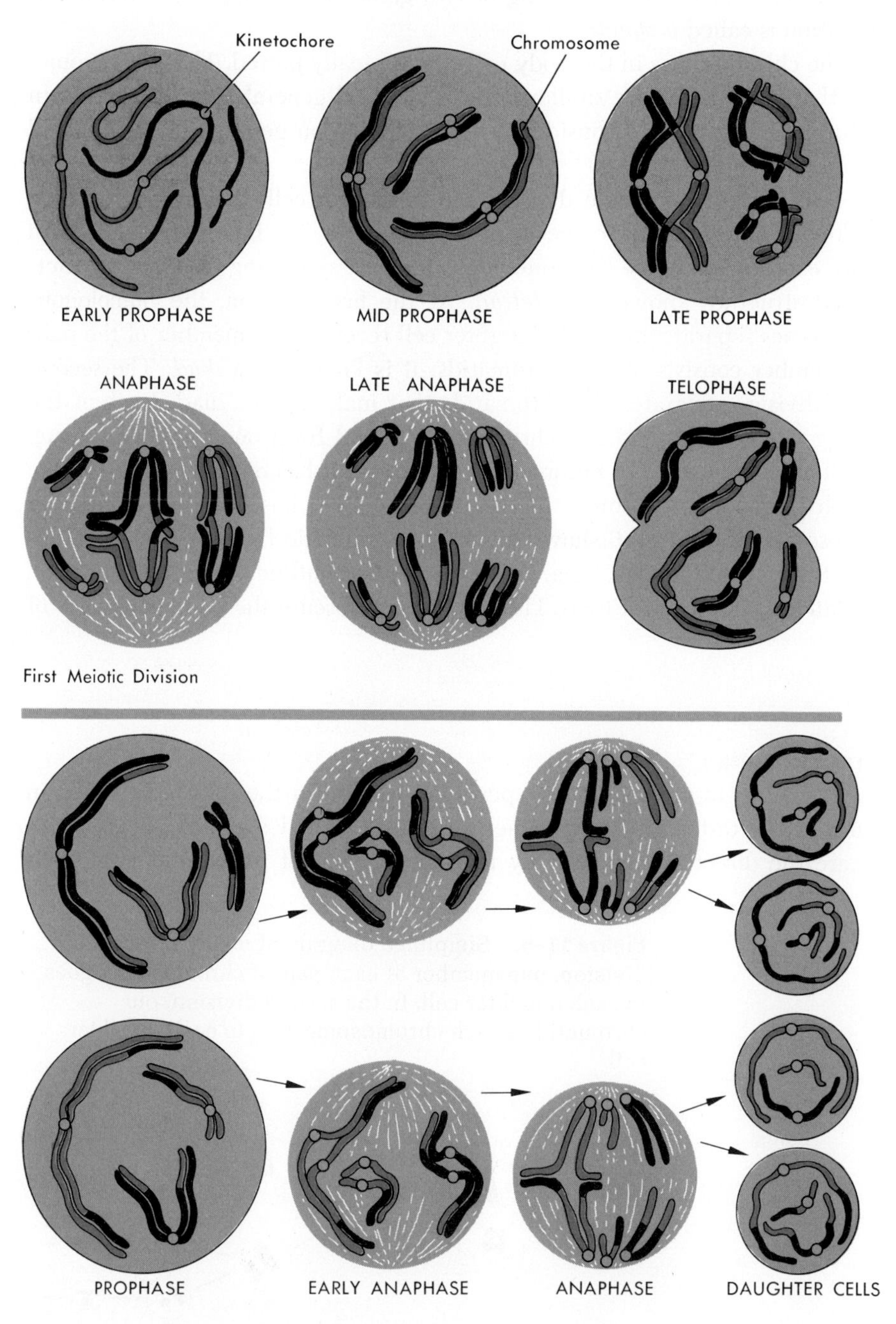

that were present in the early embryo. Spermatogonia continue to divide mitotically and ultimately give rise to cells called *primary spermatocytes.*

First Meiotic Division. The primary spermatocyte undergoes the first meiotic division. The chromatin material of the cell resolves itself into long, slender threads, which are even longer than in ordinary mitosis. The threads arrange themselves into units corresponding to pairs of chromosomes, and the members of each pair come to lie side by side, that is, they go into synapsis. As the threads shorten and thicken, it can be seen that each is double; the four chromatids of the two homologous chromosomes form the tetrad. The kinetochores of the homologous chromosomes move apart while the chromatid strands wrap around each other and seem to be exchanging parts. This results in a cross-shaped figure called a *chiasma.* The significance of this exchange of parts by the strands and its effect on variation are discussed in Chapter 15.

In the meantime, a spindle has formed and the tetrads move to take up the metaphase position. During anaphase, the two chromatids that are connected by a kinetochore (they form a diad) separate from the other two members of the tetrad. The two diads that were once part of a single tetrad move toward opposite poles.

There is usually no nuclear reconstruction during the first telophase in meiosis. The primary spermatocyte divides and each daughter cell, known as a *secondary spermatocyte,* proceeds directly into the second meiotic division. The period between the two meiotic divisions is designated by the special term *interkinesis.* There is no doubling of the chromatin material during interkinesis as there is during interphase, because this doubling took place before the first meiotic division.

Second Meiotic Division. The secondary spermatocytes undergo the second meiotic division. During prophase, the chromatids of each diad are still connected at their kinetochore although they are divergent elsewhere along their length. Again a spindle forms and the diads move to the equatorial plate. With the division of the kinetochores and the separation of the chromatid pairs, each chromatid now becomes a separate and complete chromosome. The chromosomes move to the poles, the nuclei are reconstructed, and the cell divides. The nucleus of the primary spermatocyte has now given rise to four daughter nuclei, each of which contains one strand from each pair of homologous chromosomes.

Development of Sperm. The four cells that result from the two meiotic divisions are the *spermatids.* Each develops without further division into the male gamete, the sperm cell. A sperm looks like a very long-tailed tadpole. The enlarged head contains the nucleus and a very little cytoplasm, the small body is rich in mitochondria, which provide energy for the lashing of the long, whip-like flagellum by which the sperm swims. The general process of sperm cell formation from the original spermatogonium is called *spermatogenesis.*

Gamete Formation in the Female

The process of meiosis is the same in both sexes, but gamete formation differs in detail. The sperm contributes little besides its nucleus to the zygote; the egg must provide both nuclear material and stored food for the developing embryo. The *oögonium,* which corresponds to the spermatogonium, undergoes a period of growth so that the *primary oöcyte* is much larger than the primary spermatocyte. In the male four functional sperm are formed from each primary spermatocyte, whereas in the female only one egg develops from each primary oöcyte. The distribution of the cytoplasm is quite unequal in the meiotic divisions that give rise to the egg cell. Although the nuclear divisions follow the same pattern as in the male, the stored cytoplasmic food material each time goes practically all to one cell, leaving the other cell with a nonfunctional nucleus and a minimum of cytoplasm. This little structure is called a *polar body.* If the first polar body formed at the first meiotic division undergoes a second meiotic division (it may not), three polar bodies are formed for each egg. Thus four daughter cells are produced, as there are in spermatogenesis, but only onc is functional.

Figure 11–8 gives a comparison of spermatogenesis and oögenesis. This should help emphasize the basic similarity of gamete formation in the two sexes. Figure 11–9 illustrates gamete formation in diagrammatic form, together with its natural outcome, fertilization.

Methods of Ensuring Fertilization

A few species of animals and some plants are self-fertilizing, that is, the gametes that unite to form the zygote are produced by a single parent. Much more frequently, animals are bisexual. The male produces the sperm, the female produces the egg, and cross-fertilization is mandatory. Cross-fertilization confers the advantage of variability on the species, but it does present one complication not encountered in asexual reproduction or self-fertilization. It requires the cooperation of two individuals.

Animals have developed a wide variety of techniques for ensuring that the egg and sperm encounter each other. Because you are probably more familiar with the backboned animals, the vertebrates, we take our examples from this

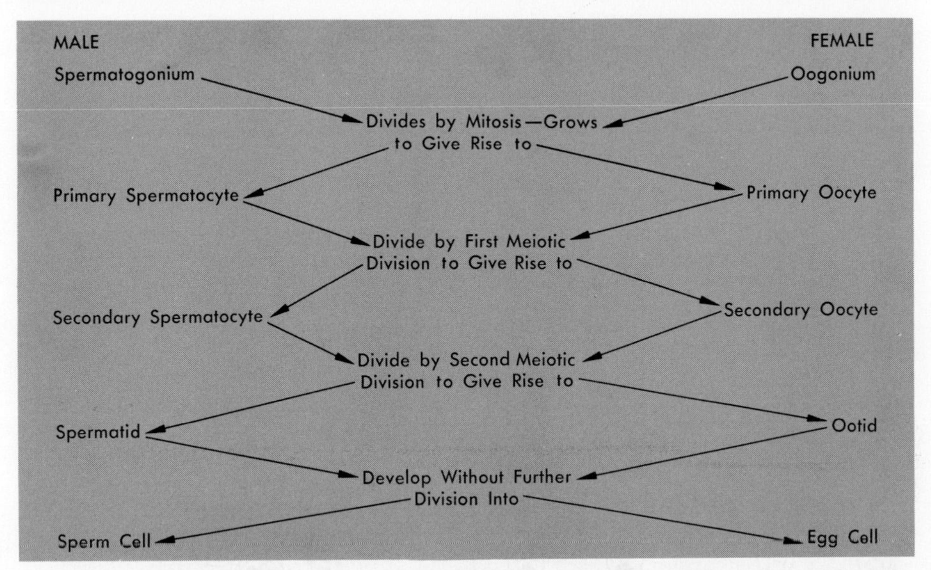

Figure 11–8. The stages of gamete formation in the male and female.

group. The various patterns of breeding behavior can be grouped into four major categories: fertilization outside the body without actual contact between male and female; fertilization outside the body with some sort of sexual embrace or *amplexus;* fertilization inside the female's body without copulation; and fertilization in the female's body involving sperm transfer by copulation.

External Fertilization Without Amplexus. Some fishes, though by no means all of them, use this method of fertilization. The fish come together in large numbers during the breeding season. The females simply release their eggs into the water, the males release their sperm at about the same time, and eggs and sperm must come together. The eggs drift about; if sperm pass close enough by they may be attracted to an egg, swim to it by means of their long flagella, and one will fertilize it. This approach calls for tremendous numbers of both eggs and sperm because there is an enormous loss of gametes that never take part in fertilization. The ocean sunfish has been reported to produce up to 28 million eggs in a single season.

External Fertilization with Amplexus. The male comes into some sort of bodily contact with the female, usually an embrace. This type of breeding is carried on by some fishes and frogs and toads. The frogs that call from ponds and ditches in the spring are males advertising their presence to the females.

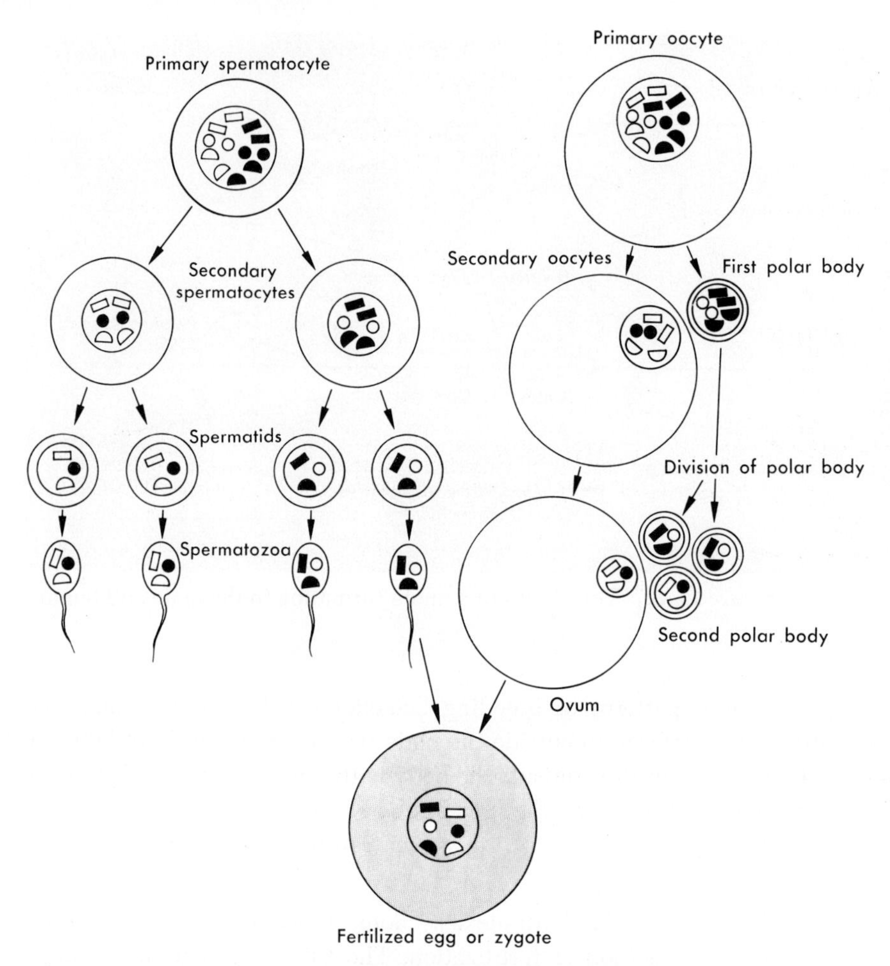

Figure 11–9. Diagram of gamete formation and fertilization.

When a female approaches a male, he embraces her. This stimulates her to release her eggs and he releases his sperm at the same time. Figure 11–10 shows two breeding pairs of American toads. Because they are in an embrace, the sperm and eggs are released in the same place and have a better chance of encountering one another than when there is no sexual contact. Amplexus usually occurs in water, but a few kinds of frogs may embrace in damp places on land. Amplexus is a more efficient procedure than external fertilization without bodily contact, but most frogs deposit large numbers of eggs. Figure 11–11 shows egg masses laid by several wood frogs. Females of this species lay up to 3,000 eggs at one time.

Figure 11–10. Breeding American toads (*Bufo americanus*) on Carnegie Museum's Powdermill Nature Reserve. (Carnegie Museum—M. Graham Netting photo.)

Internal Fertilization Without Copulation. The type of breeding known as internal fertilization without copulation is not widely practiced among the vertebrates, but in the amphibians called salamanders the male deposits sperm-filled gelatinous capsules called *spermatophores,* which the female finds and takes into her reproductive tract. Usually the male first stimulates the female by a rather complex courtship prelude. Internal fertilization results in much less wastage of the gametes, and salamanders usually lay far fewer eggs than frogs.

Internal Fertilization with Copulation. Internal fertilization with copulation is the type of sexual behavior practiced by some fishes and by all the higher terrestrial vertebrates, the reptiles, birds, and mammals, including man. Here the sperm are transferred directly into the body of the female, usually by a specialized intromittent organ. In mammals this organ is the penis.

Figure 11–11. Egg masses of the Eastern wood frog (*Rana sylvatica*) on Carnegie Museum's Powdermill Nature Reserve. (Carnegie Museum—M. Graham Netting photo.)

SUGGESTED READINGS

Burke, J. D.: *Cell Biology,* Williams & Wilkins Company, Baltimore, 1970.

Francoeur, R. T.: *Utopian Motherhood: New Trends in Human Reproduction,* Doubleday & Company, Inc., New York, 1970.

Fraser, A.: *Heredity, Genes, and Chromosomes,* McGraw-Hill Book Company, New York, 1966.

Orr, R. T.: *Vertebrate Biology,* 3rd ed., W. B. Saunders Company, Philadelphia, 1971.

CHAPTER 12

Development of the Individual

Once the sexual behavior of the adults has ensured the coming together of the gametes, fertilization normally ensues. It is the resultant zygote that, by repeated mitotic cell divisions, will give rise to the adult of the next generation. This is one of the most amazing facts of biology—that a single cell, by dividing and dividing and dividing again, can give rise to sometimes billions of cells of many different kinds and yet so precisely arranged that the resultant organism is not only a smoothly functioning individual but an individual of the same kind as the parent. It is also one of the least understood aspects of biology. Biologists can follow the course of embryonic development, but they are just at the threshold of an understanding of the mechanisms involved.

THE ZYGOTE

The primordial germ cell is *diploid,* that is, the chromosomes are present in pairs. As a result of the meiotic divisions of the germ cells, the gametes are *haploid;* they contain only one member from each of the original pairs of chromosomes. Fertilization is not simply the penetration of the egg by the sperm but an actual fusion of the two haploid nuclei of the gametes. The zygote, or fertilized egg, now has a single diploid nucleus, in which the chromosomes are again paired. Each gametic nucleus has brought in one member of each pair. The zygote thus has the same number and same kind of chromosomes as the parents.

Because the zygote must develop not just into an adult organism but also into an organism of a specific kind, the zygotes of different kinds of organisms follow different paths of development. A plant does not develop the way an animal does, an invertebrate, such as a worm or insect, does not develop the way a vertebrate does. It is impossible to discuss the development of all organisms, or even of all multicellular animals, except in terms so broad that no concrete picture would emerge. Because man is a vertebrate, we limit our discussion to the developmental patterns of the vertebrates.

It has long been recognized that the early embryonic stages of related animals resemble each other and that divergencies appear as development proceeds. This is true for the vertebrates with one major exception. The very earliest stages of embryonic development are more different than some of the later stages. This results largely from differences in the amount of stored food material (yolk) in the eggs of the different kinds of vertebrates. Mammalian embryos are typically nourished from a very early stage by food material derived from the tissues of the mother. They need little in the way of a reserve food supply, and the egg has almost no yolk. The egg is so small it is hardly visible to the naked eye. Amphibians and some fishes have only a moderate amount of yolk. Their eggs vary in size but are often about two millimeters (one twelfth of an inch) in diameter. Other fishes and all reptiles and birds have very large amounts of yolk—the yolk of a hen's egg is a single cell, enormously swollen by masses of stored food.

A zygote is polarized. Even though it is round, it is already orientated so that the nucleus and an area of clear cytoplasm are located near one pole. This is known as the *animal pole* and is destined to be the center of the earliest visible embryonic activity. The opposite pole, called the *vegetal pole,* marks the site of the greatest concentration of yolk.

EMBRYONIC DEVELOPMENT—THE EARLIEST STAGES

It is easier to understand the basic pattern of the early stages of vertebrate development by looking first, not at a vertebrate but at a small, marine relative of the vertebrates, a little, fishlike animal called *amphioxus.* Like mammal eggs, amphioxus eggs have very little yolk.

Early Development of Amphioxus

Early development is divided into two periods, first a period of rapid, synchronous cell divisions and then a period in which, while cell division continues, there is a moving and folding of cell masses.

Cleavage. The series of divisions that mark the beginning of development are known as *cleavages*. They are mitotic divisions that differ from the mitotic division of a single-celled animal in two ways. First, the cells do not separate but remain clinging together. Second, there is no growth period between successive divisions. The result is that the initial cell is divided into successively smaller cells known as *blastomeres*. The first cleavage extends from the animal to the vegetal pole and divides the zygote into two symmetrical blastomeres. One will produce the right side of the body, the other the left. The second cleavage is at right angles to the first. It is also meridional and divides each of the first two blastomeres; the resultant structure is composed of four blastomeres about equal in size. The third cleavage is horizontal to the first two. It is slightly above the equatorial plane so that the four cells around the animal pole are somewhat smaller than the four around the vegetal pole. Successive cleavages produce smaller and smaller blastomeres with a more pronounced difference in size between the upper and lower blastomeres. Figure 12–1 shows the early cleavage stages in amphioxus.

After the third cleavage a central space develops between the blastomeres and as cell division proceeds this cavity increases in size. The result is a struc-

Figure 12–1. Early cleavage in amphioxus. A: The zygote. B: Two cell stage. C: Four cell stage. D: Eight cell stage. E. Sixteen cell stage. F: Thirty-two cell stage.

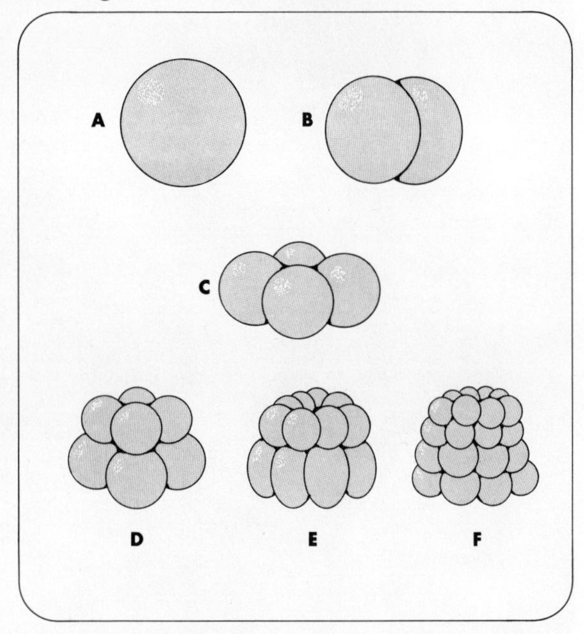

ture composed of a single layer of cells arranged in the shape of a hollow ball. This hollow ball is called a *blastula* and the cavity is called a *blastocoel* (see Figure 12–2A).

Gastrulation. The next stage of development consists of a folding in (invagination) of the blastula wall much as one might push in the side of a soft rubber

Figure 12–2. Sections through early stages of amphioxus development. A: Blastula. B–E: Stages in gastrulation. F: Gastrula.

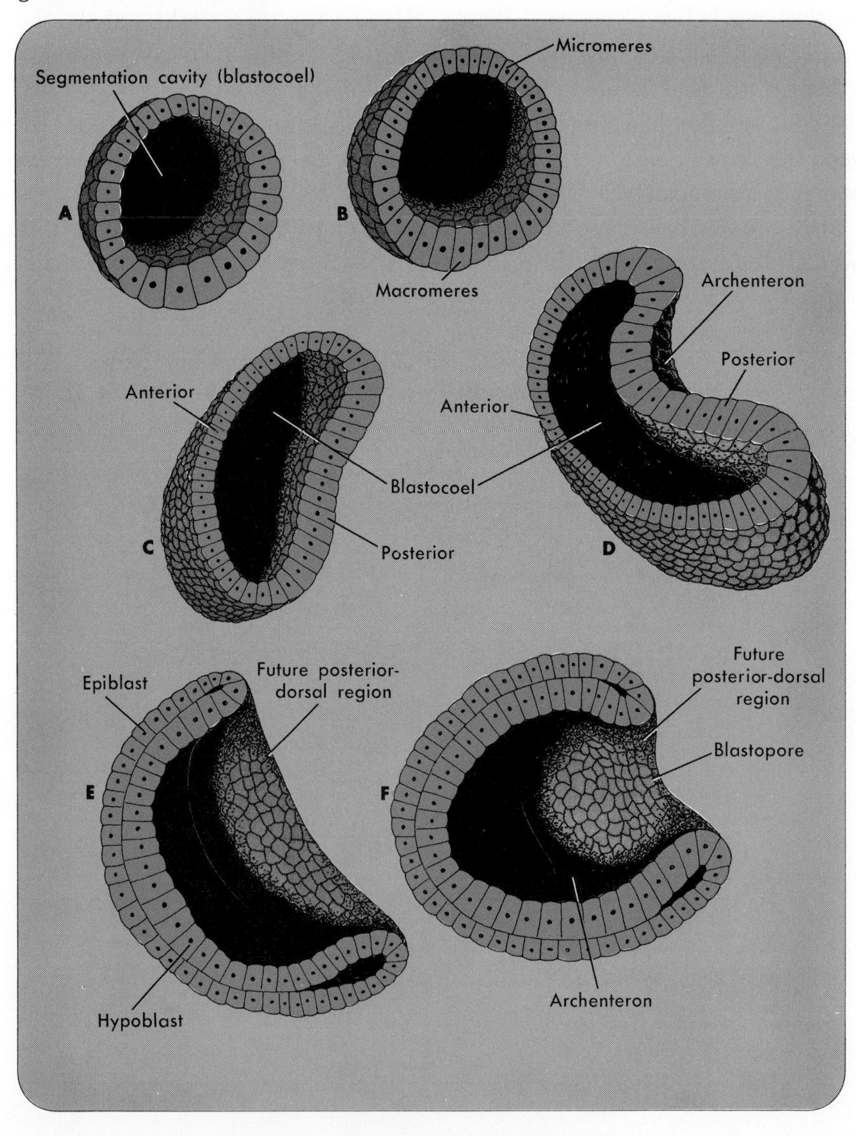

ball (see Figure 12–2B–E). Invagination carries the larger yolky cells inward and continues until the infolding layer comes in contact with the cells of the opposite wall of the blastula. Invagination essentially obliterates the blastocoel but at the same time it leads to the formation of a new cavity in the now two-cell layered, cup-shaped structure. This structure is known as a *gastrula.* Its cavity is called the *gastrocoel* or, more familiarly, the *cavity of the primitive gut* (*archenteron*). The cavity of the archenteron leads to the outside by way of the *blastopore.* With the development of the blastopore (see Figure 12–2F) a head-tail orientation of the embryo is established. The blastopore indicates the future posterior end of the animal, whereas the other end of the embryo will develop into the head. The outer layer of cells of the gastrula is called the *epiblast;* the inner layer is known as the *hypoblast.*

During gastrulation, the cells begin to grow between divisions so that the embryo starts to increase in size. In particular, it elongates, becoming sausage-shaped rather than round.

Differentiation of Epiblast and Hypoblast. Even before gastrulation is complete, the epiblast and hypoblast begin to differentiate.

EPIBLAST DIFFERENTIATION. The cells of the dorsal region of the epiblast begin to flatten and thicken to form the *neural plate.* At this stage the epiblast has differentiated into two different tissues: the *neurectoderm,* which forms the neural plate and will give rise to the nervous system, and the *skin ectoderm,* which will ultimately develop into the epidermis of the skin. As development continues, the skin ectoderm grows up over the neural plate and fuses along the midline. The neural plate itself then folds up its edges and becomes a tube, the *neural tube.* This will develop into the brain and spinal cord.

HYPOBLAST DIFFERENTIATION. At the same time that the epliblast is differentiating, the hypoblast also begins to change. That part below the neural plate differentiates from the rest and is now called *mesoderm.* The remainder of the hypoblast will become the epithelial lining of the archenteron or primitive gut and is called *endoderm.* The mesoderm lateral to the midline pinches off into a series of mesodermal pouches that grow downward to separate the ectoderm from the endoderm; the cavities in the pouches fuse to become the *coelom* or true body cavity. Along the midline, the mesoderm does not form pouches but instead gives rise to a rod of stiffening material, the *notchord,* which lies between the neural tube and the gut.

Ectoderm, endoderm, and mesoderm are oftentimes called the *primary germ layers,* and an animal that develops all three is said to be *triploblastic.* Figure 12–3 shows the differentiation of the germ layers.

Development of True Gut. As development proceeds, the blastopore is closed over by ectoderm so that the cavity of the archenteron has no opening to the

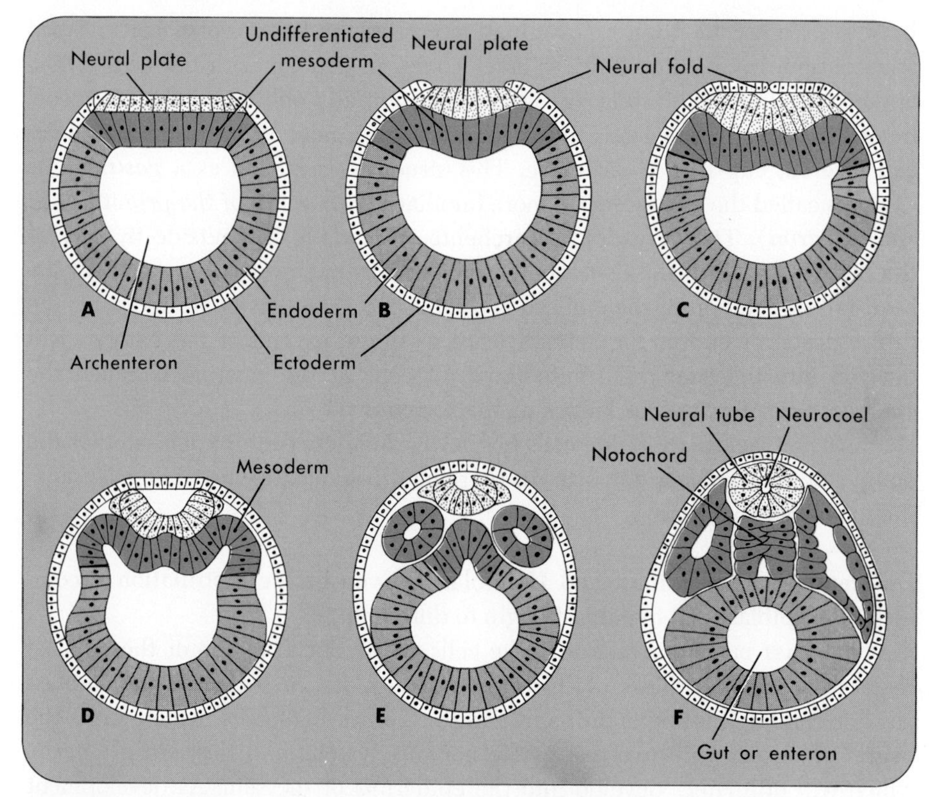

Figure 12–3. Differentiation of the germ layers in amphioxus leading to the formation of the neural tube, notochord, and mesodermal pouches. Later the mesodermal pouches grow down to surround the gut. The cavities within them become the general body cavity or coelom. The body wall is composed of ectoderm and mesoderm, the wall of the gut of mesoderm and endoderm.

outside. Shortly, though, a new opening penetrates the body wall in the region that will develop into the mouth, and another forms in the region of the obliterated blastopore to become the anus. When these openings have developed, the gut is a true tube and is called the *enteron* or *true gut* instead of the *archenteron* or *primitive gut.*

Other Patterns of Early Development

By a number of different pathways, all vertebrate embryos arrive at a stage similar to that to which the amphioxus embryo has been traced. They all become triploblastic, with an inner tube running from mouth to anus. The ectoderm is on the outside, the endoderm lines the inner tube, and the mesoderm lies between the two. The mesoderm itself is divided into two layers, an outer one lying against the inner margin of the ectoderm and an inner one covering the endoderm. The cavity between the two is the coelom. Neural tube and notochord are also present.

Development of Amphibian. The amphibian egg has considerably more yolk than the amphioxus egg. Cleavage is similar but the blastula consists of several cell layers and the cells of the animal portion are much smaller than the yolk-heavy vegetal cells. These large cells do not invaginate during gastrulation. Instead, the small cells of the animal hemisphere begin to move down over them and start to fold inward. The area where infolding begins is marked by a crescent-shaped notch. As infolding spreads, the notch becomes a circular opening, the blastopore, surrounding a central plug of yolk cells (see Figure 12–4).

Figure 12–4. Sections through early stages of frog development. A: Blastula. B–F: Gastrulation.

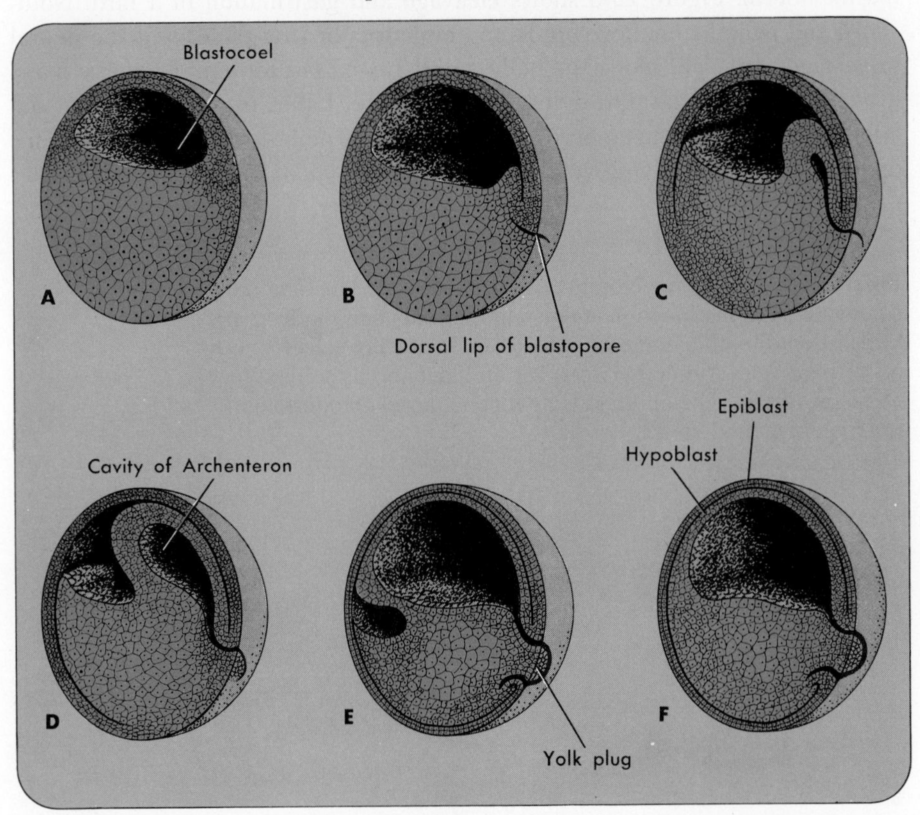

The place where the infolding first starts is known as the *dorsal lip* of the blastopore. The infolded cells spread under the outer layer (epiblast) to become the hypoblast and separate into endoderm and mesoderm. The notochord and neural tube develop. Cells both migrate and change in shape during the early stages of embryonic development. The cells of the neural plate elongate and become pinched in at their apexes to form the neural tube. Cytoplasmic microtubules are involved in elongation and microfilaments in apical constriction.

Development of Bird. The early cleavages of the bird's egg do not cut through the yolk. Instead, they simply divide a small area at the animal pole. The result is a flat disc of cells lying on top of the huge, undivided yolk. The lower cells of this disc split off to form the hypoblast whereas the upper layer becomes the epiblast. Cells from the epiblast move toward the midline to form a thickened area known as the *primitive streak.* In the center of the primitive streak is a groove, the *primitive groove.* Migrating epiblast cells pass downward through the groove to spread between the epiblast, which becomes the ectoderm, and the hypoblast, which becomes the endoderm. The cells that migrated in form the mesoderm. Figure 12–5 shows cleavage and gastrulation in a bird. Notochord and neural tube develop. If an amphioxus or frog embryo at the neural tube stage were split along the belly and flattened out on top of a yolky mass, it would resemble the bird embryo at this stage. Later, infoldings around the edges of the developing embryo serve to pinch the embryo off so that it remains attached to the underlying yolk only by a narrow *yolk stalk.*

Figure 12–5. Early development of the chick. A: The blastoderm corresponds to the blastula stage in amphioxus and the frog. B: Separation of epiblast and hypoblast. C: Migrating cells from the primitive streak spread between the epiblast and hypoblast to form mesoderm. The epiblast is now considered ectoderm and the hypoblast endoderm.

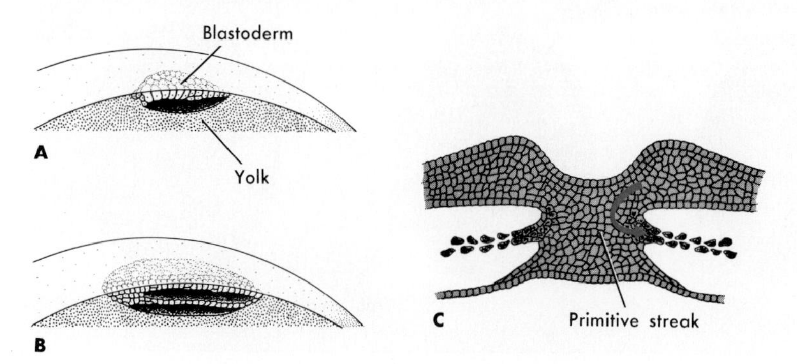

Development of Mammal. The early stages of development vary in the mammals. The account that follows is based largely on the development of the human embryo. Cleavage divisions pass all the way through the small egg, as they do in amphioxus, but the divisions are apparently not synchronous. The first divides the egg into two unequal cells. The larger cell divides first to produce a three-cell stage, and then the small cell divides. Thus, although the number of cells in the cleaving amphioxus egg increases geometrically (two, four, eight, sixteen, and so on), the number in the cleaving mammal egg increases arithmetically (two, three, four, five; see Figure 12–6). A hollow ball stage is reached, but again this differs from the blastula of amphioxus. It is known as a *blastocyst*. There is an outer layer of flattened cells called the

Figure 12–6. Cleavage and blastocyst formation in a mammal. A: Egg. B–E: Early cleavage stages. F: Solid ball stage known as a morula (= mulberry). G: Section through the morula. H and I: Formation of blastocyst.

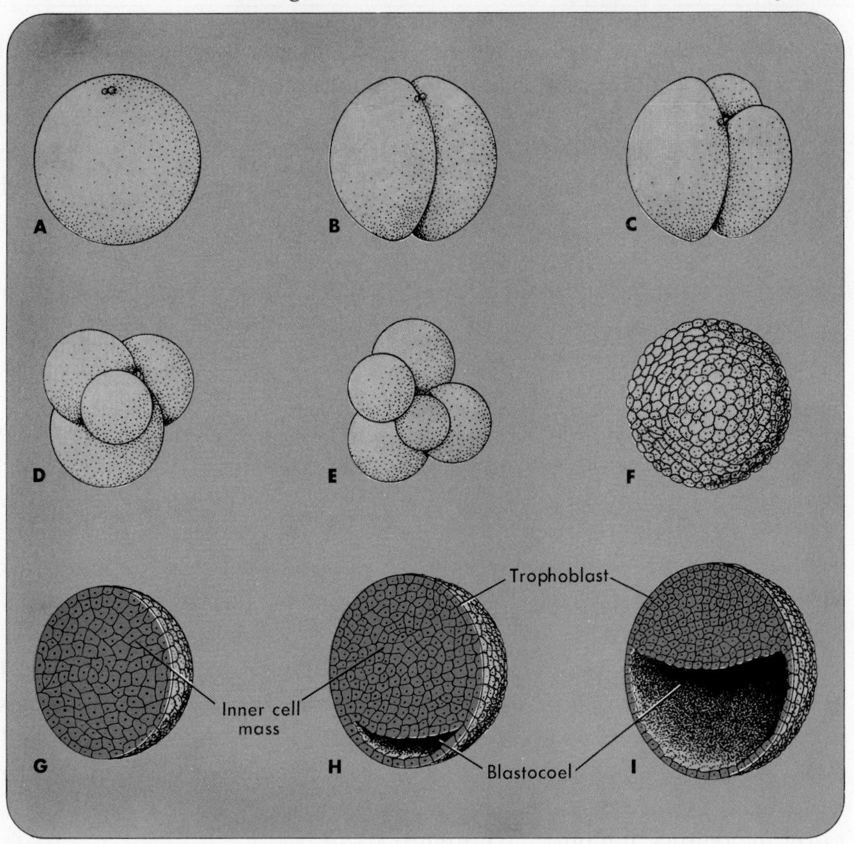

trophoblast. Pressed against one side of the trophoblast is a mass of cells, the *inner cell mass.* Figure 12–6 shows cleavage and blastulation in an advanced mammal.

The cells of the inner cell mass sort themselves out into two layers much as the cells in the developing bird embryo do. Then two cavities develop, one in the upper layer, one in the lower, so that the whole cell mass has the shape of a figure eight. The cells that surround the upper cavity will become ectoderm, those that surround the lower cavity, endoderm. The disc of cells separating the two cavities is called the *embryonic disc.* It develops into the embryo proper in very much the same way that the bird embryo develops. The further history of the other parts of the inner cell mass and of the trophoblast appears in the section on extraembryonic membranes. Figure 12–7 is a sketch of a sixteen-day human embryo, showing the embryonic disc.

EXTRAEMBRYONIC MEMBRANES

The embryo is itself a functioning organism, carrying on the metabolic processes required for growth at a very rapid rate and facing many of the same problems the adult does. The embryo must take in food and oxygen, get rid of carbon dioxide and nitrogenous waste products, and it must be protected against destructive forces in the environment. For these functions, vertebrate embryos develop one or more sacs bounded by membranes that lie outside the body of the embryo. A *yolk sac* is formed in all vertebrates. Reptiles, birds, and mammals have three additional membranes: the *amnion, chorion*, and *allantois.*

Yolk Sac

The method by which the yolk sac is formed differs in the various groups of vertebrates. In reptiles and birds, the yolk sac consists of a layer of extra-embryonic mesoderm and endoderm that grows down to surround the mass of yolk. Nutrients in the yolk are digested by the endoderm cells and transported to the body of the embryo by blood vessels that develop in the yolk sac mesoderm.

Although very little yolk is present in the human egg, it develops a yolk sac just as all vertebrates do. A layer of cells differentiates from the inner layer of the trophoblast to become the extraembryonic mesoderm. It surrounds the inner cell mass and forms a body stalk connecting this to the trophoblast. Thus, the cavity in the inner cell mass lying below the embryonic disc is surrounded by layers of mesoderm and endoderm to form a sac that corresponds to the yolk sac of the reptiles and birds (see Figure 12–7).

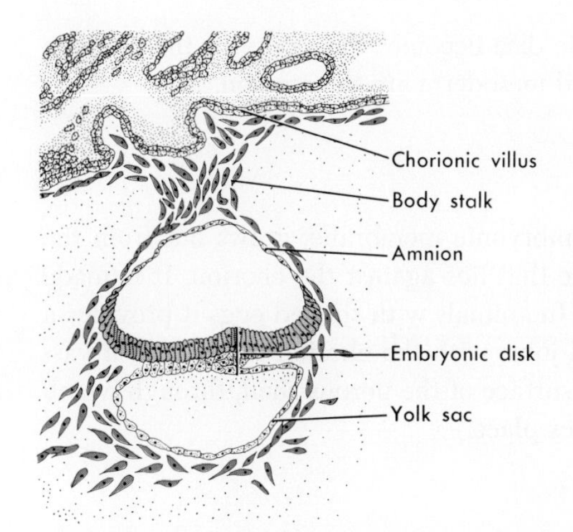

Figure 12–7. Sketch of a 16-day human embryo. (Migrating mesoderm cells are in color.)

Amnion and Chorion

Most fish and amphibian eggs are laid in water, which cushions them from jars and protects them from desiccation. Oxygen dissolved in the water diffuses into the egg, and carbon dioxide and nitrogenous wastes diffuse out. Birds, many reptiles, and a few mammals lay shelled eggs on land; other reptiles and almost all mammals develop within the body of the mother. They need additional mechanisms, both for protection and for exchange with the environment.

Folds of the extraembryonic ectoderm and mesoderm grow up to surround the developing embryo of the reptile or bird. These folds meet and fuse above the embryo to form two sacs. The inner sac surrounds a fluid-filled cavity in which the embryo is protected from mechanical shocks and also from desiccation. The membrane bounding this sac is known as the *amnion.* The outer membrane, called the *chorion,* completely surrounds the embryo and all the other membranes. Figure 12–8 shows diagrammatically the development of the extraembryonic membranes in a reptile or bird.

In man, the trophoblast with its underlying mesoderm forms the chorion.

Figure 12–8. Diagram of the development of the extraembryonic membranes in a reptile or bird.

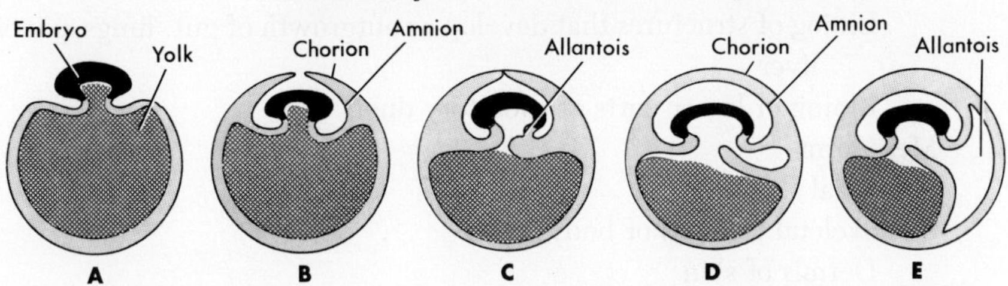

The cavity lying above the embryonic disc becomes the cavity of the amnion; its surrounding layers of ectoderm and mesoderm are the amnion.

Allantois

The allantois, the fourth extraembryonic membrane, grows out from the posterior end of the gut to form a sac that lies against the chorion. It is made up of both endoderm and mesoderm. In animals with shelled eggs it provides a storage place for waste products and joins with the chorion to produce a vascular membrane applied to the inner surface of the porous shell, through which the exchange of respiratory gases takes place.

Placenta

In most mammals, the developing young receive oxygen and nutrients and eliminate waste products by way of the mother's bloodstream through a structure called the *placenta.* This is a dual organ, formed in part from the lining of the uterus of the mother and in part from the extraembryonic membranes. In man the allantois grows out through the body stalk and fuses with a part of the chorion to form the embryonic part of the placenta (see Figure 12–9).

LATER DEVELOPMENT

The differentiation of the primary germ layers into the definitive organs that are characteristic of the adult is known as *organogenesis.* We do not follow it in detail here, but simply list the major structures derived from each germ layer.

- Ectoderm
    - Neural canal and from it the nervous system (neurectoderm)
    - Epidermis of skin (skin ectoderm)
- Endoderm
    - Lining of gut
    - Lining of structures that develop as outgrowth of gut: lungs, pancreas, liver
    - Lining of lower parts of excretory ducts
- Mesoderm
    - Axial skeleton
    - Skeletal muscles of body
    - Dermis of skin

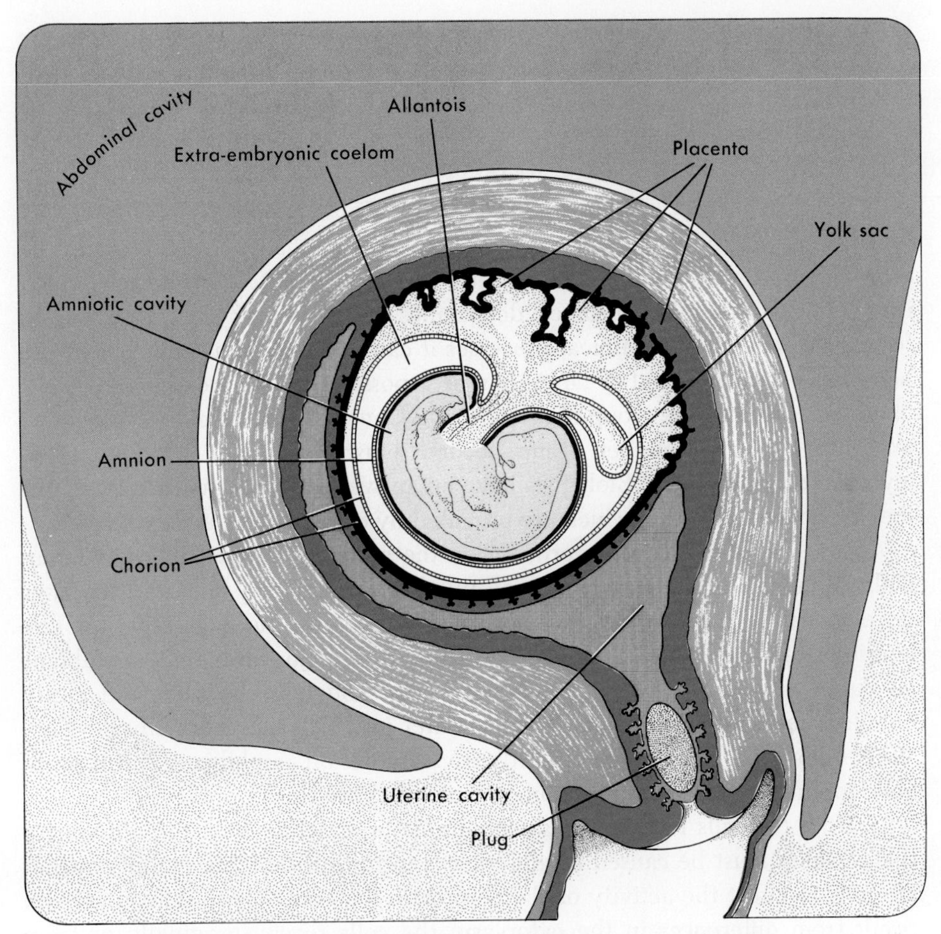

Figure 12–9. A human embryo in the uterus showing the extraembryonic membranes. The placenta includes tissues from the allantois, chorion, and uterine wall.

Excretory system
Internal reproductive organs
Lining of coelom
Circulatory system
Smooth muscles of digestive system
Other parts of lungs, pancreas, and liver

Some structures of the body are not formed directly from the subdivisions of the three primary germ layers but rather from an undifferentiated tissue, *mesenchyme*. Mesenchyme cells have the ability to migrate to where they will be needed and there differentiate. For example, mesenchyme flows out into the developing limb buds where it gives rise to the limb bones and muscles.

MECHANISMS OF DEVELOPMENT

Embryonic development proceeds by mitosis so that each cell receives the same kind and same amount of chromatin material. The DNA of the cell, which is contained in the chromosomes, determines what proteins the cell is able to form, and hence what chemical processes it can carry on. If all the cells of the body have the same kind of DNA, how is it possible for differentiation of cells to occur? How can one group of cells become bone cells capable of laying down a hard matrix and another group become muscle cells and produce actin, myosin, and myoglobin? Biologists have no answers to these questions yet, but they do have some clues, which are provided by experimental embryology.

It is very difficult to experiment on the early embryonic stages of mammals. Mammals produce only a small number of eggs at a time and the eggs develop within the body of the mother. Just keeping the eggs alive for any length of time out where they can be studied is a major problem. Amphibians lay large numbers of eggs, which are large enough to be seen and easily handled, and simple to maintain in the water in which they normally develop. What follows has been derived mostly from studies on amphibians, but it probably applies to the developing human embryo as well.

If all the cells receive the same chromatin material, then the differences that develop must be caused by differences in the surrounding cytoplasm that are able to affect the activity of the chromatin material. These differences can result from differences in the cytoplasm the cells receive originally or from differences in the location of the cells within the embryo in relation to other cell masses, or to the environment. For example, in amphioxus, cells of the hypoblast have more yolk and cells of the epiblast are in contact with the extra-embryonic environment.

Differences in Egg Cytoplasm

The constituents of the egg cytoplasm are not distributed evenly. Yolk is the most obvious example, but there are other differences as well. When the sperm penetrates an amphibian egg, a crescent-shaped area known as the *gray crescent* appears on the opposite side. The location of the gray crescent is determined partly by the point of entrance of the sperm and partly by differences in the distribution of material in the egg cytoplasm. The first cleavage division separates the gray crescent material so that each blastomere receives the same amount of material. If the two blastomeres are separated, each will develop into a complete embryo. It is possible, though, so to manipulate the egg that the cleavage division gives most of the gray crescent to one blastomere. If the two blastomeres are then separated, the one receiving the gray crescent develops normally but the other forms only an undifferentiated mass of cells. This shows that cytoplasmic constitutents can influence the future course of development of the cells. Figure 12–10 is a diagram showing the importance of the distribution of gray crescent materials.

The Organizer and Induction

The area where the gray crescent was located becomes the dorsal lip of the blastopore in the developing gastrula. Cells from this region can be dissected out from one embryo and implanted on another embryo on the region opposite to its own blastopore. These cells will migrate inward as in normal gastrulation, form mesoderm and notochord, and the adjacent cells of the host embryo will form a neural tube and other organs. Two essentially complete, conjoined embryos can develop from the two dorsal lips (see Figure 12–11).

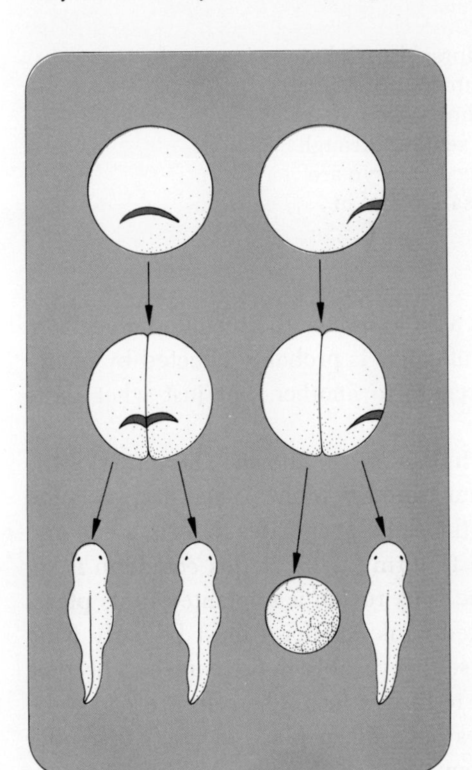

Figure 12–10. The effect of the distribution of gray crescent material. In the figure on the left the gray crescent is equally divided between two blastomeres. Each is able to develop into a normal embryo. In the figure on the right the gray crescent goes to one blastomere. It develops normally while the blastomere lacking gray crescent material does not.

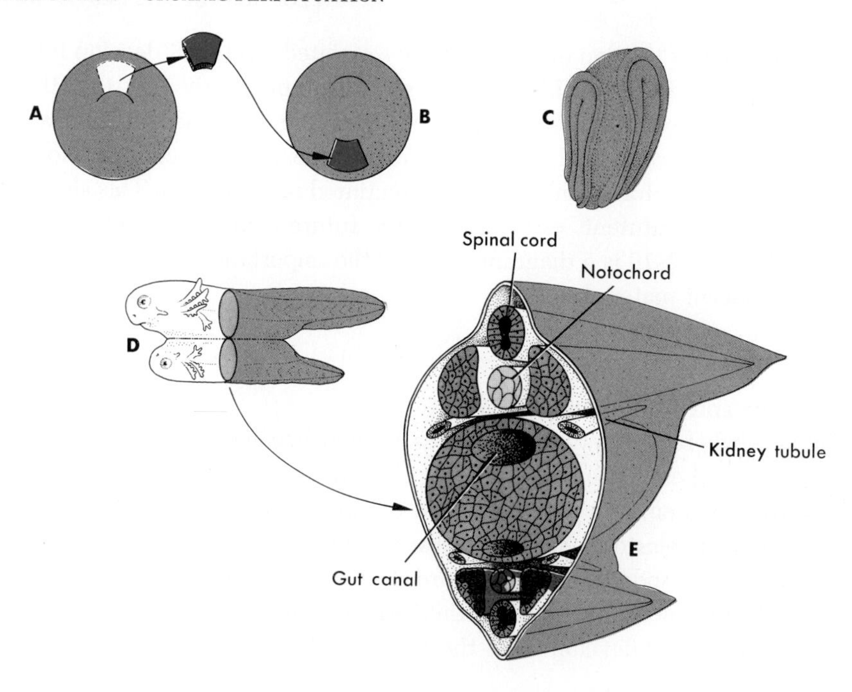

Figure 12–11. Inductive activity of the dorsal lip. Cells from the dorsal lip region of A are implanted on B. As a result, two neural plates develop (C). D shows the two conjoined embryos at a later stage. E is a section through the two embryos. The organs of the induced embryo are formed in part from the transplanted dorsal lip and in part from the cell of the host embryo.

Cells from the dorsal lip region are thus able to induce the surrounding cells to form other organs of the embryo. The influence is probably effected by chemical substances that diffuse from one region to another, but just what these substances are is not known.

Many more localized examples of induction are known. The retina of the eye develops from a stalk that pushes out from the brain. When it approaches the overlying skin ectoderm, the end of the stalk invaginates to form a cup and the ectoderm above the cup is induced to form a lens. If this ectoderm is removed and a patch of ectoderm from the flank region is implanted in its place, it too will form a lens. But the implanted flank ectoderm must be early flank ectoderm. A little later, the cells will be set irrevocably in the course to become flank skin and can no longer be induced to form a lens. The cells are said to be no longer competent. But what changes have taken place in them to render them resistant to induction is still not known.

We deferred the discussion of the reproductive system in our earlier account of the structure of the body because we believed that you would understand human reproduction better if you had some background knowledge of the major cellular processes involved and of the basic outlines of embryonic development. It is now time to turn again to the human body and consider reproduction as carried on by man.

SUGGESTED READINGS

Deuchar, E.: *Biochemical Aspects of Amphibian Development*, John Wiley & Sons, Inc., New York, 1966.

Ebert, J. D.: *Interacting Systems in Development*, Holt, Rinehart and Winston, Inc., New York, 1968.

Francis, C. C.: *Introduction to Human Anatomy*, C. V. Mosby, St. Louis, Mo., 1973.

Markert, C. L., and H. Ursprung: *Developmental Genetics*, Foundations of Developmental Biology Series, Prentice-Hall, Inc., Englewood Cliffs, N. J., 1971.

Saunders, J. W., Jr.: *Animal Morphogenesis*, Macmillan Publishing Co., Inc., New York, 1968.

Twitty, V. C.: *Of Scientists and Salamanders*, W. H. Freeman and Co., San Francisco, 1966.

Waddington, C. H.: *Principles of Development and Differentiation*, Macmillan Publishing Co., Inc., New York, 1966.

CHAPTER 13

Human Reproduction

Every species that has ever lived on the earth has had to meet two requirements in order to survive—maintenance of the individual and maintenance of the race. A failure of either of these requirements inevitably leads to the extinction of the species.

The previous discussion of the human body was concerned with the organ systems involved in the maintenance of the individual. This chapter deals with human reproduction, which leads to the maintenance of the species.

HUMAN REPRODUCTIVE SYSTEMS

Because men and women differ in the roles they play in the production of offspring, their reproductive systems also differ in structure.

Male Genital System

The functions of the male genital system include the production of sperm and hormones and the transfer of the sperm to the female.

Structure of Male System. The genital organs of the male are the *testes*, which produce both sperm and male hormones; the *vasa deferentia*, tubules that transport the sperm from the testes to the urethra; various glands, such as the *seminal vesicles, prostate,* and *Cowper's gland,* that produce a fluid medium for the sperm; and the penis, by which the sperm are transferred to the female. Figure 13–1 shows the position of these structures.

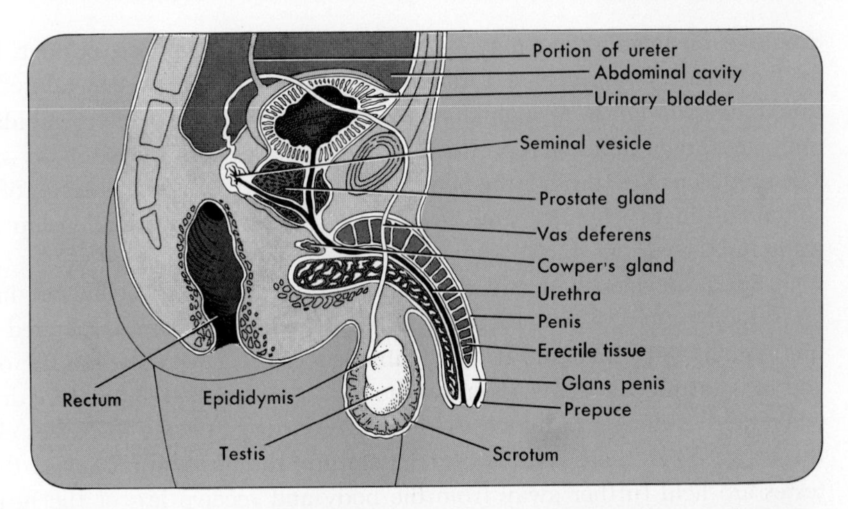

Figure 13–1. Structure of the male genital system.

The testes, or male gonads, are ovoid, somewhat flattened glands, about thirty-five to fifty millimeters (one and a half to two inches) long, and twenty-five millimeters (one inch) wide that are suspended in a musculomembranous pouch, the *scrotum.* They form originally from undifferentiated reproductive tissue in the abdominal cavity and can be recognized as male gonads by about the seventh week of embryonic life. The growth of the gonads at this time is stimulated by gonadotrophic hormones secreted by the placenta. Under the influence of these hormones, the testes secrete *androgens,* hormones that bring about the development of the other male sexual organs. About the beginning of the seventh month of development, the testes descend by a narrow opening, the *inguinal canal,* connecting the coelom with the scrotal sac. This canal normally pinches closed just prior to birth. If it should fail to do so, the individual is susceptible to a protrusion of a loop of the intestine into the canal to form an *inguinal hernia.*

The testes essentially stop growth when they are separated from the placental effect at birth and very little change takes place in them until adolescence, when active growth begins again under stimulation of the anterior pituitary gonadotrophic hormones.

The adult testis is covered with a tough, fibrous coat, the *tunica albuginea.* The tunic contains numerous branching smooth muscle cells. It has been suggested that contraction of these muscles may be responsible for the transport of sperm from the testis. The bulk of the testis is made up of a number of pyramid-shaped lobules, each of which is composed of several coiled seminiferous tubules. Each tubule is from thirty to sixty centimenters (one to two feet) long; it is estimated that there are eight hundred or more tubules in each testis. The tubules come together in a network of canals from which a series of ducts known as the *vasa efferentia* pass through the tunic and form a convoluted mass called the *epididymis.* A single duct, the *vas deferens,* leads from the epididymis to the urethra, a tube the reproductive system shares with the excretory system. The urethra passes through the penis, where it is surrounded by masses of tissue containing many vascular sinuses. Figure 13–2 shows the relationship of the testis and the various ducts.

Sperm develop best at a temperature several degrees below the internal body temperature. The scrotal sac in which the testes are suspended has a thermoregulatory function. It is a double-layered bag with skin on the outside and muscle and connective tissue on the inside. When the temperature drops or when there is danger of injury, the muscles contract to bring the testes close to the body. When there is a rise in temperature, the scrotum relaxes so that the testes are held further away from the body and receive less of the heat that radiates from it.

Spermatogenesis in man begins at puberty and usually continues throughout life. The spermatids that result from the second meiotic division are attached for a while to special *Sertoli cells,* which are believed to nourish them in some way. They then metamorphose into the spermatozoa (see Figure 13–3) and are transported into the vasa efferentia and epididymis, where they are stored.

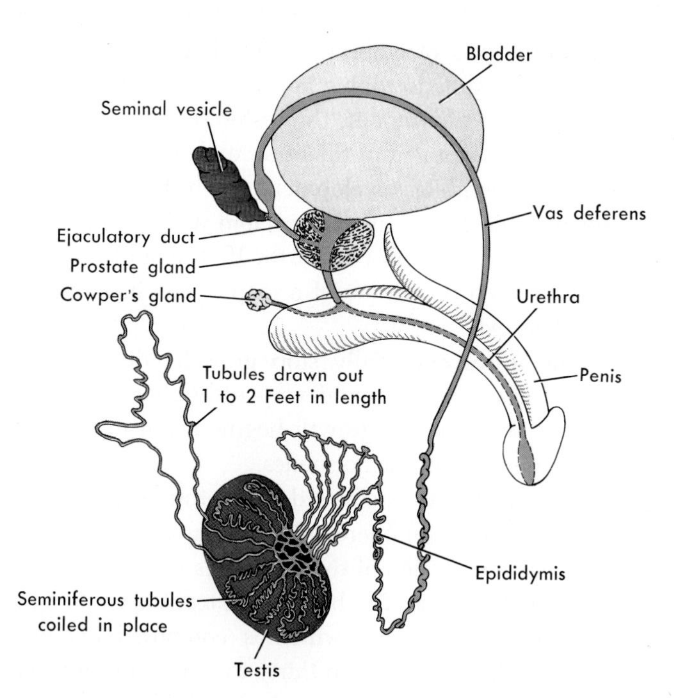

Figure 13–2. The testis and associated ducts of the human male.

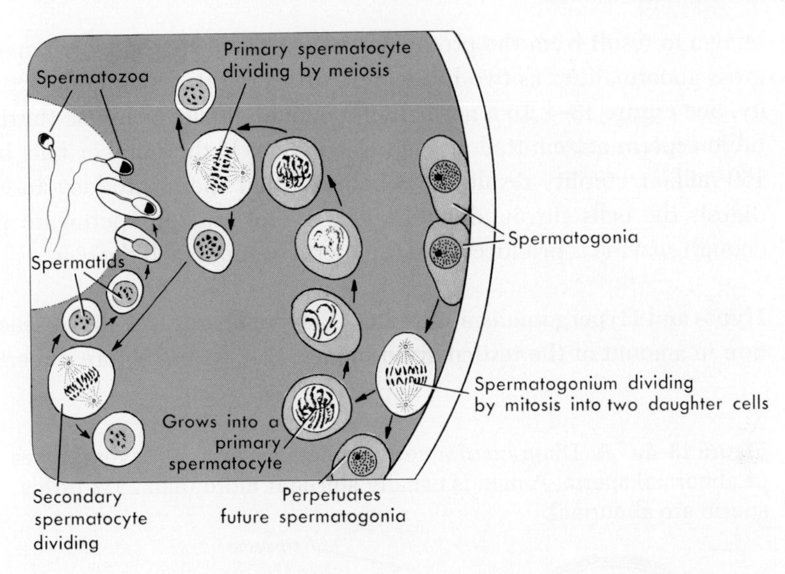

Figure 13–3. Spermatogenesis in a seminiferous tubule.

Transfer of Sperm. Usually the vascular sinuses in the penis are devoid of blood and the penis is flabby. On sexual stimulation, blood engorges the tissue of the penis, causing erection of that organ. If stimulation continues, it results in peristaltic waves that pass from the epididymis through the vas deferens and carry the sperm along with them. Alkaline fluids from the seminal vesicles, prostate, and Cowper's gland are poured over the sperm to form the semen or seminal fluid. The semen reaches the tip of the penis and is there forced out by muscular contraction of the penis—the process of ejaculation. The sperm, which were immobile in the acid conditions of the vasa deferentia, become mobile in the seminal fluid; if the penis is inserted in the vagina of a woman at the time of ejaculation, the sperm are carried up her reproductive tract, apparently partly by their own motion and partly by rhythmic contractions of the tract.

Causes of Sterility. Sterility is not uncommon in man. If the testes fail to descend into the scrotum (*cryptorchidism*) spermatogenesis is inhibited by the high temperature of the body. It has been suggested that sterility can be brought on by wearing tight-fitting shorts, which hold the testes close to the body, or by some occupations that involve exposure to high temperatures for long periods of time. Sterility can also be caused by a previous infection of the genital ducts by such diseases as typhus or mumps. Other cases of sterility are known to result from the production of abnormal sperm, which may have such gross abnormalities as two heads, but are more often simply defective in motility. See Figure 13–4. In a normally functioning male, from one-third to one-half billion sperm are emitted at each ejaculation. If the number falls below about 150 million, sterility results. It is believed that sperm produce an enzyme that digests the cells surrounding the egg. If not enough sperm are present, not enough enzyme is produced and the sperm cannot reach the egg.

Hypo- and Hypergonadism in Males. Hypogonadism is the absence or reduction in amount of the testicular hormones. If it occurs before puberty, it results

Figure 13–4. A: Diagram of a normal human sperm. B: Several types of abnormal sperm. A man is usually sterile if more than 25% of his sperm are abnormal.

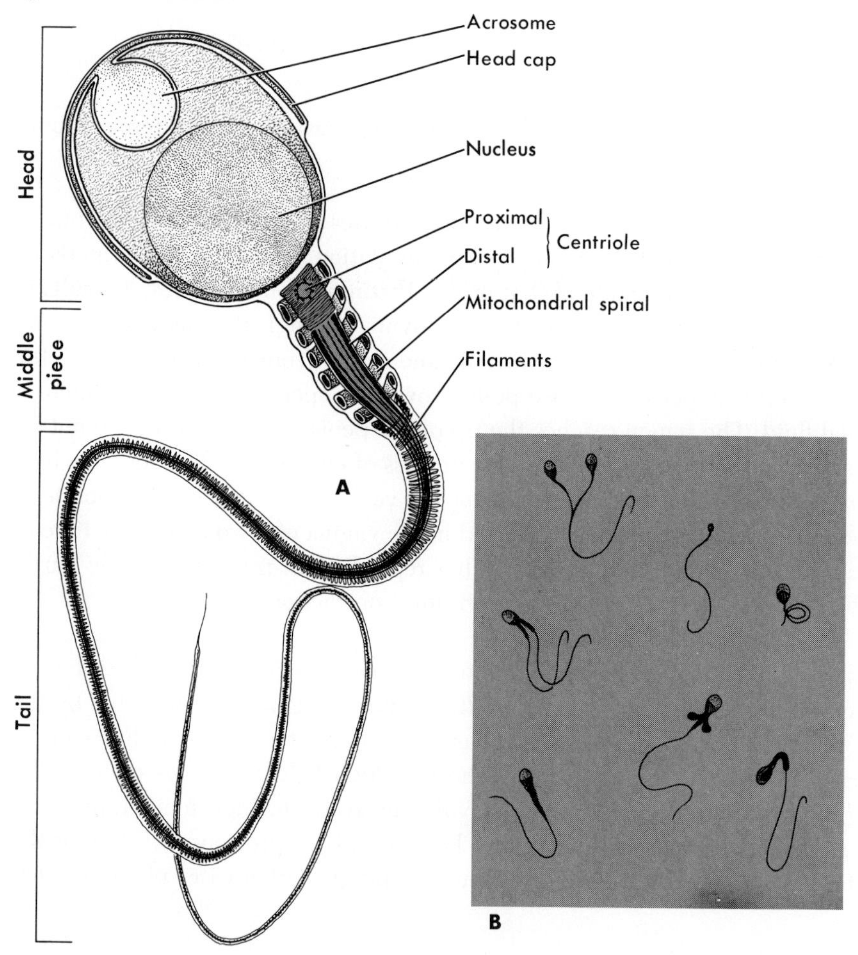

in excessive growth of the long bones, a feminine build, and reduced body hair. True *eunuchism,* in which the testes are completely destroyed either by disease or castration, is marked by a typical eunuch build. The legs are long in comparison to the rest of the body and fat is distributed in the female pattern, that is, in the mammary glands, on the belly below the umbilicus, around the hips, and in the axillary region. The beard does not develop, and the penis, prostate, seminal vesicles, and vasa deferentia remain infantile (see Figure 13–4). If eunuchism occurs after maturity is reached, the results are much less marked.

Hypergonadism is the overproduction of male hormones. In boys it results in the excessive development of the genitalia and secondary sexual characteristics; the body is usually short because of premature closure of the epiphyses. Hypergonadism developing after puberty is not generally accompanied by clinically recognizable symptoms.

Female Genital System

The female system not only produces ova and hormones and receives the sperm from the male but also must provide for the development of the embryo.

Structure of Female System. The reproductive structures of the female are generally grouped as external and internal organs. The external organs include two folds of tissue (*labia majora* and *labia minora*) that surround the genital opening; the *hymen,* a thin sheet of tissue that nearly covers the opening, which usually ruptures either before or during the first intercourse; a small, sensitive structure called the *clitoris;* and certain mucous and sebaceous glands.

The internal organs are the vagina, uterus, Fallopian tubes, and ovaries (see Figure 13–5).

The *vagina* is a muscular tube about twelve centimeters (five inches) long; it is lined with a membrane containing many mucous glands that secrete a lubricating fluid upon sexual stimulation. Circular and longitudinal layers of muscle surrounding the vagina undergo rhythmic contractions at the climax of sexual intercourse.

The *uterus* is made up of two parts—a broad triangular body and a lower, tubular neck, the *cervix,* which projects into the vagina. The wall of the uterus consists of three layers: an outer layer of visceral peritoneum that holds it in place; a middle layer of thick masses of smooth muscle; and an inner glandular coat called the *endometrium.*

The *Fallopian tubes,* or *oviducts,* are narrow tubes about ten centimeters (four inches) long that extend from the upper corners of the uterus to terminate about twenty-five millimeters (one inch) from the ovaries. The end of each of these tubes is expanded and divided into a number of finger-like projections, the

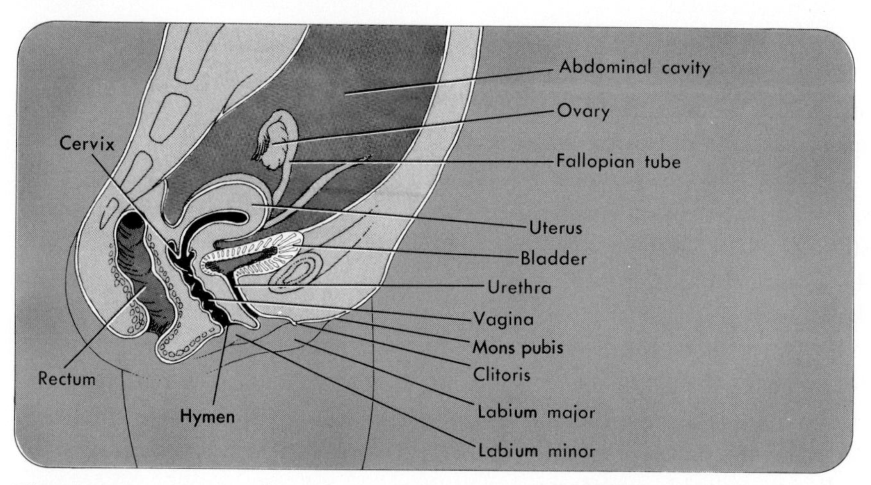

Figure 13–5. Section through the abdominal cavity of the human female showing the reproductive organs and their relationship to other abdominal organs.

fimbriae. The Fallopian tubes, like the uterus, are made up of three layers; the inner mucous layer is ciliated. Beating of the cilia carries a constant stream of fluid from the abdominal cavity into the tubes. When the ovary expels an egg, the egg is swept along with the fluid into one of the tubes and moved by ciliary action down into the uterus.

The *ovaries* are flattened, somewhat ovoid structures that lie deep in the pelvic cavity, one on either side of the uterus. Each ovary is about thirty-seven millimeters (one and a half inches) long, and both ovaries together weigh about as much as an American nickel. The ovaries do not hang free but are attached to the uterus by an ovarian ligament. Each ovary is made up of an outer fibrous coat and an inner *stroma*. The stroma is composed of an outer layer of germinal tissue and an inner mass of connective tissue, blood vessels, and some smooth muscle fibers.

In addition to producing eggs, the ovaries make two kinds of hormones, *estrogens* (mainly *estradiol*) and *progesterone*. The estrogens not only contribute to the control of the female sexual cycle but are also necessary for the development of the female sexual characteristics. If the ovaries are removed before puberty, development of the other sex organs is inhibited as is the development of the breasts, sexual desire, and such feminine traits as the deposit of fat around the hips. If the ovaries are removed after puberty, menstruation ceases, sexual desire diminishes, the body becomes more masculine in form, and hair may develop on the chest and face.

Ovulation. During embryonic development, about 400,000 *primary follicles* are formed from the germinal tissue of the ovaries. Each of these follicles consists of an oögonium surrounded by a layer of epithelial cells. Because a woman may ovulate over a span of about thirty-eight years (ages twelve to fifty), and because during this time she produces an average of one mature ovum each twenty-eight days, it can be calculated that only about 500 of these primary follicles ever reach maturity. The rest degenerate.

Under stimulation of the follicle-stimulating hormone (FSH) from the anterior pituitary at puberty, a primary follicle starts to develop into a mature *Graffian follicle.* The primary follicle is buried in the stroma, and the oögonium increases in size and accumulates fat globules and cytoplasm to become a primary oöcyte. Meanwhile a thin transparent structure, the *zona pellucida,* develops between the oöcyte and the surrounding follicular cells. These cells become granular and proliferate so that ultimately there is a very thick layer of *granulosa cells* surrounding the oöcyte. A cavity forms in this layer and becomes filled with *follicular fluid.* The follicle starts its development deep in the stroma but winds up near the surface of the ovary, where it forms a blister-like protuberance.

While the follicle has been developing, meiosis has been taking place. It begins about the time the zona pellucida forms and reaches the stage of interkinesis about the time the ovum erupts from the ovary. The outer follicular wall disintegrates and the follicle ruptures, releasing the ovum and the surrounding several layers of granulosa cells, together with the follicular fluid, into the body cavity. This process, known as *ovulation,* normally takes place about midway between the menstrual periods, but there is much variation. Maturation of the ovum continues in the oviduct, but the second polar body is not extruded until fertilization occurs. Figure 13–6 shows the development of a follicle in the ovary and ovulation.

Sterile women are sometimes treated with FSH (pregnancy pills) in order to stimulate follicular growth. Frequently, more than one follicle is stimulated, several ova are released at the same time, and a multiple pregnancy occurs. This also happens naturally, of course. Twins that develop from different ova are known as *fraternal twins,* whereas if they develop from a single ovum that divides early in development to give rise to two embryos, they are known as *identical twins.*

Uterine Changes. While the follicle is developing it secretes increasing amounts of a hormone of its own, estradiol, one of the estrogens. This brings about a thickening and vascularization of the endometrial layer of the uterus. Under stimulation from the luteinizing hormone (LH) of the anterior pituitary, the parts of the follicle remaining on the surface of the ovary after ovulation

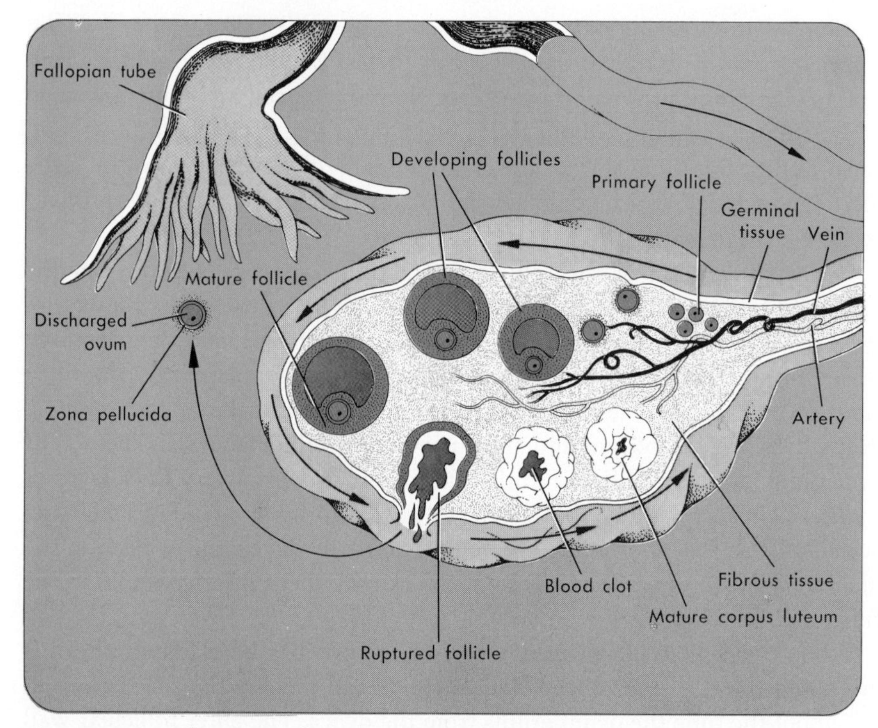

Figure 13–6. The ovary showing successive stages in the development of a follicle, ovulation, and the formation of the corpus luteum.

develop into a rounded, yellowish body, the *corpus luteum.* The corpus luteum produces both estradiol and the hormone progesterone. Progesterone increases the thickening of the endometrium, prepares it for implantation of the embryo, and also functions to maintain the vascular condition of the uterus during the early months of pregnancy. If fertilization does not occur, the corpus luteum begins to degenerate, with a corresponding reduction in the production of progesterone. The sudden drop in the level of progesterone in the blood apparently halts secretion by the glands of the endometrium. The outer layer degenerates and is sloughed off. This material, together with blood from the underlying layer, is discharged through the opening of the vagina and constitutes the *menstrual flow.*

Summary of the Menstrual Cycle. At the end of menstruation, the output of both progesterone and estradiol from the ovary is very low. Both of these hormones have an inhibitory effect on the anterior pituitary. Released from this inhibition, the pituitary increases its production of FSH, which stimulates the development of a follicle. The follicle produces estradiol in increasing amounts up to about the fourteenth day, when ovulation occurs. In addition to stimulating the growth of the endometrium, estradiol inhibits FSH production and stimulates LH production by the pituitary. As the level of FSH falls that of LH increases. Just before ovulation, there is a sudden rise in the level of LH and this is apparently what triggers the rupture of the follicle. LH then stimulates the development of the corpus luteum, which produces, in addition to estradiol, large amounts of progesterone. Besides its stimulatory effect on the endometrium, progesterone inhibits the production of LH by the pituitary. If implantation of an embryo in the uterus does not take place, the corpus luteum degenerates, thereby reducing the amount of progesterone circulating in the blood. Lacking progesterone to sustain it, the vascular wall of the uterus degenerates, with a corresponding loss of blood. Unlike other blood in the body, this blood has lost the ability to clot, and instead flows from the uterus through the vagina to the outside. With the degeneration of the corpus luteum and the reduction in the production of progesterone and estradiol, the pituitary begins to secrete FSH and the whole cycle begins over again. Figure 13–7 illustrates the chief features of this cycle.

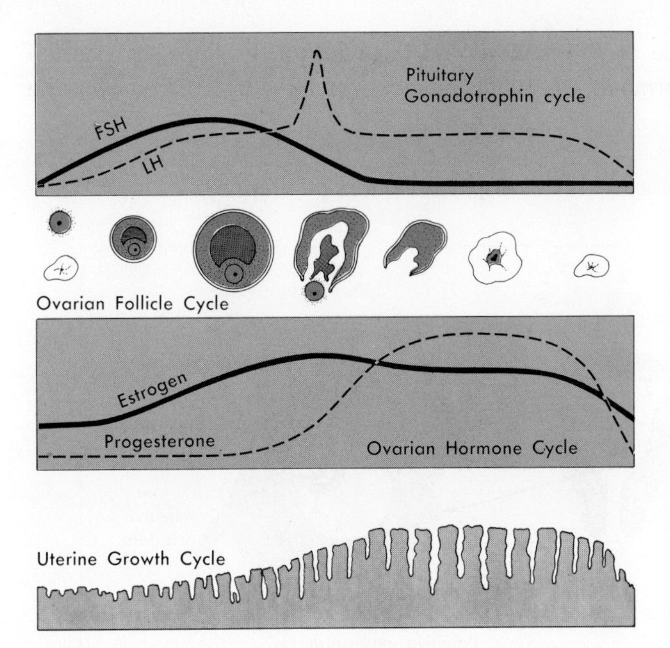

Figure 13–7. The human menstrual cycle. Increased production of FSH by the pituitary stimulates follicular growth in the ovary. Estrogens formed by the follicle inhibit FSH production and stimulate LH production. Increase in the amount of LH causes ovulation and the formation of the corpus luteum, which produces progesterone. Progesterone inhibits LH production. Both progesterone and estrogens stimulate growth of the uterine wall. When the corpus luteum degenerates, production of the ovarian hormones declines and part of the uterine wall is sloughed off (menstruation). The decrease in estrogen production by the ovary allows increased production of FSH by the pituitary to start the cycle again.

PREGNANCY AND BIRTH

Biologically speaking, a normal outcome of the extrusion of an ovum by the ovary is that it will be fertilized and a pregnancy will ensue, followed by the birth of a baby.

Pregnancy

Although it takes some seventy-two hours for an egg to make the journey from the ovary to the uterus, evidence seems to indicate that the egg can only be fertilized during the first twenty-four hours or so after ovulation. Because of this timing, fertilization normally takes place in a Fallopian tube. If intercourse has occurred during the appropriate interval, the sperm enter the cervix, cross the uterus, and travel up the Fallopian tube to fertilize the egg. The zygote then starts early cleavage and makes its way to the uterus.

Implantation. From the best of the scanty information available about very early human development, it seems that the trophoblast tissue from the developing embryo starts to grow into the endometrium about seven to ten days after ovulation. This is known as *implantation* (see Figure 13–8).

Hormonal Maintenance of Pregnancy. As the embryo develops, its chorion begins to secrete a hormone, *chorionic gonadotrophin,* which takes over the

Figure 13–8. Fertilization, early development, and implantation of the human embryo.

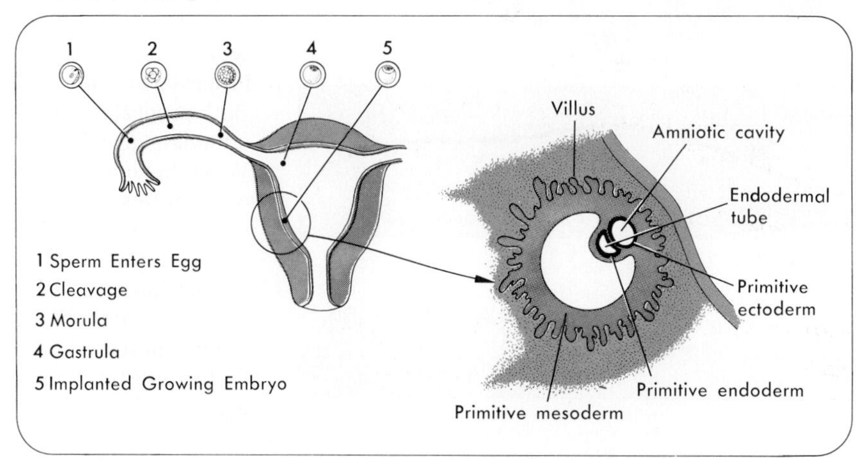

function of the now waning LH in sustaining the corpus luteum. Production of progesterone is thus maintained, degeneration of the endometrium is prevented, and the uterus remains in a condition to sustain the developing embryo. The corpus luteum also secretes estradiol, which suppresses the production of FSH and prevents the growth of other follicles. After the placenta is well formed, it produces increasing amounts of progesterone and, somewhere between the fourth and seventh month, takes over this function entirely from the corpus luteum, which then degenerates.

The Placenta. As pregnancy progresses, the growing embryo bulges out from the uterine wall. It eventually fills the entire uterine cavity and is attached to the placenta by the narrow *umbilical cord,* through which course the blood vessels that transport materials between the embryo and the placenta. The umbilical cord represents the body stalk, and the blood vessels develop in the allantois. These blood vessels do not fuse with the vascular spaces in the maternal part of the placenta, but remain separated from them by tissue known as the *placental barrier.* Small molecules are able to diffuse across this barrier. Larger molecules are generally excluded, but some, such as the antibodies that confer immunity on the mother, are able to pass. The newborn child is immune to the diseases to which his mother is immune, but this immunity lasts only a few months. Then the child must begin to manufacture his own antibodies.

The placental barrier usually functions to protect the child from harmful substances that may be present in the mother's bloodstream. Sometimes, though, it fails. The German measles virus is able to pass and may cause serious damage to the developing heart and nervous system of the embryo. Certain drugs are also able to pass the barrier. A tragic example of such a drug is the sedative thalidomide, which some years ago caused the birth of many babies with incompletely developed limbs.

Gestation Period. The period of development from fertilization to birth is known as the *gestation period.* In humans this period is generally about 275 days. Birth may be expected around 280 days after the end of the last menstrual period. During the first two months of pregnancy the developing human is called an *embryo* but thereafter it is known as a *fetus.* Although there are variations, the fetus at the end of the gestation period is usually head downward in the uterus in a position known as *universal flexion,* with the head and limbs folded in close to the trunk. This is because the fetus develops strong flexing muscles before it develops the antagonistic extensor muscles, and because intermittent contractions of the uterus during gestation tend to keep the fetus confined in as small a space as possible.

Birth

In normal childbirth, the fetus together with its membranes is moved down the *birth canal* by muscular contractions of the uterus. The birth canal is made up of the cervix of the uterus, the true pelvis, and the soft tissues that line the canal. The true pelvis, that part of the pelvic girdle through which the fetus must pass, has an inlet and a somewhat larger outlet (see Figure 13–9). Because the inlet is the narrowest part of the canal, if the head of the fetus can pass through the inlet it can traverse the rest of the canal and normal birth can take place.

Actual childbirth normally begins with rhythmic contractions of the muscular wall of the uterus. This indicates the onset of *labor,* which is divided into three stages.

The first stage of labor is called the *stage of dilation.* When labor begins, the outlet of the true pelvis is almost completely closed by soft tissue, the *pelvic floor.* The increasingly severe contractions of the uterine wall exert a considerable pressure on the amniotic sac. Because the pelvic floor is the weakest part of the fetal chamber, the amnion tends to push down through the floor. This results in a dilation of the cervix. After the cervix if fully dilated, the pressure is held only by the wall of the amnion. At this point the amnion usually ruptures, releasing the amniotic fluid and bringing the first stage of labor to a close (see Figure 13–10).

The second stage of labor is *expulsion.* The movement of the fetus from the uterus through the birth canal to the outside is brought about by stronger

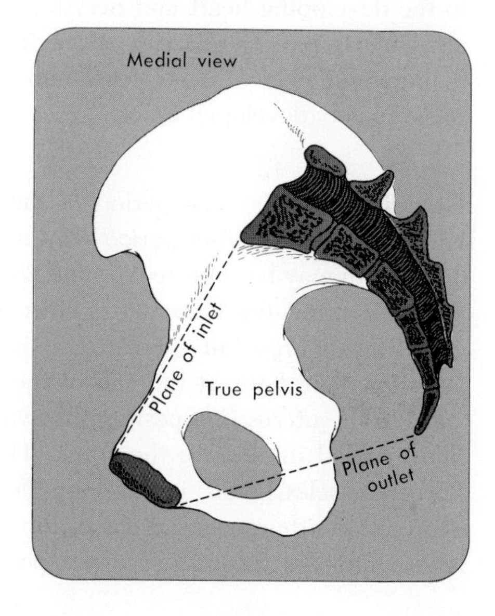

Figure 13–9. One half of pelvic girdle showing inlet and outlet.

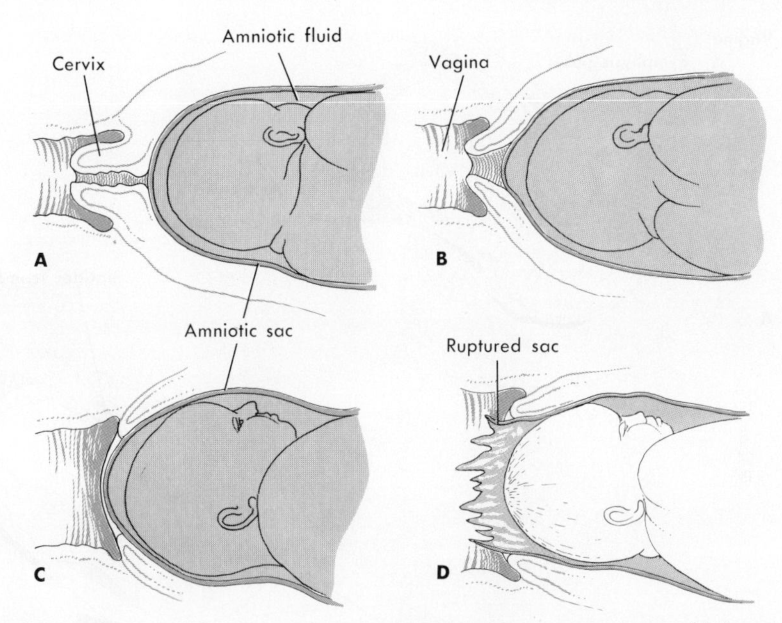

Figure 13–10. First stage of labor.

contractions of the uterus at more frequent intervals (see Figure 13–11). Unless the mother is under anesthesia or heavy sedation, voluntary contractions of her abdominal muscles aid in the expulsion of the fetus. The second stage ends when the fetus is outside the body of the mother.

During the third stage of labor the contraction of the uterine muscles results in the separation of the placenta from the uterine wall and its expulsion as the afterbirth. Constriction of the blood vessels in the walls of the contracting uterus prevent serious hemorrhage.

Changes at Birth

As long as the fetus is attached to the mother by way of the placenta and umbilical cord, it is dependent on her for oxygen. When birth occurs, the attachment is broken and the infant has to depend on its own respiratory system. The shift from one system to the other must take place promptly, and is made possible by some drastic changes in the infant's circulatory system.

The circulatory system of the fetus differs in three major ways from that of the child. It is connected with the placenta by way of the umbilical blood vessels; the umbilical vein flows into the inferior vena cava, which carries blood to the right atrium. An artery, the *ductus arteriosus*, leads from the pulmonary artery into the dorsal aorta so that blood from the right ventricle can bypass the

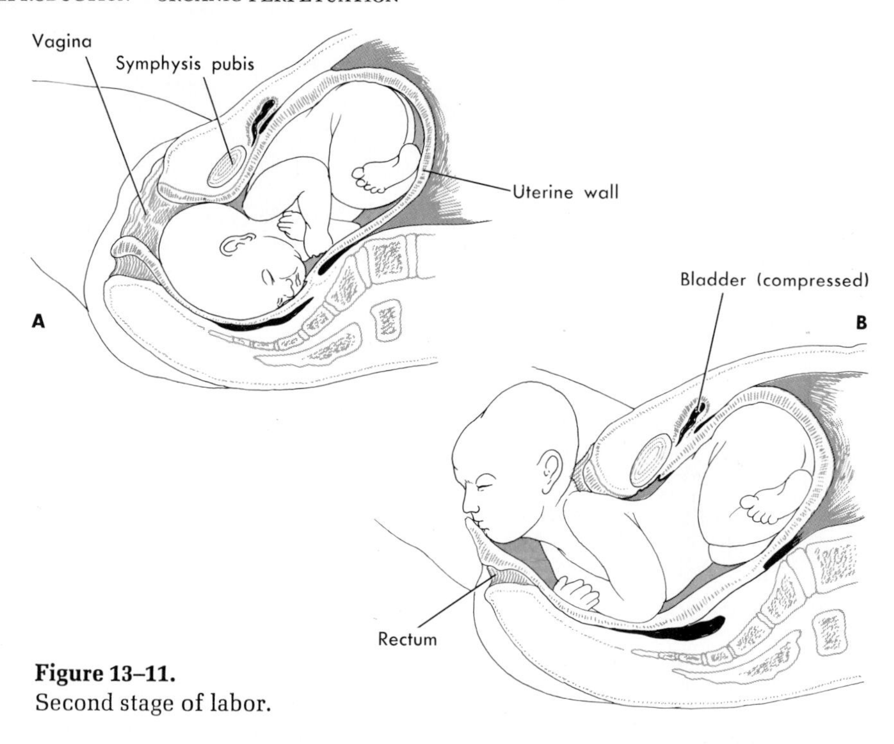

Figure 13–11.
Second stage of labor.

lungs and enter directly into the systemic circulation. Finally, there is an opening between the right and left atria, the *foramen ovale*. This opening is covered by a flap on the left side that functions as a valve to let blood flow from the right to the left atrium but not in the opposite direction. These characters of the fetal circulation are shown in Figures 13–12 and 13–13.

While the fetus is in the uterus the lungs are collapsed and are not functioning. Venous blood from the body and the placenta enters the right atrium. From there it passes either directly through the foramen ovale to the left side of the heart or to the right ventricle. From the right ventricle the venous blood enters the pulmonary artery and most of it is shunted by way of the ductus arteriosus into the dorsal aorta and systemic circulation. At birth the placental circulation is halted, the level of oxygen in the blood drops, the respiratory center in the brain is stimulated, and the baby takes his first breath. The lungs fill with air and the chest expands. This permits abundant blood to flow through the pulmonary circulation and restricts the flow through the ductus arteriosus. Blood now passes from the right ventricle to the pulmonary arteries to the lungs to the pulmonary veins to the left atrium. This additional flow of blood into the left atrium raises the pressure there and pushes the flap covering the foramen ovale against the interatrial wall, closing the hole so that a full double circulation is established. In a few months the ductus arteriosus withers, and generally during the first year of life the flap in the left atrium fuses to the interatrial wall so that the separation of the two atria becomes complete.

Sometimes either the ductus arteriosus or the foramen ovale remains open. This means that unoxygenated blood is entering the systemic circulation. The result is a "blue baby." The skin appears bluish and, because insufficient oxygen is reaching the tissues, the child is incapable of normal activity. Formerly such children usually died young but now, with modern surgical techniques, these defects can be corrected.

AGING

Birth is followed by a period of continued growth, maturity, and senescence. In spite of much research, we still are not sure of what causes aging. One suggestion is that aging results from the differential ability of cells to divide. Some cell types, such as muscle cells and neurons, lose this ability

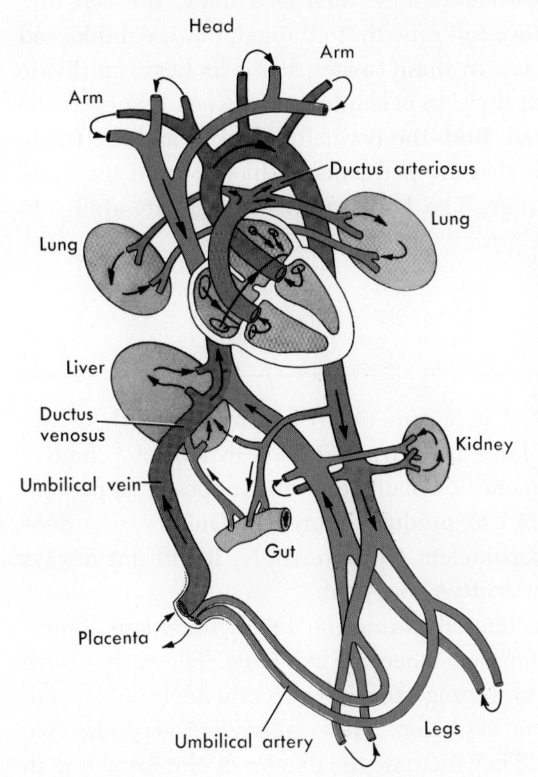

Figure 13–12. Fetal circulation. The umbilical arteries carry blood to the placenta where waste products are discharged and oxygen and nutrients enter the bloodstream. The umbilical vein returns blood from the placenta. The ductus arteriosus allows blood to pass from the pulmonary artery to the dorsal aorta.

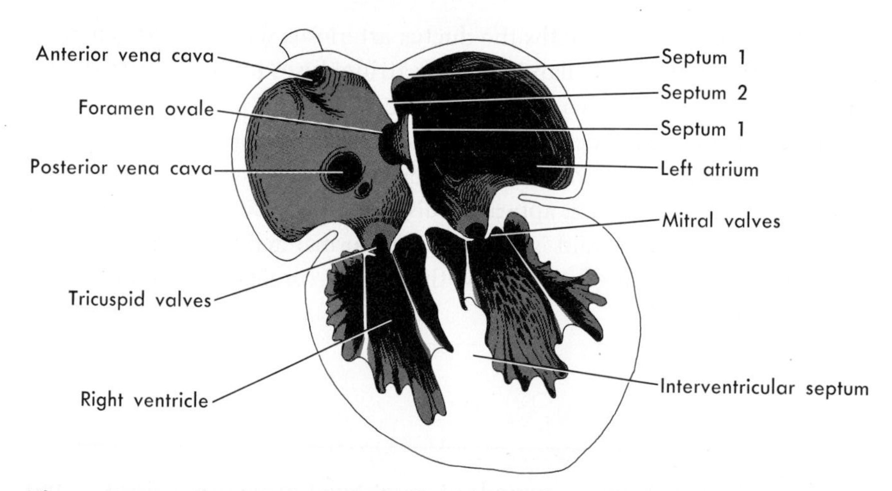

Figure 13–13. Section through the fetal heart showing the foramen ovale between the atria and the septa that later grow together to close it.

during the period of growth. In other tissues, such as kidney, the cells are normally blocked from dividing once full growth it attained, but are unblocked when part of the tissue is destroyed. In these tissues, the cells begin to divide again to replace the lost cells. Cell division is continuous in such tissues as the epidermis and the intestinal mucosa. Both the normally blocked and the freely dividing tissues contain some cells that are permanently blocked and the number of these cells increases with age. The body gradually loses its ability to repair itself and replace lost tissue.

BIRTH CONTROL

Man has long been interested in means of preventing the birth of unwanted children. Many methods have been tried. The herb lore of primitive societies frequently includes the names of plants that are supposed to prevent conception and others that are said to produce abortions. Quinine was once used in the rural South as an abortifacient. Unfortunately, it did not always work and when it failed the baby was often born deaf.

Interest in contraception accelerated during the 1960's when oral contraceptives containing synthetic sex hormones became available. The earlier birth control pills contained estrogen and progestogen (a synthetic form of progesterone) and acted by inhibiting ovulation. These pills were very effective but could have serious side effects. They increase the danger of clot formation in the blood vessels and may elevate the blood pressure with a concomitant risk of strokes. A newer pill, which contains only progestogen, apparently acts by preventing the sperm from traveling up the oviduct to reach the ovum. It maintains the mucus at the opening into the uterus in a state that inhibits sperm migration. This pill is slightly less effective than the earlier pills in preventing conception, and its complete safety has yet to be established. The same thing can be said of other methods of introducing sex hormones into the body, including injections and the insertion of hormone-coated devices into the uterus (IUD's). One reason that it takes so long to test the safety of birth control methods that depend on synthetic hormones is that some of these substances have been implicated as causes of cancer, but in humans the cancer may not develop for twenty years or more.

A successful male oral contraceptive is still not available. Male sterilization by tying off or cutting the vasa deferentia (*vasoligation* or *vasectomy*) is relatively simple, but is usually irreversible. However, the forced retention of semen in the body may also have serious side effects. It may set off an autoimmune response, the testes may degenerate, and cysts and tumors may form. A completely safe, effective, reversible means of artificial birth control is yet to be developed.

SUGGESTED READINGS

Francis, C. C.: *Introduction to Human Anatomy,* C. V. Mosby, St. Louis, Mo., 1973.

Frye, B. E.: *Hormonal Control in Vertebrates,* Macmillan Publishing Co., Inc., New York, 1967.

Grollman, S.: *The Human Body,* 3rd ed., Macmillan Publishing Co., Inc., New York, 1974.

Hardin, G., ed.: *Population, Evolution, Birth Control,* W. H. Freeman and Co., San Francisco, 1964.

Kimber, D. C., et al.: *Anatomy and Physiology,* 16th ed., Macmillan Publishing Co., Inc., New York, 1972.

Inheritance— the Mechanisms of Variation

We are truly heirs of all the ages;
but as honest men it behooves us to
learn the extent of our inheritance . . .
JOHN TYNDALL, *Matter and Force*

CHAPTER **14**

Some Elementary Genetics

"You have your father's nose and your mother's coloring." Everyone has heard comments like this, perhaps made them. They show that most people already know several basic facts of inheritance. It is common knowledge that physical characters are transmitted from parent to child and also that the child differs from either of his parents as well as from his brothers and sisters. Everyone knows that like begets like; dogs beget dogs, not cats or canaries. Everyone also knows that except for identical twins, all the children of the same parents are not identical and that the pups in a litter differ from one another.

From such common knowledge has grown the science of genetics, which seeks a precise understanding of the transmission of biological properties from generation to generation.

THE WORK OF MENDEL

Man must have been aware that offspring both resemble and differ from their parents ever since he became a thinking creature. He used his knowledge in trial-and-error breeding experiments to improve his domestic animals and plants. Genetics as a science, though, had a very clear-cut beginning. It started in the midnineteenth century with the work of an Austrian monk, Gregor Johann Mendel.

Mendel, an avid gardener, became interested in the transmission of certain characteristics of garden peas from parent to offspring. The pea plant is normally self-fertilizing; it has a flower so constructed that pollen grains (which act as sperm) from other flowers cannot reach the female part of the flower and initiate fertilization. The offspring receive all their characters from a single parent. But a careful experimenter can remove the stamens, which produce the pollen, from the flower, artificially introduce pollen from another plant, and so make controlled crosses between different plants.

Early Experiments

Mendel started with true-breeding plants, that is, plants from stocks that, when allowed to fertilize themselves, always produce offspring that were just like the parent. These are called *pure lines*. In some of his earlier experiments, Mendel crossed plants with alternative characters and then let the offspring of such plants fertilize themselves to produce the next generation. Among the crosses he made were plants having smooth seeds with ones having wrinkled seeds, plants having green seeds with ones having yellow seeds, plants having short stems with ones having long stem, and plants having flowers at the tips of the stems with those having flowers at the twig axils (the angle between the twig and the stem).

Mendel's Results. Table 14–1 gives the results of some of Mendel's crosses. All of these crosses had similar results. For one thing, the offspring of each original cross all looked alike and were like just one of the parents. The expression of one trait prevented the other from appearing. When these offspring were allowed to fertilize themselves, some of their progeny showed the trait of one of the original parents and some showed the trait of the other. But these traits were not present in equal numbers. The trait that appeared in the first generation was about three times as common in the second generation as was the trait that did not appear in the first.

Mendel's Interpretation. Mendel was astute enough to realize that the key to the situation lay in the fact that the traits that were not expressed in the first generation were not lost, but simply failed to appear. With this in mind he was

Table 14–1. Results Reported by Mendel for Some of His Crosses

Appearance of Parents	Appearance of 1st Generation	Appearance of 2nd Generation		Ratio
Smooth seeds × wrinkled	All smooth	5,474 smooth	1,850 wrinkled	2.96:1
Yellow seeds × green	All yellow	6,022 yellow	2,001 green	3.01:1
Long stems × short	All long	787 long	277 short	2.84:1
Axial flowers × terminal	All axial	651 axial	207 terminal	3.14:1

able to visualise a rather simple mechanism to account for the 3:1 ratio in the second generation. Assume that for a given pair of alternative characters each plant of the first generation receives one determinant from each parent. If we let X represent the determinant that is expressed and x represent the determinant that is not expressed, each of these plants would be designated Xx. Now suppose that each of the gametes produced by these plants contains only one of these determinants, but that both kinds are equally represented in the gametes produced, that is, half the gametes have X and half of them have x. If it is a matter of chance which pollen grain contributes the sperm to which egg, the resultant combinations would be as shown in Figure 14–1.

In summary, ¼ of the offspring would have XX, ²⁄₄ of them would have Xx, and ¼ would have xx. This gives a ratio of $1XX$:$2Xx$:$1xx$. Because in the first generation the character represented by X is expressed but that represented by x is not, one would expect that in the second generation the Xx individuals would also show the character represented by X. Thus, the second generation would show this trait in a 3:1 ratio; i.e., 3 ($1XX$ plus $2Xx$) to 1 ($1xx$).

Now apply this idea to one of Mendel's basic crosses. The cross of smooth versus wrinkled seeds would trace through as shown in Figure 14–2. Notice two things. The hereditary determinants exist in pairs in the parents as well as in subsequent generations, and the actual results shown in Table 14–1 do not fit the 3:1 ratio precisely. Such deviations from predicted ratios are no more than would be expected if it is chance that determines which sperm fertilizes which egg, as it seems to be.

Confirmation of Mendel's Results. Mendel made other crosses to confirm the validity of this interpretation and since his day similar crosses have been made with many other kinds of organisms, from bacteria to mammals. A great many pairs of contrasting characters are known that are inherited just as seed shape is inherited in peas. For example, coat color in sheep is transmitted in this way. If a white sheep from a pure line is mated with a black sheep, it has only white offspring. In brother-sister matings between those offspring, both white and black sheep are produced in the ratio of three white to one black. Figure 14–3 shows this cross.

Egg Cells	Sperm From Pollen Grain: X	x
X	X X	X x
x	x X	x x

Figure 14–1. The theoretical basis of Mendel's results. The offspring produced by each of the four possible combinations of egg and sperm are shown in the boxes.

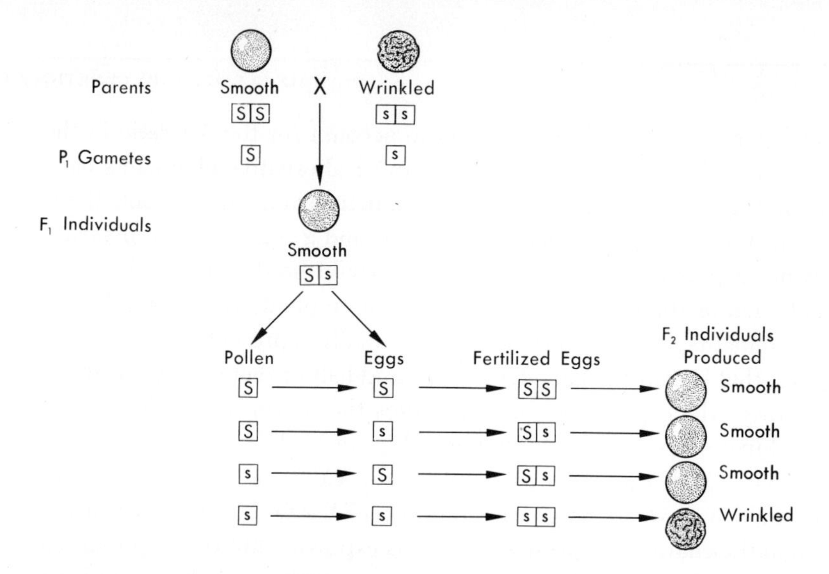

Figure 14–2. Inheritance of seed shape in garden peas. When a plant having smooth seeds is crossed with one having wrinkled seeds, the first generation offspring (F_1) all have smooth seeds. Three out of four individuals of the second generation (F_2) have smooth seeds and one out of four have wrinkled seeds.

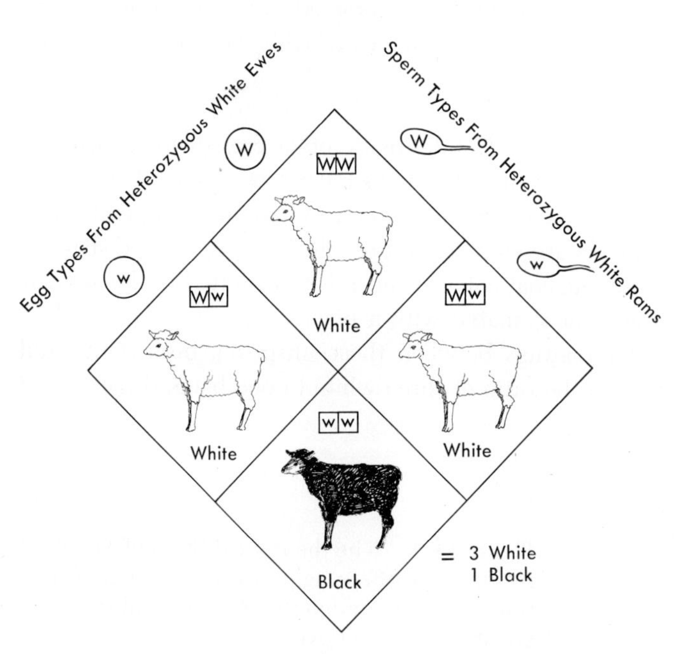

Figure 14–3. Inheritance of coat color in sheep.

Some Necessary Terminology

At this point you should become familiar with the precise meaning of some words used in genetics. The study of the mechanism of transmission of hereditary characters is a sort of problem-solving science; if you understand the concepts underlying the terminology, you will be able to solve your problems and will not have to memorize the results of each particular cross.

In genetic crosses, *parents* means the individuals involved in the original cross. They are often referred to as the parental, or P_1 generation. Their offspring make up the *first filial generation,* abbreviated F_1. The offspring of the F_1 are the *second filial generation* or F_2.

The determinants of heredity characters are called *genes.* Geneticists have a much better understanding today of just what a gene is than was possible even a few years ago. We take up the nature of genes in Chapter 16, but for the present, just think of a gene as the unit of heredity that controls the expression of a particular character.

The members of a pair of such units that control alternative characters are called *alleles.* For example, the *W* and *w* that control coat color in sheep are alleles. An individual in which both members of a pair of alleles are the same, that is, *AA* or *yy*, is said to be *homozygous* for that trait. If the alleles are unlike, *Aa* or *Yy*, the individual is *heterozygous.*

A gene that prevents the expression of its allele is said to be *dominant,* whereas the allele whose expression is prevented is said to be *recessive;* in seed color in peas, yellow is dominant to green, and green is recessive to yellow.

In genetic terminology, the offspring of a cross between individuals that differ in at least one heritable trait are called *hybrids.* If the cross involves only one pair of alternative characters (tall versus short in peas for example), it is called a *monohybrid cross.* A cross involving two sets of traits (tall–short stems and yellow–green color in peas) is a *dihybrid cross.* If the cross involves three sets, it is a *trihybrid,* and so on.

The *phenotype* of an individual refers to the characters that are expressed in it. A pea plant can have a long stem and smooth, green seeds. On the other hand, *genotype* refers to the genetic constitution of the individual, the particular combination of genes that it possesses. Individuals that have the same phenotype may have different genotypes. The genotype of a pea plant with smooth yellow seeds can be either *SSYY* or *SsYY* or *SSYy* or *SsYy*.

In the crosses thus far considered a capital letter has been used to stand for the dominant allele and a lower case of the same letter to stand for the recessive one; the use of this convention is continued in the following discussion of genetics.

Back Crosses

If F_1 individuals are self-fertilized or inbred, their offspring show a 3:1 phenotypic ratio. What will be the results if these individuals are backcrossed to their parents? Figure 14–4 shows that the F_1 sheep, if crossed with its recessive (black) parent, has white and black offspring in a 1:1 ratio rather than 3:1. On the other hand, if it is backcrossed with the white (dominant) parent (see Figure 14–5) it has all white offspring. The backcross with the white parent does not show whether the F_1 is *WW* or *Ww*, but backcrossing with the black parent demonstrates that the F_1 individual must be *Ww*, because it produces gametes carrying the alternative genes in equal numbers. For this reason, a cross of an F_1 with its recessive parent or with some other recessive individual is known as a *test cross*. Notice that some backcrosses may be test crosses and some test crosses may be backcrosses, but the terms are not identical in meaning.

The Dihybrid Cross

Mendel did not stop with doing monohybrid crosses and working out the backcross; he also tried some dihybrid crosses. Again his results have been confirmed many times. A pure line tall sweet pea plant with red flowers was crossed with a dwarf plant with white flowers. (Tall is dominant to dwarf and red is

Figure 14–4. Backcross of a heterozygous F_1 white sheep to its black parent. White and black offspring are produced in a 1:1 ratio.

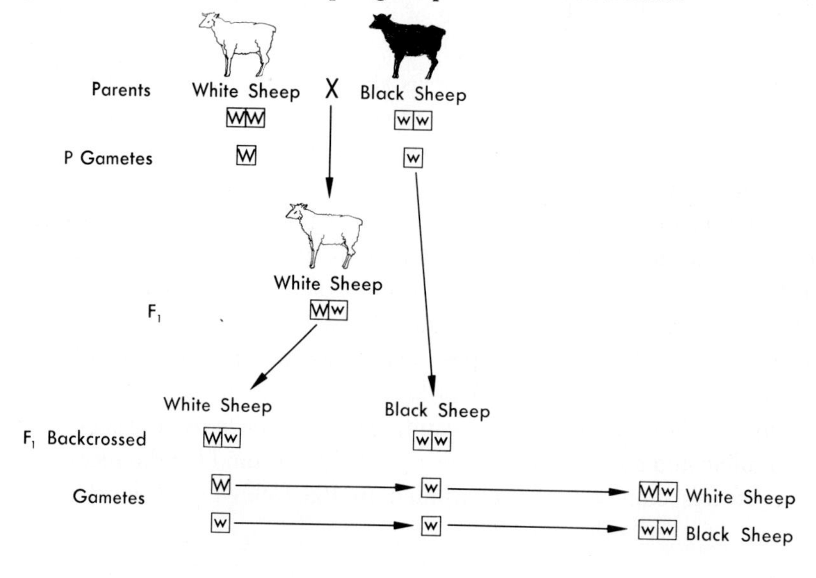

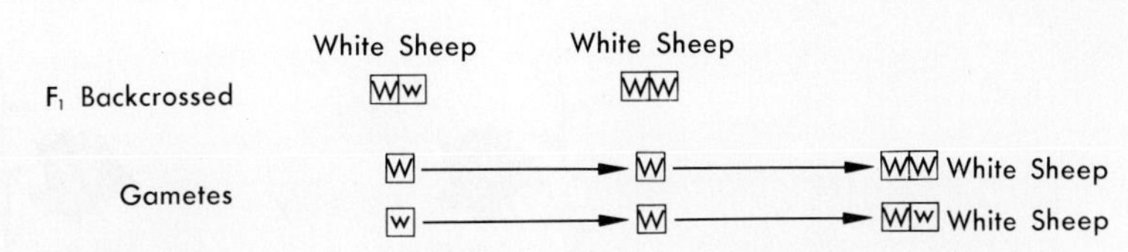

Figure 14–5. If the heterozygous F_1 white sheep is backcrossed to its white parent, all the offspring are white.

dominant to white.) In the monohybrid cross one member of the pair of alleles is passed to each gamete, and in the dihybrid cross one member of each pair of alleles is present in each gamete. This cross is pictured in Figure 14–6.

What kind of offspring are to be expected in the F_2 generation? The F_1 will produce four kinds of gametes in equal numbers: *TR*, *Tr*, *tR*, and *tr*.

An easy way to calculate the possible results of fertilization is to make a checkerboard and list the gametes from one parent along one side and those from the other parent along an adjacent side. One can then calculate the possible combinations by filling in the squares of the checkerboard. This is shown in Figure 14–7.

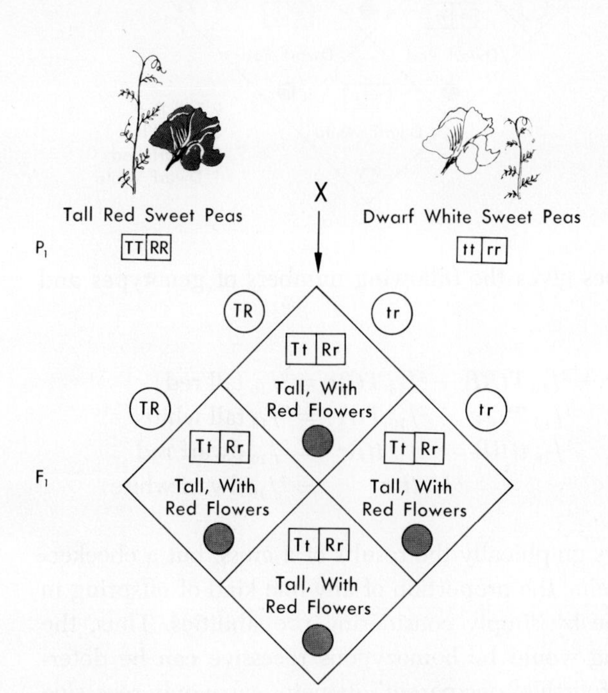

Figure 14–6. The F_1 generation of a dihybrid cross.

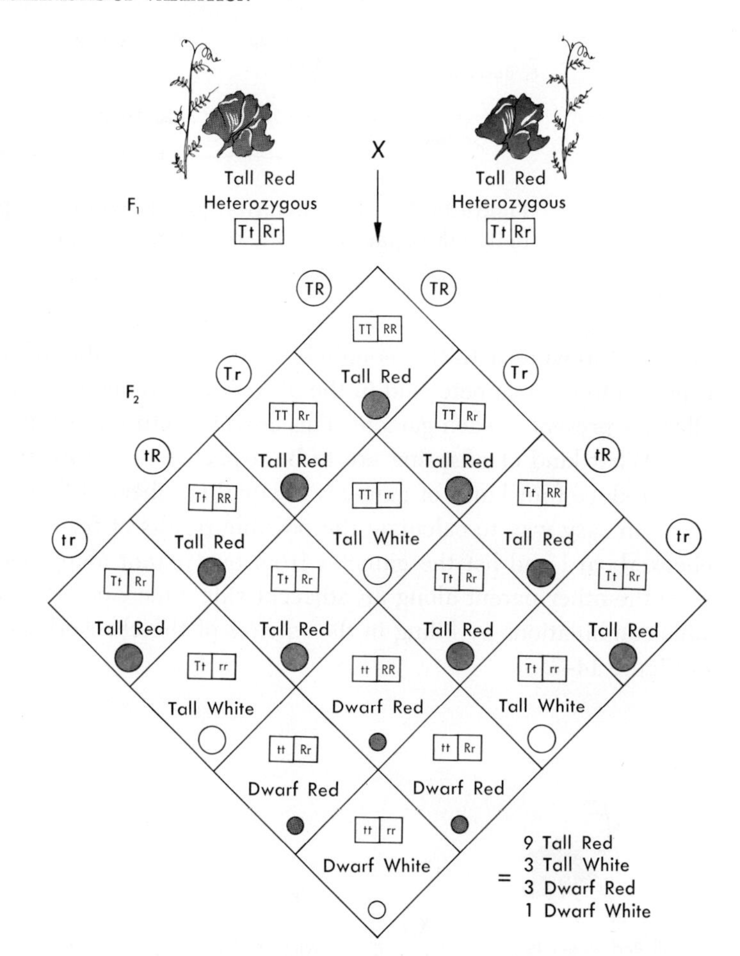

Figure 14–7. The F_2 generation of the cross shown in Figure 14–6.

Collecting like genotypes gives the following numbers of genotypes and phenotypes:

$$\begin{aligned}
{}^1/_{16}\,TTRR + {}^2/_{16}\,TTRr + {}^2/_{16}\,TtRR + {}^4/_{16}\,TtRr &= {}^9/_{16}\text{ tall red}\\
{}^1/_{16}\,TTrr + {}^2/_{16}\,Ttrr &= {}^3/_{16}\text{ tall white}\\
{}^1/_{16}\,ttRR + {}^2/_{16}\,ttRr &= {}^3/_{16}\text{ dwarf red}\\
{}^1/_{16}\,ttrr &= {}^1/_{16}\text{ dwarf white}
\end{aligned}$$

The checkerboard shows graphically the results of a cross, but a checkerboard is not needed to determine the proportion of any one kind of offspring in a given cross. It can be done by simply considering probabilities. Thus, the chance that any one offspring would be homozygous recessive can be determined by figuring that if one fourth of one parent's gametes carry only recessive genes and one fourth of the other parent's gametes carry only recessive genes, then the probability that any particular offspring would have only recessive genes is

$$\tfrac{1}{4} \times \tfrac{1}{4} = \tfrac{1}{16}\ ttrr$$

In other words, one out of every sixteen offspring should be homozygous recessive.

This is simply an example of a basic law of probability, which says that the frequency with which two or more independent events occur together is equal to the product of the frequencies with which each event occurs independently. Considering each pair of alleles separately, the probability that a *TtRr* individual crossed with another *TtRr* individual will produce tall offspring is three fourths ($Tt \times Tt$) and the probability that red offspring will be produced is also three fourths ($Rr \times Rr$). Then the probability that tall red offspring will be produced is

$$\tfrac{3}{4} \times \tfrac{3}{4} = \tfrac{9}{16} \text{ tall red offspring}$$

The Trihybrid Cross

One of the most elaborate of Mendel's experiments was a trihybrid cross involving three pairs of alternative characters: yellow and green seeds, smooth and wrinkled seeds, and colored and colorless seed coats. The results are given in outline form as follows: the phenotypic ratio in the F_2 here is 27:9:9:9:3:3:3:1.

Outline of Trihybrid Cross

P_1 phenotypes	yellow-smooth-colored × green-wrinkled-colorless
P_1 genotypes	*YY SS CC* *yy ss cc*
P_1 gametes	*YSC* *ysc*
F_1 genotype	*Yy Ss Cc*
F_1 phenotype	yellow-smooth-colored
F_1 male gametes	*YSC YSc YsC ySC Ysc ySc ysC ysc*
F_1 female gametes	*YSC YSc YsC ySC Ysc ySc ysC ysc*

The F_2 generation will consist of:

27 distinct genotypes	8 distinct phenotypes
1 *YYSSCC*	27 yellow-smooth-colored
2 *YYSSCc*	

2 *YYSsCC*
2 *YySSCC*
4 *YYSsCc*
4 *YySSCc*
4 *YySsCC*
8 *YySsCc*

1 *YYSScc* — 9 yellow-smooth-colorless
2 *YYSscc*
2 *YySScc*
4 *YySscc*

1 *YYssCC* — 9 yellow-wrinkled-colored
2*YYssCc*
2 *YyssCC*
4 *YyssCc*

1 *yySSCC* — 9 green-smooth-colored
2 *yySSCc*
2 *yySsCC*
4 *yySsCc*

1 *YYsscc* — 3 yellow-wrinkled-colorless
2 *Yysscc*

1 *yySScc* — 3 green-smooth-colorless
2 *yySscc*

1 *yyssCC* — 3 green-wrinkled-colored
2 *yyssCc*

1 *yysscc* — 1 green-wrinkled-colorless

MENDEL'S LAWS

From his various experiments Mendel derived two generalizations, now known as Mendel's laws. The first of these is that each individual of a sexually reproducing population has determinants in pairs, that the members of a pair separate (segregate) when gametes are produced, and that pairs are formed again at fertilization. This is known today as the law of segregation and recombination.

Mendel's second law is that a pair of alleles controlling the development of a trait are sorted out and inherited independently of any other pair of alleles (law of independent assortment). In the dihybrid cross diagrammed previously, it is simply chance that determines whether or not the gene for long stem winds up in the same individual as that for red flower; taking the characters by themselves, the F_2 contains 3/4 tall plants and 1/4 dwarf, and also 3/4 red flowers and 1/4 white. In other words, insofar as these two properties are concerned, there is an independent assortment.

In the next chapter we discuss the many exceptions to this law. A great deal of luck is involved in most successful scientific research and Mendel was especially lucky in that all the characters he investigated in peas are inherited independently. Otherwise his results would have been much less clear-cut.

MODIFICATIONS OF MENDELIAN INHERITANCE

Mendel's work was simply ignored by other biologists at the time of his investigation, and it was not until after the turn of the century that it was rediscovered and its significance realized. Since then a great many crosses have been made and Mendelian inheritance has been amply confirmed. But many crosses do not yield the expected ratios. In the rest of this chapter we describe some of these perplexing results and show how they can be explained.

Incomplete Dominance

For every character that Mendel studied, the effect of one gene completely dominated that of its allele. Frequently, though, the heterozygous F_1, instead of having the phenotype of one of the parents, is somewhat in between.

If a homozygous red-flowered snapdragon (*RR*) is crossed with a homozygous white-flowered one (*rr*), the heterozygous F_1 (*Rr*) is neither red nor white but is instead pink. If the F_1 is self-fertilized, it gives rise to offspring in the ratio of 1 red (*RR*): 2 pink (*Rr*): 1 white (*rr*). Similar situations occur in animals. In shorthorn cattle, a cross between a white cow and a red bull gives an offspring of an intermediate type known as *roan*. This cross is shown in Figure 14–8. Intermediate forms result when a gene fails to show complete dominance.

Lethal Genes

A number of genes bring about the death of the individual in which they occur in the homozygous state. (Why not in the heterozygous individual?) A

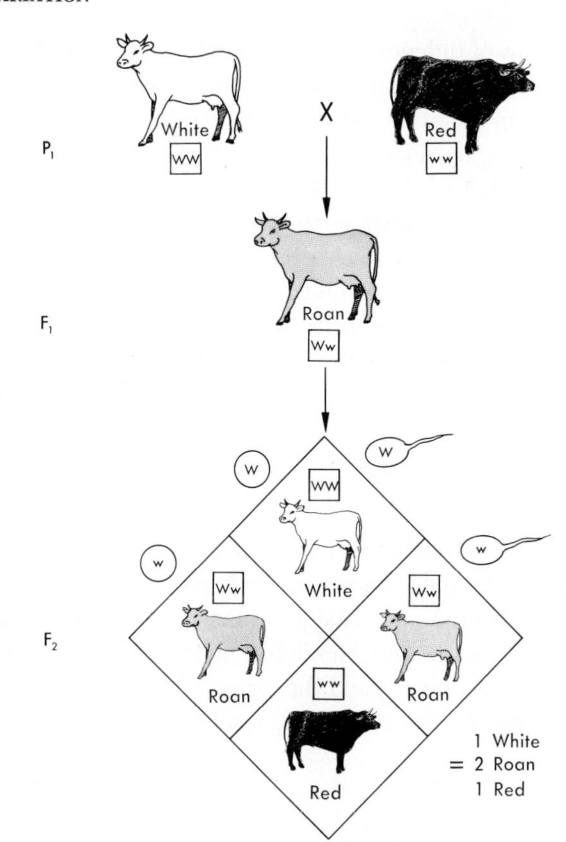

Figure 14–8. Incomplete dominance of coat color in cattle.

gene is known in corn that, if present in the homozygous state, prevents the development of chlorophyll. If a seed homozygous for this condition is planted, it can germinate and actually grow into a seedling plant, but as soon as all the stored food material in the corn grain is used up the seedling dies, for it lacks the ability to manufacture its own food.

Some mice have a yellow coat color instead of the normal gray. If two yellow-coated mice are mated, they have offspring in the ratio of two yellow to one gray. It has now been shown that this just represents the 2 *Yy*:1 *yy* part of a 1 *YY*:2 *Yy*:1 *yy* ratio, and that the homozygous *YY* embryo dies because it does not become properly attached to the uterine lining of the mother.

Two things should be noticed about this cross. One is that the gene *Y* has an effect on more than one character. It determines both coat color and the ability of the embryo to attach to the uterus. Such a gene is said to be *pleiotropic.* (Probably most, if not all, genes are pleiotropic to some extent.) Also, a single pleiotropic gene may act both as a dominant and as a recessive. Thus *Y* is dominant in respect to coat color but is recessive in determining attachment to the uterus.

Gene Collaboration

Comb shape in chickens is determined by two pairs of genes. With one pair, if the dominant allele P is present, it causes the development of a type known as *pea comb*. There is another gene, R, which causes the development of a rose comb if present as a dominant. It neither dominant P nor dominant R is present, the comb is the normal single type. Should both dominant P and dominant R occur in the same individual, a comb known as *walnut* develops. This cross is shown in Figure 14–9.

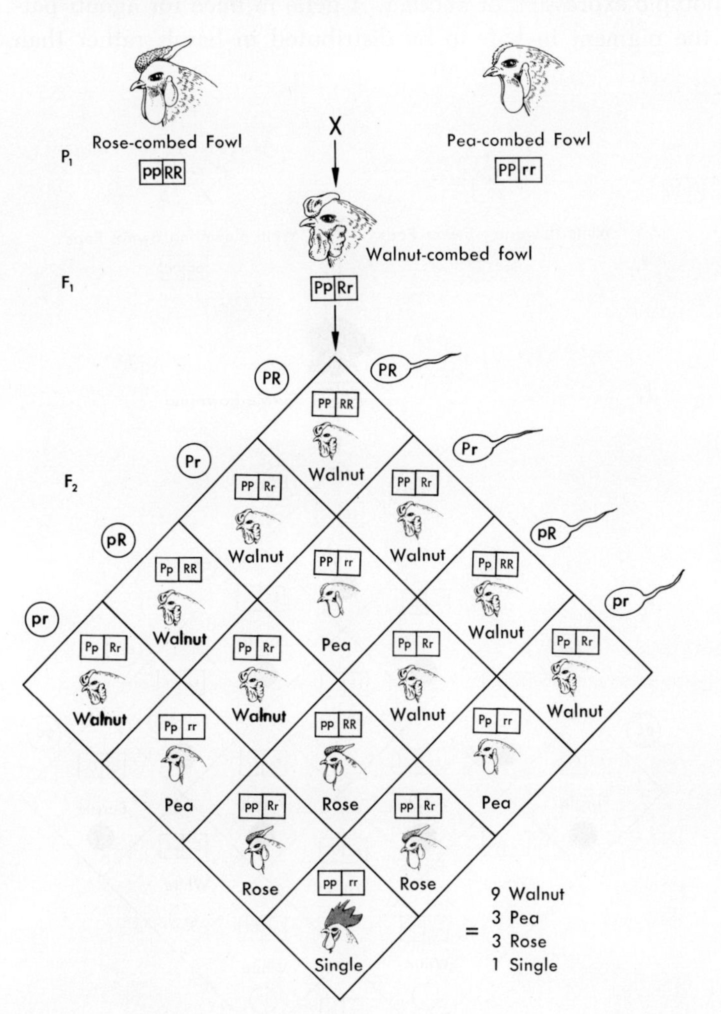

Figure 14–9. The inheritance of comb shape in chickens.

Complementary Genes

In sweet peas a gene, *C*, brings about the development of *chromagen* (a colorless pigment precursor). Another gene, *E*, causes the development of an enzyme that turns chromagen purplish red. If either chromagen or the enzyme is present by itself, the flower is white, but if both *C* and *E* occur in the same individual, the flower is purple (see Figure 14–10).

Modifying Genes

Another type of gene behavior occurs when the presence of one gene modifies the phenotypic expression of another. A gene in mice for agouti pattern (*A*) causes the pigment in hair to be distributed in bands rather than

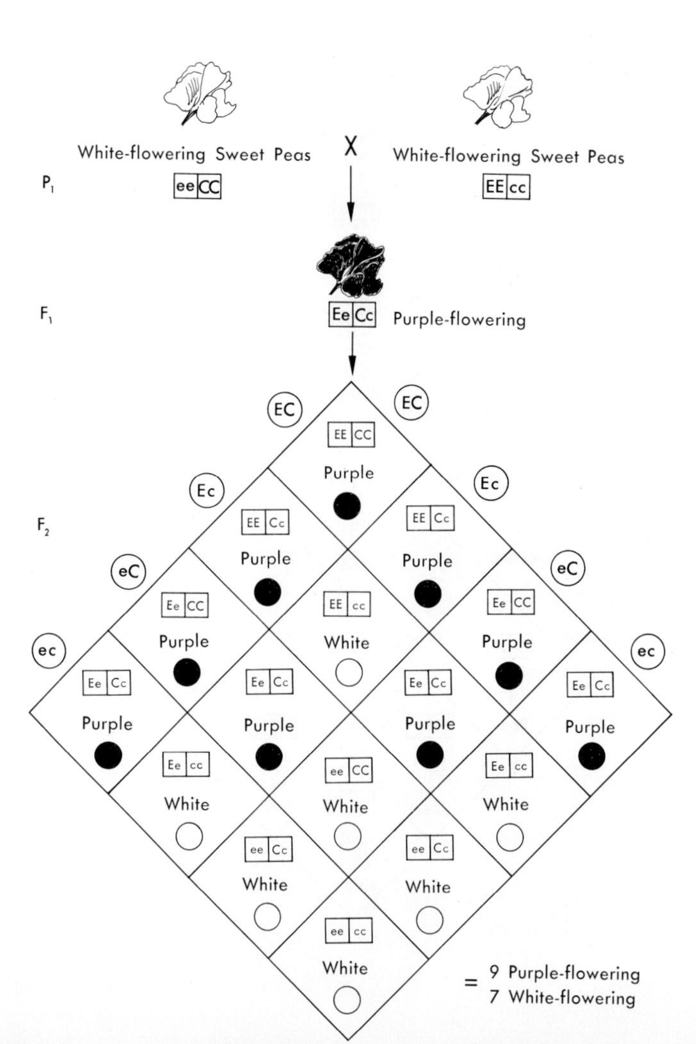

Figure 14–10. Complimentary genes in the inheritance of flower color.

uniformly along the hair, giving the individual a sort of salt-and-pepper effect known as *agouti*. Another gene (C) is responsible for the development of pigment. If the dominant gene for agouti occurs in a mouse in which there is no pigment, the mouse is, of course, just white. Figure 14–11 shows the results when a mouse homozygous dominant for black pigment and homozygous recessive for agouti is crossed with a mouse homozygous recessive for pigmentation and homozygous dominant for the agouti gene.

Multiple Alleles

In all the crosses discussed so far, each gene has had only a single alternative allele. Frequently, though, three or more different forms of alleles occur. Consider a well-known example in humans.

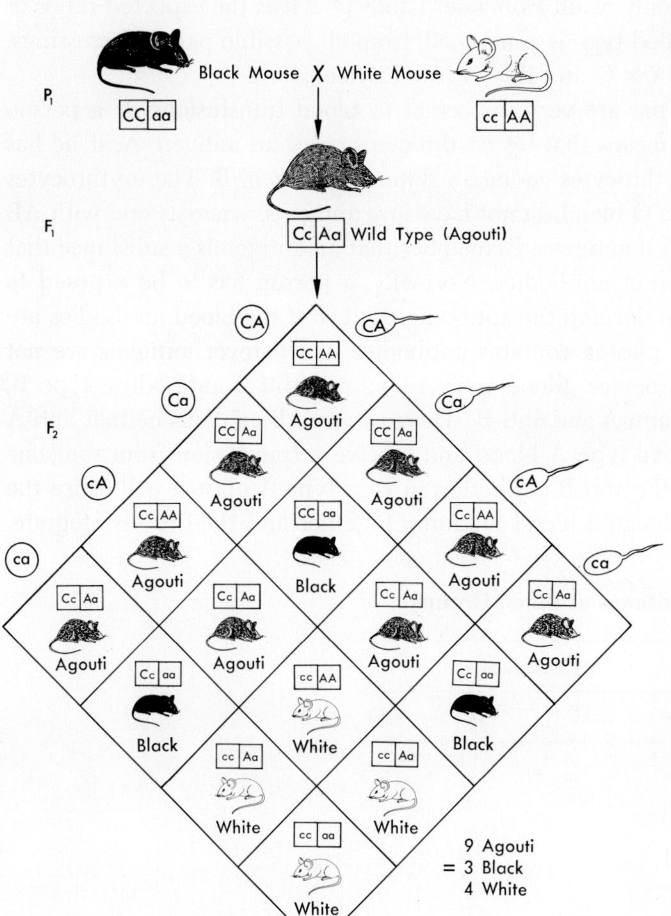

Figure 14–11. Modifying genes in the inheritance of coat color in mice.

On the metal tag worn by each American serviceman is stamped either A, B, AB, or O. These refer to his blood type. The four basic blood types are determined by allelic genes that occur in three different forms: i, I^A, I^B. I^A and I^B are dominant to i but lack dominance for each other; if both occur in the same individual a blood type known as AB is produced. Because genes occur in pairs, any one individual can have any combination of one or two of the three alleles. The possible genotypes of the four basic blood groups are:

A	*B*	*AB*	*O*
I^AI^A	I^BI^B	I^AI^B	ii
I^Ai	I^Bi		

Type A and type B can each result from two different genotypes, whereas types AB and O can only result from one. Table 14–2 lists the expected ratios of children, so far as blood type is concerned, from all possible parental crossings. Why are the crosses A × O and B × O listed twice?

These blood types are very important in blood transfusions. If a person has type A blood, it means that his erythrocytes carry an antigen, A; if he has type B blood, his erythrocytes contain a different antigen, B. The erythrocytes of a person with type O blood do not have any antigens, whereas one with AB blood has both A and B antigens. Remember that an antigen is a substance that causes the production of antibodies. Normally, a person has to be exposed to an antigen in order to develop the antibodies to it. But the blood antibodies are different. The blood plasma contains antibodies to whatever antigens are not present on the erythrocytes. Blood type A contains anti-B antibodies; type B, anti-A; type O, both anti-A and anti-B; whereas type AB contains neither anti-A nor anti-B. If you have type A blood and receive a transfusion from someone who has B antigens, the anti-B antibodies in your type A plasma will cause the erythrocytes in the donated blood to clump together and then to disintegrate.

Table 14–2. The Inheritance of Blood Groups

	Children			
Parents	A	AB	B	O
1. AB × AB	1/4	1/2	1/4	—
2. AB × O	1/2	—	1/2	—
3. A × O	1/2	—	—	1/2
4. A × O	All	—	—	—
5. B × O	—	—	1/2	1/2
6. B × O	—	—	All	—
7. O × O	—	—	—	All

Table 14–3. Possible Donor-Recipient Relationships

Blood Type	Can Give Blood to	Can Receive Blood from
O	O, A, B, AB	O
A	A, AB	O, A
B	B, AB	O, B
AB	AB	O, A, B, AB

The clumps may block small blood vessels. More importantly, as the erythrocytes break up, they release large amounts of hemoglobin into the blood stream. In the passage of the filtrate through the kidney tubules, the hemoglobin may be concentrated to the point where it precipitates out, blocking the tubules and causing a fatal failure of kidney function. You can receive small amounts of blood from an O donor, however, because his blood plasma is usually so diluted by yours that his anti-A antibodies will cause little clumping and destruction of erythrocytes. Table 14–3 shows the possible donor-recipient relationships of the blood types.

Environmental Control of Phenotype

The expression of a trait determined by a gene depends not only on the presence of the gene itself but on the right environment for the expression of that gene. The environmental factor concerned may be either external or internal.

Color in a variety of the Chinese primrose is inherited in simple monohybrid fashion with incomplete dominance. Grown in the garden, if a pure line red flower is crossed with a white flower, all of the F_1 are pink. If the F_1 are self-fertilized, the F_2 occur in a typical 1 red:2 pink:1 white ratio. But, if the offspring of red or pink flowers are grown in a hothouse at temperatures of 35°C (95°F) or above, they are all white. No matter how many generations are reared in the hothouse, they will all be white, but as soon as the plants are replaced in a garden the *RR* flowers turn out red and the *Rr* flowers, pink. In Chapter 13 we pointed out that if eunuchism occurs early enough in a man, he will not develop a beard. The color and type of beard are inherited, but their development is dependent on the proper hormonal environment. Thus, although a woman or a eunuch has the genes for a particular type of beard, it will not develop in the absence of the hormones produced by the testes.

Mendel assumed that genes occur in pairs in the zygote, singly in the gamete. The assumption provided the simplest explanation for his results, but because he shared with other biologists of the time a complete ignorance of what actually happens during germ cell formation, it remained an assumption. Mendel could not point to a physical basis for the behavior of the genes. Perhaps this is part of the reason that his results were ignored for so long. The next great advances in genetics came when that physical basis was found.

SUGGESTED READINGS

Burns, G. W.: *The Science of Genetics,* 2nd ed., Macmillan Publishing Co., Inc., New York, 1972.

Fraser, A.: *Heredity, Genes, and Chromosomes,* McGraw-Hill Book Company, New York, 1966.

Herskowitz, I. H.: *Principles of Genetics,* Macmillan Publishing Co., Inc., New York, 1973.

Srb, A. M., R. D. Owen, and R. S. Edgar: *General Genetics,* W. H. Freeman and Co., San Francisco, 1965.

Stern, C.: *Principles of Human Genetics,* 3rd ed., W. H. Freeman and Co., San Francisco, 1973.

Strickberger, M. W.: *Genetics,* Macmillan Publishing Co., Inc., New York, 1968.

CHAPTER **15**

Cytogenetics

Between 1866, when Mendel published his work, and 1900, when its significance was first realized, great advances were made in the field of *cytology,* the study of cells. During this time special techniques were developed for fixing and staining cells to make it easier to examine them microscopically, and major improvements were also made in the microscopes themselves.

SUTTON-BOVERI HYPOTHESIS

One thing that was being studied very actively during these years was the formation of germ cells by meiosis. About 1902, W. S. Sutton, an American cytologist who had been studying meiosis, realized that there is a striking parallel between the observed transmission of chromosomes from generation to generation and the theoretical passage of genes from generation to generation. He saw that if the genes are in the chromosomes, this would provide a precise explanation for

1. The segregation of the genes at gamete formation.
2. The recombination of the pair of genes when the chromosomes pair up again at fertilization.
3. The independent assortment of genes controlling different traits so long as each pair is in a different pair of chromosomes.

The idea that the genes are in the chromosomes came to be known as the Sutton-Boveri hypothesis after Sutton and T. Boveri, a cytologist who contributed much to our knowledge of the details of the maturation of the germ cells. The gradual acceptance of this idea led to a merging of the fields of cytology and genetics to form a new subscience—*cytogenetics*. About the same time, something else happened that had a major effect on the progress of genetics. Mendel had to wait from one year to the next to see the results of his crosses of garden peas. In the early years of this century, T. H. Morgan began using the little fruit fly *Drosophila* in genetic studies. Fruit flies are easy to raise in the laboratory in enormous numbers, they take up much less space than pea plants, and the time between one generation and the next is only about twelve days. All these factors contributed to a tremendous speedup of genetic research. And *Drosophila* proved to be ideal for cytogenetic studies. For one thing, it has only four pairs of chromosomes, which are easy to tell apart. Biologists owe much of their present knowledge of cytogenetics to the little fruit fly.

THE NATURE AND NUMBER OF CHROMOSOMES

In the light microscope a chromosome in an interphase cell appears to be a threadlike structure called the *chromonema* (plural: *chromonemata*). Under the electron microscope, each chromonema shows up not simply as a solid body but rather as a mass of very fine fibers. It must be composed either of a bundle of smaller fibers or of a single long fiber that is coiled and recoiled. The chromonema itself coils and uncoils during cell division, which accounts for the changes in the length and thickness of the chromosomes during the various stages of division.

Chemical Composition of Chromosomes

Chemically the chromosomes are made up of deoxyribonucleic acid (DNA) aggregated with a special group of proteins known as *histones*. It is these DNA-histone complexes that are referred to as *nucleoproteins*. (We describe the chemical structure of DNA in the next chapter.) In addition to these major components, chromosomes also contain some other proteins and some ribonucleic acid (RNA).

Chromosome Configuration

Although all chromosomes are similar in chemical composition, they differ markedly in appearance.

Location of Kinetochore. Generally there is in each chromosome a single differentiated place along its length that acts as the point where force is exerted to separate the chromatids of a dividing chromosome. This spot is known as the *kinetochore*. If it is near the center of the chromosome, the chromosome appears V-shaped during anaphase, with the kinetochore leading the way along the spindle fiber to the pole and the chromosome arms streaming out behind. If the kinetochore is near one end, the short and long arms give the migrating chromosome a J-shape (see Figure 15–1). It is doubtful whether the kinetochore ever occurs at the very end of a chromosome under normal conditions, although it sometimes may be so close to the end that the anaphase chromosome appears almost I-shaped.

Chromomeres and Knobs. Under high magnification a chromosome usually does not seem to be uniform along its length, but rather to have a series of enlargements and constrictions. This is apparently caused by differential coiling of the chromonema in different regions. The smaller of the enlargements are known as *chromomeres;* the larger ones are simply called *knobs.* The location of these enlargements is different in different chromosomes.

Nucleolus. There is usually a nucleolus (or several nucleoli) in the nucleus of an interphase cell. It contains tiny bodies like the *ribosomes* found in the cytoplasm and is rich in the particular kind of RNA found in ribosomes. Apparently ribosomes are formed in the nucleolus and pass out to the cytoplasm through pores in the nuclear membrane. The nucleolus is attached to a particular region of one of the chromosomes called the *nucleolar organizer*. This region is thought to contain the genes that are responsible for the formation of ribosomes.

Chromosome Complement

Each species has a typical number of chromosomes (see Chapter 11), but this number varies from species to species. In most organisms, the chromosomes are grouped into homologous pairs. The members of a pair are identical; they are the same length, have the kinetochore in the same place, and show the same pattern of knobs and chromomeres. But the members of each chromosome pair differ in these characteristics from every other pair. These differences make it possible to study individual chromosomes. The different kinds of chromo-

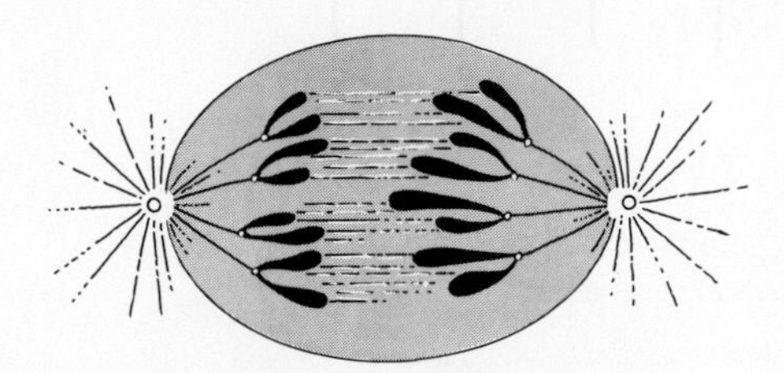

Figure 15–1. The shape a chromosome assumes during anaphase is determined by the location of the kinetochore.

somes in an organism are designated by numbers, starting with 1 for the longest pair and ending with the highest number for the shortest pair. Thus, in the somatic cells of corn, which has twenty chromosomes, there are two homologous number 1 chromosomes, two number 2, and so on (see Figure 15–2).

It is a convention of cytogenetics to use the letter n to indicate the number of kinds of chromosomes. Somatic cells, which have two of each kind, are said to have $2n$ chromosomes. In corn, then, there are $2n = 20$ chromosomes in the somatic cells, but in the gametes, in which there is only one of each pair, the chromosomal complement is $n = 10$. Table 15–1 should help make this relationship clear.

When the chromosomes occur in homologous pairs in a cell, the cell (or organism of which it is a part) is said to be *diploid.* If there is just one of each kind of chromosome present, as in most gametes, the cell is *haploid.* Most organisms, including man, are diploid.

CHROMOSOMES AND INHERITANCE

The arrangement of chromosomes in pairs, which is the normal result of sexual reproduction, allows for a recombination of the characters of the two

Figure 15–2. Diagram of the ten different chromosomes in corn. The location of the kinetochore is indicated by a dash across the chromosome. The open circles represent knobs. The black circle on chromosome 6 is the nucleolus organizer. (After Longley, Botan. Rev., 7:266, 1941.)

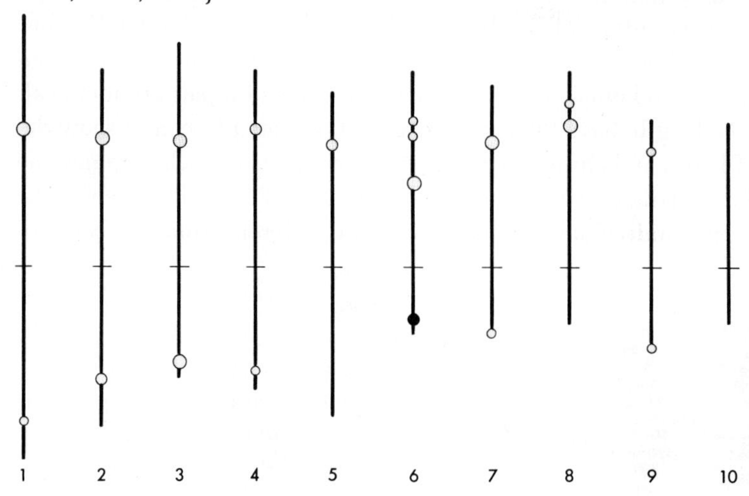

Table 15–1. Chromosome Complement of Some Organisms

Organism	Chromosomes in Gametes (n)	Pairs of Chromosomes in Somatic Cells	Chromosomes in Somatic Cells (2n)
Man	23	23	46
Mouse	20	20	40
Corn	10	10	20

parents. Recombination is not confined to genes that are located on different chromosome pairs.

Linkage and Crossing Over

Shortly after the Sutton-Boveri theory was proposed, some of its adherents suggested that if two different kinds of genes should reside in the same chromosome, they would not be independently assorted, but would instead be linked with one another. One of the first crosses that had a bearing on this idea was made between a sweet pea homozygous for purple flowers and elongated pollen grains and one homozygous for red flowers and round pollen grains. The F_1 all had purple flowers and elongated pollen grains. When these F_1's were allowed to self-fertilize, the F_2 showed a ratio that could not be explained either by independent assortment or by complete linkage. A total of 6,952 F_2 progeny were obtained. Table 15–2 shows the actual numbers of the different phenotypes, together with the expected numbers.

After several years, during which some careful cytogenetic work and some astute reasoning were applied to the problem, an explanation was found. Such results would be expected only if the genes for flower color and pollen grain shape are in the same chromosome and if the homologous chromosomes sometimes exchange parts during meiosis. Remember that during prophase 1 the

Table 15–2. Inheritance of Flower Color and Pollen Grain Shape in Sweet Peas

Phenotype of F_2	Expected F_2 Numbers with Independent Assortment	Expected F_2 Numbers with Complete Linkage	Actual F_2
Purple-elongate	9/16 or 3,910.5	3/4 or 5,214	4,831
Purple-round	3/16 or 1,303.5	none	390
Red-elongate	3/16 or 1,303.5	none	393
Red-round	1/16 or 434.5	1/4 or 1,738	1,338

chromatids form *chiasmata* (crosses); these are the places where this exchange takes place. Figure 15–3 shows how crossover may occur. The position of the hypothetical genes *A* and *B* and their alleles *a* and *b* are indicated on the chromatids. If a break occurs at the point where the chiasma is formed, and the end of the chromatid part bearing *B* is joined to the end of the part bearing *a*, while the end of the part bearing *b* joins the end of the part bearing *A*, you can see that gene *b* is now linked with *A* and gene *B* is linked with *a*. It works out, then, that in addition to gametes *AB* and *ab*, gametes *Ab* and *aB* will also be formed in direct proportion to the percentage of times that crossover occurs. Gametes *AB* and *ab* always occur in equal numbers as do gametes *Ab* and *aB*.

Drosophila has yielded much information about crossing over. (For some reason crossing over does not occur in male *Drosophila*. This seems to be peculiar to this group and is not characteristic of organisms in general.) In these flies, gray body (*G*) is dominant to black body (*g*), and normal wings (*N*) is dominant to vestigial wings (*n*).[1] The genes for body color and wing type occur in the same chromosomes. If a homozygous gray-bodied, vestigial-winged fly is crossed with a homozygous black, normal-winged one, the F_1 will be heterozygous gray-bodied, normal-winged flies. If a female F_1 is mated in a test cross

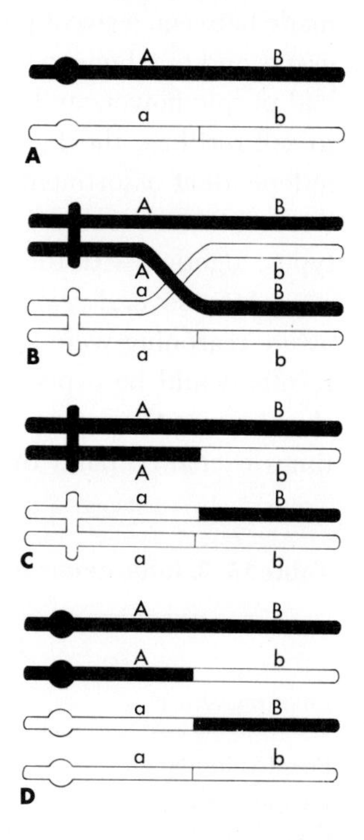

Figure 15–3. Crossing over during meiosis. A: The two homologous chromosomes of the diploid cell. One carries genes *A* and *B*, the other their alleles *a* and *b*. B: Chiasma formation during prophase I. The two chromosomes have duplicated and are in synapsis. Two of the chromatids are exchanging parts. C: The homologous chromosomes are separated at the end of the first meiotic division. D: At the end of the second meiotic division each of the four gametes has received a chromosome with a different complement of genes (*AB*, *Ab*, *aB*, *ab*).

[1] These are not the symbols used by geneticists for these traits, but for the sake of simplicity the convention established in the last chapter is continued.

with a homozygous recessive male, she will have 41.5 per cent black, normal-winged offspring, 41.5 per cent gray, vestigial-winged, 8.5 per cent black, vestigial-winged, and 8.5 per cent gray, normal-winged. If no crossing over occurred, the gametes of the female would be either *Gn* or *gN*. With crossover, they would be *gn* or *GN*. Because it is obvious that 8.5 per cent of the gametes were *gn* and 8.5 per cent were *GN*, then it follows that 17 per cent (8.5 plus 8.5) of her gametes resulted from crossing over and 83 per cent (41.5 plus 41.5) did not. See Figure 15–4.

When appropriate test crosses are made, it is possible to calculate the percentage of crossover by observing the numbers of various phenotypes in the offspring. Also, because the crossover ratio is constant for any given cross, once

Figure 15–4. Backcross of a female *Drosophila* heterozygous for body color and wing length to a recessive male. The genes are on the same chromosome. The presence of four classes of offspring indicates that crossing over has occurred.

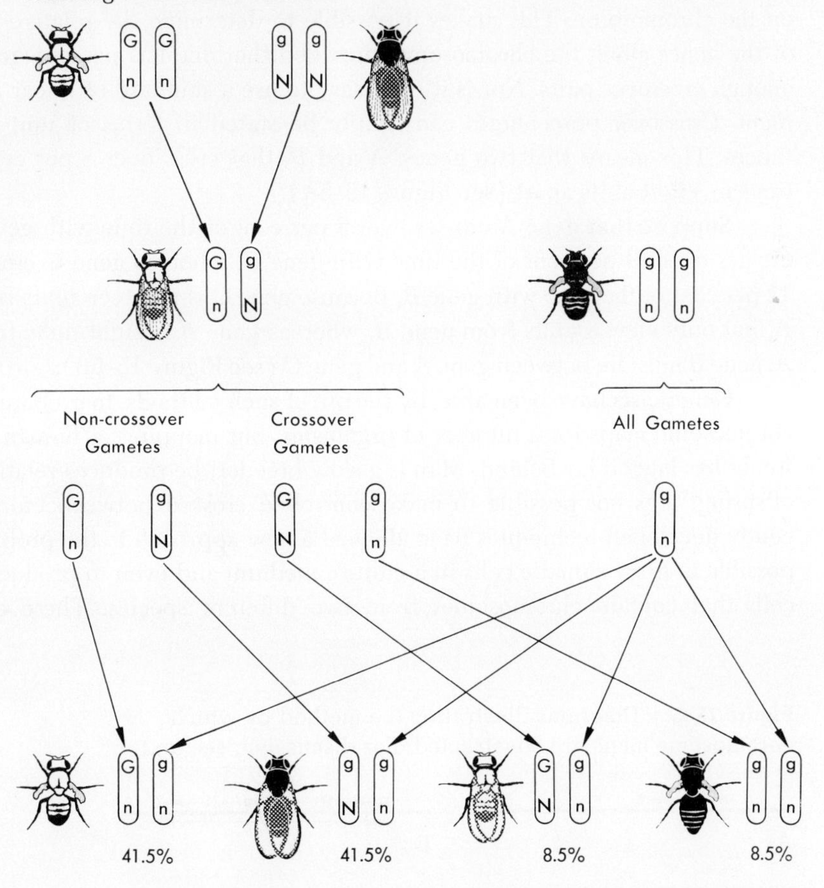

this ratio is known, the phenotypic ratio of the offspring of future crosses can be predicted.

Crossing over is not a rare aberration but is a normal part of meiosis in almost all diploid organisms. It provides for recombinations between genes that are in the same chromosome and so is an important source of variation among the offspring.

Chromosome Mapping

Crossing over has provided geneticists with an important tool by which they can determine the position of the genes in the chromosomes.

The genes are arranged in a linear order in the chromosomes. In *Drosophila,* the genes for white eye and for yellow body cross over 1 per cent of the time, whereas the genes for white eye and cut wing (which gives a scalloped effect to the edge of the wing) cross over 20 per cent of the time. The percentage of times that two linked genes cross over is a function of their distance apart on the chromosome. This makes it possible to determine the relative positions of the genes along the chromosome, provided that one has percentage data on enough crossover pairs. Nor is it necessary to use a standard of linear measurement. Crossover percentages can simply be stated in terms of unit-map distances. This means that two genes, *A* and *B*, that cross over 8 per cent of the time are eight units apart (see Figure 15–5A).

Suppose that gene *A* crosses over 8 per cent of the time with gene *B*, but crosses over 19 per cent of the time with gene *C*, whereas gene *C* crosses over 11 per cent of the time with gene *B*. Because gene *C* is nineteen units from gene *A*, but only eleven units from gene *B*, whereas gene *B* is eight units from gene *A*, gene *B* must lie between gene *A* and gene *C* (see Figure 15–5B).

Geneticists have been able, by the use of such methods, to prepare detailed chromosome maps for a number of organisms, but mapping of human chromosomes has lagged far behind. Man is a slow breeder; he produces relatively few offspring; it is not possible to make controlled crossed between humans. Recently developed techniques have allowed a new approach to the problem. It is possible to grow somatic cells in a culture medium and even to produce hybrid cells that contain chromosomes from two different species. These cells will

Figure 15–5. Diagrams illustrating the method by which chromosome maps are constructed. For discussion, see text.

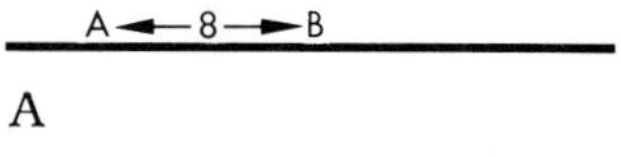

A ←8→ B

A

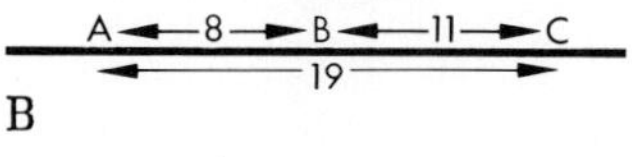

A ←8→ B ←11→ C
←19→

B

continue to divide and give rise to lines of daughter cells. In somatic-cell hybrids between mouse and human or hamster and human, the human chromosomes are progressively lost. It is possible to test these cells for the presence of specifically human enzymes, which are the products of genes. If a given human chromosome is lost from a cell line, and at the same time certain enzymes are lost, then we can say that the genes for those enzymes reside in that chromosome. Eventually we will have comprehensive chromosome maps for humans as well as for fruit flies.

Sex and Chromosomes

There are exceptions to the rule that chromosomes are always in homologous pairs in the somatic cells of diploid organisms. An important exception, at least so far as man and many other organisms are concerned, has to do with the determination of sex.

Sex Determination. In man, *Drosophila,* and many other species, one pair of chromosomes is alike in the female but unlike in the male. One of the pair in the male resembles the members of the female pair, the other is different. The latter is frequently smaller, but may simply be different in shape. These chromosomes, in both the male and the female, are called *sex chromosomes;* other chromosomes, which form identical pairs in both sexes, are called *autosomes.* The sex chromosomes are conventionally labeled X and Y chromosomes, X for the ones that are alike in the female and Y for the different one in the male. Figure 15–6 shows the sex chromosomes and autosomes of *Drosophila* as they might appear on the equatorial plate during mitosis.

In both man and *Drosophila,* sex is determined at conception. The female produces only one type of gamete; each contains one of each pair of autosomes and an X chromosome. In the male, half the gametes contain one of each pair of autosomes and an X chromosome, the other half contain one of each pair of autosomes and a Y chromosome. If the egg is fertilized by a sperm containing the X chromosome, this brings two X chromosomes together and the zygote de-

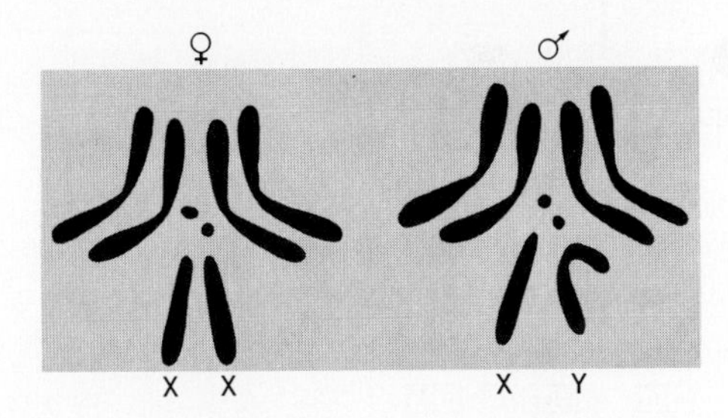

Figure 15–6. Chromosomes in *Drosophila.* The sex chromosomes are the ones marked X in the female (indicated by the symbol ♀) and X and Y in the male (indicated by the symbol ♂).

velops into a female. If the egg is fertilized by a sperm containing the Y chromosome, this brings an X and a Y chromosome together and the zygote develops into a male.

Sex Linkage. A number of genes on the X chromosome seem to be lacking on the Y chromosome. This means that the female has two of this particular gene, but the male has only one. Such genes are said to be *sex linked.*

A familiar example of a sex-linked trait in man is color blindness. The genes for this trait are carried on the X chromosome and the gene for normal vision (N) is dominant to the gene for color blindness (n). Figure 15–7 shows that more men than women are color blind because if a man has the recessive gene he is *ipso facto* color blind, whereas if a woman has the recessive gene, she may have the dominant allele with it.

Other Patterns of Sex Determination. Although in man the female is XX and the male XY, this is not the only type of sex determination known. In barnyard

Figure 15–7. Sex-linked inheritance of color blindness. Females are indicated by ♀, males by ♂. If a woman homozygous for normal vision marries a color-blind man, all the children will have normal vision but the daughters will be heterozygous. If a daughter marries a normal man, all her daughters will have normal vision but half the sons will be color-blind. If a color-blind woman marries a normal man, the daughters will be normal heterozygous, the sons color-blind. If a heterozygous woman marries a color-blind man, half the daughters and half the sons will be color-blind.

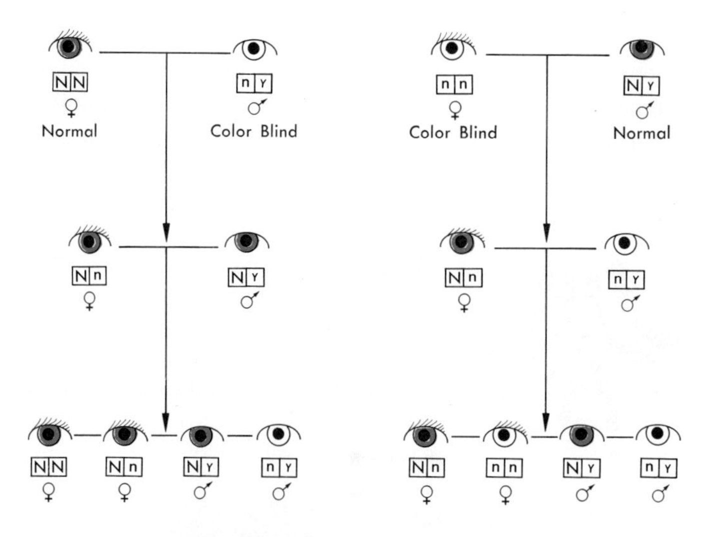

fowls (as in all birds so far tested) the rooster is XX and the hen is XY. Some insects have XX females, but the male is simply X, for there is no Y chromosome at all. One of the most striking situations is found in the honeybee. Here the female has a diploid number of thirty-two and produces haploid eggs with sixteen chromosomes. The male is haploid to begin with and produces haploid sperm by a mitotic type of cell division. Normally all eggs laid by the queen develop. If they have been fertilized, they develop into diploid females. If they have not been fertilized, they develop into haploid males (drones).

CHROMOSOME MODIFICATIONS

Every kind of organism has a characteristic set of genes, its *genome*. Because genes are carried on the chromosomes, in diploids each gamete normally has one complete genome and each somatic cell has two. This typical condition can be modified in a number of ways.

Ploidy

One major class of modifications involves the addition or deletion of whole chromosomes or sets of chromosomes, which is known as *ploidy*. The two basic types of ploidy are *aneuploidy* and *euploidy*.

Aneuploidy. The majority of the chromosomes are arranged in pairs, but one or more chromosomes may lack a partner, or one or more pairs may have an extra chromosome in the set. This can result from the failure of the two members of a pair to separate during meiosis. If both members go to one pole, one daughter cell will have an extra chromosome and the other will lack one.

Aneuploidy occurs in humans, often with tragic effects.

DOWN'S SYNDROME. Down's syndrome is unfortunately a relatively frequent abnormality in man; in lay terms it is called *Mongolian idiocy*. Affected individuals are mentally deficient and physically retarded. Down's syndrome is always associated with, and seems to be the result of, aneuploidy in the form of an extra chromosome in set 21 (see Figure 15–8). The frequency of children of this type seems to increase with the age of the mother at childbirth. It is about a hundred times more common in children whose mothers are past forty-five when they are born than it is in children whose mothers are under twenty-eight. Why this is so and why the addition of one small chromosome causes such a drastic effect in the child are not known.

TURNER'S SYNDROME. This rather rare syndrome results from the absence of a sex chromosome. One X is present but there is no other X or Y to go with it, so the individual is X0. A person with Turner's syndrome has female external

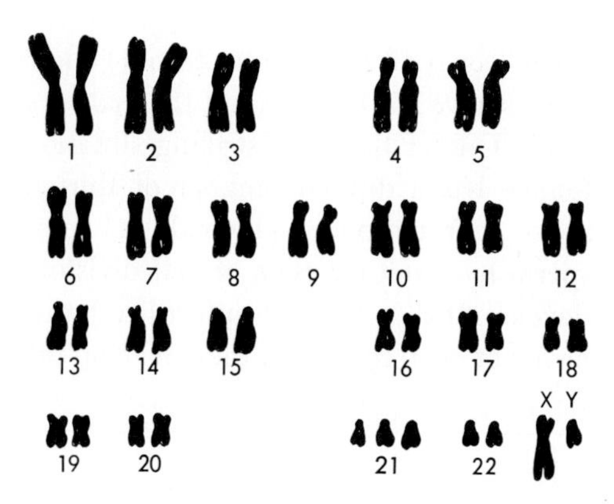

Figure 15–8. The chromosome complement of an individual with Down's syndrome. Note the extra chromosome in set 21.

and internal genitalia, but the uterus and breasts are infantile and the gonads are vestigial or absent entirely. Intelligence is usually impaired.

KLINEFELTER'S SYNDROME. The genitalia are male, but the testes are abnormally small; few if any sperm are produced; body hair is reduced; the breasts tend to be enlarged as compared to those of normal males. The aneuploidy here is the addition of an X chromosome so that the sex chromosomes of the individual are XXY.

Euploidy. Euploidy is merely an extension of aneuploidy in which whole sets of chromosomes, rather than just one or two members of a set, are involved. The chromosomes may be in sets of three (triploidy), four (tetraploidy), or even more. *Polyploidy* is the general term used for having one or more extra sets of chromosomes. *Monoploidy,* only one set of chromosomes in a normally diploid organism, has been reported in *Drosophila,* but the condition is apparently very rare. It is probably almost always fatal at a very early stage of development.

Polyploidy is much more common in plants than it is in animals. This is because the presence of extra sets of chromosomes usually interferes with the regular distribution of the chromosomes to the daughter cells during meiosis and thus prevents the formation of normal gametes. Plants that are able to reproduce asexually can pass the polyploid condition on to their offspring; the higher animals, which normally must reproduce sexually, cannot. Several species of salamanders and lizards are known, however, in which all the individuals are triploid females. The eggs are formed without meiosis, and there is no union of male and female nuclei at the beginning of development.

Polyploid plants are sometimes fertile, and recently a number of species of frogs that reproduce sexually have been found to be polyploids. The mechanisms by which normal gamete formation can be combined with polyploidy are the province of advanced genetics and are not considered here.

Polyploid plants are frequently larger and more vigorous than their diploid counterparts (see Figure 15–9). Certain chemicals, such as *colchicine,* can

Figure 15–9. Comparison of tetraploid (on left) and diploid (on right) snapdragons. (Burpee Seeds photo.)

induce the formation of polyploids; this is now a common practice in horticulture. Some fancy fruits, such as Baldwin apples, are triploids.

Very rarely a human triploid embryo is formed. It is always very abnormal and is usually aborted early. If it does survive to birth, it dies in a few hours. For some reason, if a woman stops taking birth control pills and conceives during the next seven months, the chances of triploidy are somewhat increased.

Chromosome Aberrations

Modifications of gene ratios and gene action can also be brought about by aberrations within the chromosomes. Such aberrations can result from a breaking of the chromosomes and a rejoining of the parts. If a chromosome is thought of as a long, stringlike structure with genes arranged in linear sequence, and the genes are identified by the letters of the alphabet, the chromosome can be pictured thus:

A B · C D E F

A duplication occurs when a segment of the chromosome is repeated:

A B · C D C D E F

Here the segment containing C and D is duplicated.

In a deletion, a part of the chromosome has been lost:

A B · E F

The section containing C and D is missing.

Deletions and duplications resemble aneuploidy, except that parts of chromosomes, rather than whole chromosomes, are involved.

An inversion can be represented thus:

A B · E D C F

The segment containing genes C, D, and E has simply been turned around. Inversions can have a genetic effect because the activities of genes can apparently be modified by their position in relation to other genes in the chromosome.

Sometimes part of a chromosome breaks off and becomes attached to another chromosome in the set. This is a translocation.

The frequency of chromosomal aberrations, and their attendant genetic abnormalities, is increased by exposure to certain chemicals, such as mustard gas and LSD, and also by X-ray and atomic radiation. The same factors can also cause cancer, and broken chromosomes are often found in cancer cells.

SUGGESTED READINGS

Fraser, A.: *Heredity, Genes and Chromosomes,* McGraw-Hill Book Company, New York, 1966.

Herskowitz, I. H.: *Principles of Genetics,* Macmillan Publishing Co., Inc., New York, 1973.

Lerner, L. M.: *Heredity, Evolution, and Society,* W. H. Freeman and Co., San Francisco, 1968.

Markert, C. L., and H. Ursprung: *Developmental Genetics*, Foundations of Developmental Biology Series, Prentice-Hall, Inc., Englewood Cliffs, N. J., 1971.

Srb, A. M., R. D. Owen, and R. S. Edgar: *General Genetics,* W. H. Freeman and Co., San Francisco, 1965.

Stern, C.: *Principles of Human Genetics,* 3rd ed., W. H. Freeman and Co., San Francisco, 1973.

Strickberger, M. W.: *Genetics*, Macmillan Publishing Co., Inc., New York, 1968.

Swanson, C. P.: *Cytology and Cytogenetics,* Prentice-Hall, Inc., Englewood Cliffs, N. J., 1957.

CHAPTER 16

The Nature of Genes and Mutations

The Mendelian geneticists discovered that there are such things as genes, and the cytogeneticists found that the genes are arranged in linear order in the chromosomes. But it remained for the molecular geneticists to provide an acceptable functional model of a gene. There are still questions to be answered, and some current ideas concerning the gene may have to be modified in detail, but the outline is clear at last. It is now known what genes are, how they function, and how they are transmitted to the next generation.

Like the cytogenetists, the molecular genetists found favorable organisms on which to work: molds and particularly bacteria, the tiniest of living things. Bacteria can be raised in astronomical numbers in a small laboratory, and under favorable conditions their generation time can be as short as twenty minutes.

DNA

Although DNA is found in the chromosomes, it is not confined to them—it is present also in such cytoplasmic organelles as mitochondria, chloroplasts, and centrioles. These are all structures that, like the chromosomes, have the ability to replicate themselves. But the vast majority of the cell's DNA is in the chromosomes, it is the stuff of which genes are made.

Structure of DNA

DNA, like protein, is a polymer; that is, a long chain of repeated units. DNA differs from protein in that the units, instead of being amino acids, are *deoxyribonucleotides* (*nucleotides* for short).

Nucleotides—The Subunits. Each nucleotide is made up of a five-carbon sugar submolecule, a phosphate (PO_4) submolecule, and a submolecule of one of four compounds, two of which are known as *purines* and two of which are known as *pyrimidines.* Pyrimidines are six-carbon rings in which two of the carbons have been replaced by nitrogen, whereas purines consist of a pyrimidine ring linked to a five-member ring, which also contains two nitrogens. These ring compounds are *adenine* and *guanine* (the purines) and *cytosine* and *thymine* (the pyrimidines). Because these compounds act chemically as bases, they are often referred to collectively as *bases;* individually they are indicated simply by their initial letters: adenine (A), guanine (G), cytosine (C), and thymine (T). For the structural formulas of these compounds, see Appendix A.

How Nucleotides Combine. A series of nucleotides are bound together as indicated in Figure 16–1, in which S represents the sugar unit, P the phosphate unit, and B the base.

A DNA molecule is made up of two such chains of nucleotides, joined together by hydrogen bonds between the bases of one chain and the bases of the other. The result is a ladder-like structure, with the uprights formed by the sugar-phosphate groups, and the rungs by the bases. Because of the particular configuration of the individual bases, adenine is able to form bonds only with thymine and guanine forms bonds only with cytosine. Because adenine and guanine are purines and thymine and cytosine are pyrimidines, each rung of the ladder consists of one purine and one pyrimidine and all the rungs are the same size. There is no restriction on the order of the bases within a single chain. With this information in mind, such a ladder can be diagramed, as shown in Figure 16–2.

The DNA ladder is twisted so that it forms a helix and the two chains are thus coiled together like the strands of a rope (Figure 16–3).

DNA and Chromosomes

The DNA of a chromosome is believed to be a single, enormously long molecule. Indeed, DNA molecules are the largest biological compounds known. The chromosomes of the trillium (the largest known) contain on the average twenty-six billion nucleotides. How the DNA and proteins are arranged to form the chromosome is not known.

Replication of DNA

The model of how the DNA molecule is put together was first proposed by J. D. Watson and F. H. C. Crick in 1953, and has since been amply confirmed. This model suggests an answer to one of the most basic questions in genetics—how genes are passed on from one generation to the next.

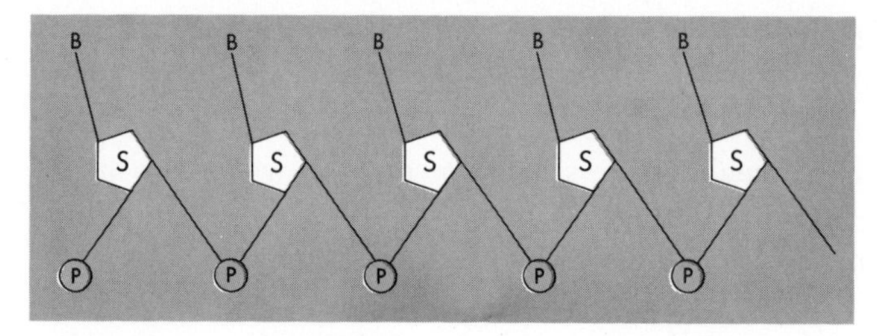

Figure 16–1. Diagram of a portion of a chain of nucleotides. The nitrogenous bases extend from the phosphate-sugar backbone.

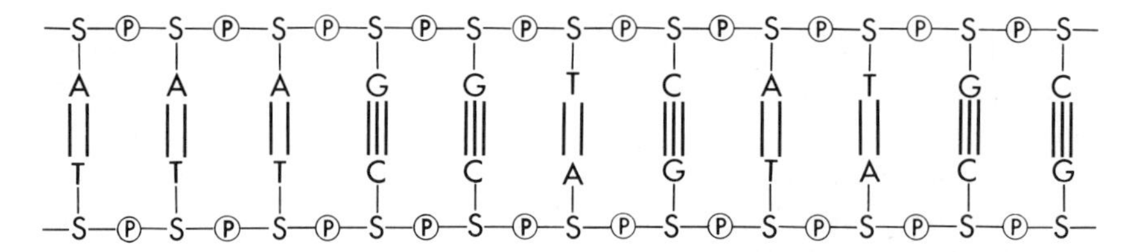

Figure 16–2. Diagram of a part of a DNA molecule. The two chains are held together by hydrogen bonds between the bases.

Figure 16–3. The DNA molecule is twisted to form a double helix.

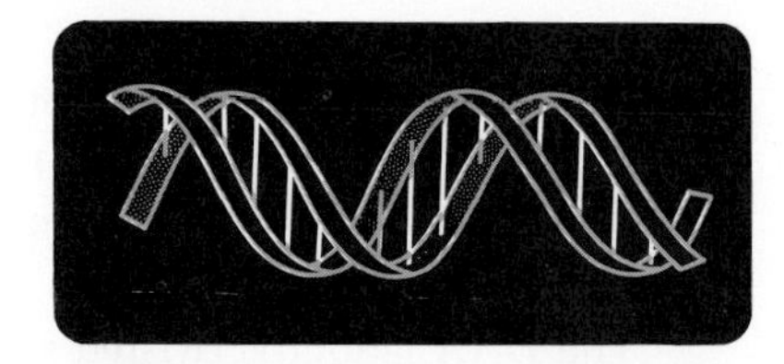

During interphase, the DNA in the nucleus doubles in amount. It is believed that the twin chains separate through a breaking of the hydrogen bonds between the pairs of bases forming the rungs of the ladder. The chains unwind and each then builds for itself a new partner. There are free nucleotides in the nucleus that form new hydrogen bonds with the bases in the chains. Bonds then form between the phosphate and sugar units of the adjacent newly attached bases. This is shown in diagrammatic fashion in Figure 16–4, in which chain II separates from chain I and chain I reconstructs a new chain II out of free nucleotides. Because A will bond only with T, T only with A, C only with G, and G only with C, it follows that the nucleotides in the new chain II are arranged in the same order as they were in the original chain II. At the same time, old chain II is making a new chain I. Thus, each chain acts as a template for the formation of a complementary chain. The result is two duplicate,

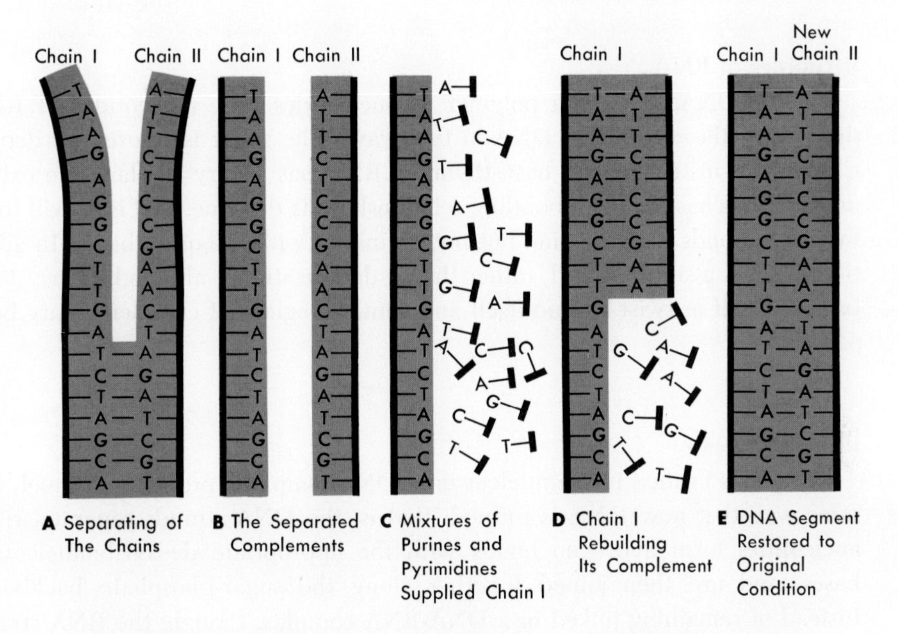

Figure 16–4. Model of the replication of DNA. Chain I is shown serving as a template for the formation of a new chain II. At the same time, the original chain II acts as a template for the formation of a new chain I (not shown).

double-stranded molecules of DNA, each of which has one strand of the original molecule and one new strand. When the chromatids separate during cell division, each contains one of these molecules.

RNA

The Watson-Crick model for DNA shows how it can act as a template so that exact copies of the genetic material can be passed on from generation to generation. But this does not answer the question of how the DNA is able to determine the characteristics of the organism. These characteristics are determined by chemical processes that take place in the cytoplasm, processes that are mediated by enzymes. Because enzymes are proteins, DNA must exert its effect by determining what proteins the cell is able to produce. Proteins are synthesized on the ribosomes found on the endoplasmic reticulum or free in the cytoplasm. What is the link between the DNA in the chromosomes of the nucleus and the ribosomes in the cytoplasm? The answer lies in the second kind of nucleic acid, *ribonucleic acid* or RNA.

Structure of RNA

Like DNA, RNA is a polymer of nucleotides. The nucleotides of RNA differ from those found in DNA in two ways. The sugar is *ribose,* not deoxyribose. And instead of the base thymine, RNA has a very similar base called *uracil,* which has the same bonding relationships as thymine; that is, it will form hydrogen bonds with adenine but not ordinarily with the other bases. In addition, RNA is a single strand, rather than a double strand, although it may loop back on itself or twist around itself and contain regions of complementary base pairing.

RNA Synthesis

RNA is formed in the nucleus on a DNA template, probably in much the same way that new DNA is formed; that is, the DNA strands separate, ribonucleotides form hydrogen bonds with the appropriate deoxyribonucleotide bases, and are then joined together along the sugar-phosphate backbone. Instead of remaining linked in a DNA-RNA complex, though, the RNA strand peels off and passes out into the cytoplasm. Another difference between DNA and RNA synthesis is that apparently only one strand of the two-stranded DNA acts as a template for RNA. It is obvious that the sequence of RNA nucleotide bases is complementary to the sequence in the strand of DNA copied, and similar to the sequence in the strand not copied. Biologists do not know how it is decided which strand is copied, nor do they know what determines when a cell should stop making RNA and begin making DNA in preparation for cell division.

Kinds of RNA

There are at least three kinds of RNA.

Ribosomal RNA. Ribosomal RNA (rRNA) makes up the bulk of the RNA in the cell, yet less is known about it and how it functions than about the other RNA's. Actually there are three different kinds of rRNA molecules that differ in molecular weight. They combine with proteins to form the ribosomes and are thus designated as rRNA. This RNA is quite stable—once formed it remains functional and does not have to be constantly replaced. Only a small part of the nuclear DNA is transcribed into rRNA.

Messenger RNA. Messenger RNA (mRNA) is less stable than rRNA. It combines with the ribosomes, usually linking two or more of them into a polyribosome. It is on the mRNA-polyribosome complex that protein synthesis actually takes place.

Transfer RNA. Transfer RNA (tRNA) comprises small RNA molecules, each containing about seventy-five to eighty nucleotides. Transfer RNA's act as acceptors for amino acids. There is at least one kind of tRNA molecule for each kind of amino acid normally found in protein.

THE GENETIC CODE AND GENES

Before proceeding it would be helpful to tie together some of the threads of the genetic story we have told so far:

1. A chain of DNA can serve as the template for the formation of a new chain of DNA.
2. A chain of DNA can also serve as the template for a chain of RNA.
3. RNA can migrate from the nucleus to the ribosomes.
4. Ribosomes are the sites of protein synthesis in the cell.
5. Proteins (including enzymes) are polymers of amino acids.
6. DNA (and RNA) are polymers of nucleotides.
7. DNA and RNA each contain four types of nucleotides.
8. Proteins are made up of about twenty kinds of amino acids.

An outline of what takes place during protein synthesis can now be put together. One of the chains of DNA acts as a template to bring free nucleotides into a chain of mRNA. The mRNA moves out to the ribosomes. Each amino acid has a specific activating enzyme that attaches an ATP molecule to the acid and uses the ATP energy to transfer the amino acid to a specific tRNA molecule. The tRNA then takes up its proper place on the mRNA. The juxtaposed amino acids are bonded together by peptide bonds and the polypeptide chain peels off (see Figure 16–5).

Polypeptides are, of course, the structural units of proteins.

Altogether this makes a rather clear picture of what takes place and how DNA functions. But one problem remains: how can four bases direct the positioning of at least twenty different amino acids?

The Nature of the Code

Obviously there cannot be a one-to-one correspondence between a base and an amino acid, because this would allow only four amino acids to be incor-

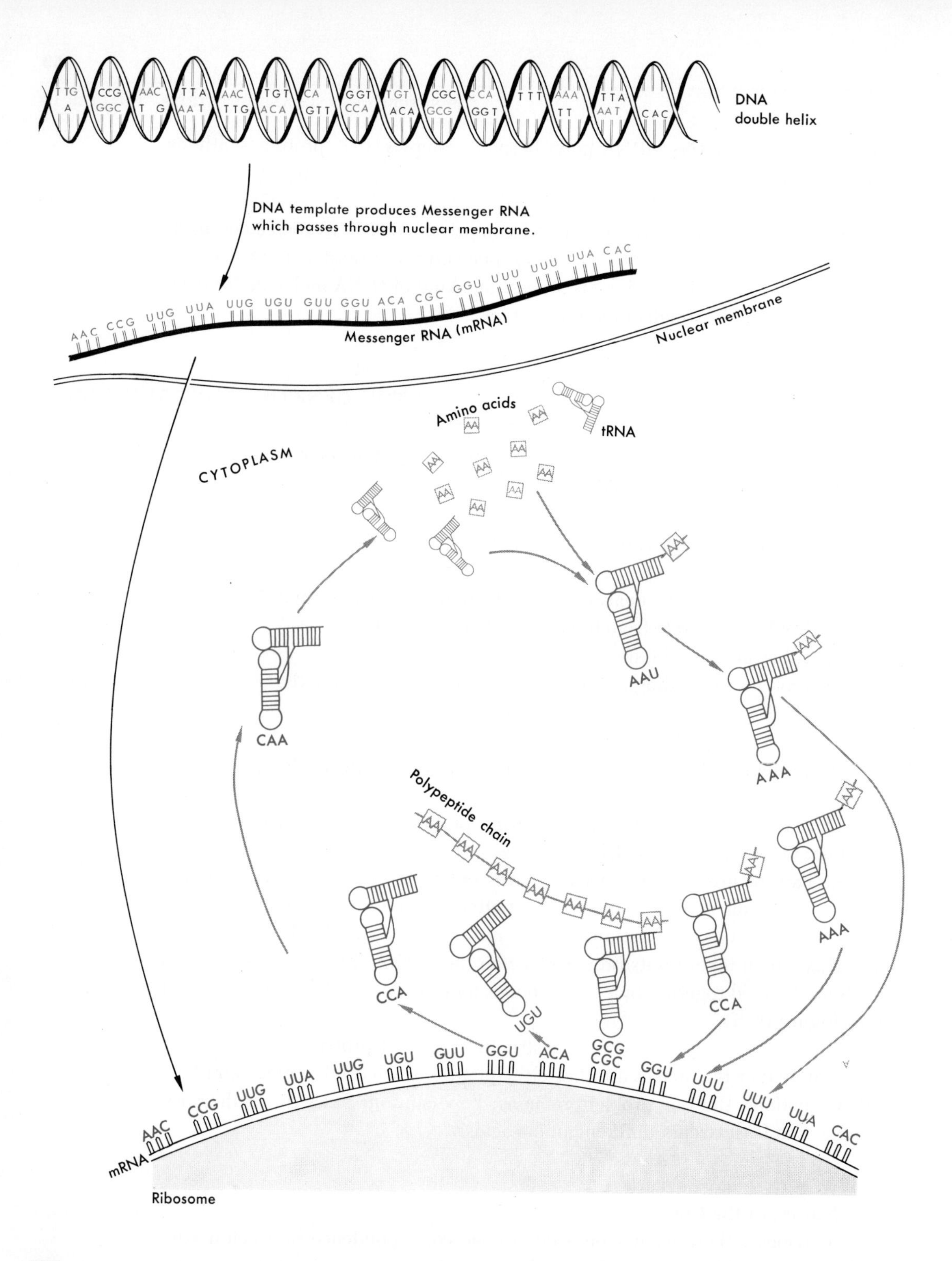

Figure 16–5. Protein synthesis.

porated into the polypeptide chain. Through a combination of inspired research and rigorous reasoning, it has been found that some group of three bases in a row (AGU, GCA, ACG, UCA, and so on) determines the position of a single amino acid, and it has also been possible to discover which triplets are specific for which amino acids. The amino acids and their triplet codes are shown in Table 16–1.

One thing is immediately apparent from Table 16–1. Most of the amino acids are coded by more than one triplet. This is not surprising because there are sixty-four permutations possible from four bases, using three at a time. There is a striking regularity in the distribution of the triplets among the amino acids. Consider the five that have four triplets each (valine or alanine for instance). In every case the first two bases are the same, and apparently it does not matter what the third base is. It has been suggested that this is a remnant of a more primitive system in which only fifteen amino acids were included in proteins and combinations of two of the four bases were sufficient to code for them. More amino acids were then incorporated into the system and it became necessary to use the third member of the triplet to distinguish between them. Thus histidine and glutamine share the same first two bases, but differ in the third,

Table 16–1. The Amino Acids and the RNA Triplets Assigned to Them

First Base	Middle Base				End Base
	U	C	A	G	
U	Phenylalanine	Serine	Tyrosine	Cysteine	U
	Phenylalanine	Serine	Tyrosine	Cysteine	C
	Leucine	Serine	(Terminator)	(Terminator)	A
	Leucine	Serine	(Terminator)	Tryptophan	G
C	Leucine	Proline	Histidine	Arginine	U
	Leucine	Proline	Histidine	Arginine	C
	Leucine	Proline	Glutamine	Arginine	A
	Leucine	Proline	Glutamine	Arginine	G
A	Isoleucine	Threonine	Asparagine	Serine	U
	Isoleucine	Threonine	Asparagine	Serine	C
	Isoleucine	Threonine	Lysine	Arginine	A
	Methionine	Threonine	Lysine	Arginine	G
G	Valine	Alanine	Aspartic acid	Glycine	U
	Valine	Alanine	Aspartic acid	Clycine	C
	Valine	Alanine	Glutamic acid	Glycine	A
	Valine	Alanine	Glutamic acid	Glycine	G

and the same is true for asparagine and lysine and for aspartic acid and glutamic acid.

Three of the triplets, UAA, UGA, and UAG, have not been assigned to any amino acid. They are sometimes, inappropriately, known as "non-sense" triplets. Far from being nonsense, they seem to serve as periods. They prevent an amino acid from falling into line and so bring an end to a particular polypeptide chain.

How the Code Works

Figure 16–6 is a diagram of a hypothetical sequence of nucleotides in a segment of a chain of mRNA. Messenger RNA can be visualized as a long series of nucleotide triplets, called *codons*. On each different kind of tRNA there is an *anticodon,* a triplet that is complementary to the codon for the specific amino acid that tRNA will accept. Thus a tRNA that will accept phenylalanine has someplace a sequence of either AAA or AAG. During protein synthesis, temporary hydrogen bonds form between the codon and the anticodon, and so the amino acids are lined up in the sequence specified by the mRNA, which in turn was specified by the DNA in the nucleus.

What Is a Gene?

Now another of the basic questions of genetics can be answered: what is a gene? A gene is a segment of DNA that specifies the order of nucleotides in RNA. The order of nucleotides in mRNA in turn specifies the order of amino acids in polypeptide chains. The order of amino acids in a polypeptide chain determines the shape that the protein molecule takes, and this in turn determines the activity of the enzyme. So it can be seen that it is indeed the DNA that ultimately determines what chemical activities the cell is able to carry on.

Every organism has many more genes than it has chromosomes. If the DNA of a chromosome is a single molecule, then each DNA molecule must comprise many genes. But it is not a string of unique genes. Some genes are represented in the genome once or a few times, but the genome also contains thousands of copies of other genes. These include the genes for tRNA and the cytochromes, molecules needed in abundance by every cell. The multiple copies of genes for these molecules probably insure their rapid production.

Figure 16–6. Diagram of a hypothetical sequence of nucleotides in a segment of mRNA.

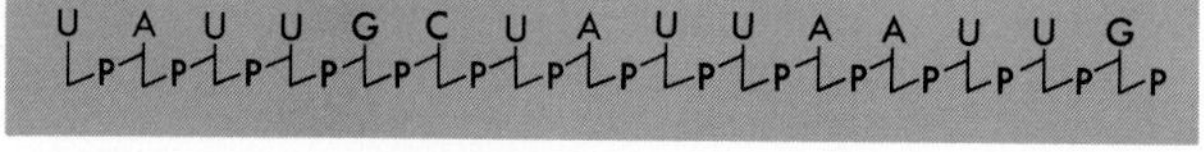

U A U U G C U A U U A A U U G

When it was first discovered that DNA is the genetic material, it was logical to assume that more complex organisms would have more DNA per nucleus than simpler ones do. In a very broad sense this is true. Complex, multicellular plants and animals have more than single-celled ones, which in turn have more than bacteria. But the generalization does not hold true for higher animals and plants. The amount of DNA per cell differs greatly from one species to another; the most complex organisms have less than some simpler ones. All salamanders have more DNA per cell than man does, and one species of salamander has twenty-five times as much DNA as man. Most of the DNA in higher organisms is not transcribed into RNA, or if it is transcribed, it is not translated into proteins. It has been estimated that at least 90 per cent of the DNA in man belongs to this silent class. We consider the possible significance of this inoperative DNA in the section on evolution.

THE NATURE OF MUTATIONS

The word *mutation* has undergone a long evolution, but it is usually taken to mean an abrupt, heritable change in the genetic material. Figure 16–7 is a family tree showing the first appearance and subsequent inheritance of a dominant mutation causing chondrodystrophic dwarfism.

A mutation may involve simply a gene or part of a gene, a part of a chromosome, or even an entire chromosome or sets of chromosomes. Mutations can be classified into two major categories:

Chromosome Mutations
- Change in chromosome number
- Change in chromosome structure

Gene Mutations
- Change in an individual gene

We discussed chromosomal mutations in Chapter 15. They are of great significance in the production of new combinations of genes, but it is the mutation of individual genes that produces new genes, the basic raw material of evolution.

Gene Mutations

The elucidation of the chemical structure and mode of action of the gene has provided geneticists with a much clearer understanding of what a gene mutation is. There are two fundamental ways in which genes can be altered.

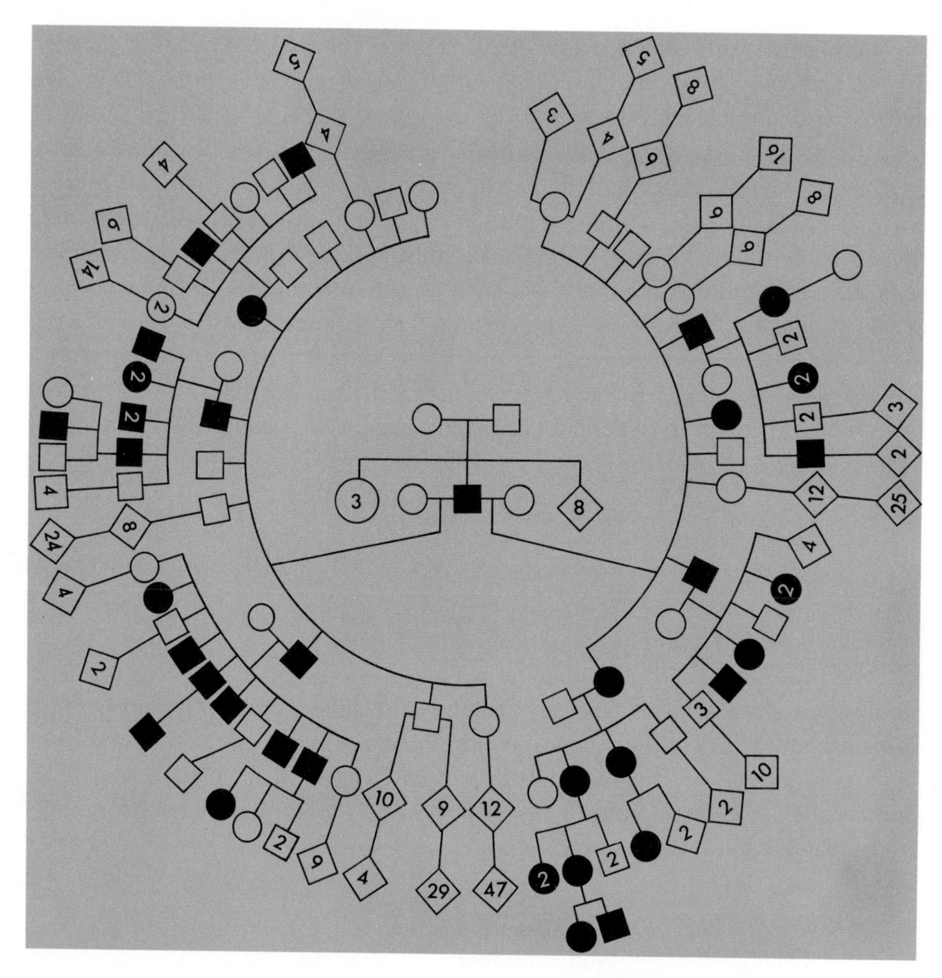

Figure 16–7. The inheritance of chondrodystrophy in man. Females are represented by circles and males by squares. A chondrodystrophic individual is indicated by a black circle or square. The original parents are shown connected by a horizontal line in the top center. Their offspring, shown in the line below, included three normal daughters, indicated by ③, eight normal offspring whose sex is not recorded, indicated by ⟨8⟩, and one chondrodystrophic son. This son was married twice to normal women, represented by the two white circles connected to the black square by horizontal lines. By these two wives he had twenty-two children, seven of them chondrodystrophic. Six of the seven married normal individuals. Between them, they had twenty-four chondrodystrophic and twenty-five normal offspring. Their grandchildren and great grandchildren are also shown. (Data from F. E. Stephens, "An Achondroplastic Mutation and the Nature of its Inheritance," J. Hered. **34,** No. 8, pp. 229–235, 1943, by permission of American Genetic Association. Chart from *Principles of Human Genetics* by Curt Stern. W. H. Freeman and Company. Copyright 1960.

Changes in Individual Nucleotides. One type of change in an individual nucleotide results from a tautomeric shift. This means that certain atoms in a base assume different arrangements; it then becomes possible for the base to bond with one it normally will not bond with. For example, an atomic rearrangement is known for adenine that permits it to bond with cytosine rather than with its usual compatriot, thymine. As a result, a nucleotide containing cytosine will occupy the site of a nucleotide containing thymine when a new chain of DNA is formed.

Nucleotide modification can also be induced chemically. Nitrous acid (HNO_2), when applied to DNA, causes the removal of NH_2 (that is, it *deaminates* both purines and pyrimidines). This also modifies the bonding pattern so that deaminated A will pair with C and deaminated C will pair with A.

A change in a single codon in mRNA can be produced in either of these ways. Such a change may or may not affect the resultant protein. If a codon for phenylalanine changes from UUU to UUC, phenylalanine will still be incorporated in its proper place. This obviously would not result in a mutation. But if the change is from UUU to CUU, leucine will be incorporated instead of phenylalanine. The results of a difference of a single amino acid in a protein are very varied. If the substitution does not seriously alter the folded structure of the protein, the change may be imperceptible. Insulin is a small protein consisting of two polypeptide chains. Beef insulin has alanine as the thirtieth amino acid in one chain, human insulin has threonine. Beef insulin, injected into a human with diabetes, apparently works in exactly the same way as human insulin.

On the other hand, the change of a single amino acid is sometimes disastrous. There is a serious, often fatal, kind of anemia in which the hemoglobin tends to precipitate when it is deoxygenated. The red blood cells are distorted; they appear sickle-shaped and are easily broken. Remember that hemoglobin is a protein formed of two pairs of polypeptide chains. It has been found that sickle-cell anemia is caused by the substitution of valine for glutamic acid as the sixth amino acid from the end in one of the pairs of chains. A comparison of the codons for valine and glutamic acid shows how this could have been brought about by a shift in a single base—from GAA to GUA.

Changes in Sequence. The second type of change in a gene is an alteration in the sequence of the nucleotides on the chain. This can result from the addition or removal of nucleotides from the chain. Figures 16–8 shows how the deletion or addition of a single nucelotide will not just bring about the formation of a new triplet but will cause a change in the entire chain from that point to its end.

If the deletion or addition occurs near the beginning of the chain, practically the whole sequence of amino acids in the resultant polypeptide will be altered. If it occurs near the end, only a few amino acids will be changed and the polypeptide may retain all or most of its normal function. If both a deletion

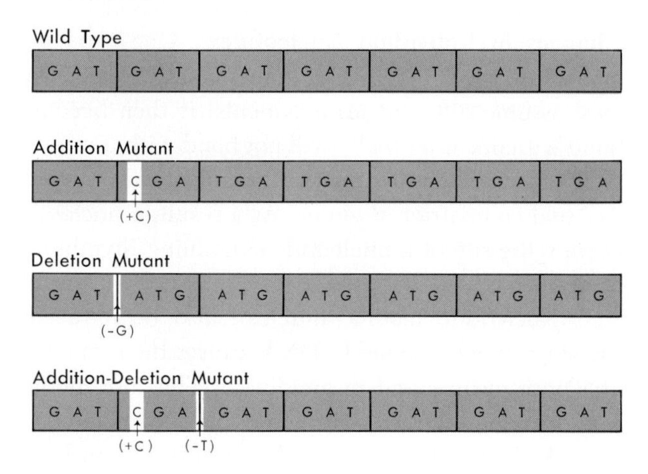

Figure 16–8. Model showing how mutations can be caused by the addition or deletion of nucleotides from a DNA chain. The normal chain (wild type) is shown at the top. If a nucleotide is added to or dropped from the chain, the codons from that point on are modified. If both an addition and a deletion occur, only the part of the chain between is modified.

and an addition occur close together, only a few amino acids will be changed and again there may be little loss of function.

Some Characteristics of Mutations

Even before the chemical nature of the gene was elucidated, geneticists had accumulated enough information about mutations to permit them to make a number of generalizations.

1. Most mutations are detrimental. Because selection has, over thousands of generations, developed a series of genes that cause the development of a well-integrated functional group of proteins, it is logical that any alteration would probably interfere with this system. Most mutations result in the loss of ability of some enzyme to function normally and so alter the normal chemical activity of the cell. It is only by a very rare chance that such an alteration represents an improvement.
2. Mutations affect the genes of all cells; they are not restricted to germ cells but occur in somatic cells as well. Obviously, though, in sexually reproducing organisms, only mutations affecting germ cells will be passed on to the next generation.
3. Spontaneous mutations occur in a species at a rate that can be predicted. Under natural conditions, the rate of spontaneous gene mutations in *Drosophila* is such that one in every twenty gametes contains a mutation. In mice, about one in every ten gametes contains a mutation and in man about one in every five. This may seem a large number, but remember that each gamete contains many thousands of genes. Genes in fact are remarkably stable. The rate of occurrence of spontaneous mutations is actually lower than would be predicted in the light of

present knowledge of the factors that may cause mutations—of the rate of occurrence of tautomeric shifts in nucleotides, for example. This suggests that the enzymes involved in DNA synthesis may be able to recognize and reject "wrong" nucleotides. There also seem to be repair enzymes that can mend breaks in the DNA chain.

4. Certain physical and chemical agents, called *mutagens*, can increase the mutation rate as much as 150-fold in cells exposed to them. Many chemicals, such as mustard gas, peroxides, and nitrous acid, are mutagenic, as are all high-energy radiations. These same factors also cause chromosomal breakage and aberrations. The activities of modern man have greatly increased the number of mutagenic agents in the environment to which all organisms are exposed.

REGULATION OF GENE ACTIVITY

The major problem of genetics today is: how are genes turned on and off? A muscle cell makes myosin but does not make hemoglobin, although it surely has the genes for hemoglobin production. Genetics here begins to merge with embryology. What apparently happens during tissue differentiation is that certain genes in the cells of a given tissue are permanently blocked and others are activated; which genes are blocked and which are activated differs from cell type to cell type. Other genes can apparently be turned on and off repeatedly.

In all except the most primitive organisms (bacteria and related forms) the chromosomes contain proteins and RNA as well as DNA. The most numerous of these proteins belong to a group known as *histones*. The histones form complexes with segments of DNA and inhibit its transcription into RNA. It is suggested that the other proteins and the RNA in the chromosome can interact with the histones and alter their activity, thereby releasing one segment or another of DNA from inhibition. The mechanism for this interaction is unknown. Some hormones apparently exert their effect by releasing specific genes from inhibition, thereby allowing the production of specific mRNA's. Again the mechanism is unknown. The question of how gene activity is regulated is one of the most important, and one of the most elusive, in biology.

SUGGESTED READINGS

Asimov, I.: *The Genetic Code*, The New American Library of World Literature, Inc., New York, 1963.
Crick, F. H. C.: "The Genetic Code: III," *Sci. Amer.*, Vol. 215, No. 4, 1966.

De Busk, A. G.: *Molecular Genetics,* Macmillan Publishing Co., Inc., New York, 1968.
Fraser, A.: *Heredity, Genes and Chromosomes,* McGraw-Hill Book Company, New York, 1966.
Herskowitz, I. H.: *Principles of Genetics,* Macmillan Publishing Co., Inc., New York, 1973.
Jukes, T. H.: *Molecules and Evolution,* Columbia University Press, New York, 1966.
Levine, R. P.: *Genetics,* 2nd ed., Holt, Rinehart and Winston, Inc., New York, 1968.
Nirenberg, M. W.: "The Genetic Code: II," *Sci. Amer.,* Vol. 208, No. 3, 1963.
Smith, H. H., ed.: *Evolution of Genetic Systems,* Brookhaven Symposia in Biology, No. 23, Gordon and Breach, New York, 1972.
Srb, A. M., R. D. Owen, and R. S. Edgar: *General Genetics,* W. H. Freeman and Co., San Francisco, 1965.
Stern, C.: *Principles of Human Genetics,* 3rd ed., W. H. Freeman and Co., San Francisco, 1973.
Strickberger, M. W.: *Genetics,* Macmillan Publishing Co., Inc., New York, 1968.
Watson, J. D.: *The Molecular Biology of the Gene,* 2nd ed., W. A. Benjamin, Inc., New York, 1968.

CHAPTER 17

Genetics and Populations

It has probably occurred to you that many traits that seem to be inherited do not fall into obvious, discrete phenotypic groups. Tall parents usually have tall children and short parents short children, but people as a whole cannot be separated into tall and short groups like Mendel's garden peas. Similarly, breeders of race horses know from empirical experience that they have a better chance of getting a winner if they crossbreed two champions or near champions than if they crossbreed a near champion with a dray horse. But horses cannot be classified into discrete groups on the basis of running speed. In fact, most characters show a gradient from the least developed to the most developed, with the majority spaced somewhere between the two extremes. This holds true for all sorts of traits, for all sorts of organisms, ranging from the amount of milk produced by dairy cows to the length of needles in pine trees. The distribution of height in a company of soldiers is shown in Figure 17–1. Such traits are said to show quantitative inheritance.

QUANTITATIVE INHERITANCE

It seemed for a time that the inheritance of quantitative characters would have to be accounted for on a different basis than the discrete determinants proposed by Mendel. However, further work has shown that such characters can be interpreted on a Mendelian basis.

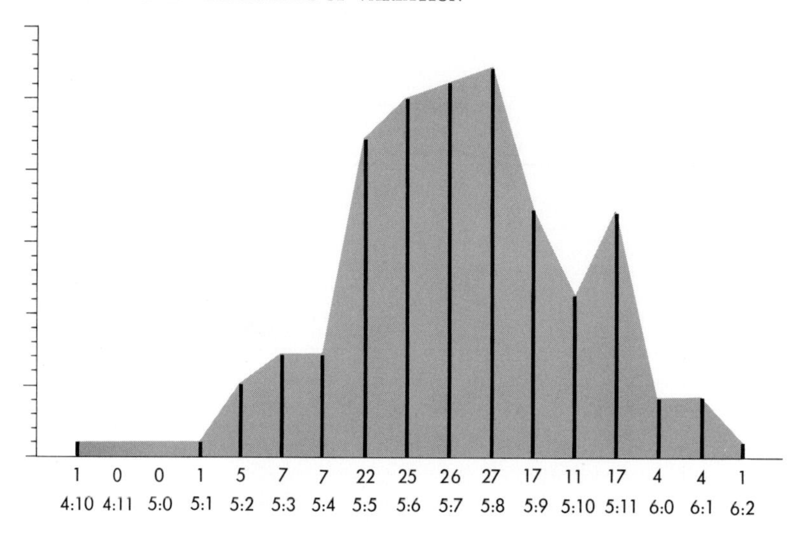

Figure 17–1. Distribution of height in a company of soldiers. The bottom row of figures indicates the height in inches, the top row the number of individuals in each group. (1 inch = approximately 25 mm.)

Nilsson-Ehle Hypothesis

One of the first cases of this type of inheritance to be worked out in detail was that of grain color in wheat. About a decade after the rediscovery of Mendel's work, H. Nilsson-Ehle found that if he crossed a red strain with a white strain, the F_1 was pink. However, when the F_1 plants were self-fertilized, the F_2 showed a great deal of variety. When he sorted the grains out, though, he could distinguish seven classes of color. This led Nilsson-Ehle to suggest that grain color in wheat is determined by three independent pairs of genes. *AA, BB,* and *CC,* and that the presence of any one of these contributes one sixth of the total pigmentation. The colors range from the red homozygous dominant (*AA, BB, CC*) to the white homozygous recessive (*aa, bb, cc*). A homozygous dominant crossed with a homozygous recessive yields a completely heterozygous F_1 (*Aa, Bb, Cc*) that contains three genes for pigmentation and is pink. If two of these F_1's are then crossed, the F_2 will, if all combinations occur in the right proportions, fall into twenty-seven different genotypes and seven phenotypes as shown in Table 17–1.

Polygenic Inheritance

The inheritance of quantitative characters that can be accounted for on the basis of the segregation of a number of supplementary pairs of allelic genes

Table 17–1. Inheritance of Grain Color in Wheat, Showing Genotypic and Phenotypic Ratios of F_2

Genotype	Phenotype
1 *AA BB CC*	1 individual in 64 with 6 genes for red
2 *AA BB Cc* 2 *AA Bb CC* 2 *Aa BB CC*	6 individuals in 64 with 5 genes for red
4 *AA Bb Cc* 4 *Aa BB Cc* 4 *Aa Bb CC* 1 *AA BB cc* 1 *AA bb CC* 1 *aa BB CC*	15 individuals in 64 with 4 genes for red
8 *Aa Bb Cc* 2 *AA Bb cc* 2 *AA bb Cc* 2 *aa BB Cc* 2 *aa Bb CC* 2 *Aa BB cc* 2 *Aa bb CC*	20 individuals in 64 with 3 genes for red
4 *Aa Bb cc* 4 *Aa bb Cc* 4 *aa Bb Cc* 1 *AA bb cc* 1 *aa BB cc* 1 *aa bb CC*	15 individuals in 64 with 2 genes for red
2 *Aa bb cc* 2 *aa Bb cc* 2 *aa bb Cc*	6 individuals in 64 with 1 gene for red
1 *aa bb cc*	1 individual in 64 with 0 genes for red

is known as *polygenic inheritance,* and the series of genes whose effects are added one on the other are known as *polygenes.*

As the number of pairs of polygenes in such crosses increases, not only does the number of phenotypic classes increase but relatively fewer of the F_2 tend to reach the extremes of the parental types. This can be seen from Tables 17–2 and 17–3, which show the distribution of the F_2 in crosses involving four and five pairs of polygenes.

Analysis of Polygenic Inheritance

As more genes are added to a polygenic series, the number of classes increases and soon becomes too large to handle. It is difficult to analyze such

Table 17–2. Phenotypic Ratios of F_2 in a Cross with Four Polygenes

1 individual out of 256 with 8 genes
8 individuals out of 256 with 7 genes
28 individuals out of 256 with 6 genes
56 individuals out of 256 with 5 genes
70 individuals out of 256 with 4 genes
56 individuals out of 256 with 3 genes
28 individuals out of 256 with 2 genes
8 individuals out of 256 with 1 gene
1 individual out of 256 with 0 genes

Table 17–3. Phenotypic Ratios of F_2 in a Cross with Five Polygenes

1 individual out of 1,024 with 10 genes
10 individuals out of 1,024 with 9 genes
45 individuals out of 1,024 with 8 genes
120 individuals out of 1,024 with 7 genes
210 individuals out of 1,024 with 6 genes
252 individuals out of 1,024 with 5 genes
210 individuals out of 1,024 with 4 genes
120 individuals out of 1,024 with 3 genes
45 individuals out of 1,024 with 2 genes
10 individuals out of 1,024 with 1 gene
1 individual out of 1,024 with 0 genes

crosses by making standard Mendelian crosses or by the study of pedigrees. But they can be analyzed by special statistical methods.

Plotting the Curve. If the phenotypic classes in a hypothetical polygene cross are plotted on a graph in which the base line represents the number of classes and the vertical represents the number of individuals in each class, and if the points are connected by a line, the result is a curve. The greater the number of polygenes involved, the smoother is the curve. This is shown in Figure 17–2.

Description of the Curve. For the correct interpretation of such a graph, four mathematical terms must be clearly understood: *mean, median, mode,* and *standard deviation.*

MEAN. The mean is simply the arithmetic average of a series of measurements. You determine the mean by adding all of the variants together and dividing the sum by the number of variants. The sum of the series of numbers 8, 8, 9, 11, 12, 14, 17, 18, 20 is 117. Because there are nine members of the series, 117 divided by 9 gives a mean of 13.

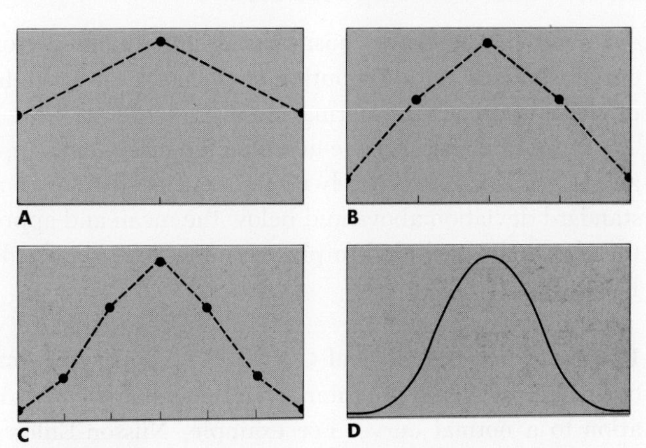

Figure 17–2. Theoretical distribution curves of phenotypes where different numbers of genes are involved. A: A single gene and allele give three classes (AA: 2 Aa: 1 aa). B: Two separate pairs of allelic genes give five classes. C: Three separate pairs of alleles, seven classes. D: An infinite number of pairs of alleles would give a smooth curve.

MEDIAN. The median is that variant that has the same number of variants both above and below it. In the series given, the number 12, which has four variants above and four below, is the median.

MODE. The mode is simply that variant that occurs the largest number of times. In this series the mode is 8.

STANDARD DEVIATION. It is often necessary to know how much variability there is in a series of numbers; that is, how much spread there is around a mean. This is usually expressed as the standard deviation. To obtain the standard deviation, you first figure the amount by which each number in the series differs from the mean. Thus, 12 is one less than the mean (-1) and 17 is four above the mean $(+4)$. Each deviation is then squared (which makes them all positive, of course). All of these squared deviations are averaged, and the square root of the average is taken. By a line of reasoning more complex than is necessary here, mathematicians can show that in calculating the average of the squared deviations it is better to divide by one less than the total number of variants $(n-1)$ rather than by the actual number (n). In summary, then, the standard deviation is the square root of the average of the squares of all the deviations. Mathematically, the standard deviation, σ, is expressed as:

$$\sigma = \sqrt{\frac{\Sigma\,(d^2)}{n-1}}$$

The Normal Curve. The curve showing the distribution of an infinite number of pairs of alleles, as shown in Figure 17–2, makes what is known as a *normal* *curve.* Obviously, no organism has an infinite number of pairs of genes, but the curves showing the distribution of phenotypes determined by only a few pairs of alleles approach this normal curve.

A normal curve is one in which the mean, median, and mode are the same, and in which approximately 68 per cent of the classes are included within one standard deviation above and below the mean and approximately 95 per cent of the classes are included in plus or minus two standard deviations (see Figure 17–3).

Polygenes and the Normal Curve. We can now examine briefly a few cases of known polygenic inheritance and compare the distribution of the F_2 generation to a normal curve. For example, Nilsson-Ehle's data on grain color in wheat are plotted on the curve shown in Figure 17–4. A similar type of distribution can be found in the numbers of rows of grains on ears of corn. In a cross between Golden Glow, which has sixteen rows per ear, and Black Mexican, which has only eight rows per ear, the F_1 average is twelve rows per ear. The F_2 form a nearly normal curve with the numbers of rows ranging from eight to eighteen per ear (see Figure 17–5). Other cases of polygenic inheritance have been worked out, including ear length in corn, fruit size in tomatoes, flower size in tobacco, and ear length in rabbits.

Polygenic Inheritance in Man. One great advantage of the application of statistics to genetic analysis is that it permits the study of the inheritance of many traits in humans. Even simple Mendelian factors are not easy to study in man. It is not possible to make controlled matings with people, or develop pure lines, or carry out backcrosses. Some information can be gathered from the study of pedigrees (see Figure 16–7), but such data are hard to acquire and they accumulate slowly. People do not breed as rapidly as fruit flies or guinea

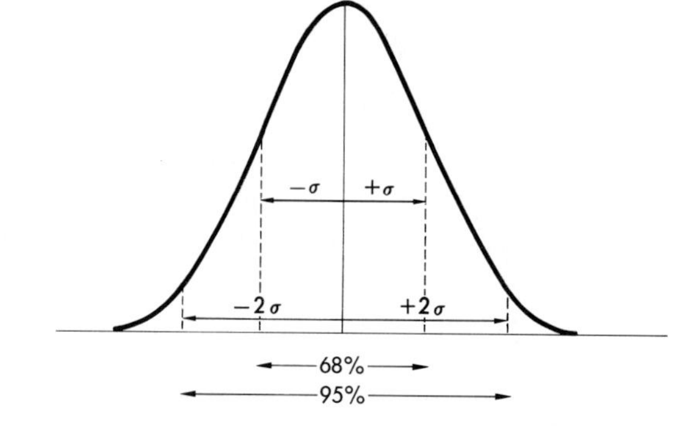

Figure 17–3. A normal curve.

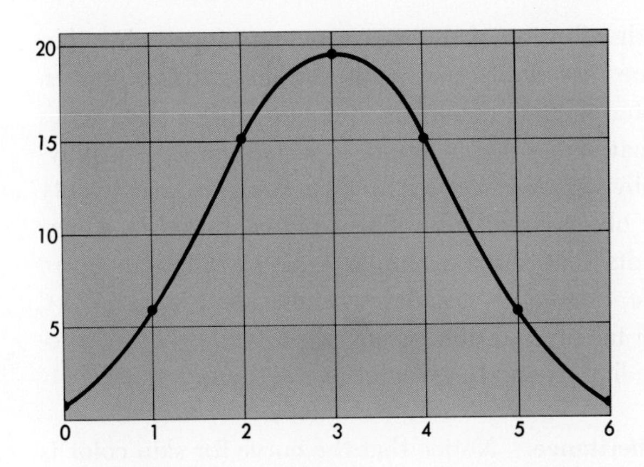

Figure 17–4. Distribution curve of the classes of grain color in wheat.

pigs, and the number of offspring per human couple is usually small. The principles of genetics might not yet have been discovered if man had confined himself solely to the study of man. With the introduction of statistical methods, though, it is possible to study polygenic inheritance without making controlled crosses. A geneticist can measure a sample of the population, and if his sample

Figure 17–5. Distribution of number of rows on the ear in the F_2 of a cross between Golden Glow corn with sixteen rows and Black Mexican with eight rows. (From L. H. Snyder and P. R. David, *The Principles of Heredity* © 1957 by D. C. Heath and Company, Boston.)

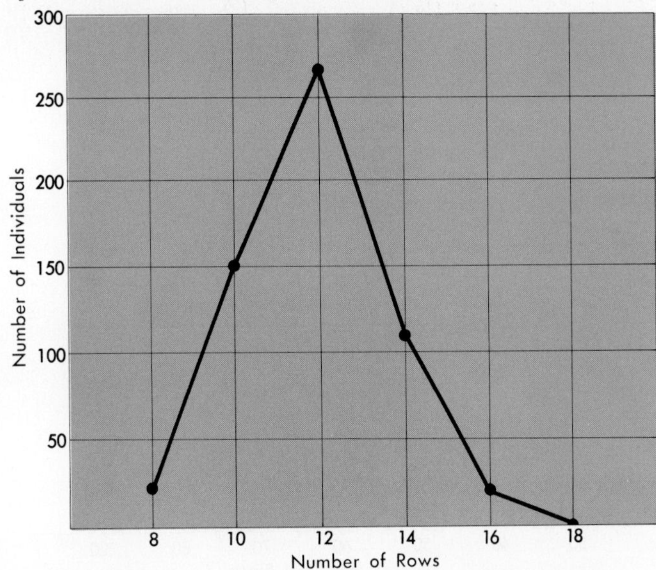

is properly chosen, and if the distribution of the measurements approaches that of the normal curve, he can be reasonably sure he is dealing with polygenes. Thus, although it is still a matter of some debate as to how many pairs of genes are involved, it is apparent that skin color in humans is determined by polygenes. Figure 17–6 shows the hypothetical curves for two, four, six, and twenty polygenes compared with the observed distribution of pigmentation in American blacks. Other human traits that show a similar type of inheritance are longevity, resistance to disease, degree of vascular tension (see Figure 17–7), scores on mental tests, musical talent, and dimensions of particular body structures, such as the length of the fingers and the weight of the thyroid gland.

Modifications of Polygenic Inheritance. Notice that the curve for skin color is not actually a normal curve. Many factors can modify a curve. For one thing, characters that are inherited in a polygenic fashion are usually also susceptible to environmental modifications. Inadequate nutrition during childhood can

Figure 17–6. Frequency of the observed distribution of skin pigmentation in the American Negro and the expected distribution based on multiple factors at 2, 4, 6, and 20 loci. (Data from Curt Stern, in *Acta Genet. Stat. Med.* **4,** 1953. By permission S. Karger, A. G. Basel. From Curt Stern, *Principles of Human Genetics* (2nd ed.), W. H. Freeman and Company. Copyright 1960.)

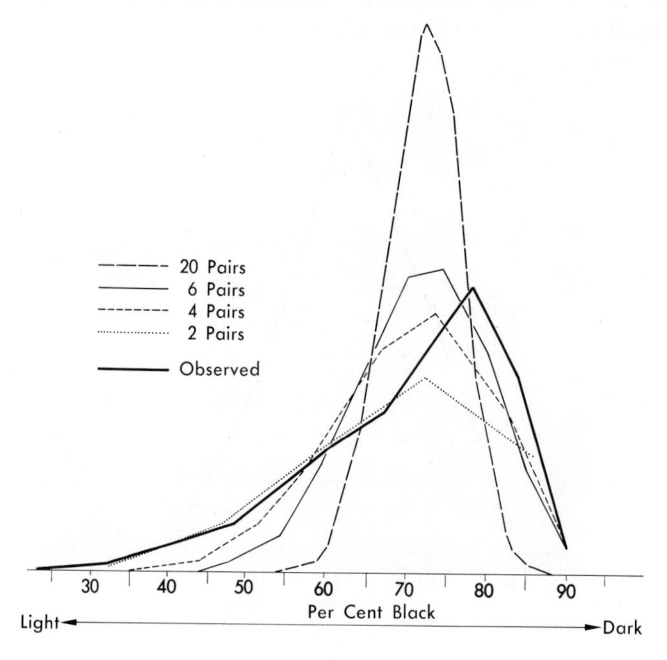

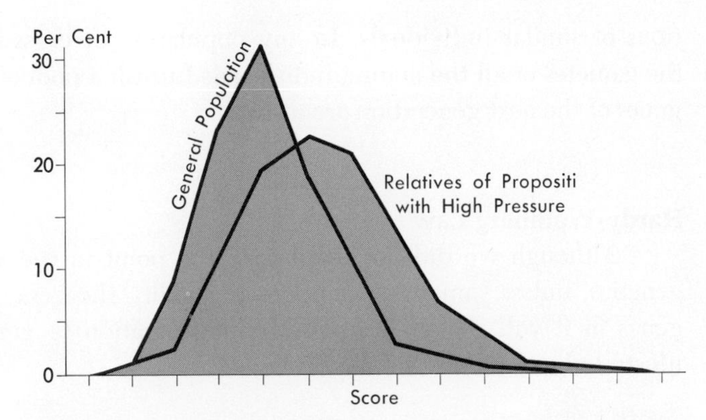

Figure 17–7. The distribution of blood pressure classes in the general population and in the relatives of individuals known to have high blood pressure (the propositi, or individuals on whom the study was based). This graph indicates that blood pressure is determined (in part) by polygenes and that the alleles causing high pressure are more common in relatives of individuals showing this condition than in the general population. (After Hamilton, Pickering, Roberts, and Sowry, *Clinical Science*, **13**, 1954.)

stunt growth, and poor education can affect the scores made on intelligence tests. All of the genotypes may not react in the same way to given environmental factors. A person who has very little pigment in his skin probably responds to a moderate dose of sunshine by turning red and peeling, whereas his friend, who has somewhat more pigment to begin with, tans; that is, he turns darker than his genotype would indicate. Even when care is taken to minimize the effects of environmental differences, it is usually impossible to tell how many genes are involved in a given case of polygenic inheritance in man.

THE GENE POOL

By the application of the sampling techniques of statistics, the study of genetics can be extended to populations. The importance of population genetics goes far beyond the demonstration of the polygenic nature of the inheritance of many characters. If genetic change is the raw material of evolution, in order to understand evolution it is necessary to understand how genes spread through a population.

Individuals of sexually reproducing organisms are members of populations of similar individuals. In any population of cross-fertilizing individuals, the gametes of all the mating individuals furnish a pool of genes from which all genes of the next generation are drawn.

Hardy-Weinberg Law

Although we did not emphasize the point in the account of Mendelian genetics, unless something happens to modify the gene pool, the ratio of the genes in it will remain constant from generation to generation. This can be illustrated in a simple monohybrid cross:

P_1	$AA \times aa$	$A{:}a = 1{:}1$
F_1	$Aa \times Aa$	$A{:}a = 1{:}1$
F_2	AA, Aa, Aa, aa	$A{:}a = 1{:}1$

In this monohybrid cross, of course, the number of *A* genes is equal to the number of *a* genes. This equality would be most unexpected in a natural population, but the principle of the constancy of the ratio of the genes in the gene pool still holds. Let us see how it works. Among the human blood antigens are two, designated M and N, that are produced by allelic genes. Individuals are either MM, MN, or NN, and the three types can be recognized by a simple blood test. Suppose that of 1,000 people 280 have only M antigens (and hence have *MM* genotypes), 500 have both M and N antigens (*MN* genotypes), and 220 have only N antigens (*NN* genotypes). This means that M individuals will contribute 28 per cent of the genes, all *M*, to the gene pool from which the next generation will develop, and MN individuals will contribute 50 per cent, only half of them *M*. The ratio of *M* genes in the pool is thus:

$$0.28 + \frac{0.50}{2} = 0.53$$

Then the ratio of *N* genes is 0.47. Now we can construct a checkerboard and add to it the ratios of the genes.

	0.53 *M*	0.47 *N*
0.53 *M*	0.28 MM	0.25 MN
0.47 *N*	0.25 MN	0.22 NN

This chart shows that the next generation will consist of 28 percent MM individuals, 50 per cent MN, and 22 per cent NN. The ratio of genes, and of genotypes, has remained unchanged.

The constancy of these relationships makes it unnecessary to construct a checkerboard each time. In any population the number of total alleles in a series must equal unity.

$$A + a = 1$$

In the F_2 this becomes

$$AA + 2\,Aa + aa = 1$$

This is simply an expanded binomial equation:

$$(A + a)^2 = A^2 + 2\,Aa + a^2$$

Suppose that 0.36 of the population are recessive:

$$a^2 = 0.36$$

Therefore,

$$a = 0.6$$

Since

$$A + a = 1$$

$$A + 0.6 = 1$$

Therefore,

$$A = 0.4$$

$$A^2 = 0.16$$

$$2\,Aa = 2(0.6 \times 0.4) = 0.48$$

Thus,

36 per cent of the population are *aa*,
16 per cent of the population are *AA*,
and 48 per cent of the population are *Aa*.

These relationships, discovered in 1908 by a mathematician (G. H. Hardy) and a physician (W. Weinberg), gave rise to the generalization known as the *Hardy-Weinberg law.* In its simplest form it states that as long as the frequencies of alleles *A* and *a* remain unchanged in a large randomly breeding population, the genotypic ratios will remain the same from generation to generation. A population in which the gene pool and the genotypic ratios do stay the same from generation to generation is said to be in *equilibrium.*

There are methods of testing to see whether a population is in equilibrium. If the heterozygotes can be recognized phenotypically, and if data on a sufficient number of individuals for several generations are available, it is possible to determine whether the population is in equilibrium simply by seeing whether the gene and genotypic ratios remain unchanged.

Frequently, though, the geneticist does not have data covering several generations, especially when he is dealing with the problems of a human population. As long as the heterozygotes can be identified phenotypically,[1] he can still decide whether a population is in equilibrium by determining whether the ratio of the genotypes in the population is what would be expected under the Hardy-Weinberg law. Because of genetic segregation and random recombination, the three genotypes occur in quadratic proportions in populations at equilibrium. Table 17–4 shows the results of a study of the M and N blood groups in

Table 17–4. Frequencies of *M* and *N* Genes and Genotypes in Six Populations*

No. of Individuals	Population	Allele Frequencies		Percentage of Blood Groups			
		M	*N*		MM	MN	NN
6,129	Whites (U.S.)	0.540	0.460	Obs.	29.16	49.58	21.26
				Exp.	29.16	49.68	21.16
278	Negroes (U.S.)	0.532	0.468	Obs.	28.42	49.64	21.94
				Exp.	28.35	49.89	21.86
205	Indians (U.S.)	0.776	0.224	Obs.	60.00	35.12	4.88
				Exp.	60.15	34.81	5.04
569	Eskimos (E. Greenland)	0.913	0.087	Obs.	83.48	15.64	0.88
				Exp.	83.35	15.89	0.76
504	Ainus	0.430	0.570	Obs.	17.86	50.20	31.94
				Exp.	18.45	49.01	32.34
730	Australians (aborigines)	0.178	0.882	Obs.	3.00	29.60	67.40
				Exp.	3.17	29.26	67.57

* Data from A. S. Wiener, *Blood Groups and Transfusions*, 3rd ed., 1943. Courtesy Charles C. Thomas, Publisher, Springfield, Illinois, and Dr. A. S. Wiener.

[1] If the heterozygotes cannot be recognized phenotypically, it is still possible to test for equilibrium, but it involves more complicated calculations than would be useful here.

six different populations. Although different for each population, the observed and expected ratios are very close, so it is reasonable to assume that for these blood groups the populations are in equilibrium.

Suppose, though, that a geneticist should find a population in which 44 per cent of the individuals are MM, 12 per cent MN, and 44 per cent NN. This would mean that half of the genes in the population are M

$$M = 0.44 + \frac{0.12}{2} = 0.5$$

and half are N. If the population is in equilibrium, then the ratio of the genotypes should be

$$(0.5\,M + 0.5\,N)^2 = 0.25\,MM + 0.50\,MN + 0.25\,NN$$

But, this is not the ratio that exists in the population. Something must have happened to modify it.

Once he discovers that a population is not in equilibrium, the geneticist can then begin to look for the factor that is modifying the ratio.

Modification of Genetic Equilibrium

In discussing the Hardy-Weinberg law, we made two assumptions. One was that breeding is random; the other, that the ratio of genes in the gene pool remains unchanged. If either of these assumptions does not hold, equilibrium will be upset.

Nonrandom Breeding. Random breeding simply means that every member of one sex in a population has an equal chance of mating with any member of the other sex. If a genetic trait has no bearing on the selection of mates, breeding will be random for that trait. The chances are very good that most people do not even know whether they are M, MN, or N, and they are certainly not going to choose their marriage partners on the basis of blood type. For this trait, breeding seems to be completely random. Suppose, though, that a trait does influence the selection of mates. Suppose that blue-eyed people tended to marry other blue-eyed people and that brown-eyed people preferred brown-eyed mates. Because blue eyes are recessive to brown eyes, some brown-eyed couples would still have blue-eyed children, but they with their genes would now become part of the blue-eyed preference group. This would lead to a steady increase in the homozygotes and a decrease in the heterozygotes. The relative proportions of the genes in the gene pool would remain the same, but the genotypes would not appear in quadratic proportions. Any factor—physical, ecological, geographical, or psychological—that interferes with random mating tends to bring about a shift in the genetic equilibrium.

One of the most common types of nonrandom breeding is inbreeding, or breeding with relatives. In humans, this is known as *consanguinity*. An important factor leading to this type of breeding is simply the availability of possible mates. Table 17–5 shows that people from small towns or country districts are more likely to marry their cousins than are people from large cities where there is more choice.

Because closely related individuals are more likely to have alleles in common than are unrelated individuals, the genetic result of inbreeding is to increase homozygosity. This tendency toward homozygosity is shown in Figure 17–8.

It is often thought that inbreeding is biologically detrimental per se, but this is not necessarily so. All that inbreeding does is increase homozygosity. If there are deleterious recessive alleles in the population (and there are in most individuals), inbreeding will increase the number of offspring showing the undesirable traits. Inbreeding will just as readily bring about the homozygosity of desirable traits. Inbreeding cannot bring about an increase in the number of deleterious genes in the gene pool or cause them to appear if they are not there to start with. Plant and animal breeders take advantage of the increased homozygosity of inbred lines, and many of their most successful products are highly inbred.

Table 17–5. First-Cousin Marriages in Different Populations*

Population	Period	No. of Marriages	First-Cousin Marriages (%)
Austria (urban)	1929–1930	31,823	0.53
Brazil (urban)	1946–1956	1,172	0.42
Spain (urban)	1920–1957	21,570	0.59
United States (urban)	±1935–1950	8,000	0.05
Brazil (rural)	1950–1951	179	19.55
Fiji Islands (rural)	±1850–1895	448	29.7
India (Caste of Parsees)	1950	512	12.9
Spain (rural)	1951–1958	814	4.67
Sweden (rural)	1890–1946	191	6.80
Switzerland (rural)	±1890–1932	52	11.5

* Data from Curt Stern, *Principles of Human Genetics*, W. H. Freeman and Co., San Francisco, 1960.

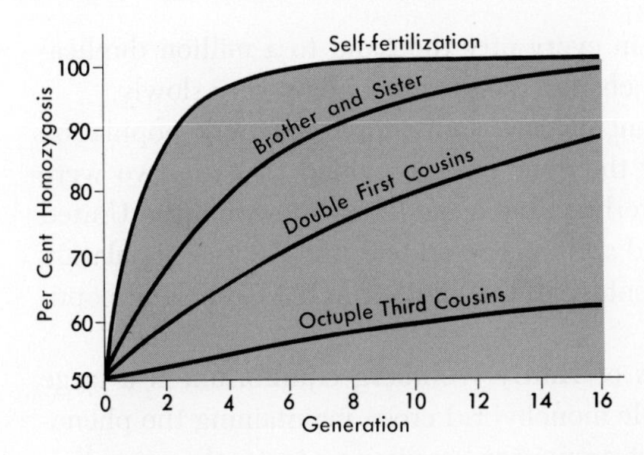

Figure 17–8. The percentage of homozygous genes present in succeeding generations in populations with different degrees of inbreeding. (After Wright, *Genetics*, **6**:172, 1921.)

Although in many parts of the world consanguineous unions are frowned on socially or even legally banned, a consideration of the human population mathematically shows that many consanguineous marriages must have occurred in the ancestry of all of us. If none had occurred, the number of ancestors of each individual would double each preceding generation. He would have two parents, four grandparents, eight great-grandparents, and so on. In other words, the number of ancestors would be two to the *n*th power, where *n* is the number of generations involved. A calculation of the number of ancestors an individual has had over the past thousand years, allowing three or four generations per century, gives anywhere from a billion to a trillion ancestors. Because the number of individuals on this earth a thousand years ago was certainly much less than this figure, there simply weren't enough people around a thousand years ago to have been his ancestors had no consanguinity occurred. As Curt Stern has pointed out, the brotherhood of man is not only a spiritual and political concept but a genetic reality.

Changes in the Gene Pool. The other condition for Hardy-Weinberg equilibrium is that the ratio of genes in the gene pool should remain unchanged. It is most unlikely that a gene pool will remain constant for many generations. Changes can be caused by

1. Mutation pressure.
2. Immigration.
3. Drift.
4. Selection pressure.

MUTATION PRESSURE. If any gene repeatedly mutates into some particular allele, then the mutant form will tend to increase in the gene pool at the expense of the mutated allele. Because the rate of mutation for any particular gene is very low, occurring perhaps once in every fifty thousand to a million duplications, mutation pressure alone will change the gene pool only very slowly.

IMMIGRATION. The movement of new individuals into any population from another population can alter the gene pool, provided that the two were different to begin with and that interbreeding occurs. If whites from the United States were to migrate to Greenland and interbreed with the Eskimo population there, this would increase the percentage of the allele N in the Greenland population (see Table 17–4).

DRIFT. One of the requisites of Hardy-Weinberg equilibrium is a large population. In the model of a simple monohybrid cross, maintaining the phenotypes and genotypes in their correct proportion requires a constantly expanding population. This is shown in Figure 17–9, in which there are two individuals in the F_1, four in the F_2, and sixteen in the F_3. Now as a matter of fact, there can be no such thing as either an infinitely large or an infinitely expanding population on this earth, although some populations are large enough so that one can apply the Hardy-Weinberg law without worrying about population size.

The point that not all individuals of a population can, under any circumstances, live and reproduce was vividly made by Nathaniel Shaler when he was professor of geology at Harvard. He calculated that if all the progeny of a plant louse should live and reproduce, at the end of the year there would be a column of plant lice having a diameter equal to that of the distance from Quincy to Brattle Street in Cambridge, Massachusetts, and thrusting itself upward through spacc at three times the speed of light.

In any natural population, then, certain individuals are bound to be removed before they can reproduce. Their removal may result from lack of food, water, room, or shelter, from disease, or from plain accidental death—but removed they will be. This removal by chance will of mathematical necessity sooner or later bring about a shift in the gene pool. If we imagine the genealogy shown in Figure 17–9 on a blackboard and simply arbitrarily remove some individuals with an eraser, we will sooner or later by chance remove more of one genotype than another (see Figure 17–10). Once this happens, the popula-

Figure 17–9. Genetic equilibrium in a constantly expanding population.

P	AA × aa	A:a=1:1
F_1	Aa × Aa	A:a=1:1
F_2	AA Aa Aa aa	A:a=1:1
F_3	AA AA Aa Aa Aa Aa aa aa AA AA Aa Aa Aa Aa aa aa	A:a=1:1

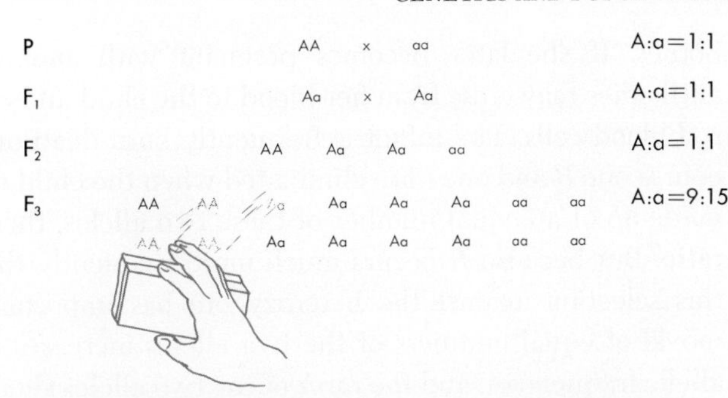

Figure 17–10. Chance removal of some individuals from a population results in a shift in gene ratios.

tion will start to drift toward homozygosity. You can see from the figure that after the four individuals (*AA, AA, AA, Aa*) are removed, another *Aa* individual will have more opportunity to breed with an *aa* individual than with an *AA* individual simply because there are more *aa*'s in the population. This phenomenon of drift toward homozygosity as a result of the chance removal of individuals is known as the *Sewall Wright effect,* after the geneticist and mathematician who first worked it out.

The larger the population is, or the more rapidly it is expanding, the less will be the effect of drift on the population. In a small population, though, drift can bring about homozygosity in relatively few generations. Because drift can work against selection pressure and bring about a shift toward the homozygosity of deleterious genes as well as of neutral or beneficient genes, its effect can be particularly devastating in small populations.

SELECTION PRESSURE. Any genotype that in any way reduces the reproductive potential of an organism will tend to be reduced or eliminated in the population, whereas any genotype that increases the reproductive potential will tend to become more abundant. Think of reproductive potential as not only the number of viable offspring the individual is able to produce but also the ability to become mature and attain reproductive capacity.

A striking case of selection pressure in humans is that of *erythroblastosis fetalis*, an often fatal anemia of the newborn. Among the blood antigens found in man and other primates is the Rh factor, named for the rhesus monkey in which it was first discovered. It is controlled by a dominant gene, *R*. Individuals who have the Rh factor (are Rh-positive) are either *RR* or *Rr*. The genotype of Rh negative individuals is *rr*. If an Rh-negative woman marries an Rh-positive man, some or all of her children may be Rh-positive (*Rr*). It sometimes happens that late in pregnancy the Rh factor in the fetus crosses the placental barrier and enters the mother's bloodstream. She then begins to produce anti-Rh antibodies. If she later becomes pregnant with another Rh-positive child, the antibodies may cross from her blood to the child and cause the destruction of its red blood cells. The infant is frequently born dead or dies soon after birth. Of course one *R* and one *r* are eliminated when the child dies. If the gene pool were made up of an equal number of these two alleles, there would be no change in ratio. But because *R* occurs much more frequently than *r* in most populations, this selection against the heterozygote has important consequences. The removal of equal numbers of the two alleles increases the imbalance of the two allelic frequencies, and the rarer of the two alleles should ultimately be lost. The elimination of *r* seems to have been reached in the Mongoloids, for in this race only the *RR* genotype is known. In the other races of man, both *R* and *r* still exist, with *R* being the predominant allele in most populations. If *r* is the most numerous gene, as it is in the Basques, then selection will be against *R*. In the chimpanzee, selection against the heterozygote has apparently eliminated *R* and in them only the genotype *rr* is known.

It is now routine practice to test the blood of a pregnant woman. If there is reason to suppose the child will be born with erythroblastosis, the doctor prepares to give it a massive blood transfusion at birth. In this way, many babies are saved that would otherwise have died. What effect will this have on selection?

Selection should not be thought of as solely a process that induces radical changes; it may instead promote stability. In a population in which mutation pressure tends to increase the percentage of some deleterious gene, selection may tend to eliminate it at the same rate and thus keep the population stable.

The net result of genetic processes continuing over millions of generations has been the development of myriads of different kinds of organisms, of different ways of meeting the problems of existence.

The major ones are considered in the next chapters.

SUGGESTED READINGS

Dobzhansky, T.: *Evolution, Genetics and Man,* John Wiley & Sons, Inc., New York, 1955.

Emmel, T. C.: *Genetics and Evolution,* Kendall/Hunt Publishing Company, Dubuque, Iowa, 1970.

Fraser, A.: *Heredity, Genes, and Chromosomes,* McGraw-Hill Book Company, New York, 1966.

Hamilton, T. H.: *Process and Pattern in Evolution,* Macmillan Publishing Co., Inc., New York, 1967.

Herskowitz, I. H.: *Principles of Genetics,* Macmillan Publishing Co., Inc., New York, 1973.

Lerner, I. M.: *Heredity, Evolution, and Society,* W. H. Freeman and Co., San Francisco, 1968.

Smith, H. H., ed.: *Evolution of Genetic Systems,* Brookhaven Symposia in Biology, No. 23, Gordon and Breach, New York, 1972.

Solbrig, O. T.: *Evolution and Systematics,* Macmillan Publishing Co., Inc., New York, 1966.

Srb, A. M., R. D. Owen, and R. S. Edgar: *General Genetics,* W. H. Freeman and Co., San Francisco, 1965.

Stebbins, G. L.: *Processes of Organic Evolution,* Prentice-Hall, Inc., Englewood Cliffs, N.J., 1966.

Stern, C.: *Principles of Human Genetics,* 3rd ed., W. H. Freeman and Co., San Francisco, 1973.

Strickberger, M. W.: *Genetics,* Macmillan Publishing Co., Inc., New York, 1968.

[illegible] L. M., [illegible] W. H. Freeman and Co., San Fran[illegible]

[illegible] New York, 1972.

[illegible] Macmillan Publishing [illegible] New York, [illegible]

[illegible] and Co., [illegible]

[illegible]

[illegible] Macmillan Publishing Co., [illegible] York, 1968.

The Diversity of Life

There are more things
in heaven and earth, Horatio, than
are dreamt of in your philosophy.
Hamlet, ACT I, SCENE 5

CHAPTER **18**

Viruses and Monera

"Animal, vegetable, or mineral?" So children ask when they play guessing games and never doubt the absolute distinctness of the three categories. It is surely one of the oldest classifications in the world. But when men came to study the minute living things revealed by the microscope, the distinctions between animal and vegetable began to break down. Some organisms can be classified as either plant or animal, and others, although obviously living, seem to be neither. Even worse, some things do not fit the classical definition of organisms, and yet they show characteristics of life. They cannot by themselves carry on metabolic activities, they do not grow, they do not respond to stimuli, and they can be crystallized. But at the same time, they are formed of biochemical molecules, they are able to reproduce, they show genetic properties, and they can mutate and recombine. These are the *viruses*. Whether one calls them organisms or not is really a matter of semantics.

VIRUSES

Many different kinds of viruses have been found, all of them parasites of either animals, flowering plants, molds, or bacteria.

Viral Structure

As seen under the electron microscope, viruses may appear as little balls, or bricks, or rods (see Figure 18–1). Viruses vary in size—the largest are as large

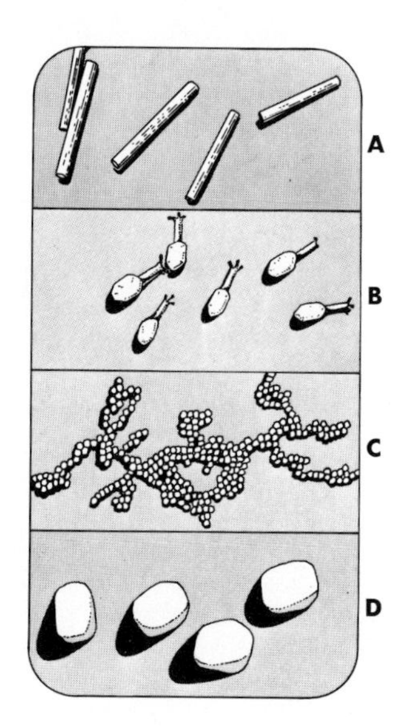

Figure 18–1. A variety of viruses. A: A rod-shaped virus that causes a disease in tobacco plants. B: A "tailed" virus, one of the kinds that attack bacteria. C: Clusters of the virus that causes poliomyelitis. D: The brick-shaped virus that causes cowpox.

as small bacteria, the smallest the size of large protein molecules (see Figure 18–2). All viruses have a central core of nucleic acid, either DNA or RNA but never both. The DNA and RNA may be either single-stranded or double-stranded. All plant and most animal viruses contain RNA, whereas a few animal viruses and most of those that attack bacteria have DNA. Surrounding the nucleic acid is a coat of protein, the *capsid,* a complicated structure composed of subunits that may be arranged in a helix or in a polyhedron (see Figure 18–3). Bacterial viruses frequently have tail-like structures that make them look a little like tadpoles. (A bacterial virus is called a *phage,* short for *bacteriophage,* meaning "eater of bacteria.") Some animal viruses are surrounded by a membrane containing lipids, proteins, and carbohydrates. Usually, but not always, this membrane is similar in composition to the cell membrane of the host.

Viral Life History

Apart from a living cell, a virus is inert. It is incapable of independent movement, but it is so small that it can diffuse through a fluid like a large molecule. It is called a *virion* rather than a *virus.*

How Viruses Enter Cells. When a virion comes in contact with the right kind of cell, it attaches to the cell membrane. This is a purely chemical reaction. For a virion, the right kind of cell is one that has a pattern of negative charges on its surface that is complementary to a pattern of positive charges on the surface of

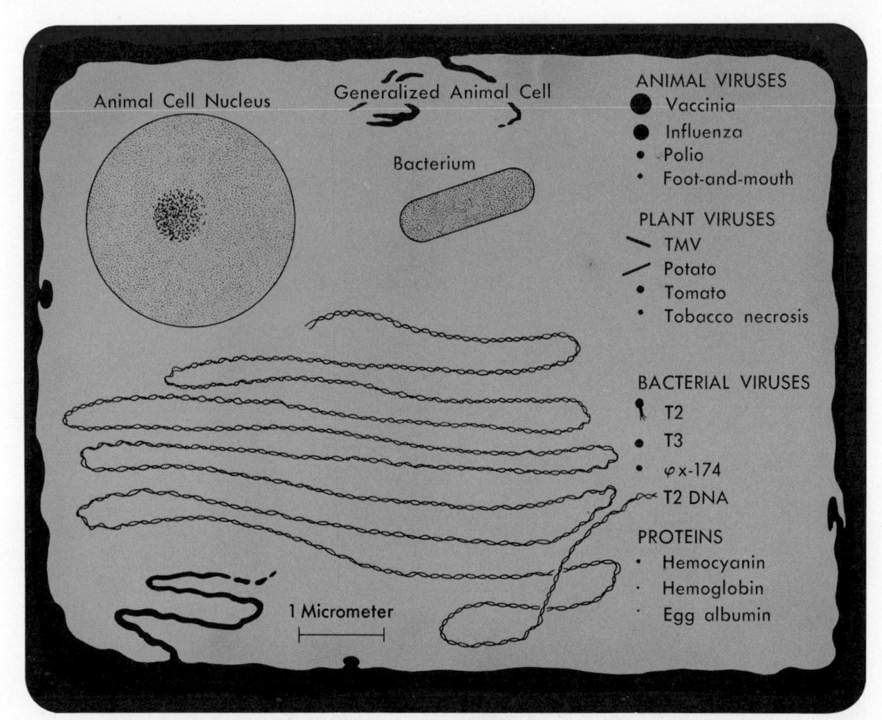

Figure 18–2. Comparison of the size of a typical animal cell, a bacterium, a number of viruses, and representative proteins. The DNA molecule of the T_2 virus is shown drawn out and greatly exaggerated in width. One micrometer is about twenty-five thousandths of an inch. (Adapted from Dean Fraser, *Viruses and Molecular Biology, Current Concepts of Biology Series*, Macmillan Publishing Co., Inc., 1967.)

the virion so that ionic bonds form between them. The virion, or at least the nucleic acid part of it, must now get inside the cell. The tail of a phage is an elaborate structure with a hollow core (see Figure 18–4). It serves as a mechanism for injecting the nucleic acid into the host bacterium while the protein coat remains outside. Most animal virions apparently stimulate the cell to which they attach to take in the whole virion by phagocytosis. When the virion is inside the cell it is called a virus. The protein coat is attacked by the cell's enzymes and the viral nucleic acid is freed. What happens next varies, depending on both the virus and the host cell.

Virulent Infection. In a virulent infection, the viral nucleic acid subverts the cell's machinery. By using the cell's ATP, free nucleotides, and enzymes, viral DNA makes mRNA after its own pattern. Viral RNA apparently acts as its own mRNA. These mRNA's attach to some of the cell's ribosomes and begin making

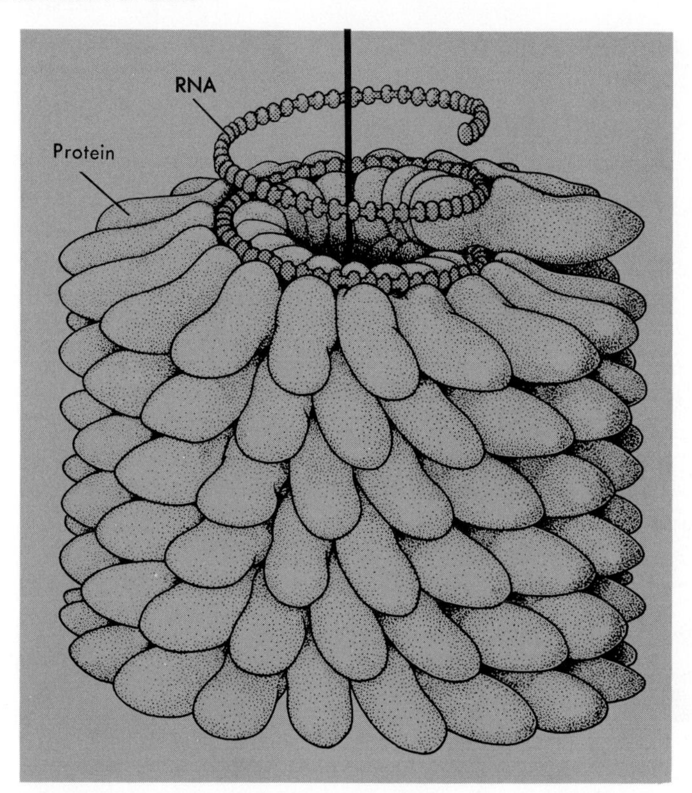

Figure 18–3. The structure of the tobacco mosaic virus showing the inner core of RNA and the protein subunits of the capsid. (Adapted from Dr. D. L. D. Caspar, as in Dean Fraser, *Viruses and Molecular Biology, Current Concepts of Biology Series*, Macmillan Publishing Co., Inc., 1967.)

proteins. Some of them interfere with the host cell's synthetic machinery. Others are the enzymes and structural proteins needed to make new viral nucleic acid and new capsids. The viral nucleic acid replicates a number of times, and the proteins and nucleic acids are assembled to form a few to several hundred new viral particles. Some viruses produce enzymes that rupture (*lyse*) the host cell so that the virions are released to find new cells to make more viruses. Other viruses do not destroy the cell; they are apparently moved out by budding, a sort of reverse pinocytosis. The virion may carry off part of the cell's plasma membrane as a protective coat that presumably shields it from antibody formation by the host.

Lysogenesis. Some phages are said to be temperate because they can establish a mutually beneficial association with one or another strain of bacterium. The

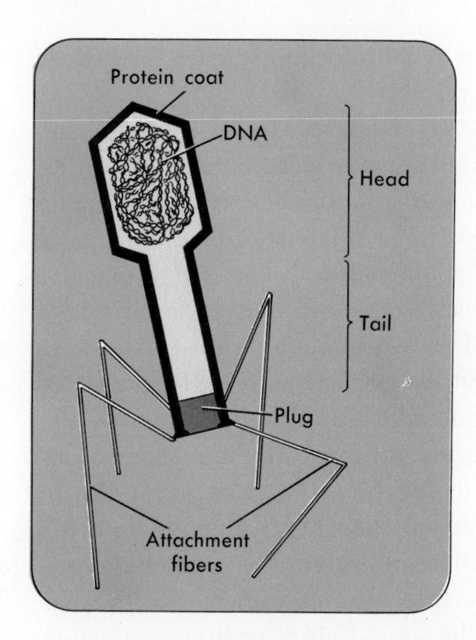

Figure 18–4. Diagram of a phage virus showing the tail. The attachment fibers carry the charges by which the virus bonds to the surface of the cell.

phage DNA enters the bacterial cell but does not disrupt it. Instead it fastens to the host's chromosome and indeed becomes part of the genome. It is then called a *prophage*. Every time the bacterial DNA replicates, the phage DNA replicates and so is passed down from generation to generation of daughter cells. The bacteria are said to be lysogenized and are resistant to infection by other similar phages. Represser proteins formed by the first phage both prevent the prophage from multiplying at the expense of its host and also suppress transcription of the DNA of other phages.

Occasionally, a prophage escapes from inhibition and becomes lytic, that is, it behaves like a virulent phage, multiplies, and ruptures the host cell. The released phage particles can then attack other bacteria and either set up new lysogenic lines or continue their destructive course. Radiation and chemicals, such as hydrogen peroxide, can switch a phage from the lysogenic to the lytic state.

Temperate Infections. Many, perhaps most, viruses do not cause virulent infections. They simply exist in the host and multiply while causing no, or only minor, signs of illness. They may be temperate in one host and virulent in another. The encephalitis virus is present in many birds but seldom makes them sick. It is transmitted from one bird to another by mosquitoes. But if the carrier mosquito should happen to bite a horse or a man, the transferred virus causes an often fatal disease. Or a virus may be temperate in one kind of cell in the host and virulent in another. Polio viruses usually multiply harmlessly in cells of the human intestinal mucosa. Rarely, the polio infection reaches the cells of the central nervous system with disastrous results.

Temperate viruses often confer benefits on the host. Many viruses are antigenic. They stimulate the production of antibodies, which confer more or less long-lasting immunity on the host. Sometimes the antibodies resulting from infection by a temperate virus protect the host from a related, virulent form. Long before anyone had heard of viruses, it was common folk knowledge that if a milkmaid with a cut on her hand milked a cow with cowpox sores on the udder, she might develop a similar sore and would thereafter be safe from smallpox. In 1796, the English doctor Edward Jenner put the matter to a test. He inoculated pus from a cowpox sore into the arm of an eight-year-old boy. Three months later he inoculated the boy with pus from a smallpox sore and thus showed that the child was immune to smallpox.

The production of antibodies is not the only benefit a temperate virus can bestow. There is also interference. The presence of one virus can sometimes prevent infection of a cell by another virus. The first virus may alter the cell's membrane so that it lacks attachment sites for the second virus, or it may somehow keep the second virus from multiplying within the cell. The influenza virus can multiply in some cells without causing obvious damage and can at the same time interfere with the multiplication of the encephalitis virus. Some viruses stimulate the cell to produce small proteins called *interferons* that can move from one cell to another and, inside the cell, protect it from a wide range of viruses. Interferons are short-lived. They seem to represent a temporary defense mechanism that gives protection until the host's antibody–forming system can take over.

Viral Genetics

Knowing that the active part of a virus is nucleic acid, you would expect to find that viruses are able to mutate. Many mutations of viruses have indeed been found. Some of these mutations make a virus more virulent or allow it to invade a new host. Other mutations make the virus less virulent. When the introduced European rabbit ran wild in Australia and effectively took over large parts of the available range land, sheepmen introduced a mosquito-borne viral disease, *myxomatosis*. The epidemic that followed was spectacular—fully 90 per cent of the rabbits died. But as time went on, the rabbits began to increase in number again. More rabbits were recovering from the disease, and those that died did not succumb so rapidly. A more temperate mutant had appeared and had replaced the original virus. When an animal dies, its viruses also die. The more quickly a rabbit died, the less chance there was for it to be bitten by a mosquito, whereas the longer a rabbit lived with the disease, the more likely it was that it would pass the virus, via mosquito, to another rabbit. Thus, the more temperate mutant spread at the expense of the virulent form. This is a normal course of evolution for a virus or for any parasite.

That a virus can mutate is not surprising. But most biologists were surprised by the announcement that viruses can also recombine. When a bacterial cell is infected with two different phages, the viral particles released may be of several different kinds, two like the parent strains and others that represent new combinations of the parental genotypes. It is very reminiscent of (although not the same as) chromosomal crossover in plants and animals, and it may be an important mechanism in viral evolution.

Viruses and Cancer

At first glance it might seem a long way between viruses and cancer. Viruses cause the destruction of cells, cancer is an enormous overgrowth of cells. But it seems more and more probable that viruses will be found to be involved in at least some human cancers. Many animal cancers are known to be caused by viruses, and viruses have been reported to be associated with a few human cancers. Viral antibodies may be present in cancers in which the viruses themselves cannot be recognized. Some viruses cause a low incidence of cancer when injected into test animals, but if the animals are then irradiated or treated with carcinogenic chemicals, the incidence of cancer is very much higher than the combined effect of either virus or carcinogen alone. Remember that X-rays and chemicals can change temperate prophages into virulent lytic phages. Our knowledge of prophages is solely a result of the fact that they occasionally become lytic. Perhaps there are many viruses that are normally present in a masked form but, when roused by carcinogens, they change the genetic constitution of the cell so that the factors that normally inhibit cell division in differentiated tissue no longer operate. Sharks, which seem to be immune to viral diseases, are also practically immune to cancer.

KINGDOM MONERA

Unlike the viruses, the two groups placed in the major subdivision of living things called *Monera* are at the cellular level of organization. They exist as cells or groups of cells. They have both DNA and RNA. They are able to take in materials from outside, to synthesize their own proteins, and to make ATP. But the cellular structure of the Monera is far less complex than that of even the single-celled plants and animals. They do not have a nuclear membrane separating the DNA from the rest of the cell, nor do they have mitochondria, an endoplasmic reticulum, a Golgi apparatus, or lysosomes. Their chromosome is a single DNA molecule, without the protein found in the chromosomes of plants and animals. Cells of this kind are said to be *procaryotic.*

Bacteria

Bacteria are minute; the smallest are no larger than a large virus, the largest are less than 0.025 millimeters (a thousandth of an inch) long (see Figure 18–2). Fifty million bacteria can exist in a single drop of water. They are omnipresent in the soil, in fresh and salt water, and in the bodies of plants and animals. They are found floating in the air, either free or attached to a particle of dust.

Kinds of Bacteria. Bacteria are divided into three major groups, depending on their shape. If a bacterium is round, it is called a *coccus;* if rod-shaped, a *bacillus;* if spiral, a *spirillum.* Sometimes the cells stay together after cell division to form chains or clumps. Thus cocci may appear as a single cell (*monococcus*), two cells (*diplococcus*), a string of cells (*streptococcus*), or a grapelike cluster (*staphylococcus*). Figure 18–5 shows a number of different kinds of bacteria.

Protection from Environment. Bacteria usually live in a hypo-osmotic solution and are in danger of hydrolysis. They form around themselves a rigid, nonliving cell wall. The inward movement of water molecules by osmosis creates a pressure within the cell and pushes the cell membrane up against the cell wall. This is called *turgor pressure.* The cell wall, although permeable to water, is rigid and resists the outward push. The back push of the wall on the cell is called *wall pressure;* it prevents hydrolysis of the cell. The increased pressure within the cell results in an increased outward flow of water molecules so that the cell remains in equilibrium with its environment. Plant cells also have cell walls, which is one of the reasons bacteria were formerly classified as plants, but the bacterial cell wall differs chemically from the plant cell wall. It is composed of polypeptides and *mucopolysaccharides* (sugars that have an amino group) and sometimes includes large amounts of lipids, whereas plant cell walls are made up largely of the carbohydrate cellulose. Penicillin inhibits the development of the materials of the bacterial cell wall. When the bacteria divide, the daughter cells have poorly developed cell walls and are readily hydrolyzed by the fluids of the host. This is why penicillin is such a useful antibiotic against many kinds of bacteria but is worthless in virus-caused diseases.

In addition to the cell wall, many bacteria develop a slimy coat, the *capsule.* It seems to protect disease-causing bacteria from the destructive activity of white blood cells.

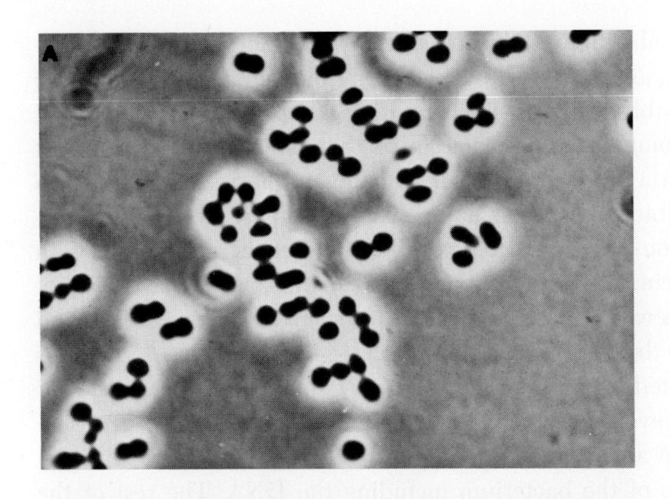

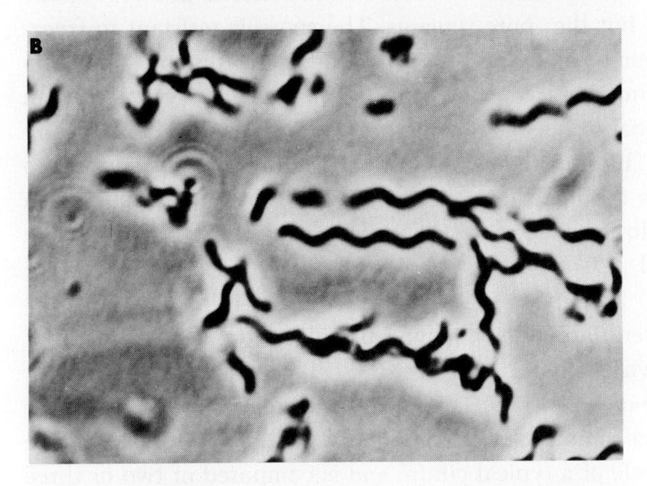

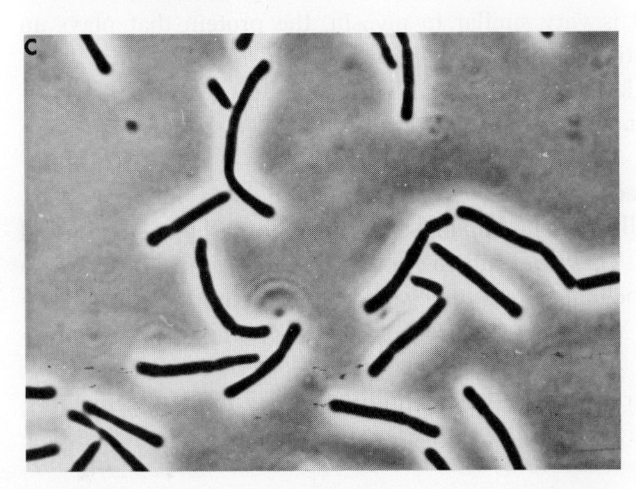

Figure 18–5. Bacterial cell types. A. *Streptococcus*, a coccus type. This species is said to be involved in dental caries. B: *Spirillum itersonii*, a spiral type. This particular species lives in water and is harmless. C: *Bacillus subtilis*, a bacillus type taken from a hay infusion. (Photos courtesy Paul H. Smith.)

Some bacteria are able to produce and secrete substances that are destructive to other micro-organisms (antibiotics). From the point of view of the bacterium, this is probably a means of eliminating competitors for something needed from the environment. Man has adapted this ability to his own needs and now cultures bacteria in huge tanks, harvests the antibiotics, and uses them as medicine. Some antibiotics have been chemically analyzed and can be produced synthetically. You have almost surely been dosed at some time with streptomycin, terramycin, neomycin, or perhaps an antibiotic that was not even heard of when this was written. Mutant strains of bacteria have developed that are resistant to the drugs that are in current use. As the nonresistant forms of bacteria are eliminated, the mutants spread very rapidly, so the search for new antibiotics seems never-ending.

Some bacteria are able to form *endospores*. A tough, resistant coat develops around a section of the bacterium including the DNA. The rest of the bacterium disintegrates, but the spore remains. The spore is resistant to freezing, to boiling for an hour or more, and to disinfectants. In this state, the bacterium can withstand unfavorable conditions, sometimes for years. It can emerge from the spore coat to begin growing and dividing when conditions are again right. Fortunately, most of the spore-forming bacteria are not harmful to man, but a few, such as the one that causes tetanus, are. Tetanus spores can remain dormant in the dirt for years, only to revive and multiply rapidly when introduced into a wound.

Movement. Many bacteria are able to move from place to place by means of flagella, which may be widely distributed over the cell or restricted to a tuft or a single flagellum at one or both ends. The bacterial flagellum is much simpler in structure than the flagella and cilia of animal and plant cells. It is about the size of a single microtubule of a typical cilium and is composed of two or three strands of a protein that is very similar to myosin, the protein that plays an active role in the contraction of muscles.

Some bacteria that lack flagella can still move by gliding over a surface—how they do this is not known.

Bacterial Nutrition. Green plants use energy from sunlight to build biochemical molecules. Specifically, plants can make sugar from carbon dioxide and water (see Chapter 29). They are said to be *autotrophic*. Animals cannot do this but must get their biochemical molecules ready-formed by the plants. They are called *heterotrophs*. Most bacteria are heterotrophs, but a few are autotrophs.

HETEROTROPHIC BACTERIA. Heterotrophic bacteria are either *saprophytes*, feeding on dead organic remains, or *parasites*, feeding on living hosts. The saprophytes are the ones that cause food spoilage. More importantly, they bring about the decay of dead animals and plants. Consider for a moment what the world would be like if there were no saprophytes. Some of the carbon built into food by plants is returned to the air by the respiration of both plants and animals. But much of this carbon remains locked in their bodies. If there were no organisms of decay, no decomposers to break down organic remains and free the carbon dioxide, it would probably long ago have become so scarce that plants would no longer be able to survive. They would have died, and the animals with them. If it were not for the saprophytes, life as we know it on this earth would soon come to an end.

The bacteria that cause disease are not all parasites. They may be saprophytes but secrete a deadly poison. This is true of the bacteria that cause botulism, an often fatal food poisoning. The diphtheria bacterium lives harmlessly in the human throat unless it is lysogenized by a certain strain of virus. Then the bacterium begins to secrete the toxin that causes the symptoms of diphtheria. Other bacteria do not secrete their poisons. But when they die and disintegrate, toxins are released that produce the symptoms of such diseases as cholera, typhoid fever, and whooping cough.

AUTOTROPHIC BACTERIA. A few of the bacteria that are able to synthesize their own biochemical molecules use the energy of sunlight, as the green plants do, and so are called *phototrophs.* Other bacteria, called *chemoautotrophs,* obtain the energy for synthesis by oxidizing inorganic materials, such as sulfur or iron. Most important of these are the *nitrifying bacteria.* Nitrogen is essential for all living things. Both nucleotides and amino acids contain nitrogen; without it neither nucleic acids nor proteins could be formed. The air is nearly 80 per cent nitrogen (N_2), but plants, animals, and most bacteria are not able to use atmospheric nitrogen. One of the products of organic decay is ammonia (NH_3). Some nitrifying bacteria obtain their energy by oxidizing ammonia in the soil to nitrite (NO_2) and others oxidize the nitrite to nitrate (NO_3). The nitrate released by nitrifying bacteria is the main source of nitrogen for plants. They build it into compounds that in turn are the nitrogen source for animals.

Bacterial Respiration. Both autotrophic and heterotrophic bacteria, like all cells, carry on cellular respiration, the process by which food molecules are oxidized and the released energy is stored in ATP. During this process, hydrogen atoms, or electrons, are transferred by a series of oxidation-reduction reactions to a final acceptor. In aerobic respiration by both the higher plants and by animals, the final acceptor is oxygen and water is formed. Many bacteria carry on aerobic respiration. If free oxygen is not available, some bacteria, the *denitrifying bacteria,* use nitrate as the oxygen source. The nitrogen is returned to the air as nitrogen gas.

In anaerobic respiration, the final acceptor is not oxygen. If it is an organic molecule, lactic acid (in animals) or alcohol (in plants) is formed. Bacteria are also capable of anaerobic respiration. The compounds they form, mostly organic acids and alcohols, are used by man in making cheese, sauerkraut, vinegar, and cattle feed.

Botanists often use the term *fermentation* for the type of respiration in which the final hydrogen acceptor is an organic molecule. They restrict the term *anaerobic respiration* to a process carried on by some bacteria and a few other organisms. In this process, an inorganic molecule other than oxygen is the final acceptor of hydrogen. The nitrogen-fixing bacteria can use free nitrogen as the acceptor and with the ammonia thus formed can synthesize biochemical nitrogen-containing compounds. This is a major way in which the nitrogen lost to the organic world by the activity of the denitrifying bacteria is replaced.

Bacterial Reproduction. Bacteria usually reproduce by simple binary fission. Under optimum conditions the daughter cells grow very rapidly and may be ready to divide again in as soon as twenty minutes. At this rate a single bacterium could have over a million descendants in less than seven hours.

It used to be thought that bacterial division was a more or less haphazard affair, because there is no obvious nucleus nor chromosomes nor mitotic apparatus. Now we known that in at least some bacteria the DNA forms a long, circular, double strand, that it replicates as the DNA in the chromosomes of higher organisms does, and that one daughter strand goes to each daughter cell. For convenience, the DNA strands are called chromosomes, although they lack the protein found in the chromosomes of plants and animals.

Bacterial Genetics. Bacteria are haploid, but occasionally they undergo a process that is much like sexual reproduction. There are two mating types, "sexes," in the common intestinal bacterium, *Escherichia coli.* When colonies of the two types are mixed, the individuals come together in pairs and a bridge forms between the pair partners. Then the male chromosome starts to flow into the female. Usually the pair separates before the transfer is complete. The male is left with an incomplete set of genes and soon dies; the female for a short time is diploid for the genes received from the male. This state does not last—something like crossing over takes place and some of the male genes replace their counterparts in the female chromosome. So it has been possible to construct chromosome maps for bacteria, far more detailed maps than are available for any mammal. Bacterial recombination is shown in very diagrammatic form in Figure 18–6.

Mutation and recombination are not the only way that variation can be introduced into the bacterial chromosome. The history of lysogenic phages has already been discussed. Sometimes when such a phage becomes lytic, it carries

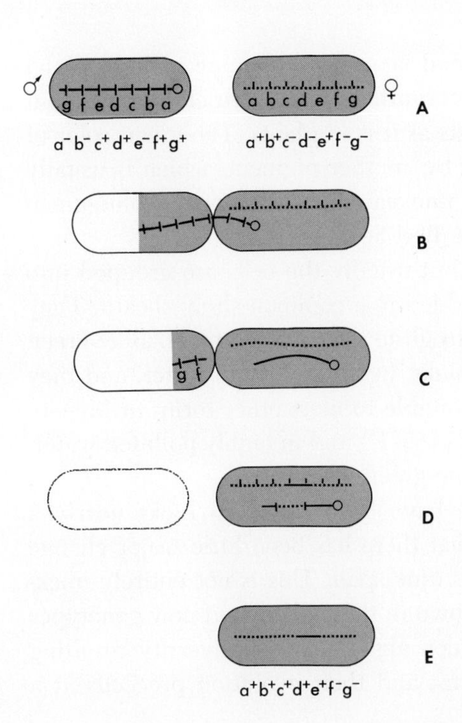

Figure 18–6. Diagram of recombination in bacteria. A: The two mating types. Their chromosomes carry different alleles, indicated by + and −. B: The male chromosome is being transferred to the female. C: The male chromosome breaks before transfer is complete. D: Crossing-over has taken place between male and female chromosomes in the female. The male dies. E: The female now has a different complement of alleles.

with it a bit of the chromosome of the bacterium with which it first established a lysogenic relationship. If later it becomes incorporated into another bacterial chromosome, it transfers to that bacterium one or more genetic characters from the original bacterium. This is called *transduction.*

There is yet another process (*transformation*) that is perhaps an artificial event, happening only in the laboratory. A bacterium that causes pneumonia has a thick capsule; colonies of it appear smooth. These bacteria quickly kill a mouse into which they are injected. A related form of this bacterium lacks the capsule and its colonies look rough; it is not virulent. But when heat-killed smooth bacteria and live rough bacteria, neither of which is deadly alone, are injected together, the mouse dies and can be shown to contain live smooth bacteria. The rough bacteria have somehow picked up the genetic information to form a capsule. The first real proof that the genetic material is DNA, not protein as most biologists had thought, came when it was shown that the rough bacteria take up (probably by pinocytosis) part of the DNA in the cellular debris of the disintegrating dead smooth bacteria and incorporate it into their chromosome.

Blue-Green Algae

The blue-green algae are less numerous and less important to man than the bacteria. Consequently, less is known about them. Like bacteria, blue-green algae lack nuclear membranes, mitochondria, endoplasmic reticulum, Golgi apparatus, and lysosomes. They are photosynthetic and have chlorophyll, but the pigment is not included in chloroplasts as it is in plants. The green color of the chlorophyll in these algae is masked by another pigment, which is usually blue but is occasionally red. (Periods of unusual abundance of red blue-green algae give the occasional red color to the Red Sea.)

Blue-green algae are single-celled, but usually the cells are grouped into clusters or filaments, with the cells embedded in a common slimy sheath. They lack flagella, but some species are able to glide over wet surfaces. Blue-green algae are found on moist rocks or tree trunks, in the soil, or in water, and they are able to live in places that are uninhabitable to most other forms of life—in hot springs with temperatures up to 71°C (185°F) and in highly polluted water. Figure 18–7 shows some characteristic blue-green algae.

Fossils of typical blue-green algae have been found in rocks nearly a billion years old. These fossils indicate that there has been little major change in blue-green algae during this enormous time span. This is not entirely unexpected, since sexual reproduction is unknown in these algae and new genotypes are produced only by mutation. Blue-green algae lack the diversity resulting from recombination found in sexual forms, and their evolution proceeds at a slower rate.

The blue-green algae contribute to the soil both by the simple addition of organic material and by the ability of many of them to fix nitrogen. Infestations

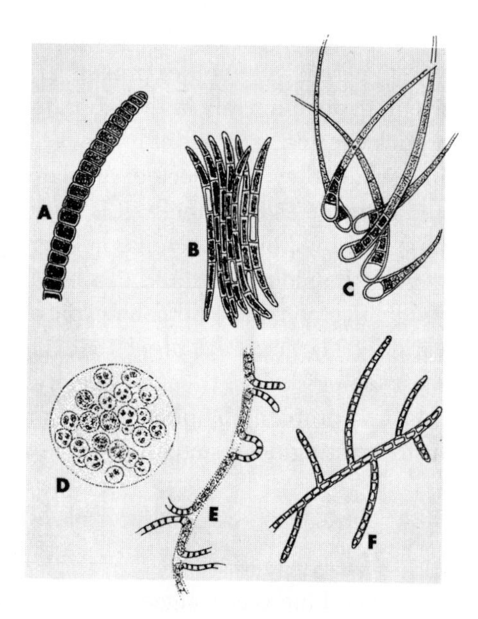

Figure 18–7. Some blue-green algae. A: *Oscillatoria*. B: *Aphanizomenon*. C: *Rivularia*. D: *Microcystis*. E: *Scytonema*. F: *Hapalosiphon*.

of these algae in city reservoirs sometimes create a nuisance by discoloring the water and giving it a fishy taste and odor.

SUGGESTED READINGS

Fraser, D.: *Viruses and Molecular Biology,* Macmillan Publishing Co., Inc., New York, 1967.

Greulach, V. A.: *Plant Function and Structure,* Macmillan Publishing Co., Inc., New York, 1973.

Krueger, R. G., et al.: *Introduction to Microbiology,* Macmillan Publishing Co., Inc., New York, 1973.

Muller, W. H.: *Botany: A Functional Approach,* 2nd ed., Macmillan Publishing Co., Inc., New York, 1969.

Volk, W. A. and Wheeler, M. F.: *Basic Microbiology,* J. B. Lippincott Co., Philadelphia, 1973.

CHAPTER 19

The Algae and Fungi

You have probably never seen a bacterium—no one had until about three hundred years ago when Leeuwenhoek ground his marvelous lenses and began looking through them at scrapings from the inside of his mouth and drops of pond water. But you are probably familiar with most kinds of organisms that are discussed in the next four chapters. They are the plants, animals, and fungi that make up such a large and important part of man's environment. They all have cells that are more or less like the "typical" cell discussed in Chapter 1, with a well-defined nucleus enclosed in a nuclear membrane, mitochondria, and endoplasmic reticulum. Such cells are said to be *eucaryotic*.

Organisms that have this kind of cell are usually considered to be either plants or animals, but it is sometimes hard to decide with which group a given organism should be classified. This is because there is no absolute criterion, no trait that is found in all plants and only in plants or in all animals and only in animals. The important difference is that plants are food manufacturers (autotrophs) and animals are heterotrophs that have to get their food from other organisms, living or dead. But some plants are more or less heterotrophic. The cells of most plants have a cell wall, usually composed largely of the carbohydrate cellulose, but some autotrophs lack a cell wall. Plants are usually not able to move about under their own power, but some autotrophs are active swimmers. These anomalous forms are usually classified with the plants. Some of them seem closely related to undoubted plants, others have simply traditionally been studied by botanists rather than zoologists. They are all included in this chapter, along with the fungi, which were formerly classed as plants but are now thought to belong to a separate kingdom.

STAGES OF PLANT COMPLEXITY

The living plants do not form a smooth sequence from the simplest to the most complex, but instead show a steplike series of stages of complexity. These stages are

1. Cellular structure.
2. Multicellular structure.
3. Cellular specialization.
4. Development of an embryo.
5. Development of specialized transport systems.
6. Development of seeds.

Botanists arrange the species of plants into thirteen distinct divisions. Each division is thought to be an assemblage of species that have a common ancestor; they have characteristics in common that are interpreted as indicating a major, distinct evolutionary line. Not every division is named or illustrated here (this classification can be found in Appendix C). Instead, attention is focused on groups that illustrate the stages of complexity.

ALGAE

For convenience, the plant divisions can be grouped into algae, bryophytes, and vascular plants. Algae are plants that are structurally simple; some are single-celled, and some are multicellular, but the somatic cells are all very similar. The plant body, whether it is composed of one cell or many, is called a *thallus*. Algae exemplify the first and second of the stages of plant complexity.

Nature of Algae

All algae are essentially aquatic. They are the seaweeds of the oceans, the pond scum of fresh waters (see Figure 19–1). A few are able to live in moist soils or on other organisms. They may turn up in unexpected places. Some flourish on snow fields, which are thereby turned decidedly pink. Other algae grow on the hairs of the three-toed sloth. They give the animal a green color that is probably a useful camouflage in the jungle trees in which it lives. Still others are responsible for the sulfur-colored underside of the sulfur-bottomed whale. Algae range in size from microscopic single-celled organisms to the giant

Figure 19–1. Masses of the green alga, *Spirogyra*, at Erie, Pa. (Carnegie Museum—O. E. Jennings photo.)

kelps, seaweeds in which the thallus may reach lengths of over sixty meters (two hundred feet). Most algae contain chlorophyll but many of them do not appear green because other pigments mask the green color of the chlorophyll.

Algae are of great economic importance; they are the basic food producers of the sea and make possible man's extensive fisheries. Some algae are eaten by people, especially in Oriental regions. They are rich sources of iodine and provide commercially important substances that are used to give body and substance to such familiar products as shaving cream and ice cream. By their photosynthetic activity they release oxygen and indeed are a major source of the oxygen needed by all animals.

Few algae are harmful to man. One species occasionally multiplies in enormous numbers in offshore waters, turning the water red—the notorious Red Tide. These algae produce a toxin that, in sufficient concentration, kills fishes by the millions. When the dead fishes are washed up on resort beaches, they can cause economic havoc. Another algal species, found mostly on the Pacific Coast of North America, also produces a virulent poison. The algae may be present in enormous numbers during the summer months when they make up a major part of the diet of marine mussels. The poison does not hurt the shellfish but accumulates in their bodies and if they are then eaten by men, the result may be serious illness or even death.

Some algae, called *diatoms,* have their cell wall impregnated with silica to form a hard shell marked with an elaborate pattern of ribs, beads, and pores as intricate as a snowflake (see Figure 19–2). In past ages diatoms flourished in the seas in uncounted billions just as they do today. When they died, their cell walls sank to the ocean floor to accumulate in layers many meters thick. Later some of these beds were elevated above the sea to form rich deposits of diatomaceous earth. It is mined commercially and used in the production of abrasives and polishes and in the manufacture of insulating materials.

Because they reproduce rapidly, consume carbon dioxide, and release oxygen, algae are being considered as occupants of spacecrafts—both for the production of food for astronauts and for the regulation of the gas content of the air.

At least six distinct lines of eucaryotic algae probably evolved from the procaryotic blue-green algae.

Structure of Algae. Algal cells are basically similar, whether they exist as separate organisms or in a multicellular body.

CELLULAR STRUCTURE. The most conspicuous organelle in an algal cell is usually the chloroplast. It is surrounded by a membrane and contains *lamellae*

Figure 19–2. A diatom, *Coscinodiscus.*

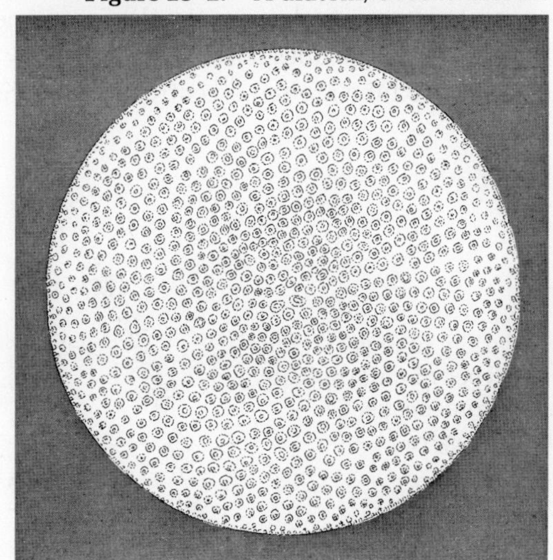

on which the chlorophyll granules are arranged. The chloroplast may be like a ribbon spiraling around close to the cell membrane or a network or a large, lobed structure, or there may be a number of small, discrete *plastids* (see Figure 19–3). Most algae have a cell wall, and some also have a gelatinous outer sheath or hard shell (sometimes considered part of the cell wall).

In common with the cells of the higher plants, some algal cells have a large, inner, membrane-lined *vacuole* filled with water containing various dissolved substances—salts, sugars, and nitrogenous compounds. The vacuole serves as a storage place and helps maintain the osmotic equilibrium of the cell.

Many algae have one or more flagella and are active swimmers. The flagella are similar structurally to those of animals cells—they have a central pair of tubules and nine double tubules arranged in a circle around the periphery, all surrounded by an extension of the cell membrane. Algae that are able to swim sometimes lack cell walls and frequently have a red "eye spot" (*stigma*), a light-sensitive organelle by which they are able to orient toward light. They may have a *contractile vacuole,* one that is able to swell up and then contract, discharging its content to the outside. This helps rid the cells of excess water that has moved in by osmosis. Some of the flagellates are heterotrophic; they are often classed with the animals and may be dangerous parasites. African sleeping sickness is caused by a flagellate. Figure 19–4 shows a flagellated alga.

MULTICELLULAR STRUCTURE. Algae are considered simple plants because they show little cellular differentiation. Yet when algal cells remain together after cell division, they produce a remarkable diversity of forms. Sometimes the cells clump together to form colonies, flat masses or balls of cells usually embedded in a gelatinous matrix. The cells in a colony may be completely separate or connected by protoplasmic strands. Sometimes the nucleus divides many times but the cytoplasm does not. These forms may take the shape of a multinucleate, cytoplasmic tube around a central vacuole. Sometimes cell division takes place

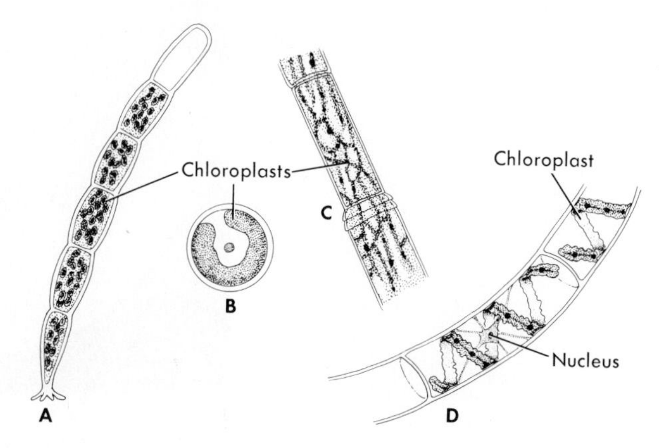

Figure 19–3. A variety of algae showing different types of chloroplasts. A: *Tribonema.* B: *Protococcus.* C: *Oedogonium.* D: *Spirogyra.*

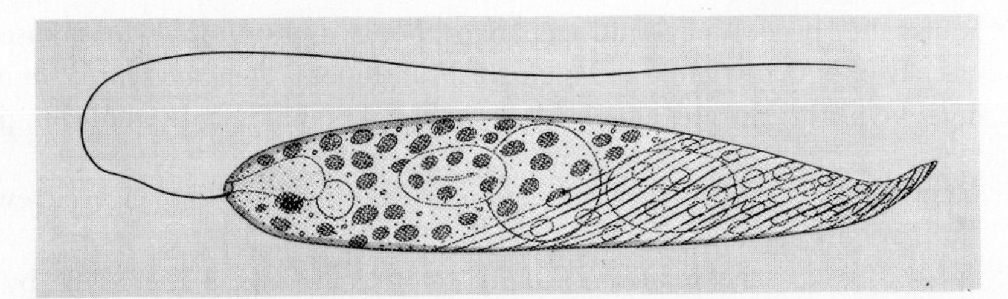

Figure 19–4. A flagellated alga, *Euglena.*

in a single direction to form a long chain of cells, a filament. Sometimes the cell divisions are in several directions so that a membranous bladelike structure several cell layers thick is formed. The basal cell of a filamentous or membranous alga is frequently modified to form a holdfast that anchors the plant in place. The large algae seem to mimic the higher plants, with a rootlike holdfast, stemlike filaments, and leaflike blades. Some multicellular algae are shown in Figure 19–5.

Nutrition. Algae absorb the materials they need directly from their surroundings and generally lack special organs of intake. Most are autotrophic, building carbohydrates from carbon dioxide and water, but a few are heterotrophic and must get their carbon in the form of biological compounds. For the other ele-

Figure 19–5. Multicellular algae. A: *Ulothrix.* B: *Macrocystis,* a large kelp. C: *Hydrodyctium.* (These are not to scale. The kelp is much reduced.)

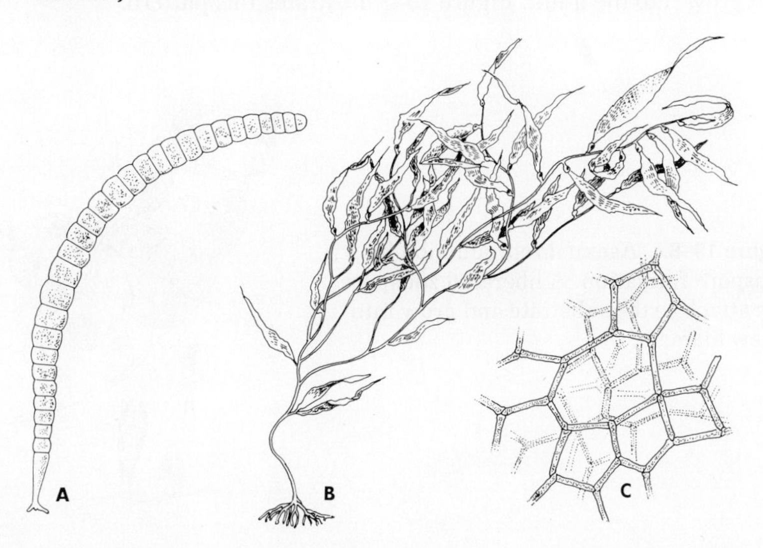

ments they need, such as nitrogen, phosphorus, and sulfur, algae can use a wide variety of both inorganic and biological substances. Thus the source for nitrogen may be nitrite, nitrate, ammonia, urea, certain amino acids, and even proteins.

Reproduction. Asexual reproduction is the only kind known in a few algae, but most are capable of sexual reproduction as well.

ASEXUAL REPRODUCTION. Many single-celled algae reproduce by simple binary fission. Multicellular algae may reproduce by fragmentation, in which parts of the colony or filament break away to form new plants. Sometimes special reproductive cells called *spores* develop. A cell in a filamentous alga may divide several times to give rise to a number of flagellated cells resembling some of the unicellular, motile algae (see Figure 19–6). Such a spore (called a *zoospore*) can swim away from the parent filament to settle down somewhere else, lose its flagella, develop a holdfast, and begin to divide to form a new filament. The spore may lack flagella but be encased in a hard covering in which it can lie dormant when conditions are unfavorable to algal growth.

SEXUAL REPRODUCTION. There are three basic kinds of sexual reproduction in the algae. Remember that in animal reproduction, division and growth take place only between fertilization and a subsequent meiosis, never between meiosis and fertilization. The animal gamete must either fuse with another gamete to make a zygote or perish. The animal body is made up of diploid cells and only the gametes are haploid. Reproduction in some algae resembles animal reproduction. This pattern is illustrated by the word diagram in Figure 19–7.

In other algae the situation is reversed. The cells of the plant body are haploid. Gametes are produced by mitosis. Two gametes fuse to form a diploid zygote, but the zygote then undergoes meiosis to produce haploid cells (spores) that grow into the adult. Figure 19–8 illustrates this pattern.

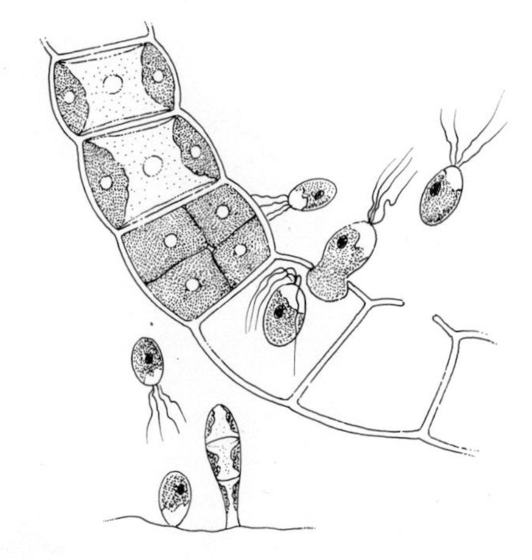

Figure 19–6. Asexual reproduction by zoospore formation. A liberated zoospore can attach to the substrate and grow into a new filament.

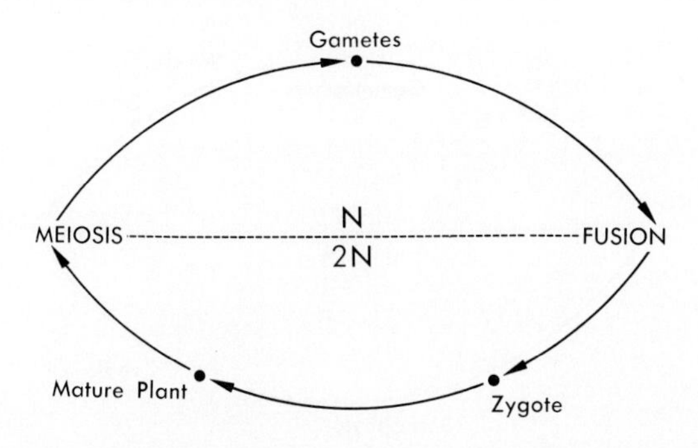

Figure 19–7. Sexual reproduction in some algae resembles animal reproduction. Only the gametes are haploid.

The third type of algal reproduction is the most complex; it is also the most interesting to evolutionists because the higher plants apparently evolved from algae showing this pattern. In a way it is a combination of the other two types. Cell division and growth occur both between fertilization and meiosis and between meiosis and fertilization. This is known as *alternation of generations.* The diploid plant body that develops from the zygote is called a *sporophyte.* It produces haploid spores by meiosis. The plant body that develops from the spore is called a *gametophyte.* It produces gametes. But a gametophyte plant is itself haploid because it developed by mitotic divisions from a haploid spore; hence it necessarily produces its gametes by simple mitotic division. This type of reproduction is summarized in Figure 19–9.

The life history of a well-known green alga illustrates alternation of generations as it actually occurs in a plant. Sea lettuce is a familiar plant along the seacoast. It is a leaf-shaped alga with a large thallus two cell layers thick and a holdfast that attaches it to rocks or pilings along the coast. A mature sporophyte plant produces zoospores by the meiosis of a special group of cells called *sporangia.* Each zoospore then settles down and grows by mitotic division to form a new thallus that is indistinguishable by the naked eye from the sporophyte plant. This new plant is the haploid gametophyte. Special groups of cells

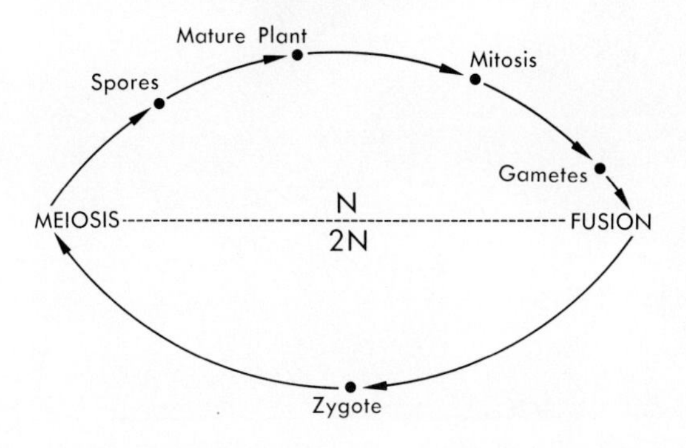

Figure 19–8. Another pattern of sexual reproduction in algae. The mature plant is haploid, gametes are formed by mitosis, the zygote (the only diploid stage) divides by meiosis to give rise to haploid spores that grow into the mature plant.

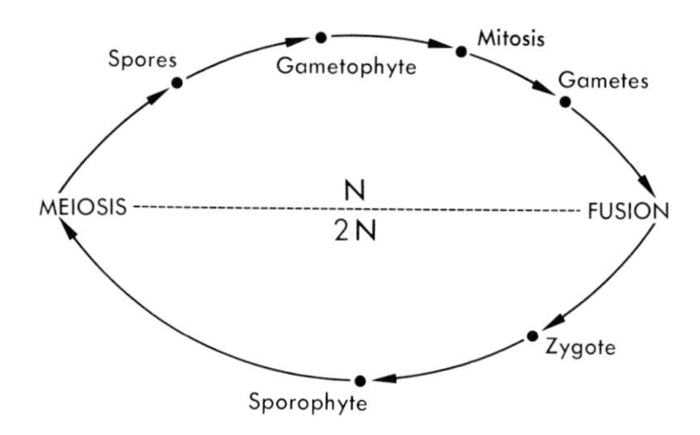

Figure 19–9. Sexual reproduction with alternation of generations. The mature diploid plant (sporophyte) gives rise to haploid spores by meiosis. The spore grows into a haploid plant (gametophyte), which produces gametes by mitosis. The gametes fuse to form the diploid zygote, which grows into the sporophyte.

called *gametangia* divide by mitosis to produce haploid gametes. All the gametes of the sea lettuce look alike, they cannot be separated into eggs and sperm, and are thus called *isogametes*. They also look like the zoospores but are smaller and have only two flagella, whereas the zoospores have four. Two isogametes from different thalli fuse to form a zygote, which, by mitotic divisions, develops into a sporophyte thallus. In summary, the gametophyte plant develops from the haploid zoospore, whereas the sporophyte plant develops as a result of the fusion of two haploid gametes. Reproduction in the sea lettuce is shown in Figure 19–10.

KINGDOM FUNGI

The slime molds and fungi differ so much from the typical algae that they are now usually placed in a separate kingdom.

Figure 19–10 Alternation of generations in the sea lettuce (*Ulva*). A: Sporophyte. B: Zoospores. C: Gametophyte. D: Gametes.

Slime Molds

Of all the groups included in this chapter, the slime molds are the most anomalous. They feed like animals and reproduce like plants. They are widespread in moist soil rich in organic material, in rotting trees and stumps, and in decaying leaf mold. There are several groups of slime molds which are probably not very closely related. In one group, the vegetative body is a mass of slimy material sometimes several feet in diameter. It usually appears as a fan-shaped network, often brightly colored. This slimy body mass, called a *plasmodium*, creeps over and into the material on which it is feeding, engulfing bacteria and other solid food particles all the while. After a time it comes to a rest and gives rise to brightly colored spore cases containing numerous tiny haploid spores, When a spore case breaks open, the spores are spread by the wind. If one comes to rest in a suitable place, it germinates, releasing one to four haploid cells. They may move about, feed, grow, and divide for a time, but eventually two fuse to form a zygote. As the zygote feeds and grows, the nucleus divides many times mitotically but the cytoplasm does not. It is this multinucleate mass of cytoplasm that is the plasmodium. Slime molds are of little economic importance, but biologists find them fascinating and they are frequently used in research on the structure of cytoplasm and the mechanics of nuclear division.

Fungi

People are usually more familiar with fungi than with algae. They have eaten mushrooms, seen a toadstool, discarded a piece of bread because it was moldy, or been treated for athlete's foot. Fungi lack chlorophyll and so are either saprophytic or parasitic. A few fungi are aquatic but most are terrestrial. They range in size from microscopic forms to ones in which the body is several meters (yards) in extent.

Fungi share with bacteria the role of decomposers. They break down organic remains and return nutrients to the soil. And like bacteria they cause disease in man, plants, and animals. Fungi have altered the course of history. The potato blight infestation of the European potato fields in 1845–1847 caused widespread famine, especially in Ireland. This helped bring about the repeal of the notorious British Corn Laws, which were designed to keep up the price of the cereal grains, and also led to a wholesale migration of the Irish to the United States. On the other hand, the most versatile antibiotic, penicillin, is produced by a fungus. The yeasts are fungi that carry on fermentation and are used in making bread, beer, wine, and whiskey.

A fungus that attacks cereal grains, such as wheat and rye, produces purplish bodies called *ergots* in the seed heads. Eating bread made from ergot-infested grain causes ergotism, a disease marked by hallucinations, convulsions, blindness, and sometimes gangrene of the hands and feet. Epidemics of ergotism, called "Holy Fire," were common in the Middle Ages and one occurred in France in 1951. LSD was first isolated from ergot.

Structure. The body or thallus of a fungus may be single-celled or multicellular. The thallus of multicellular forms, called a *mycelium,* is composed of a tangle of long, fine, often tubular, branching filaments (*hyphae*). Some hyphae are divided into separate compartments by crosswalls. The separate compartments (cells) may each have a single nucleus or there may be several nuclei. Other hyphae lack crosswalls and simply have many nuclei scattered through the cytoplasm. The cell walls of many fungi contain *chitin,* a characteristic animal polysaccharide found in shellfish and insects.

Nutrition. The fungi usually obtain their food by secreting digestive enzymes into the substrate on which they are growing and absorbing the digested food. Parasitic fungi often produce special hyphal branches that push into the cells of the host and absorb nutrients, frequently without seriously injuring the host cells.

Reproduction. Like the algae, fungi reproduce both asexually and sexually.

ASEXUAL REPRODUCTION. Asexual reproduction in unicellular forms may be by simple cell division or by budding, in which a small outgrowth develops from the mother cell, the nucleus divides, and one daughter nucleus passes into the bud, which then pinches off. Multicellular fungi reproduce by fragmentation or by the production of spores. Usually the spores are nonmotile, but some are flagellated zoospores like those of the algae.

SEXUAL REPRODUCTION. The reproductive structures of multicellular fungi, the fruiting bodies, are varied and colorful. They project above the buried mycelium and are the parts you usually see. They may be small, so that only masses of them are apparent, like the colored spots on moldy bread and the cottony tufts that show up on some insects and fishes. Or they may be quite large, like some shelf fungi that grow on tree bark (see Figure 19–11). In Europe an especially prized form of fungus has underground fruiting bodies known as *truffles.* French peasants train pigs and dogs to hunt for them by smell.

The familiar toadstools and mushrooms are fruiting bodies. Sometimes the underground mycelium is a large mass that grows outward in all directions so that the fruiting bodies that develop around its edge form a circle—an arrangement known as a *fairy circle* (see Figure 19–12).

Toadstools are not a separate group but simply mushrooms that are poisonous. A toadstool may be much more closely related to an edible form than it is to another poisonous one. Of the thousands of species of mushrooms, prob-

Figure 19–11. Shelf fungus. (Carnegie Museum—LeRoy K. Henry photo.)

ably less than a hundred are known to be poisonous and many more are known to be edible. But the only way for a layman to tell which is which is by empirical methods and this can be dangerous. If you want mushrooms, you had better buy them through regular markets. Figure 19–13 shows the fruiting body of a toadstool, the fly amanita.

Figure 19–12. Fairy-circle fungus at Columbus, Ohio. (Carnegie Museum—O. E. Jennings photo.)

Figure 19–13. Young Fly Amanita on Carnegie Museum's Powdermill Reserve. (Carnegie Museum—M. Graham Netting photo.)

Sexual reproduction in the fungi usually resembles the algal type shown in Figure 19–8. The mycelium is haploid and the gametes are produced by mitosis. They fuse to form the zygote, which divides by meiosis so that the diploid stage of the life cycle is represented by only a single cell. There are many modifications of this basic pattern. Multicellular fungi are frequently bisexual. In some, the "female" mycelium has a large, multinucleate cell called an *ascogonium* and the "male" mycelium has a similar but smaller cell called an *antheridium.* When two mycelia grow together, a bridge forms between the sex cells and the male nuclei migrate into the female cell. But the male and female nuclei do not fuse. Instead, they both continue to divide mitotically. Hyphae develop from the ascogonium, which now has two kinds of nuclei, one male and one female. The fruiting body is composed of uninucleate hyphae from both mycelia, intermingled with sexual, binucleate hyphae. In the terminal cells of these hyphae, the two nuclei eventually fuse to form the zygote. This undergoes meiosis to produce four haploid cells, which often divide again mitotically so that eight spores are formed. Figure 19–14 shows sexual reproduction in such a fungus.

The unicellular yeasts alternate between diploid and haploid stages. The haploid cells multiply asexually by budding or fission. Eventually two cells come together to form a diploid zygote, which also multiplies by budding until

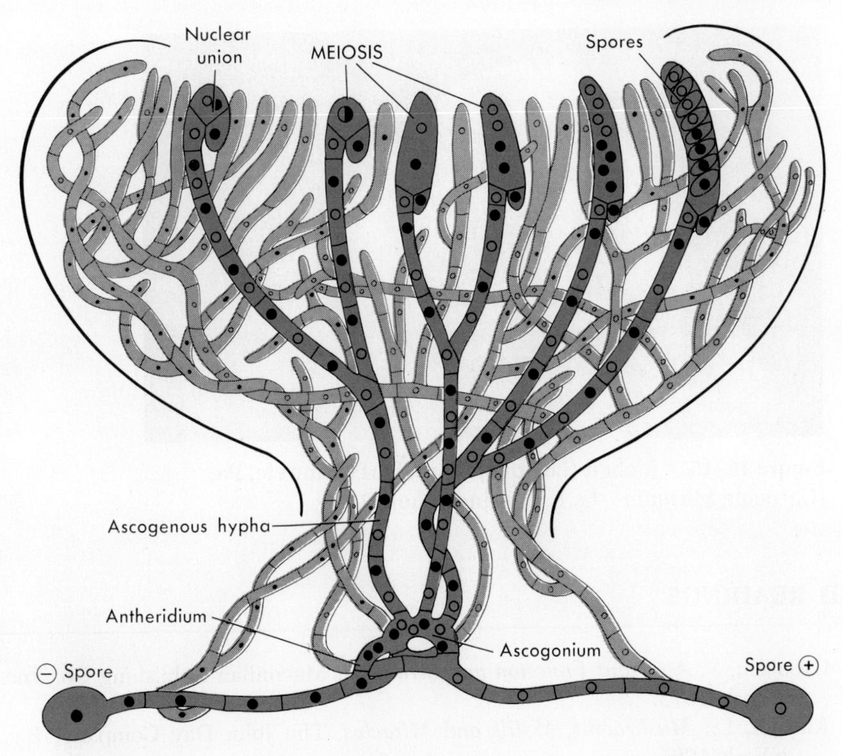

Figure 19–14. Diagram of one type of sexual reproduction in a fungus. The ascogonium and antheridium disappear before the mature spores develop.

a large number of cells are present. Then the nucleus in a cell undergoes meiosis and the four daughter nuclei become incorporated into four separate cells still contained within the body of the mother cell. Eventually the mother cell bursts and liberates the haploid spores, which begin the cycle again.

Lichens

Lichens are not a separate group of plants but rather associations between certain algae and fungi. The fungus acts as a holdfast and absorbs water and minerals from the substrate, and the algal cells live in the body of the fungus and manufacture enough food for both. If the two are separated, the alga can carry on an independent existence, but the fungus will not complete its life cycle. The whitish patches often seen on tree trunks and bare rock surfaces are lichens. See Figure 19–15.

Figure 19–15. Lichen (*Lecidea*) on rock at Ohiopyle, Pa. (Carnegie Museum—O. E. Jennings photo.)

SUGGESTED READINGS

Greulach, V. A.: *Plant Function and Structure,* Macmillan Publishing Co., Inc., New York, 1973.

Kavaler, L.: *Mushrooms, Molds and Miracles,* The John Day Company, Inc., New York, 1965.

Krueger, R. G., Gillham, N. W., and Coggin, J. H.: *Introduction to Microbiology,* Macmillan Publishing Co., Inc., New York, 1973.

Muller, W. H.: *Botany: A Functional Approach,* 2nd ed., Macmillan Publishing Co., Inc., New York, 1969.

Solbrig, O. T.: *Principles and Methods of Plant Biosystematics,* Macmillan Publishing Co., Inc., New York, 1970.

Wilson, C. L., and W. A. Loomis: *Botany,* Holt, Rinehart and Winston, Inc., New York, 1967.

CHAPTER **20**

The Higher Plants

The plants discussed so far live in water or in environments where surface water is available. Water, after all, is the natural environment of life. It is the major ingredient of protoplasm; it is necessary for photosynthesis; it provides support for the body and protection against desiccation; it contains the dissolved gases and minerals required for metabolism; it is a medium of transport for gametes and spores. In contrast, dry land is a harsh environment that could be successfully invaded only by plants that evolved the attributes needed to cope with the problems posed by such an environment.

For a plant to be truly terrestrial it must have its surface protected to prevent desiccation, but it must also have openings in the protective covering to permit gas exchange. It must have specialized surfaces for the absorption of water and minerals. If it is to grow more than a few inches tall, it must have special supportive tissues and conductive tissues to transport substances from one part of the plant to another. Finally, it must have mechanisms for bringing the gametes together and for protecting the delicate zygote during the early stages of development.

The green algae (division Chlorophyta) seem to be the one group of primitive plants having the attributes that would make possible the evolution from them of truly terrestrial plants. They have all the pigments present in higher plants. Some have alternation of generations, which is also found in the terrestrial plants. And some have large, nonmotile female gametes (eggs) and smaller, motile male gametes (sperm); this too is characteristic of the land plants.

The terrestrial plants include the bryophytes (mosses and their allies) and the ferns, conifers, and flowering plants. The last three (and a few minor forms) are sometimes grouped as ***vascular plants*** because they have special conducting tubes (vessels). The mosses and the vascular plants seem to have evolved separately from the green algae.

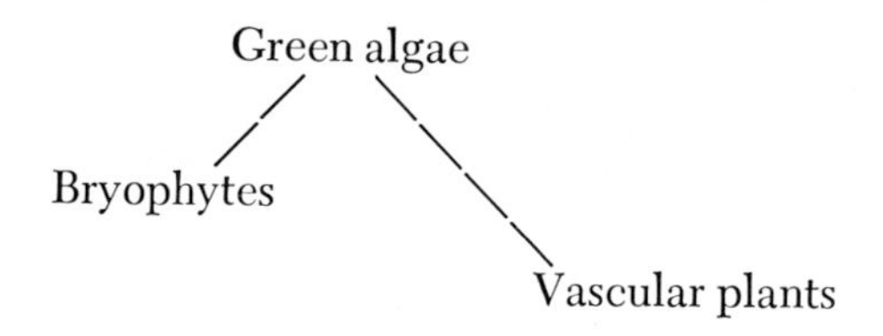

BRYOPHYTES

The bryophytes exemplify the third and fourth stages of plant complexity. Their somatic cells show more differentiation than those of the algae and fungi. They have embryos, that is, the early developmental stages are enclosed in the body of the parent plant.

Structure of Moss

The various species of the true mosses are the most familiar and most numerous of the bryophytes. The individual plants are usually small and inconspicuous, but they grow in colonies that sometimes form extensive beds (see Figure 20–1). Peat is largely made up of compacted masses of dead moss. Through their ability to absorb and hold water, mosses help protect soil from desiccation, and their dead bodies enrich the soil and make it more suitable for the growth of higher plants.

Mosses reproduce by alternation of generations. The haploid, gametophyte plants make up the beds of moss you see growing in shady, moist woodlands and on rocks around springs. The individual plant has a rootlike structure called a *rhizoid*. The rhizoid is able to take in water and minerals from the soil, but is not considered a true root because it has no special transporting tissues. The absorbed substances simply pass from cell to cell by osmosis and diffusion. The stem of a moss may lie prostrate or stand erect, but it lacks special supporting or conductive tissues; for this reason most mosses are only a few inches tall at most. The stem may show some differentiation, with the outer cells containing chloroplasts and the inner cells being elongated. Leaflike blades extend out from the stem. The cells of the outer layer of the plant produce a waxlike substance that forms a thin protective covering, the *cuticle*. There are openings in this layer through which gas exchange can take place; water and dissolved minerals can also be absorbed through these openings. The sex organs are borne

Figure 20–1. A true moss (*Stereodon*) on rock at Carnegie Museum's Powdermill Nature Reserve. (Carnegie Museum—LeRoy K. Henry photo.)

at the tips of the stems. Sometimes the leaves surrounding these organs are modified to form flower-like rosettes, often pink in color.

Reproduction of Mosses

Mosses can spread vegetatively, but sexual reproduction is probably more important. The sex organs are the *archegonium*, which produces the egg, and the *antheridium,* which produces sperm. In some species, the male and female organs are borne on the same plant, in others the male and female plants are separate. All the cells of the algal gametangia are reproductive cells, but in mosses the gametes are surrounded by a layer of protective, nonreproductive cells. Because the plants are haploid, the gametes are produced by mitosis. A mature antheridium discharges its sperm during periods of wet weather. The sperm may travel through a surface film of water or be carried by splashing rain drops to the archegonium, where they swim down a fluid-filled passage to reach the egg. Fertilization takes place, and the zygote begins to divide to form a spindle-shaped embryo within the archegonium. The embryo develops a foot that attaches to the parent plant, a stalk, and at the tip of the stalk an enlarged, spore-bearing capsule. This sporophyte plant develops chlorophyll and so is able to manufacture some of its own food, but it remains attached to the gametophyte parent from which it gets its nitrogen and other essential elements. Figure 20–2 shows a moss plant with a sporophyte growing from the gametophyte. Spores are produced by meiosis and are disseminated by air currents; when they settle in suitable places they give rise to the haploid gametophytes.

The failure of the mosses to evolve a mechanism of fertilization that is not dependent on water, coupled with their lack of conductive and supporting

Figure 20–2. A moss gametophyte with a sporophyte growing from the top.

tissues, has made the mosses less successful inhabitants of dry land than the vascular plants.

VASCULAR PLANTS

Trees and grasses, garden flowers, fruits, and vegetables, all the things most people think of when they hear the word *plant,* are vascular plants. In addition to a protective covering and an embryo, they have developed tissues that are specialized for conduction and support. Water, minerals, and nutrients can be carried from one part of the plant to another more rapidly than by diffusion from cell to cell, and the plant can grow to a considerable size. The above-ground part of the plant can rise well above the surface and the roots can extend down into the ground to reach water below the surface layer of the soil. The vascular plants are thus less dependent than mosses on the presence of water near the surface and are able to grow in many places where mosses cannot.

Plant Tissues

The specializations necessary for terrestrial life involve the differentiation of plant somatic cells into a number of different kinds of tissues.

Meristem. A meristem cell is cubical or elongated with a thin wall and without the large central vacuole found in most plant cells. In contrast to most plant cell types, meristem cells continue to divide mitotically. The cells to which they give rise differentiate to form the other types of tissues. The presence of meristem means that growth in a plant can continue as long as the plant lives. Apical meristem is present at the tips of both the twigs and the roots and makes elongation possible. Lateral meristem forms a sheet or sheets under the outer layer of roots and stems and makes possible an increase in root or stem diameter. Lateral meristem is usually called *cambium.*

Simple Tissues. Simple tissues are each composed of a single type of cell. The epidermis is a protective tissue, forming the outer layer of leaves, flowers, and young stems and roots. The cells are cubical, with large, central vacuoles (see Figure 20–3A). On the above-ground parts of the plant, epidermal cells secrete *cutin,* a waxy substance that forms the protective cuticle. The outer layers of woody stems and roots are made up of the thick walls of *cork cells* (see Figure 20–3E), which die after the walls are laid down.

Parenchyma is a tissue composed of large, rounded, thin-walled cells with

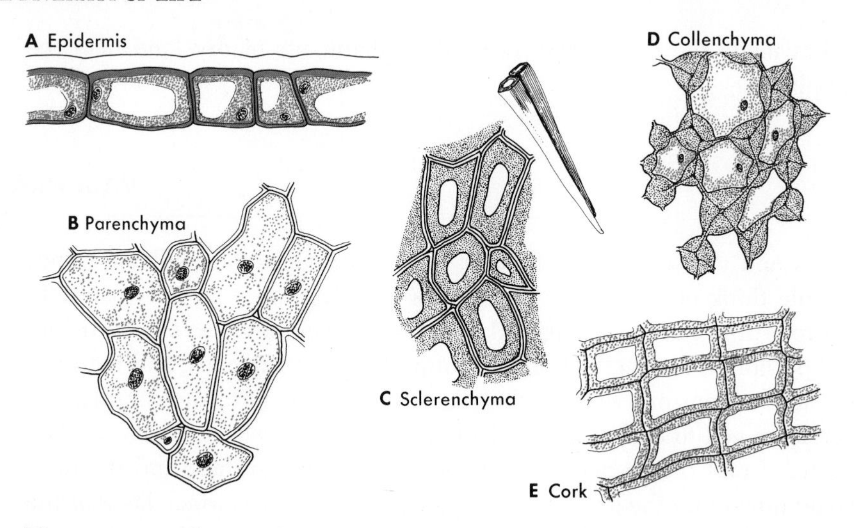

Figure 20–3. The simple tissues of a plant.

large central vacuoles. Parenchymous cells in the leaves contain chloroplasts and carry on most of the photosynthetic activity of the plant (Figure 20–3B). In the stem and root, parenchymous cells store water and nutrients. The bulk of an Irish potato is made up of parenchyma in which starch is stored. Parenchymous cells are still able to divide. When a plant is injured, the parenchymous cells may function like meristem and so play an active role in tissue repair.

Collenchyma and *sclerenchyma* are supportive tissues. Collenchyma cells (see Figure 20–3C) are similar to parenchyma cells, but their walls are thickened. Collenchyma is frequently found near the epidermis in young stems. Sclerenchyma (Figure 20–3D) is made up of thick, tough cell walls but as in cork, the cells have perished. It provides strength and mechanical support. Fibers, like those of linen, and stone cells, the hard, gritty bits in pears that make it unwise for people with false teeth to eat pears in public, are examples of sclerenchyma.

Complex Tissues. Complex tissues contain more than one type of cell. The two complex tissues (vascular tissues) are *xylem* and *phloem*. These are the conductive tissues through which flow water and food materials, the sap of the plant.

XYLEM. Xylem consists of various types of hollow conducting tubes and ducts bound together with sclerenchymous tissue. The tubes are formed of the walls of cells that die at maturity. Xylem carries water and minerals from the roots to the rest of the plant and is also a major supportive tissue. The wood in a tree trunk is xylem.

Just what causes water to move through the xylem tubes from the roots to the topmost leaves of the tallest tree has been a matter of debate for years. The current theory is that the most important factor is the cohesion of water molecules. These molecules have a very strong attraction for one another. Water is constantly being lost from the leaves by evaporation (transpiration). As water is lost by the cells, water moves into them by osmosis from adjacent cells. Some of the cells in the leaf are next to xylem tubes, and as molecules move from the tubes into the cells they pull on molecules behind them and these in turn pull molecules further down in the tube. In this manner the whole column is moved upward by a force exerted from above. Figure 20–4 shows part of a xylem tube.

PHLOEM. Phloem is a vascular tissue concerned with the transport of foods. It is a two-way system and can move food either from the leaves to the roots or from the roots to the leaves. Unlike xylem, it is a living tissue. The cylindrical cells (sieve tube elements) are connected end to end, and the cell walls that separate them (sieve plates) are perforated by holes through which protoplasmic strands connect one cell with another. The cells of the sieve tube lack nuclei but are connected by protoplasmic strands that pass through pores in the tube cell wall to adjacent companion cells that have nuclei. These companion cells may help carry on the metabolic activities of the phloem.

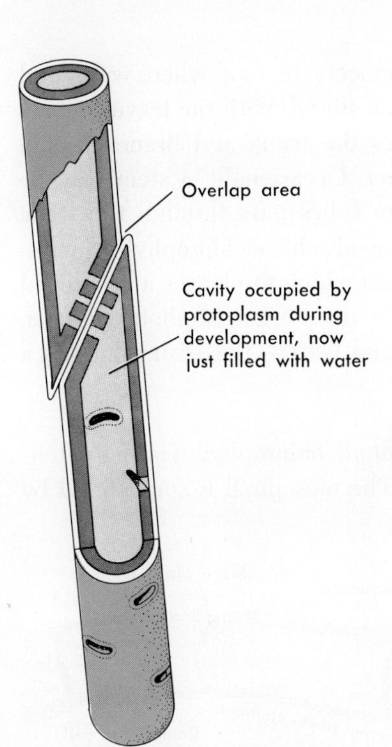

Figure 20–4. Part of a xylem tube. A section of the wall has been cut away to show the inner cavity and the openings between the end walls of the cells.

Again there is a problem of how materials are transported through the phloem, because food moves from one part to another of a plant much more rapidly than can be accounted for by simple diffusion from cell to cell. No satisfactory answer has yet been found. Figure 20–5 shows part of a phloem tube.

Vegetative Structures

The parts of the plant that are concerned with the growth and development of the individual are called *vegetative structures.* In most vascular plants they are the root, the stem, and the leaf. The arrangement of the tissues in those structures differs from one group to another, and the structures themselves show enormous variation.

Root. The root is a much-branched part of the plant that is usually underground. Its primary function is to absorb water and minerals from the soil, which it does through the root hairs, tiny extensions of epidermal cells that pass between the soil particles (see Figure 20–6). The inner part of the root consists largely of xylem and phloem tubes and parenchyma cells. Root parenchyma is often an important food storage area, as in the carrot. Roots also serve to anchor the plant.

Stem. The stem is the part of the plant that connects the root, where water and minerals are taken in and food is used and often stored, with the leaves, where food is manufactured. The stem thus includes the trunk and branches of a tree as well as the soft stem of a garden flower. Occasionally a stem may be underground (a *rhizome*). Xylem and phloem tubes pass through the stem, which supports the leaves and flowers. Stems may also have chlorophyll-containing cells and carry on photosynthesis. In cacti, in which the leaves are reduced to sharp spines (see Figure 20–7), the stem is the main organ of photosynthesis. Stems may also serve for the storage of food and water—the Irish potato is a modified storage area of an underground stem.

Leaf. The bulk of the leaf is made up of *mesophyll,* chlorophyll-bearing parenchymous cells with air spaces between them. The mesophyll is surrounded by

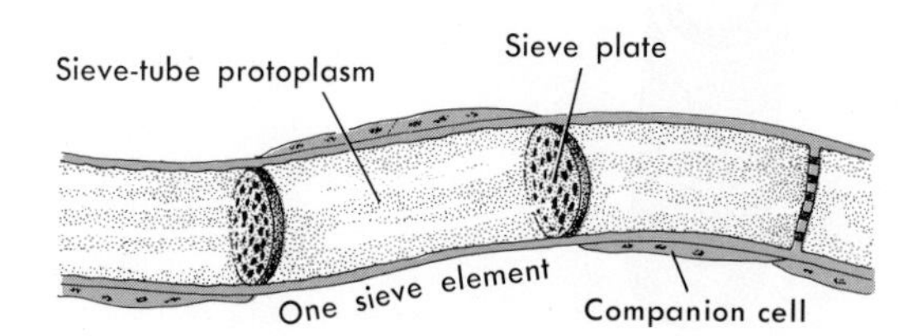

Figure 20–5. Part of a phloem tube showing sieve cells and companion cells.

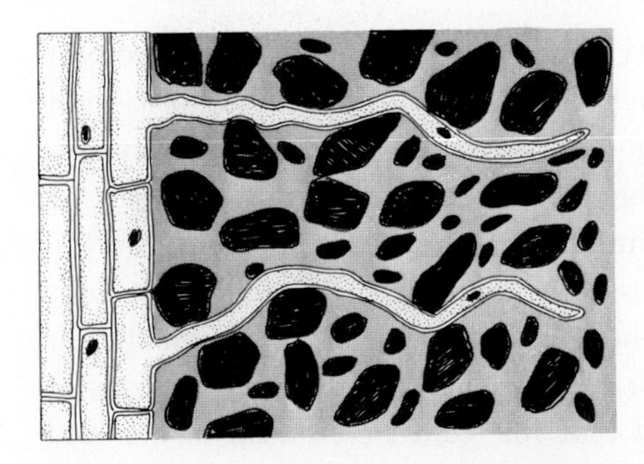

Figure 20–6. Root hairs extending out between soil particles from the root epidermal cells.

an outer layer of epidermal cells covered with an impermeable cuticle. Scattered between the epidermal cells are openings called *stomata* (singular: *stoma*) through which gas exchange and transpiration take place. Each stoma is surrounded by two kidney-shaped guard cells. Remember that a plant cell is surrounded by a cell wall and that as water moves into the cell by osmosis, the cell swells and exerts turgor pressure on the cell wall. When the back pressure exerted by the wall equals the turgor pressure, no more water enters the cell

Figure 20–7. A cactus showing the leaves modified to form spines. (Carnegie Museum—Neil D. Richmond photo.)

but it remains stiff and swollen, and is said to be *turgid.* In a guard cell, the part of the cell wall bordering the stoma is thicker than the wall on the other side of the cell. When the cell is turgid, the turgor pressure causes it to buckle outward. When photosynthesis starts in the morning, starch stored in the guard cells is converted to glucose. This increases the number of molecules in solution in the cytoplasm, which in turn increases the osmotic influx into the cells. The cells thus become turgid, the stomata open, and exchange of gases between the air and the spaces in the mesophyll can take place. When photosynthesis stops at night, the sugar is reconverted to starch and the guard cells lose turgor and become less rigid. They tend to collapse toward each other so that the stomata are more or less closed. This cuts down on the loss of water by transpiration. Figure 20–8 shows stomata in a leaf. Running through the mesophyll are bundles of vascular tissue, the leaf veins, through which water is brought to the leaf, and the food manufactured there is transferred to other parts of the plant for use or storage.

KINDS OF VASCULAR PLANTS

Six divisions are included in the vascular plants. Several are primitive groups, known mostly as fossils, in which the leaves are either lacking or repre-

Figure 20–8. Stomata and guard cells in leaf epidermis. (Courtesy of Ward's Natural Science Establishment, Inc., Rochester, N.Y.)

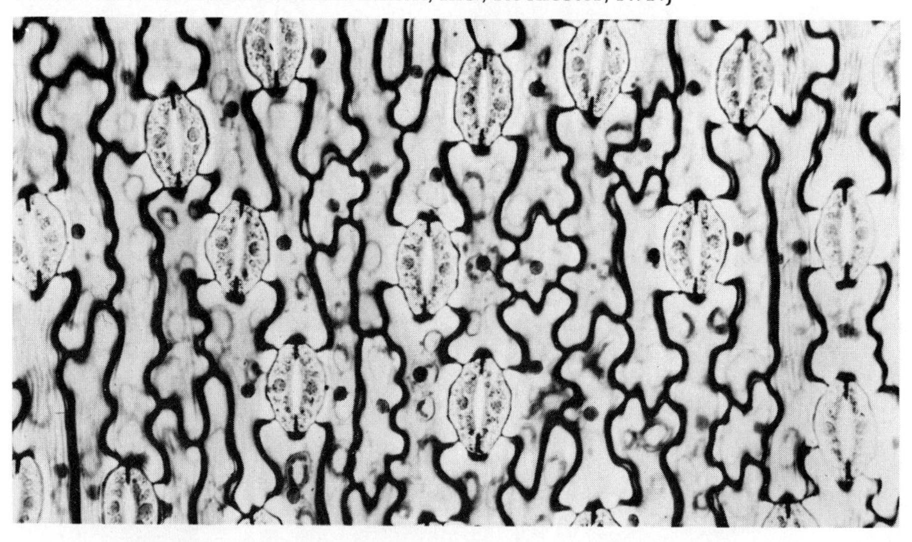

sented by small outpouchings of the stem. Survivors today include the club mosses and scouring rushes. The ferns differ from these primitive forms in having leaves that are specially modified structures with vascular bundles leading into them. The great majority of living land plants are seed plants. They have true leaves as do the ferns, and also produce seeds. The seed plants include the flowering plants, the conifers (cone-bearing trees such as the pines), the cycads, and several minor groups usually classed with the conifers.

Ferns

The ferns range from small plants to tree ferns that may reach twenty meters (sixty feet) in height. With their vascular tissues, roots, and well-developed leaves, the ferns are truly terrestrial plants. They also show adaptation to terrestrialism in their life history. The sporophyte generation is predominant and the gametophyte generation is very reduced and insignificant. The common garden fern is a good example. It has large, conspicuous leaves growing from an underground stem. At the proper season, little spore-producing bodies (*sporangia*) develop on the undersides of the leaves; each sporangium consists of a stalk terminating in a capsule. Within the capsule, spore mother cells give rise to tiny haploid spores by meiosis (see Figure 20–9). When the sporangia open, the spores are scattered and if one chances to fall on moist ground, it starts mitotic division and develops into a small, heartshaped gametophyte called a *prothallus*. This little plant has *archegonia* where eggs are produced by mitosis on one part of the body and *antheridia* where sperm are produced on another (see Figure 20–10). If the prothallus is covered with a film of water when the gametes are mature, the sperm swim to and fertilize the eggs to form the zygotes. The advantage of a small gametophyte is that the sperm have a short distance to swim; fertilization can take place even though the water film is only present for a very short period of time. On the other hand, the large sporophyte, with its differentiated tissues forming roots, stem, and leaves, may be able to carry on for a number of years, even during times of drought, until conditions are right for reproduction.

The zygote begins to grow within the archegonium and, although for a while the embryo is parasitic on the gametophyte, it soon becomes a nutritionally independent sporophyte, with its own root, stem, and leaves. Figure 20–11 shows a fern prothallus with a sporophyte growing from it.

Figure 20–12 is a word diagram of a fern life history. The major difference between it and the life history of a moss lies in the increased size and independence of the sporophyte generation, which is a self-sustaining plant, nutritionally independent of the gametophyte, whereas in the moss the sporophyte is dependent on the gametophyte.

Figure 20–9. Spore production by a fern. A: A sporophyte plant. B: Underside of a leaf. C: Part of B magnified to show sporangia. D: Sporangia more highly magnified. The thick-walled cells that form the annulus contract as the sporangium dries and this ruptures the capsule, releasing the spores.

Seed Plants

The remaining groups of vascular plants are the most successful of all, with over a quarter of a million species living today. Together they are known as the *seed plants.* A seed is an embryonic sporophyte, together with a stored food supply, encased in a tough seed coat. The embryo lies dormant within the seed coat and is capable of tolerating a long period of adverse environmental

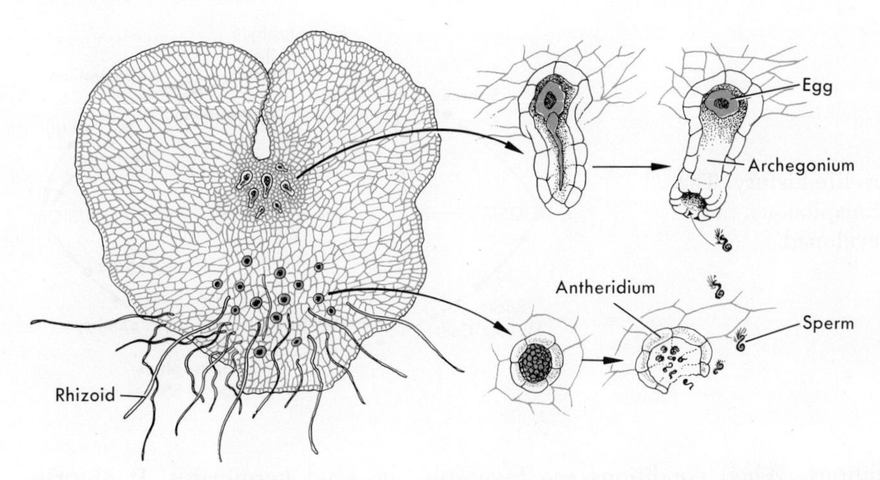

Figure 20–10. Fern prothallus showing archegonia and antheridia.

Figure 20–11. Fern prothallus with a young sporophyte growing from it. (Courtesy of Ward's Natural Science Establishment, Inc., Rochester, N.Y.)

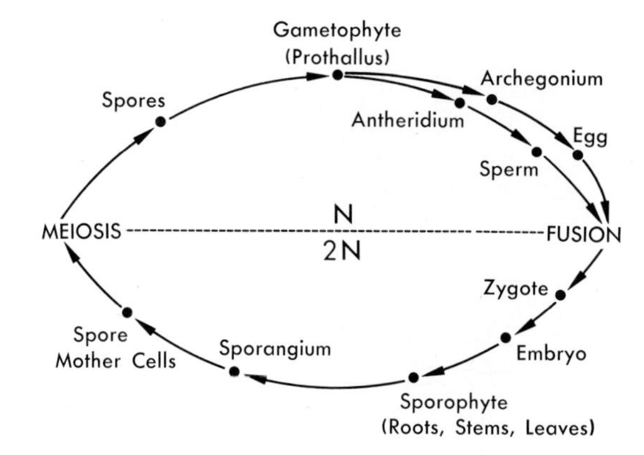

Figure 20–12. Fern life history. The gametophyte is inconspicuous, the sporophyte well developed.

conditions. When conditions are favorable the seed germinates. It absorbs water, the inner tissues swell and rupture the seed coat, and the embryo resumes growth. The evolution of the seed makes it possible for a species to survive in an area even when some catastrophe, such as drought or fire, has eliminated all the mature sporophytes. Some seed plants produce naked seeds that are usually borne on cones. The seeds of the flowering plants form within special structures, the flowers.

Conifers. This group includes such plants as pines, cedars, firs, spruces, the giant redwoods, and the ginkgo. The first five, and their relatives, include the largest trees alive today and supply about 75 per cent of the lumber used in building. The leaves of conifers are usually needles. The ginkgo is a large, broad-leaved tree that is resistant to smog and industrial gases. It has been introduced as a shade tree from China, its native home, into many of the large cities of the world.

The conifers have further modified the pattern of alternation of generations that is such a characteristic feature of plant reproduction. The gametophyte no longer exists as an independent plant. Furthermore, although the ferns have only a single type of sporangium and a single type of spore, the conifers have two.

Reproduction in the pine tree has been very thoroughly studied. *Cones* develop at the tips of some of the twigs. A cone is essentially nothing more than a central axis or stem bearing reduced, scalelike leaves called *bracts.* There are two sorts of cones, male and female.

Between the bracts of the female cone lie specialized sporangia called *megasporangia.* The *megaspore mother cell* within gives rise, by meiosis, to four haploid *megaspores,* three of which disintegrate (remember the polar bodies of the egg) (see Figure 20–13). The remaining megaspore starts to grow (germin-

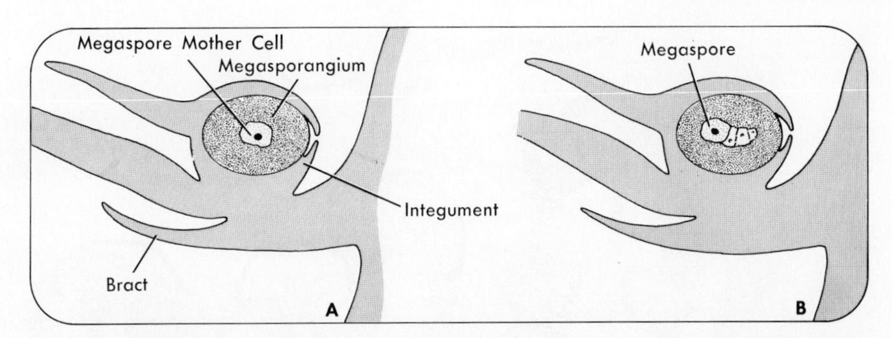

Figure 20–13. Megaspore formation in the pine. A: Megaspore mother cell in megasporangium. B: The megaspore mother cell undergoes meiosis to produce one megaspore and three small cells that disintegrate.

ate) by mitotic division to form a multicellular *megagametophyte,* which contains two to five archegonia.

Meanwhile, on another part of the tree, male cones bear *microsporangia.* Here each *microspore mother cell* undergoes meiosis to form four *microspores.* Each of these divides mitotically and gives rise to a *pollen grain,* which is a two-celled gametophyte enclosed in a capsule. One of the cells is called a *generative cell* and the other a *tube cell.* When the microsporangia rupture, large numbers of these pollen grains are set free. The capsule of the pine pollen grain has winglike extensions and the grains are easily disseminated by air currents (see Figure 20–14).

Should a pollen grain come in contact with a megasporangium in a female cone, the tube cell develops a tube, which penetrates the megasporangium. Meanwhile the generative cell divides to form two sperm cells, which migrate down the tube. Eventually one of them fertilizes an egg cell in one of the archegonia, and the other cells of the male gametophyte then disintegrate. Once a zygote is formed, the remainder of the megagametophyte is reorganized into stored food material to be used by the embryo in its development. The tissue that originally surrounded the megagametophyte becomes dry and hard to form a seed coat, a part of which may extend into a flat, winglike structure that aids in the dispersal of the seed by wind.

The word diagram in Figure 20–15 summarizes the life history of the pine. Now, instead of a single gametophyte that produces both eggs and sperm, there are two types of sporangia that produce different types of spores. Those from the megasporangium develop into a female gametophyte and those from the microsporangium develop into male gametophytes. Thus, the fusion of sperm and egg has to be preceded by the coming together of male and female gametophytes—the process known as *pollination.*

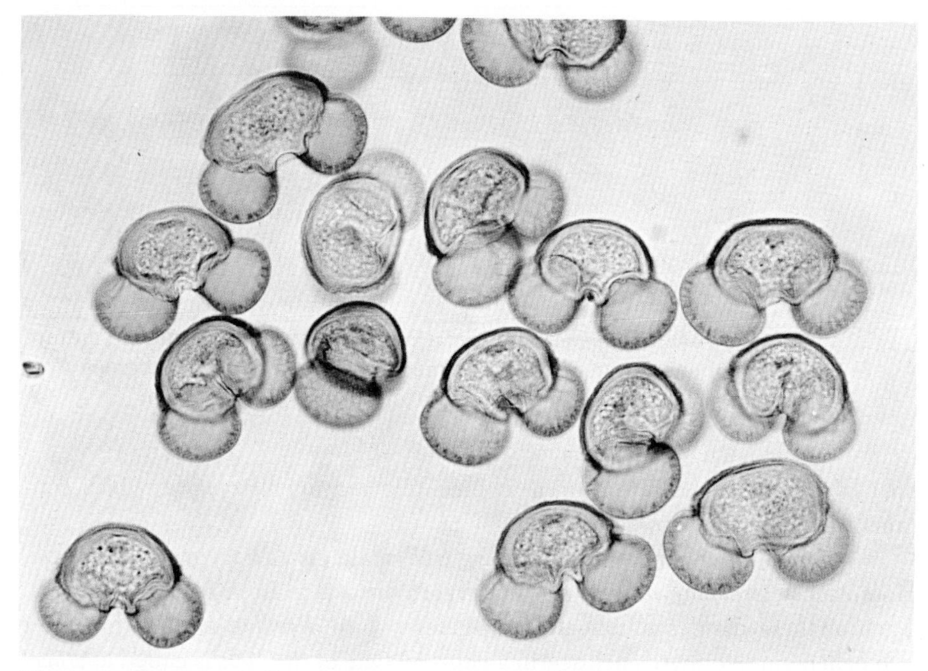

Figure 20–14. Pine pollen. (Courtesy of Ward's Natural Science Establishment, Inc., Rochester, N.Y.)

Figure 20–15. Pine life history. The two gametophyte plants are very inconspicuous and only function in reproduction.

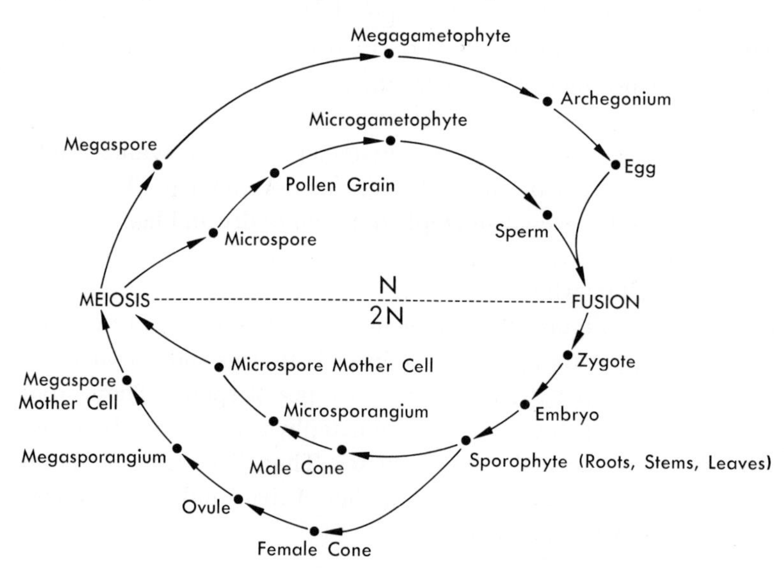

Flowering Plants. The flowering plants are placed in the division Anthophyta (formerly class Angiospermae). They are the most successful plants and are the major food source for man. Many flowering plants are large woody trees such as the oak and maple, but even more are short-lived herbaceous plants. They do not form wood, which consists of layers of xylem laid down during the growing season each year. They have shortened the life span of the sporophyte, or at least the above-ground part of it, to a point where only a brief time is needed to produce seeds. This is an advantage for plants that live in places where the growing season is very short. In the desert, enough ground water to support a mature sporophyte may be present for only a few weeks out of several years. When rains do come, the seeds of a previous generation germinate and in a few days the ground is carpeted with thousands of flowers; the desert in bloom is a glorious sight. The flowers produce seeds that can persist in the sands and rocks for several years, although the parent plants wilt and die as the desert sands dry out.

REPRODUCTION IN FLOWERING PLANTS. The possession of a flower is the outstanding feature of the anthophytes. The flower provides a protective covering for the female gametophyte and frequently serves as an attractant to insects that carry pollen from one plant to another. Many flowering plants are no longer dependent on wind pollination as are the conifers. Figure 20–16 shows that the flower develops on the enlarged end (the *receptacle*) of a flower stalk. Typically, there is an outer row of *sepals*, which are usually green and leaflike. Next

Figure 20–16. Section through a flower. Encircled anther is enlarged.

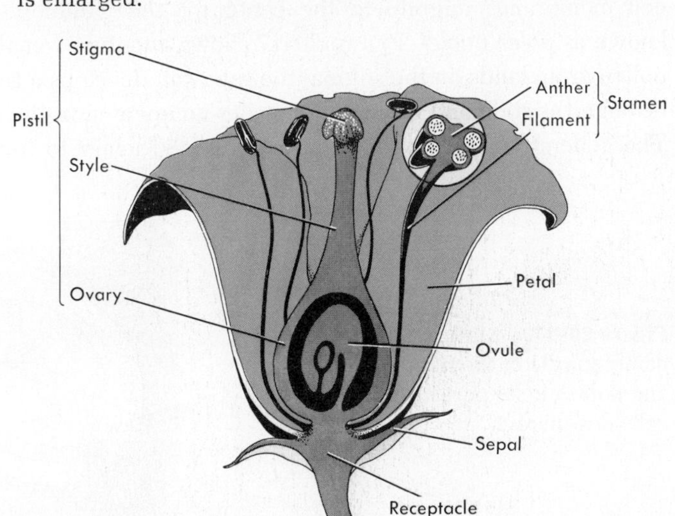

are one or more rows of *petals*, which are often brightly colored. Median to the petals is a series of male organs, the *stamens*, each of which is made up of a stalk, the *filament*, capped by a pollen-producing structure, the *anther*. Finally, in the center of the flower is the female organ, the *pistil*, which consists of a basal ovary and an elongated portion, the *style*, bearing a pollen-receiving structure, the *stigma*. The seed develops in the ovule within the ovary, and the ovary wall forms the fruit around the seed.

Some flowers do not have all flower parts. One flower may have only stamens and another only a pistil. The two types may be borne on the same plant, or one plant may have only male flowers and another only female flowers. A few flowers, such as the violet, are self-pollinating, but most flowers are cross-pollinated. Even if the stamens and pistil are in the same flower, they usually mature at different times so that self-fertilization does not take place. Cross-fertilization, it should be recalled, is an important source of variation.

Pollen grains are formed in the anther in much the same way as in the male pine cone. The microspore mother cell gives rise by meiosis to four haploid microspores. Each divides by mitosis to form the two-celled gametophyte plant that, when it develops an investing layer of protecting material, is the pollen grain. One of its cells is known as the *tube cell* and the other as the *generative cell*.

Within the ovule a megaspore mother cell produces four haploid daughter cells by meiosis. Three of these perish, and only one persists to develop. Generally, its nucleus divides mitotically three times to give rise to the female gametophyte. Cell membranes develop around six of the eight nuclei. One of these is the egg, the other five ultimately disintegrate. The two nuclei that lack cell membranes migrate to the center of the gametophyte body; they are known as *polar nuclei*. Figure 20–17 shows a mature female gametophyte. If a pollen grain lands on the stigma, the tube cell develops a tube that grows down through the style and ultimately reaches an opening in the ovule, the *micropile*. The generative cell by this time has divided again to form two sperm. They

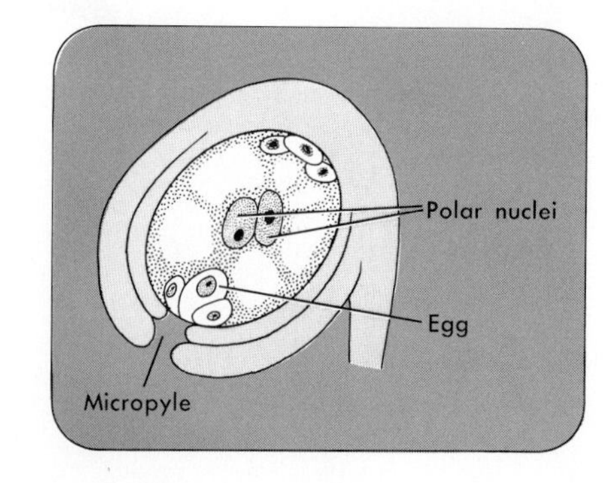

Figure 20–17. Mature female gametophyte in the ovule. Only the egg and polar nuclei persist. The other five cells degenerate.

move down the tube and one fertilizes the egg to form a zygote. The nucleus of the other fuses with the two polar nuclei to form a triploid nucleus. This divides mitotically a number of times, and cell membranes form around the daughter nuclei; the resultant structure is the *endosperm*. The zygote develops by mitotic divisions into an embryo plant; the endosperm develops into a mass of rich, stored food material; the wall of the ovule develops into a seed coat; the wall of the ovary with perhaps some additional floral parts develops into the fruit.

As the embryo plant forms within its seed coat, it gives rise to an embryonic root, stem, and leaf or leaves; the latter are known as *cotyledons*. The embryo may take nearly all of the endosperm into its own body and store it again in the cotyledons, as in a bean or pea, or it may leave most of it as endosperm until it is utilized as in a corn grain.

Figure 20–18 is a word diagram summarizing the life history of the flowering plants.

Many flowering plants also reproduce asexually. A strawberry plant sends out procumbent stems (runners) from which new plants develop. Asexual, or vegetative, reproduction is of great importance in horticulture. For one thing, it is more rapid than sexual reproduction. Some plants, such as the banana and navel orange, do not produce seeds and must be propagated vegetatively. Others, such as the triploid Baldwin apple, produce seeds but the offspring that

Figure 20–18. The life history of a flowering plant. It differs from that of the pine chiefly in the fusion of one of the sperm nuclei with the two polar nuclei to form the 3N endosperm.

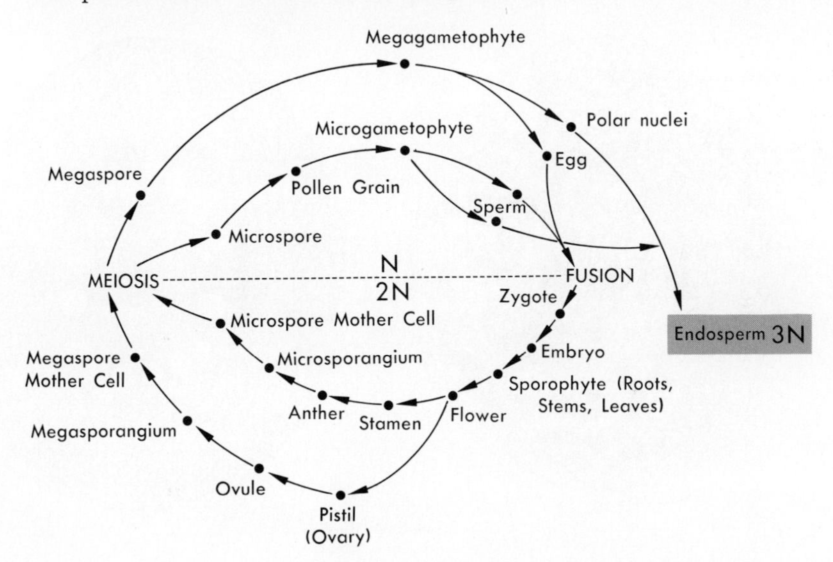

develop from them are very variable and seldom show the combination of desired characters that are found in the parent stock. Most fruits and ornamental trees and shrubs grown commercially are the products of vegetative reproduction.

DICOTS AND MONOCOTS. The anthophytes are divided into two groups, depending on whether one cotyledon is formed by the embryo (*monocotyledons,* familiarly known as *monocots*) or two cotyledons are formed (*dicotyledons* or *dicots*). In dicots, the vascular tissue in the root and stem is in the form of a cylinder and cambium (lateral meristem) is present. In the monocot the vascular tissue is usually in scattered bundles and cambium is usually absent. Figure 20–19 shows the difference between a dicot and a monocot root.

The dicots include most of the familiar trees and shrubs; such plants as oak, hickory, apple, rose, pea, and beet belong here. A rule of thumb useful in identifying dicots without having to examine the embryo is that the veins of the leaves are arranged in a network rather than being parallel, and the flower parts, the sepals, petals, and stamens, are in multiples of four or five (see Figure 20–20).

The monocots are plants with parallel veins in the leaves and with the floral parts in either threes or multiples of three (see Figure 20–21). Lilies and grasses, including the cereal grains and sugar cane, are examples.

Isaiah's dictum "All flesh is grass," is closer to the truth than many people

Figure 20–19. Section through the root of a young dicot (left) and a young monocot (right). (Courtesy of Ward's Natural Science Establishment, Inc., Rochester, N.Y.)

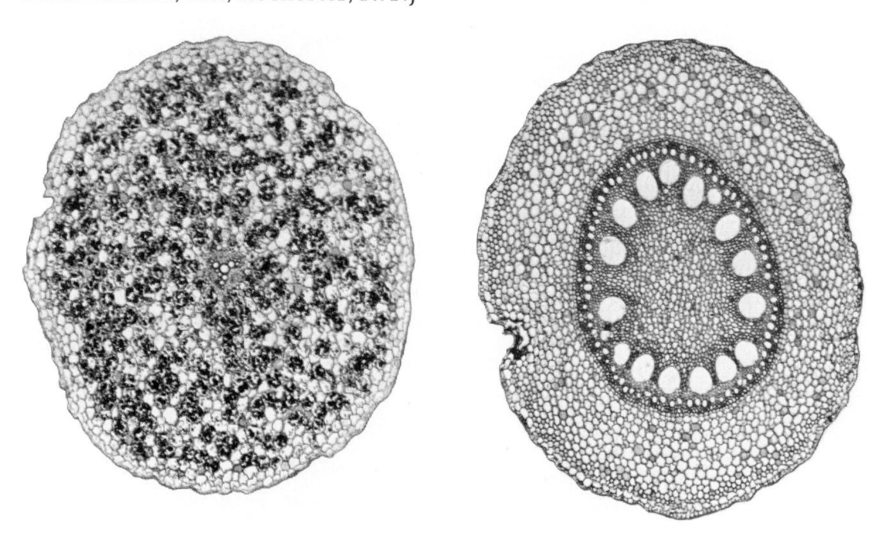

Figure 20–20. Hibiscus (a dicot) growing on Carnegie Museum's Powdermill Nature Reserve. (Carnegie Museum—M. Graham Netting photo.)

realize. The presence of such grains as rice and wheat have made possible the development of some of the most concentrated human populations in the world. Not only is man dependent on the grasses directly, but his main sources of meat (cattle, goats, sheep) are grass eaters and his chickens feed on grass seeds. Without the grasses, there would be no beef or mutton, no butter, milk, cheese, or eggs. Nor would there be bourbon, scotch, rum, or beer.

PLANT MOVEMENTS

People ordinarily think of plants as being incapable of movement, and indeed they are far less mobile than most animals. Still, they are capable of a variety of movements.

Figure 20–21. Trillium (a monocot) growing on Carnegie Museum's Powdermill Nature Reserve. (Carnegie Museum—M. Graham Netting photo.)

Turgor Movements

Some plant parts are able to move quite rapidly as a result of changes of turgor pressure in the cells. An example of such movement in plants is the closing of the stomata. The leaves of some plants, such as the bean and mimosa, fold up at night, as do many flowers. These movements result from the loss of water by special groups of cells. The cells are no longer turgid, they lose their rigidity, and they yield to the pressure of nearby turgid cells. This brings about a change in the relative positions of the parts of the leaf or flower. The leaf of the sensitive plant is divided into many small leaflets; if you run your finger along the main stalk of such a leaf, it droops immediately and the leaflets close up. Again, this results from loss of turgor by special cell groups called *pulvini*. Gradually the cells regain their turgor and the leaf opens again. Perhaps the most spectacular movements are those of carnivorous plants. Some plants that grow in nitrogen-poor soils have specially modified leaves that are able to trap insects. The leaf of the Venus's flytrap is two-lobed with a series of stiff teeth around the edge and sensitive hairs on the surface. If an insect lands on the leaf, the two lobes snap shut. Then special glands secrete enzymes, the insect is digested, and the nutrients are absorbed by the leaf. How the sudden loss of turgor pressure that results in these rapid movements is brought about is not known. Figure 20–22 is a picture of a Venus's-flytrap.

Plant Hormones

Most plant movements result from differential growth. Two processes are involved in growth: cell division and cell elongation. Meristem tissue is present

Figure 20–22. Venus's flytrap (*Dionea muscipula*). (Photograph courtesy of Dr. W. S. Justice.)

at the tip of the plant shoot (a young, growing stem or twig) and also at the tip of the root. Behind the meristem is a region where the cells are no longer dividing but are elongating. Much of the growth in height seen in plants results not so much from an increase in the number of cells as from this elongation of the cells behind the meristem region. The elongation of cells is regulated by chemical substances called *auxins,* the best known of which is *indoleacetic acid,* or IAA. It is formed in the tip of the shoot and moves down through the plant. It stimulates elongation of the cells in the region below the shoot meristem. IAA also influences cell division and cell differentiation. An auxin thus corresponds to an animal hormone; it is produced in one part of the plant and exerts its effect on another part.

Because auxins tend to move away from the light, if a plant shoot is illuminated on one side, the auxins accumulate on the other side. The cells here elongate more rapidly than those on the side exposed to the light so that one side of the plant is growing more rapidly than the other side. The result is that the plant bends toward the light. Movements such as this, caused by differential growth in response to outside stimuli, are called *tropisms.*

Curiously, the concentration of auxin that stimulates elongation in the shoot retards it in the root. If you place a germinating seedling flat on the ground so that both the shoot and the root are horizontal, the auxin accumulates in the cells closest to the ground. Those in the shoot elongate more rapidly than the overlying cells and bend the shoot upward; those in the root are inhibited and the more rapid growth of the overlying cells bends the root down into the ground. Thus, no matter in what position the seed falls, the plant develops normally, with the shoot growing up and the root growing down (see Figure 20–23).

In addition to auxins, other substances known as *gibberellins* and *kinetins* influence cell division and cell elongation in plants.

Botanists believe that hormones are also involved in flowering, although no such hormones have yet been isolated. Some plants, such as the chrysanthemum and poinsettia, will not flower when the nights are too short; they need a certain number of hours of darkness. If the dark period is interrupted by a flash of light, which may be as brief as thirty seconds, flowering will not take place. Such plants are known as *short day plants,* although *long night plants* would be a more appropriate designation, because it is the length of the dark period that is important. These plants do not flourish in the Arctic, where nights are very short during the growing season. They usually bloom in the spring or fall.

Other plants, such as the rose mallow, are called *long day plants.* These plants need a certain number of hours of daylight; they are summer bloomers. The critical length of day varies from species to species, but if they need more than about thirteen hours such plants will not bloom in the tropics, where day and night are about equal in length throughout the year.

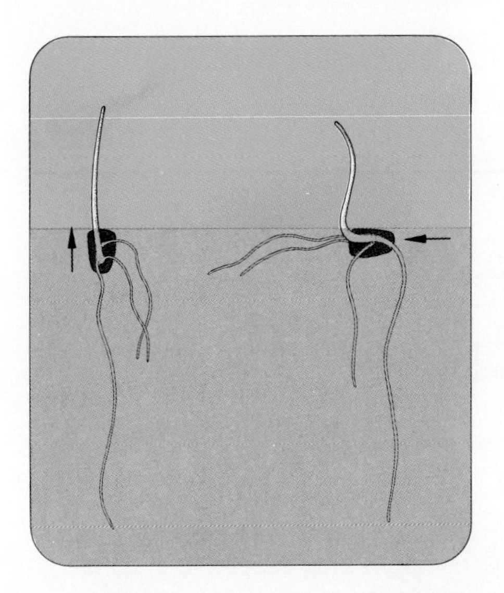

Figure 20–23. The effect of auxins on germinating seedlings. The seedling on the left was placed in an upright position. The shoot grows upward and the root downward. The seedling on the right was germinated in a horizontal position. Accumulation of auxins in the lower part caused the shoot to bend upward and the root to bend downward.

Still other plants seem indifferent to the length of night or day. If they get enough light to manufacture sufficient food, they will flower.

The receptor of the light stimulus is located in the leaves. If a leaf of a short day plant is shielded from light for the appropriate period, while the rest of the plant is exposed, the plant will flower. This indicates that something is transmitted from the leaf to the flower bud. The receptor is *phytochrome,* a protein attached to a pigment molecule. It exists in two possible forms, P_R and P_{FR}. When exposed to red light, P_R is converted to P_{FR}; in the dark, P_{FR} slowly changes back to P_R. This is reminiscent of the changes that take place in visual purple. The amount of P_R that accumulates in the leaf, then, is a measure of the duration of the dark period. But how this is related to the production of the hypothetical flowering hormone is not known.

SUGGESTED READINGS

Bierhorst, D. W.: *Morphology of Vascular Plants,* Macmillan Publishing Co., Inc., New York, 1971.

Galston, A. W., and P. J. Davies: *Control Mechanisms in Plant Growth,* Prentice-Hall, Inc., Englewood Cliffs, N.J., 1970.

Greulach, V. A.: *Plant Function and Structure,* Macmillan Publishing Co., Inc., New York, 1973.

Muller, W. H.: *Botany: A Functional Approach,* 2nd ed., Macmillan Publishing Co., Inc., New York, 1969.

Steward, F. C.: *Growth and Organization in Plants,* Addison-Wesley Publishing Co., Inc., Reading, Mass. 1968.

Wilson, C. L., W. A. Loomis, and T. A. Steeves: *Botany,* 5th ed., Holt, Rinehart and Winston, Inc., New York, 1967.

CHAPTER **21**

The Lower Animals

If the anomalous organisms are grouped with the plants, it might seem possible to list the traits that characterize the animals. But it is not. There are well over a million different kinds of animals and there is an enormous diversity among them. They are all heterotrophs and do not show alternation of generations with separate diploid and haploid stages. Animals are usually motile for at least part of the life cycle. None of these traits separate all animals from all plants, nor do all these traits taken together.

The same characters are shown by some of the flagellated organisms that are here included with the plants. Many flagellates are clearly related to other algae; they carry on photosynthesis and structurally resemble the reproductive zoospores of some of the filamentous algae and fungi. Some of these flagellates may at times lose their chlorophyll and become saprophytic or feed by engulfing solid particles. Still other flagellates never have chlorophyll; they are obligatory heterotrophs and are usually considered single-celled animals by zoologists. It seems possible that the present-day flagellates are the not greatly modified descendants of primitive organisms that gave rise both to the algae and higher plants and to the animals (see Figure 21–1).

It is thus not surprising that it is hard to know where to draw the line between animals and plants. Whether this line is drawn between the single-celled animals and the flagellates, or between the flagellates and other algae, or between the autotrophic and heterotrophic flagellates, it is bound to separate some forms that are obviously related. We have chosen to discuss the flagellates with the algae, but this is an arbitrary choice.

Zoologists call the major divisions of the animal kingdom *phyla* (singular: *phylum*), but there is no general agreement about how many phyla should be

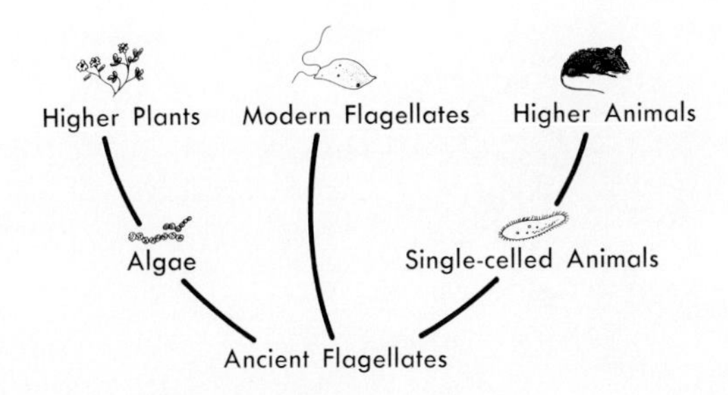

Figure 21–1. Hypothetical origin of the plants and animals from the flagellates.

recognized. Many phyla contain only a small number of species and little is known of the role they play in the economy of nature. The ones considered here are mainly those that illustrate major advances in organization.

PROTOZOA

The phylum Protozoa includes a number of diverse groups that probably evolved independently from different flagellate ancestors. They exist as single cells or cell colonies and this is about the only structural feature they have in common. They are small enough so that they can take in oxygen and get rid of carbon dioxide and nitrogenous wastes, mostly in the form of ammonia, by diffusion. Many protozoa, especially those that live in fresh water, have contractile vacuoles like those of the flagellate algae, by which they pump out the water drawn into their bodies by osmosis. Despite their small size, the protozoa make up a goodly part of the *biomass* (the total amount of living matter) on the earth, for they occur in astronomical numbers. Most of them are free-living, but a number are parasitic, causing such human diseases as amoebic dysentry and malaria. Some of the free-living protozoa contribute to the bad odors and tastes sometimes noticed in the municipal water supply; they are obnoxious rather than dangerous to man. Figure 21–2 shows a number of different kinds of protozoans.

Amoebas and Their Allies (Sarcodina)

The sarcodines have *pseudopods,* extensions of the body used in feeding and, in those forms that move about, in locomotion. The pseudopods may be blunt, finger-like processes (see Figure 21–2A and B) or thin and threadlike (see Figure 21–2C). The amoeba-like sarcodines have a constantly changing body shape (see Figure 21–2A) but others have *spherical symmetry,* that is, they are round so that any plane passing through the center of the body divides the animal into equal halves. Some of these animals are naked (see Figure 21–2A)

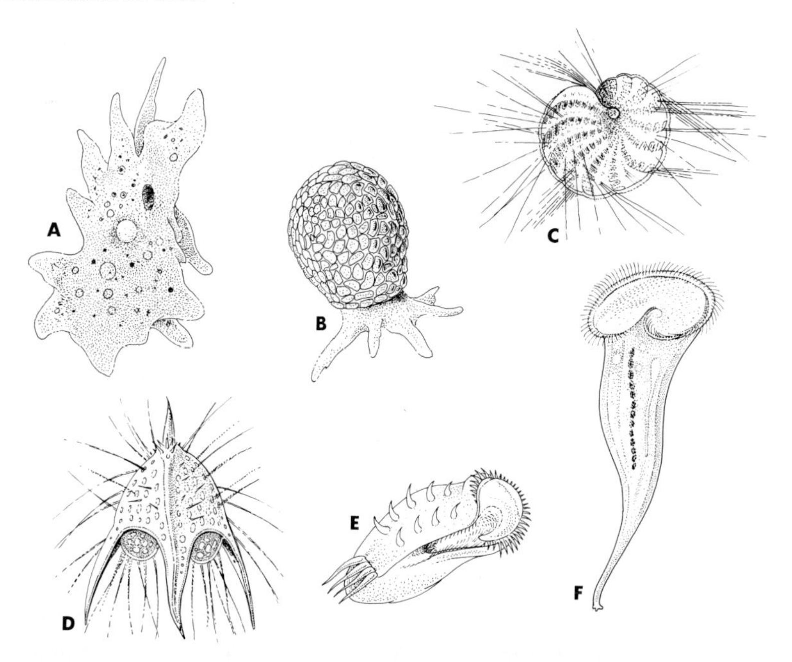

Figure 21–2. A variety of protozoans. A: A naked amoeba. B: A shelled amoeba with blunt pseudopods. C: A shelled sarcodine with threadlike pseudopods. D: A sarcodine with an internal skeleton (radiolarian). E: A free-swimming ciliate. F: An attached ciliate.

but many form elaborate shells of lime or silica (see Figure 21–2C). Others secrete about themselves a sticky matrix; they ingest sand grains and pass these out into the matrix to form the shell (see Figure 21–2B). Still others have internal supporting structures of interconnected rods and plates of silica (see Figure 21–2D). In some parts of the world so many of these animals have died that their shells, sinking to the bottom of the sea, have formed great layers of rock. The White Cliffs of Dover are composed almost completely of the shells of these protozoans. Figure 21–3 shows the structural details of a naked amoeba.

Locomotion. The naked amoebas creep over a surface by extending one or more blunt pseudopods, into which the cytoplasm of the rest of the body flows. Some of the species with threadlike pseudopods pull or drag themselves along the sea bottom with these processes. Others simply float on the surface of the water, drifting with the currents.

Nutrition. Most sarcodines feed on other animals and small plants, which they capture with their pseudopods. If an amoeba comes upon another protozoan in

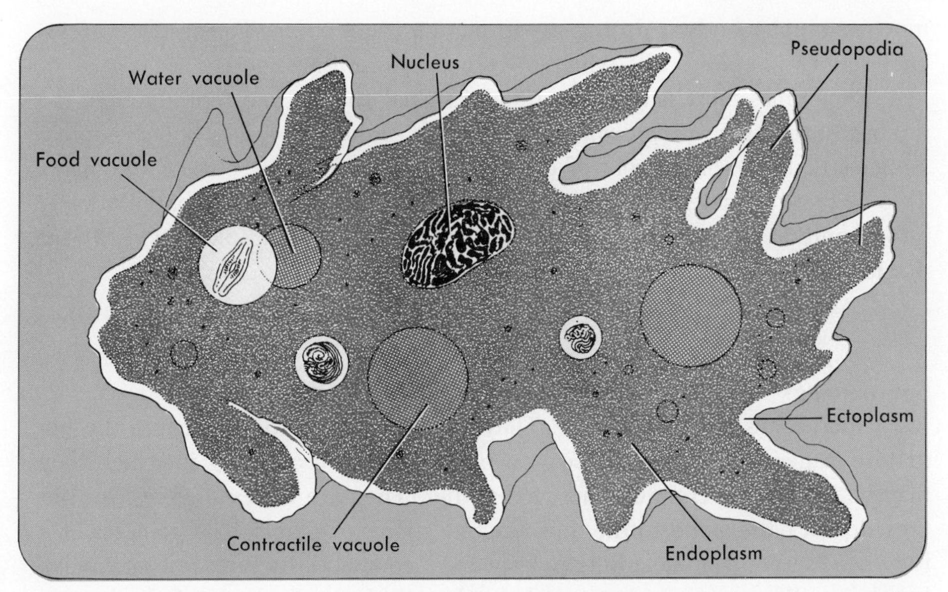

Figure 21–3. Structure of a naked amoeba.

its wanderings, it simply engulfs it by extending pseudopods around it and taking it in, along with a considerable amount of water, as a food vacuole. Enzymes present in the amoeba's protoplasm pass into the vacuole and the food is gradually digested and assimilated. Any undigested remains are cast away by a reverse process; the vacuole moves out to the cell membrane and the animal flows away from it. This is simply another example of phagocytosis, the process by which white blood cells engulf and destroy bacteria.

Sarcodines that have threadlike pseudopods use them to trap their prey. The victim becomes stuck in a mucus-like substance covering the pseudopods. This substance apparently contains enzymes that paralyze the prey and begin digestion. Then the prey is drawn into the body in a food vacuole.

Reproduction. Some sarcodines have only a single nucleus, others have many nuclei within a single cell membrane. Asexual reproduction is usually carried out by binary fission. Multinucleate species may divide by multiple fission, producing a large number of daughter cells at one time.

Sexual reproduction may also occur. One of the spherical forms (*Actinophrys*) withdraws its fine, needle-like pseudopods into its body and encysts, that is, it secretes a tough, protective coat around itself. It then divides mitotically to form two daughter cells. Next the nucleus of each daughter cell divides meiotically. At each meiotic division, one of the daughter nuclei is extruded, much as a polar body is during the formation of an ovum. The result is a cyst enclosing two haploid cells, which then fuse to form the zygote.

Sometimes there is an alternation of sexual and asexual reproduction; the diploid sexual forms (*gamonts*) produce a number of flagellated haploid gametes that are reminiscent of the flagellated gametes of some of the algae. Two of these gametes fuse to form a diploid cell, which develops mitotically into a multinucleate *schizont.* The schizont divides by multiple fission to give rise to the young gamonts.

Sporozoans

Sporozoans are all parasitic. They are called *sporozoans* because the immature stages are often enclosed in a hard, protective covering in which they may be transferred from one host to another. They are then called *spores,* but they do not correspond to the reproductive spores of plants. The gametes of a few species have flagella; adults lack special locomotor structures but may move by gliding. They absorb their nutrients directly from the body of the host in which they live.

The life history of the sporozoan is often complex and usually involves an alternation of sexual and asexual stages.

For man, by far the most important of the sporozoans are the plasmodia, which include the parasites that cause malaria. Throughout human history, probably more people have been sick with malaria than with any other disease. Each year, about 300 million new cases of malaria are contracted and almost 3 million people die of it. The life history of the malarial parasite is complicated and we give only the bare outline here. Asexual stages live in the red blood cells of a human being and reproduce by multiple fission. At regular intervals (three days, two days, or more frequently, depending on which species of parasite is involved) all the infected cells break open, liberating the parasites into the bloodstream. This brings on an attack of chills and fever in the host. The malarial parasites enter other red blood cells and continue the cycle. They cannot become sexually mature until they have been ingested by a mosquito. In the mosquito, they produce haploid gametes that fuse to form the diploid zygotes, which in turn produce by multiple fission a number of *sporozoites.* These make their way to the salivary glands of the mosquito, ready to be injected into another man. Figure 21–4 is a simplified sketch of the life history of a malarial parasite.

When a parasite must spend part of its life in one host and part in another, the one in which it reproduces sexually is called the *definitive host,* the other the *intermediate host.* So here man is the intermediate, the mosquito the definitive host.

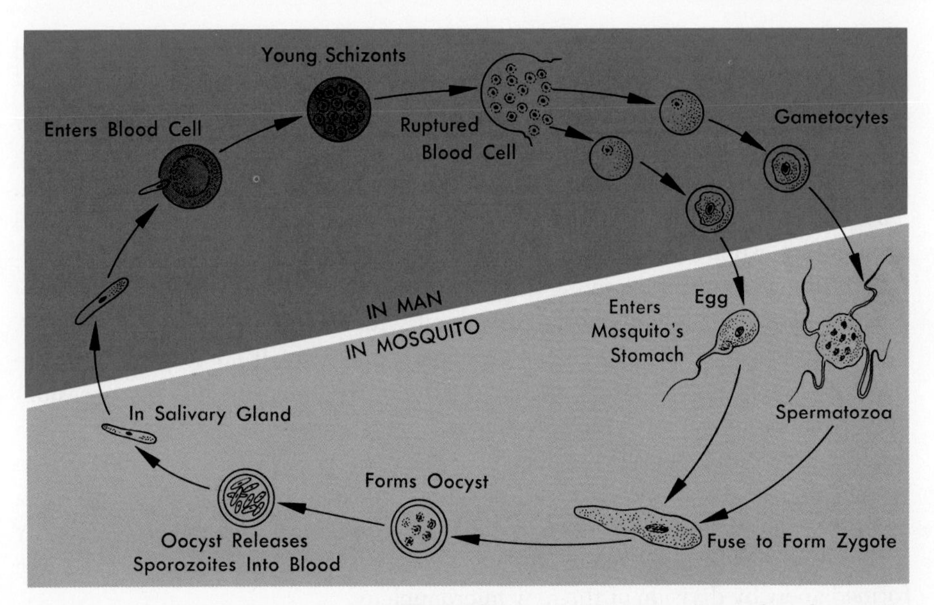

Figure 21–4. Life history of a malarial parasite.

Ciliated Protozoans

A ciliate has a definite body shape, with the rigid outer surface of the body forming a covering called the *pellicle*. All ciliates have cilia, used for locomotion or for food gathering. Most species are free-swimming (see Figure 21–2E) but some attach to the substrate by a stalk (see Figure 21–2F). Some live harmlessly in the bodies of other animals.

Among the great diversity of forms, we consider only one, the slipper animalcule (*Paramecium*) (see Figure 21–5). It is large for a protozoan, large

Figure 21–5. *Paramecium*, a ciliated protozoan.

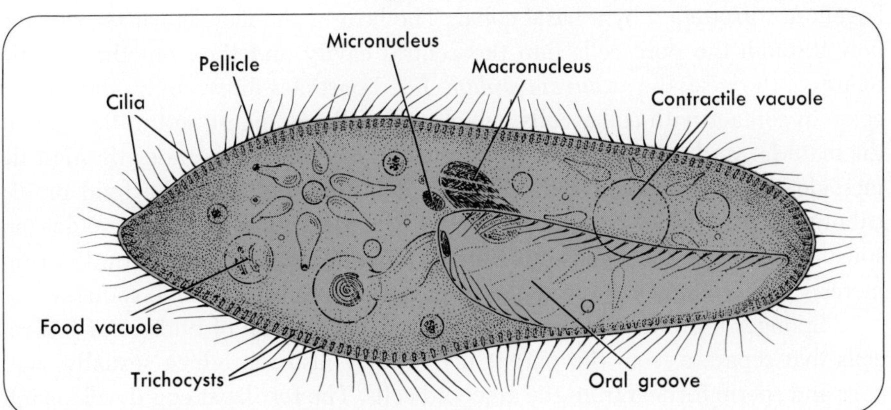

enough so that under favorable conditions it can be seen by the naked eye, and it has a definite head and tail ends. Running the length of the body are rows of cilia; the cilia in each row are connected by a band of fibers that are apparently able to transmit impulses, by which the beat of the cilia is coordinated. Along one side is an *oral groove,* leading to a mouth, where food vacuoles are formed. The food vacuole moves in a definite path through the body, and the undigested remains are expelled in a special anal region.

Ciliates are unique among animals in having two kinds of nuclei, a large *macronucleus* that is apparently concerned with the synthesis of the enzymes needed to regulate body functions and one or more small *micronuclei* that are involved in reproduction. Ciliates usually divide by binary fission, but occasionally two individuals mate and exchange nuclear material. The details of the nuclear divisions that precede and follow mating are complex and differ from species to species, but the net result is that at the completion of the process each individual has a new micronucleus that is a combination of the micronuclear material of the original mates. The macronuclei have disintegrated and are formed anew by division of the new micronucleus.

SPONGES (PORIFERA)

Sponges are the simplest multicellular animals with a differentiation of labor among the cells. Most species are marine but a few live in fresh water. Sponges may grow in various complex, folded, branching forms, but the basic construction, shown in the most primitive types, is that of a hollow vase (see Figure 21–6). The body wall is made up of outer and inner layers of cells separated by a gelatinous material called *mesoglea.* Tube-shaped *pore cells* in the body wall permit water to pass through into the inner cavity of the sponge. In the lining of the inner surface are numerous specialized *collar cells,* each with a flagellum surrounded by a basal collar. The beat of the flagella causes water to flow through the pore cells into the central cavity and then out through the mouth of the vase, the *excurrent siphon.* Food particles drawn in by the current come in contact with collar cells, which engulf them, digest and utilize part of the material, and pass the rest on to amoeba-like cells that wander through the mesoglea. These ameboid cells continue digestion and store the food or distribute it to the other cells in the body. Scattered through the mesoglea are spicules of lime or silica and sometimes protein fibers. The latter form the commercial sponges, now largely replaced in the market by synthetic sponges.

Sponges reproduce asexually through the formation of buds or of balls of cells that separate from the parent body. They also reproduce sexually, with eggs and sperm formed from the ameboid cells. The fertilized egg develops into

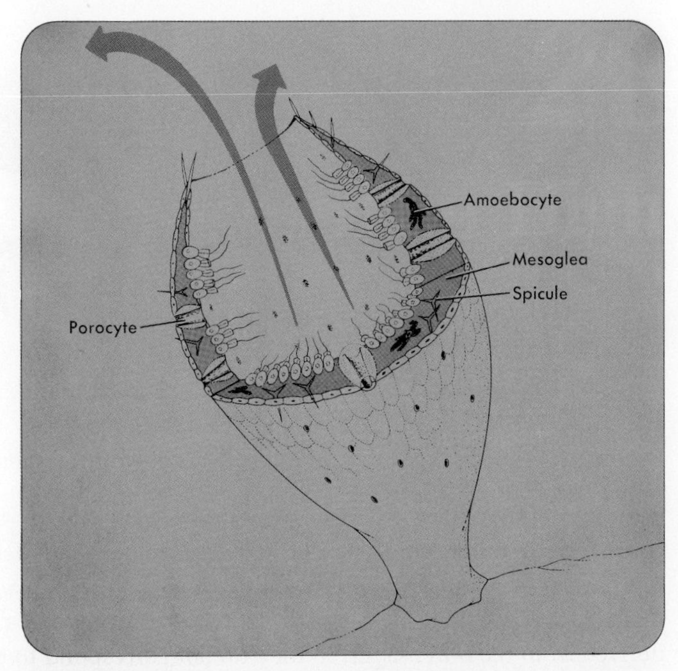

Figure 21–6. A simple sponge. Part of the body wall has been cut away to show the internal structure.

a flagellated ball of cells that swims away from the parent sponge to settle down someplace else.

In structure and in development, the sponges are quite unlike any of the other multicellular animals. They may have arisen from a different flagellate line and they almost surely did not give rise to any of the more highly evolved animals.

JELLYFISHES, CORALS, AND THEIR ALLIES (CNIDARIA)

The cnidarians are often called *coelenterates*. They are all aquatic, and most are marine, although a few simple freshwater forms are known. The individual animals range in size from the tiny freshwater hydra, perhaps six millimeters (a quarter of an inch) tall, to large marine jellyfish, nearly two meters (about six feet) in diameter. The group also includes the sea anemones and colonial, shelled forms, the corals.

The cnidarians have an inner digestive cavity called the *gastrovascular cavity*, which has a single opening to the outside, the mouth. The body wall

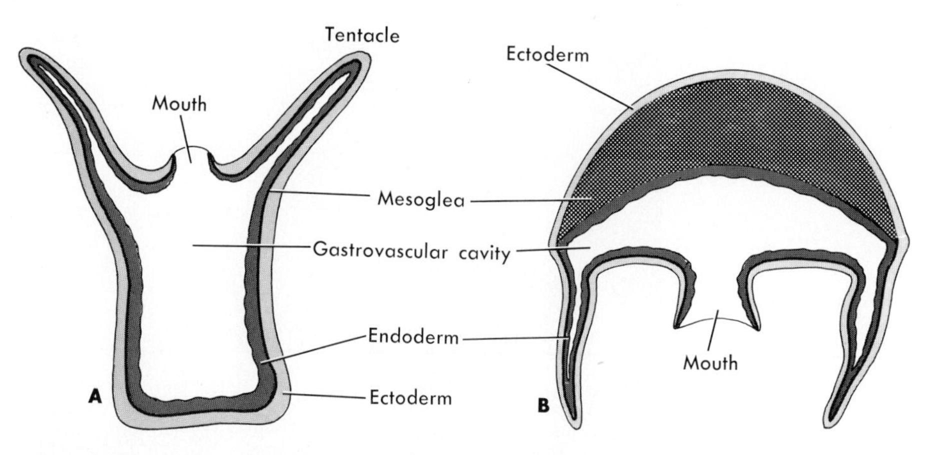

Figure 21–7. Body types in cnidarians. A: A polyp. B: A medusa.

consists of an outer, epidermal layer and an inner, endodermal layer that lines the gastrovascular cavity. Between the two layers lies a mesogleal layer. In many forms ameboid cells wander through the mesoglea, but these cells are derived from the outer layer, so they do not correspond to the mesodermal cells of the higher animals in which the mesoderm is derived from the endoderm.

There are two basic body forms. One, the *polyp,* is usually attached at the base, with the mouth end directed upward. The other, the *medusa,* is usually free-floating and resembles a flattened, upside-down polyp with a thickened mesogleal layer (see Figure 21–7). A jellyfish is a medusoid cnidarian. Both polyp and medusa show *radial symmetry,* that is, any plane passing through a central axis will divide the body into more or less equal halves.

Structure of Hydra

Hydra is a polyp. Some of the cells in its outer layer have extensions of their basal ends that contain contractile fibers so that hydra is able to change its body shape. There are also sensory cells and simple nerve cells that form a network, which enables hydra to respond to stimuli by coordinated movements of parts of the body. Surrounding the mouth are a number of highly mobile tentacles bearing stinging cells. When an animal of the appropriate size comes in contact with the tentacles, the stinging cells discharge barbed tubes that penetrate the prey and inject a paralyzing toxin. The tentacles pull the prey to the mouth, which opens to receive it. Gland cells in the wall of the gastrovascular cavity secrete enzymes and food is digested in the cavity. This is truly a digestive cavity and digestion is extracellular as in man. Some intracellular digestion, however, also takes place, for phagocytic cells in the lining of the cavity are able to engulf and digest minute food particles.

Because each cell in the body is close to or in contact with the water in which the animal lives, both respiration and excretion are carried on by direct diffusion. Remains of digested food particles are extruded through the mouth by contractions of the body (see Figure 21–8).

Reproduction

Hydra reproduces both asexually, by budding (see Figure 11–4), and sexually, by producing eggs and sperm. The zygote develops directly into a little hydra. Many cnidarians, though, show alternation of form (*metagenesis*). The polyp stage gives rise to medusas by asexual budding and the medusas produce the gametes by meiosis. This should not be confused with the alternation of generations that is so characteristic of plants. In cnidarians, only the gametes are haploid and cell division does not take place between meiosis and the formation of the zygote. Figure 21–9 is a word diagram of this pattern of reproduction. In some cnidarians, only the medusa form exists and reproduction is always sexual.

FLATWORMS (PLATYHELMINTHES)

Instead of being radially symmetrical like the coelenterates, the flatworms have true *bilateral symmetry* as do man and other higher animals. There is

Figure 21–8. Structure and function in hydra.

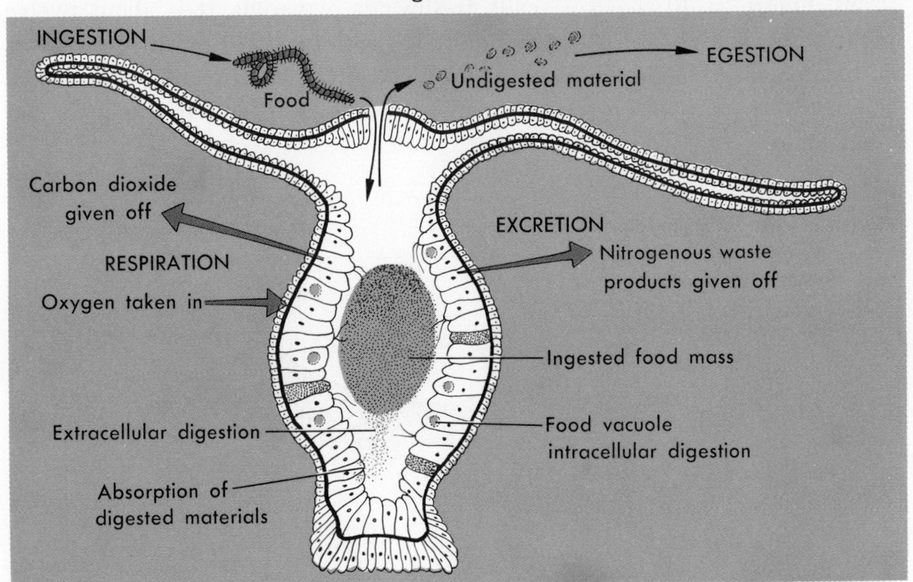

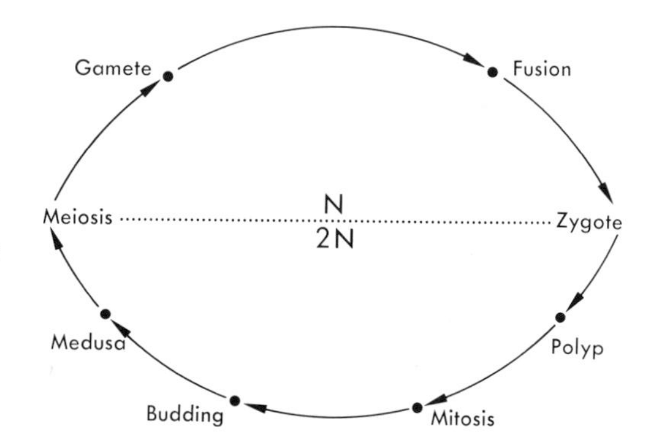

Figure 21–9. Metagenesis in cnidarians. The polyp stage reproduces asexually to form the medusas. The medusa reproduces sexually.

only one way in which a plane passing through the body can divide it into two halves (right and left) that are essentially mirror images of each other. A flatworm has a definite head where nervous tissue is concentrated and back (dorsal) and belly (ventral) surfaces. It also has a well-developed mesodermal layer and shows more tissue differentiation than a cnidarian. True organ systems are present. Some flatworms are free-living, in water or in damp ground; others are parasitic. The latter include some of the most spectacular agents of human disease.

Structure of Flatworms

Because the parasites are modified in adaptation to their specialized environment, the basic flatworm organization can best be seen in the free-living forms. Figure 21–10 shows a small, freshwater planarian. It is about twelve millimeters (half an inch) long and quite flat with two light-sensitive eye spots on the dorsal surface in the head region. Ventrally it has a mouth, from which a protrusible pharynx (*proboscis*) can project, and behind this is a minute genital pore.

Figure 21–10. A planarian.

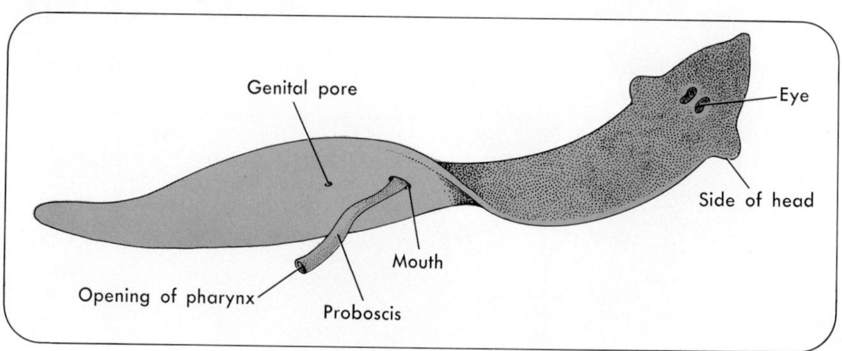

A planarian has circular and longitudinal layers of muscle cells and also diagonal muscle fibers that run dorsoventrally so that it is capable of a variety of movements. The digestive system is much branched and complex, but it has only one opening, the mouth. The planarian feeds on other small animals, living or dead, which it captures by its protrusible pharynx. The food passes into the digestive cavity where enzymatic digestion begins. When the food is partially broken up, the particles are ingested by phagocytic cells in the walls of the gastrovascular cavity. These cells complete digestion, and the nutrients then pass by diffusion to the other cells. The extensive branching of the gastrovascular cavity means that no part of the body is too far from a source of food.

The planarian is small enough to obtain oxygen and eliminate carbon dioxide by diffusion through the body surface.

Water content is regulated by a system of branched tubules, *protonephridia,* which open to the outside but end blindly in the body. The cell at the tip of the closed end is horseshoe-shaped and has a tuft of cilia on the inner curve (see Figure 21–11). When the living cells are examined under a microscope, the constant beating of the cilia makes a flickering that gives these cells the name of *flame cells.* The beating cilia set up a current that causes water to flow down the tubules and so to the outside. The system is called an excretory system, but probably most nitrogeneous wastes simply diffuse out through the whole body surface and the system is mainly concerned with osmotic regulation.

A planarian has a central nervous system with two nerve cords running the length of the body and an accumulation of nervous tissue in one end, the *cerebral ganglion.* There is also a concentration of sense organs in the head region. Not only are eye spots present but tactile and chemoreceptors are more abundant here than on other parts of the body. The concentration of nervous tissue and sense organs at one end of the body to form a head is known as *cephalization.* Most of the animals that we have thus far considered are either

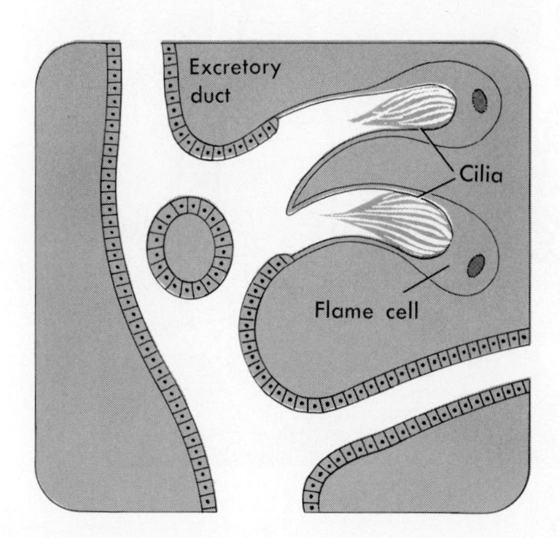

Figure 21–11. Detail of excretory system of a planarian. The beating cilia cause water to move down the excretory duct.

stationary or drift with water currents. It has been suggested that the ancestors of the flatworms adopted a mode of existence of creeping over the bottom, and that this led to the differentiation of dorsal and ventral surfaces and of a directive head end. Cephalization and bilateral symmetry are characteristics of free-moving animals.

Figures 21–12 and 21–13 show the systems of a planarian.

Reproduction

Almost all flatworms are hermaphroditic, with a single individual containing both male and female reproductive organs, but cross-fertilization from one individual to another seems to be the rule. Some, if not all, of the free-living forms are also able to reproduce asexually either by binary or multiple fission.

Parasitic Flatworms

The parasitic flatworms include the flukes and the tapeworms.

The Chinese liver fluke is a little worm about eight millimeters (one third of an inch) long. The adults live in the livers of humans and some other fish-eating mammals. The flukes tend to congregate in the upper bile ducts, particularly, those of the left lobe of the liver. The incredible number of 21,000 flukes have been recovered from the liver of a single individual on autopsy. The worms reproduce sexually and the eggs pass down the bile duct to the intestine and

Figure 21–12. Systems of a planarian. A: Digestive system. B: Excretory system. C: Nervous system.

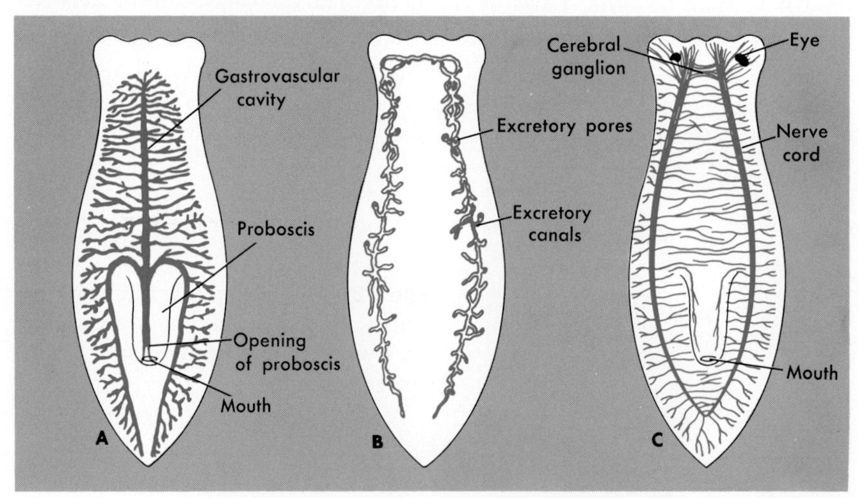

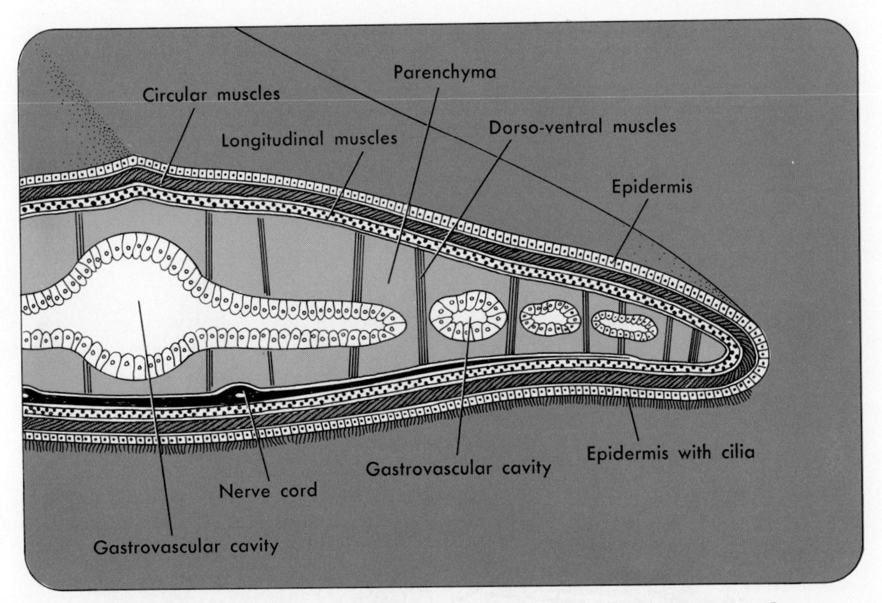

Figure 21–13. Part of a section through a planarian showing the arrangement of the muscles and the ventral nerve cord and branches of the gastrovascular cavity on one side.

thence with fecal material to the outside. Only if the eggs get into a pond or waterway where certain snails and fishes occur can the life history be continued. The use of *nightsoil* (human feces and excretory products) to fertilize crops facilitates the continuation of the cycle. If the forms that hatch from the eggs are ingested by a snail, they develop in its tissues into forms that reproduce asexually a number of times and emerge some four or five weeks later as small, motile animals called *cercariae*. These cercariae swim around until they encounter a fish, whereupon they burrow under its scales and into its flesh and there encyst. If a man or other mammal should eat such a fish, either raw or insufficiently cooked, the cysts rupture in the digestive tract of the host and the young flukes make their way up the bile duct to the liver. Light infections may cause few symptoms, but heavy infestations can result in serious liver damage. Figure 21–14 shows the life cycle of the Chinese liver fluke.

ROUNDWORMS (NEMATODA)

Roundworms are everywhere—in soil, fresh water, the sea, and the bodies of plants and animals. A spadeful of garden soil may contain a million of them. Many roundworms are microscopic, but some of the parasitic ones are nearly

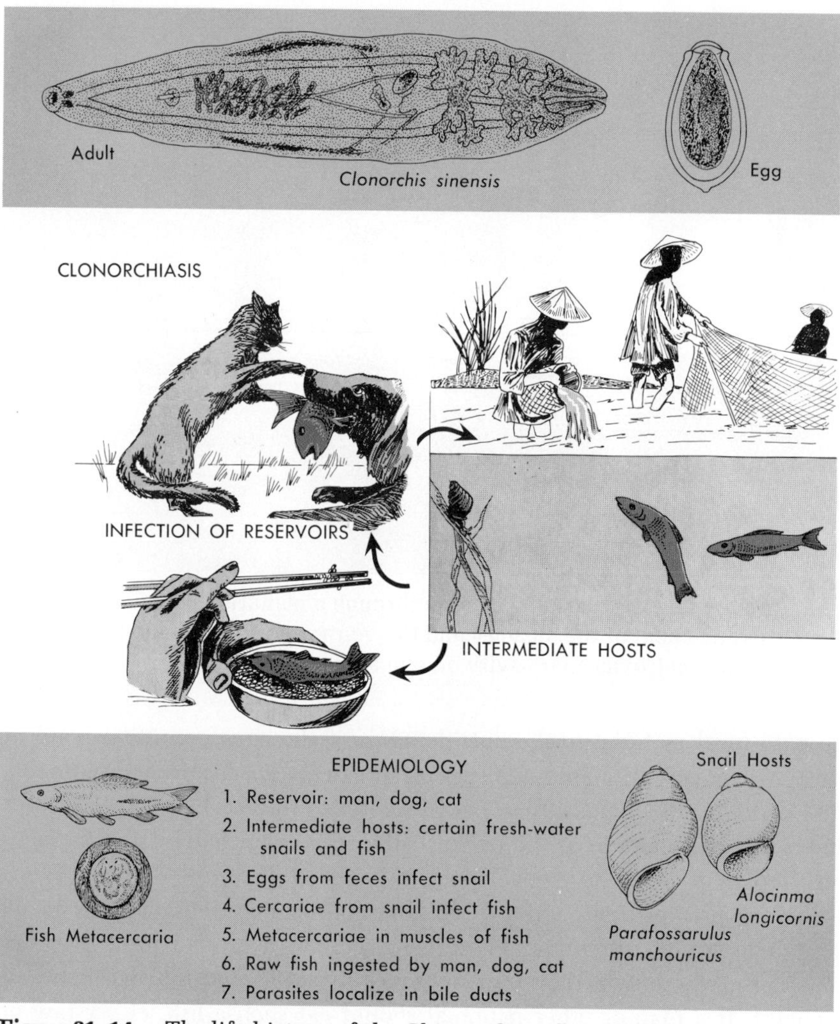

Figure 21–14. The life history of the Chinese liver fluke. (From Hunter, Frye, and Swartzwelder, *A Manual of Tropical Medicine* (4th ed.), 1966. By permission of W. B. Saunders Company and George W. Hunter.)

a meter (three feet) long. A roundworm has a slender, elongate body surrounded by a tough, noncellular cuticle. The body wall is composed of ectoderm and mesoderm. Within is a cavity, the *pseudocoelom,* in which the internal organs lie. This is not a true coelom, but is derived from the embryonic blastocoel. It lies between the mesoderm and endoderm, not within the mesoderm. There is no mesodermal layer surrounding the gut, and there are no mesenteries extending in from the body wall to support the internal organs. The roundworms have made one major advance over the flatworms; the digestive cavity has developed a posterior opening, the anus. It is now a true gut, a tube with a mouth at one end and an anus at the other. Ingestion and egestion can take place at the same time. Furthermore, the separation of the gut from the body wall means that the movement of material through the gut is not dependent on movements effected by the general body musculature. In nematodes, the fore-

Figure 21–15. Life histories of some parasitic roundworms. (From Hunter, Frye, and Swartzwelder, *A Manual of Tropical Medicine* (4th ed.), 1966. By permission of W. B. Saunders Company and George W. Hunter.)

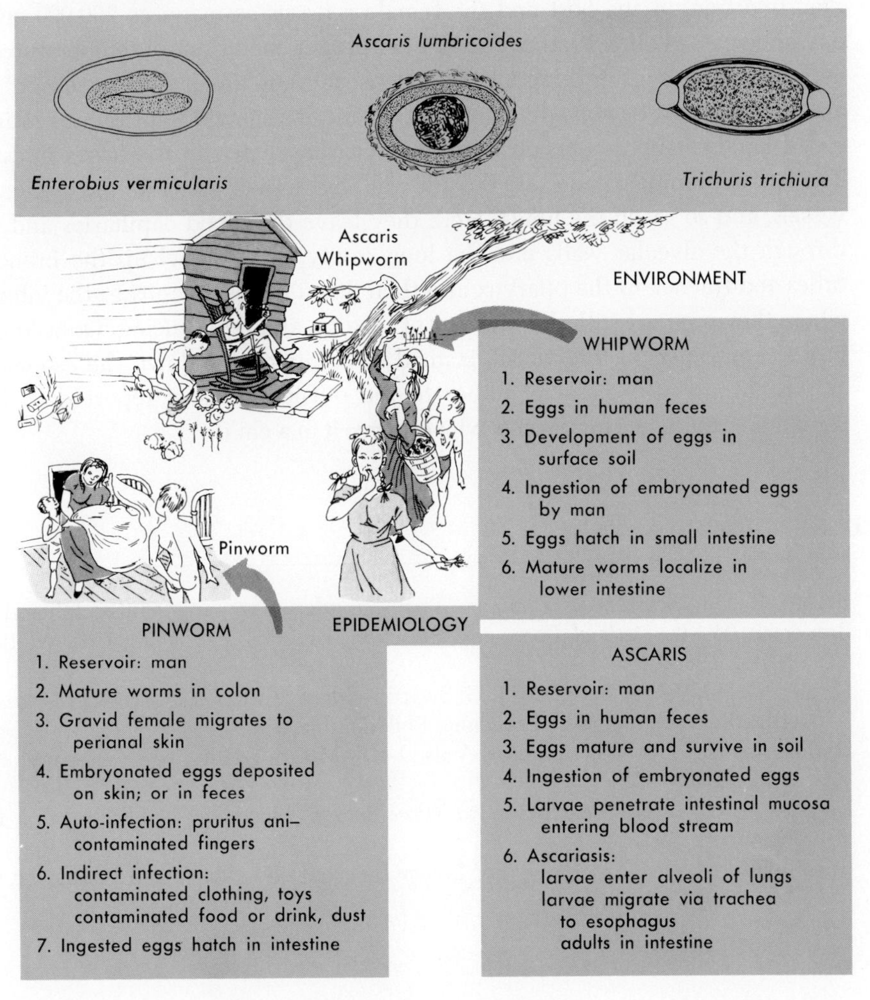

part of the gut is a muscular pharynx formed by an invagination of the ectoderm and mesoderm. Below the pharynx the gut wall lacks muscular tissue because it is composed only of endoderm. Food is forced along by the pumping action of the pharynx.

Many parasitic roundworms do little damage to the host, but some are very destructive to plants and a few cause serious diseases in man. Figure 21–15 shows the life cycles of some of the parasitic roundworms. In *elephantiasis* the worms block the lymphatic ducts; the legs and scrotum are particularly affected and swell enormously. One roundworm infection, technically known as *ascariasis*, is caused by a large intestinal roundworm. An adult female may reach 250 millimeters (ten inches) in length. The adult roundworms live in the human intestine and feed on the material eaten by the host. Mating takes place in the digestive tract of the host and the female releases as many as 200,000 eggs a day to be passed out with the feces. Under proper conditions of temperature and moisture, these eggs develop into infectious stages in about three weeks. If they are then ingested by some human on raw fruit or vegetables, or are taken in by soiled hands or by *geophagia* (eating dirt), they hatch in the host's intestine. The larvae promptly penetrate the gut wall, make their way into the circulatory vessels, and so reach the lungs. Here they leave the blood capillaries and pass through the alveolar walls into the lung cavity. They crawl up the bronchial tubes and trachea to the pharynx and then down the esophagus to the intestine where they proceed with the business of feeding and reproducing. Occasionally, in a sleeping person, the worm, instead of turning back down the esophagus, crawls through the nasal chamber to the outside. This is reported to be a traumatic experience for parents who observe it in a child.

SUGGESTED READINGS

Barnes, R. D.: *Invertebrate Zoology*, W. B. Saunders Company, Philadelphia, 1968.

Breneman, W. R.: *Animal Form and Function*, Blaisdell Publishing Co., Waltham, Mass., 1966.

Hunter, G. W., W. W. Frye, and J. C. Swartzwelder: *A Manual of Tropical Medicine*, 4th ed., W. B. Saunders Company, Philadelphia, 1966.

Hyman, L. H.: *The Invertebrates*, Vols. I-III, McGraw-Hill Book Company, New York, 1940–1955.

Kreuger, R. G., et al.: *Introduction to Microbiology*, Macmillan Publishing Co., Inc., New York, 1973.

Russell-Hunter, W. D.: *A Biology of Lower Invertebrates*, Macmillan Publishing Co., Inc., New York, 1968.

CHAPTER 22

The Higher Animals

With the development of bilateral symmetry and a "tube within a tube" body plan, you can see dimly foreshadowed the structure of man himself. All the higher animals have this basic construction and they also have a true coelom.

COELOM

A coelom is a body cavity within the mesoderm. This third germ layer gives rise to the muscular and vascular parts of the wall of the gut as well as of the body wall and forms the mesenteries and peritoneum by which the internal organs are supported and surrounded. Figure 22–1 shows diagrammatic cross sections of the acoelomate (without a coelom), pseudocoelomate (with a false coelom), and coelomate body plans.

The formation of a cavity between the gut and the body wall allowed a physical separation of the locomotor and digestive functions. Thus, each of these functions could evolve and differentiate more freely. The presence of mesodermal muscular tissue in the gut wall allowed food to be moved by peristaltic waves. The coelom provided space for the gut to enlarge and coil so that the surface available for the absorption of food could increase as the size of the body increased. Increase in size, though, necessitated further differentiation of the mesodermal tissue. Acoelomate animals are either small or very slender and flattened, or have the bulk of their bodies made up of a largely nonliving mesoglea. The actively metabolizing cells are always close enough either to the body surface or to the digestive cavity so that diffusion can take the place of special circulatory, respiratory, and excretory organs. Most coelomate animals

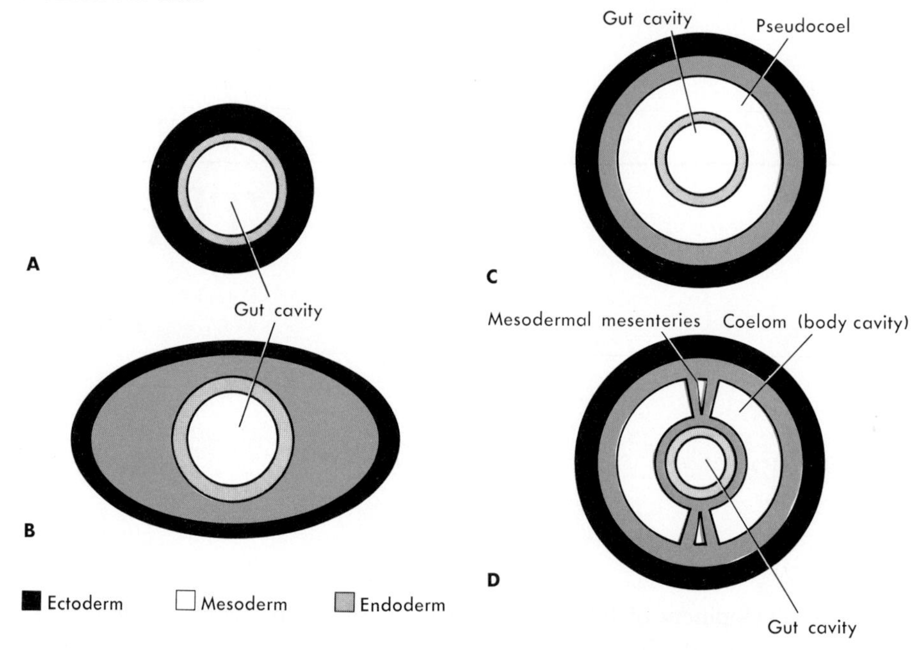

Figure 22–1. Diagram of body plans of multicellular animals. A: Acoelomate type. Ectoderm and endoderm are present but no true mesoderm and the only body cavity is the gut (gastrovascular cavity). Cnidarians show this type of organization. B: Acoelomate type with true mesoderm. Flatworms exemplify this pattern. C: A pseudocoel is present between mesoderm and endoderm. This pattern is found in roundworms. D: A true coelom develops within the mesoderm. Advanced invertebrates and all vertebrates are coelomate animals.

have a respiratory system, a well-developed circulatory system, and an excretory system that, in addition to maintaining the salt and water balance, can also remove nitrogeneous wastes. These animals are freed from the restrictions limiting size in the acoelomates. Not all coelomates are large, but all large animals are coelomates.

EMBRYONIC DEVELOPMENT

The coelomate phyla are divided into two great groups that differ in their patterns of embryonic development.

Deuterostomes

The animals whose development was discussed in Chapter 12 are all deuterostomes. Four aspects of their development are important here. First, look again at Figure 12–1 and notice that the early cleavage planes are either parallel with, or at right angles to, the axis running from the animal to the vegetal pole. The cells in the upper row lie directly above the cells in the lower row. This kind of cleavage is called *radial cleavage*. Second, the ultimate fate of the cells is not irrevocably fixed until relatively late in development (indeterminate development). Remember some of the experiments on amphibian development. For example, if cells of the dorsal lip of the blastopore are transplanted to the belly region of another embryo, they can cause the development of a complete embryo. Third, in deuterostomes the mesoderm primitively develops as outpocketings from the wall of the archenteron. And fourth, the anus develops in the region of the blastopore, whereas the mouth forms elsewhere. In addition to the vertebrates and their allies (phylum Chordata), the deuterostomes include the starfishes and sea urchins (phylum Echinodermata) and several smaller phyla.

Protostomes

The protostomes include a number of minor phyla and three major invertebrate groups: the segmented worms (earthworms, leeches), the mollusks (oysters, snails, squids), and the arthropods (lobsters, spiders, insects). Like the deuterostomes, protostomes show a diversity of developmental patterns, depending largely on the amount of yolk in the egg. Basically, though, the cleavage planes form at an angle to the central axis. In the eight-cell stage, the cells in the upper row lie in the grooves between the cells in the lower row. This pattern continues so that at each division each new cell is located in the angle between the two cells above and below it. Figure 22–2 is a diagram of the eight- and sixteen-cell stages of an egg with this type of cleavage (*spiral cleavage*). Furthermore, the fate of the cells is determined very early. Even in the four-cell stage, each cell can give rise to only a fourth of a gastrula (deter-

Figure 22–2. Spiral cleavage in a protostome. A: Eight-cell stage. B: Sixteen-cell stage.

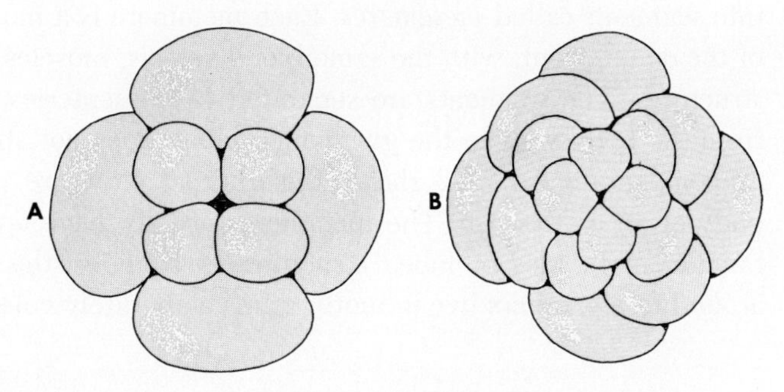

minate development). Most of the mesoderm develops as a solid sheet of cells from a single cell that can be identified in the early blastula, and the coelom forms as a split in this sheet. The mouth of a protostome develops from the region of the blastopore. This is the reason for the name *protostome*—the true mouth develops at or near the site of the "first mouth," the blastopore. (*Deuterostome* means "second mouth.")

SEGMENTED WORMS (ANNELIDA)

Did you ever walk down a sidewalk on a sunny morning after a night of warm rain and notice the bodies of earthworms? They moved out of their burrows during the hours of high humidity and desiccated before they could get back to their burrows, for the earthworms, even though they are terrestrial, are still dependent on humid surroundings. Most segmented worms are aquatic, living in either salt or fresh water. They are especially numerous in the intertidal zone. One species found on the Pacific Coast of the United States lives in colonies that average 2,500 to 3,000 individual worms a square foot.

Segmented worms range in size from tiny forms about half a millimeter (a fiftieth of an inch) long to the giant earthworm of Australia, which may reach a maximum length of about three to four meters (nine to twelve feet). The burrowing activities of earthworms are useful in aerating the soil and in mixing organic material through it, and worms form an important part of the diet of many fishes and other animals. The bloodsucking leeches may be a nuisance, particularly the large terrestrial ones found in tropical jungles. Formerly, doctors used to apply leeches to boils or other inflamed areas to draw the blood. The practice was so widespread that the doctors themselves were often called leeches.

General Structure

The most noticeable characteristic of an annelid is the division of its body into segments called *metameres*. Each metamere is a more or less faithful copy of the one in front, with the same blood vessels, muscles, nerves, and excretory structures. The segments are separated by mesenteries, called *septa*, running from the body wall to the gut, which alone does not show the segmented arrangement. Figure 22–3 shows the internal structure of the forepart of the body of an earthworm. The metameres usually have appendages, either little bundles of bristles or lobed structures bearing bristles. Some annelids move around freely, others live in more or less elaborately constructed tubes.

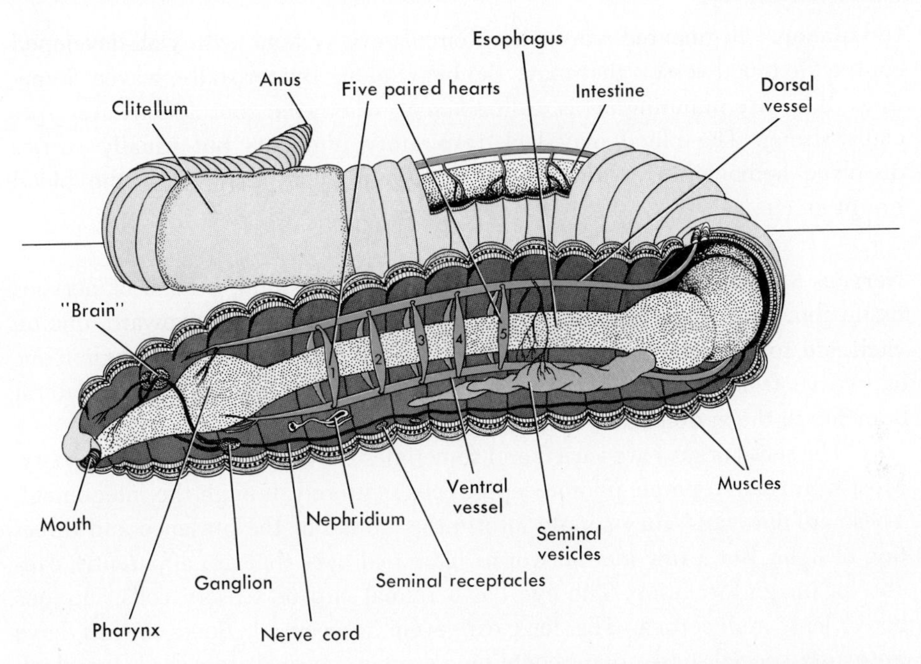

Figure 22–3. Structure of earthworm.

Nutrition. Earthworms and many other annelids live on decaying organic material in the soil, through which they literally eat their way. Digestion is extracellular. The undigested remains, mostly inorganic soil particles, form the worm castings sometimes seen on top of the ground. Other annelids are carnivorous, feeding on small invertebrates. The leeches are either carnivorous or external, bloodsucking parasites.

Excretion. Annelids have specialized excretory structures, the *nephridia,* which constantly remove nitrogeneous wastes from the coelom. A nephridium is a small, ciliated funnel opening into the coelom and leading into a coiled tubule that opens to the outside. Water is reabsorbed through the excretory tubules of the terrestrial worms, and in both terrestrial and freshwater species, salts are likewise reabsorbed.

Respiration. Many annelids lack respiratory structures and simply carry on exchange of gases through the skin. Others have a variety of special expanded external respiratory surfaces called *gills.* The gills vary greatly in shape. They may be flattened lobes or long threads, or branched and feather-like. Usually they are modified parts of the appendages.

Circulation. Segmented worms have circulatory systems with well-developed contractile blood vessels that move the blood along by peristaltic waves. Sometimes there are pumping areas, called *hearts,* where the contractions are especially strong. The blood may lack respiratory pigments but usually carries dissolved hemoglobin or another respiratory pigment that colors the blood bright green.

Nervous System. The nervous system consists of a brain, a mass of nervous tissue above the pharynx, from which a pair of nerves pass downward, one on each side, to join again in a ganglion below the pharynx. From this ganglion one or two ventral nerve cords pass backward below the gut, giving off lateral branches to the various body segments.

The sense organs are varied and sometimes show considerable complexity. Most worms have simple photoreceptor cells scattered through the integument. These are not eyes—they simply inform the animal of the presence and direction of light. But a few marine worms have real eyes that are apparently capable of image formation. The eye has a retinal cup of sensory cells, vitreous body, lens, and cornea. The lens can even be focused. Some worms have *statocysts,* special sense organs containing sand grains or diatom shells by which the animal is made aware of changes in position. There are also chemoreceptors and tactile receptors. The ability of earthworms to sense vibrations in the ground is utilized by Florida fishermen. In an area where there is *hardpan,* a dense, impervious layer close under the surface of the ground, a man drives a wooden stake in until the end rests on the hardpan. Then he saws across the stake with an iron rod. The worms respond to the vibrations thus set up by leaving their burrows and appearing on the surface around the stake in an area several meters (yards) in diameter. It is easier than digging for bait.

Reproduction. Some aquatic worms are able to reproduce asexually by budding or by fission, but most of them reproduce sexually.

Cross-fertilization usually takes place, even though some species are hermaphroditic. The gametes are shed into the coelom—in fact they usually complete maturation there. They may escape from the coelom through the nephridia or through special sex ducts. Marine worms may have separate sexual and asexual forms. The asexual form (*atoke*) either changes directly into the sexual form (*epitoke*) or gives rise to it by budding. The epitokes swim to the surface in huge swarms and then break open, releasing the eggs and sperm. Swarming is remarkably synchronized; the Samoan palolo worm epitokes swarm on the night at the beginning of the last lunar quarter in October or November. The natives of the region know when the epitokes will appear and wait for them in boats, for they are a great delicacy. The exact mechanism of synchronization is not known, but it has been shown that the formation of epitokes is inhibited by neurosecretory cells in the brain—an example of hormonal control.

ARTHROPODS (ARTHROPODA)

Three groups of organisms have made a success of terrestrial existence in a major way. They are the seed plants, the arthropods, and the vertebrates. For the autotrophic plants, the *sine qua non* of successful life on land are a tough, more or less impervious covering to prevent desiccation, a protected respiratory surface, supportive tissues, a system of transport, a terrestrially adapted reproductive system, and a means of protecting the developing embryo. The heterotrophic animals need more. They must have a locomotor apparatus for getting around and finding food, and because their food is made up of other terrestrial organisms with tough outer coverings, they must have jaws or some similar device to crush or penetrate such a covering. Not all arthropods have all these traits and not all are terrestrial, but those that are include the most successful of all land animals, the insects. Arthropods are also successful as aquatic animals; about 80 per cent of all living kinds of animals belong to this one phylum.

General Structure

The Arthropoda (meaning jointed feet) are coelomate animals with a true gut, a metameric construction, a ventral nerve cord, and jointed appendages. In most arthropods the metameres are more or less obscured by a fusion of various segments, but they are still evident in a centipede or in the "tail" (really the abdomen) of a lobster.

An arthropod is covered by a hard, outer, armor composed of *chitin,* a polysaccharide secreted by cells of the epidermis. Obviously, if the armor were a single, continuous sheet, the animal would not be able to move at all. Instead, the armor is composed of many plates connected by regions where the covering is very thin and flexible. The appendages are also covered by sheaths of chitin divided into segments to form the joints. They are variously modified for many different functions (see Figure 22–4).

The presence of the armor is correlated with changes in the musculature. The outer covering of an earthworm is a thin, flexible cuticle and the body musculature consists of layers of circular and longitudinal muscles below the epidermis. Movement of the body is accomplished by alternate contractions of these muscles. Because the arthropods must move by flexion between the plates of armor, as vertebrates move by flexion between parts of the skeleton, the

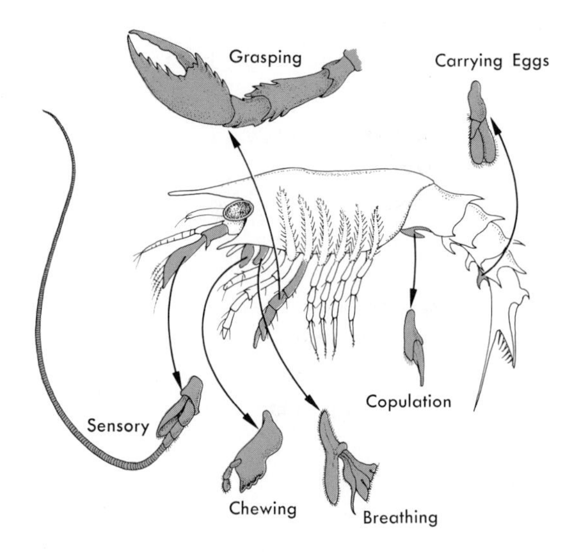

Figure 22–4. The jointed appendages of a crustacean are modified to serve a variety of functions.

former have striated skeletal muscles attached to the inner surfaces of the plates. The armor is thus really an external skeleton.

The arthropod external skeleton did not develop as an adaptation to life on land. The earliest arthropods lived in the sea, and their armor must have been useful in that environment. It does not seem to have served originally as a protection against predators because active predators were apparently very rare or absent when the shelled arthropods appeared. Perhaps the primary adaptive advantage of the jointed external skeleton with attached muscles was that it allowed a far greater diversity of movement than is possible for any of the animals discussed so far. That it also prepared the arthropods for the invasion of the land was only by chance.

The presence of an external nonliving skeleton creates special problems of growth. At intervals, a new, soft cuticle develops under the hard external skeleton. When this occurs, the outer skeleton cracks and the animal crawls out of the old skeleton. The soft-shelled crabs of the market are nothing but individuals that have just shed their old skeletons. Rapid growth takes place before the new skeleton hardens so that successive sheddings (*molts*) permit the animal to grow larger and larger. The process is regulated by hormones.

Reproduction

Asexual reproduction by budding or fission does not take place in arthropods, but a few are parthenogenetic. This means that the egg develops without being fertilized. For some species, males have never been found and may not exist. Or parthenogenesis may be seasonal. In the spring only wingless female plant lice occur. They produce diploid eggs by mitosis, which grow into females and so on for a number of generations. Toward fall, two new kinds of females appear. One lays eggs that develop parthenogenetically into winged males, the other lays eggs that develop parthenogenetically into winged females. These produce eggs and sperm by meiosis, mating takes place, and the zygote is encased in a hard shell in which it overwinters, to hatch in the spring as a wingless female.

Usually, though, arthropods reproduce sexually, and usually, too, the sexes are separate. Even in the few hermaphroditic forms, cross-fertilization normally occurs. Primitive groups show evidence of spiral cleavage, but in most arthropods, development is modified by the presence of large amounts of yolk.

Arthropods Other Than Insects

Most arthropods are insects. Another group important to man is the crustaceans, including the lobsters, shrimp, crabs, barnacles, and a number of tiny forms known as *microcrustaceans* (see Figure 22–5). The larger crustaceans are eaten by man and other animals and the microcrustaceans are major items in the diet of many fishes. Crustaceans are aquatic and usually breathe by means of gills, which are modified parts of the appendages. Most excretion takes place through the gills. The head and trunk segments are fused into a single body part, the *cephalothorax,* whereas the abdominal segments remain distinct.

The arachnids are also important to man. This group includes spiders, scorpions, daddy longlegs (also called harvestmen), mites, and ticks. Figure 22–6 shows a variety of arachnids. Like the insects, they are primarily terrestrial, but structurally they are more like crustaceans, with cephalothorax and abdomen. The spiders are better known, more feared, and less dangerous than members of the other large groups of arachnids, the mites and ticks. The widow spiders (black, brown, red, and so on) are capable of causing serious illness or even death in a human, but most spider bites are rather minor, and the group as a whole serves man well by helping to keep obnoxious insects under control. The less feared mites and ticks are not only a considerable annoyance to human beings, but are also carriers of such serious diseases as Rocky Mountain spotted fever; they also transmit diseases to domestic animals.

Insects

Insects are so widespread and abundant and touch on man's life in so many ways that no one is free from their influence, both good and bad. Many of the flowering plants, including most vegetables, fruits, and garden flowers, depend on insects for pollination. One of the finest of all cloths, silk, is spun from the

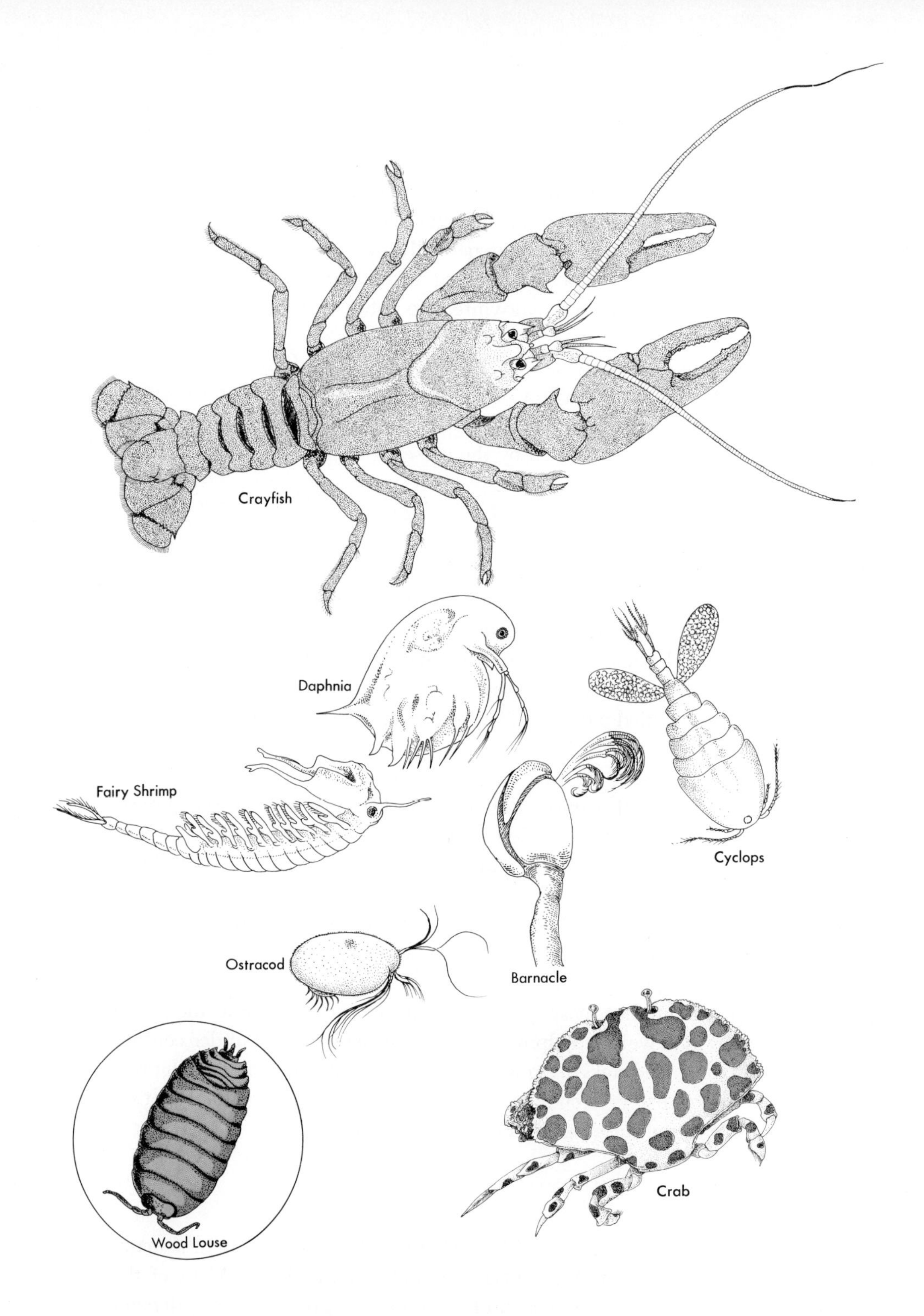

Figure 22–5. Various kinds of crustaceans (not shown to scale).

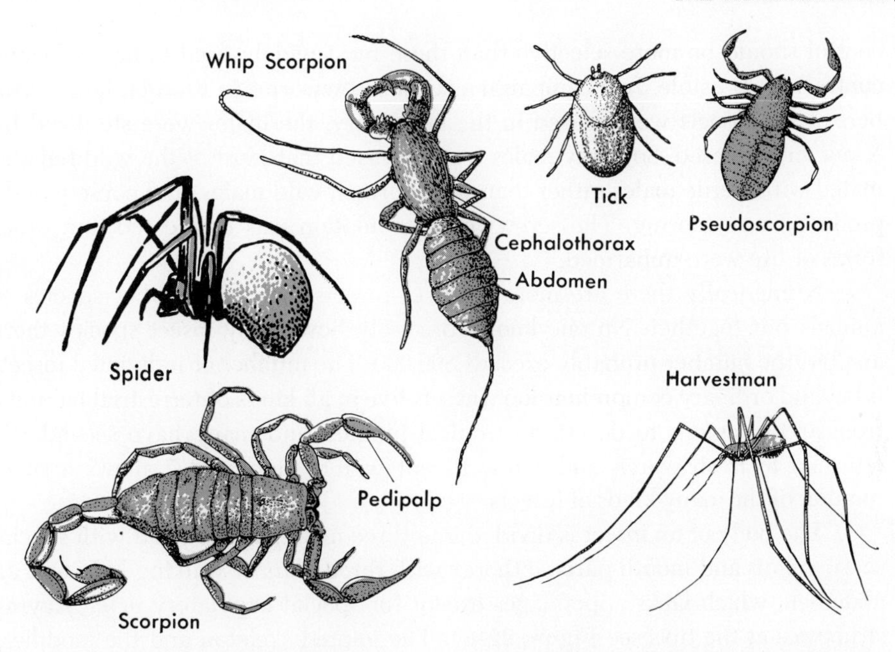

Figure 22–6. Some well-known arachnids. (From L. S. Dillon, *Principles of Animal Biology*. Macmillan Publishing Co., Inc., 1965.)

fibers of the cocoon of the silkworm. Aquatic insects are extremely important as a food source for freshwater fishes, and terrestrial insects are equally important in the diet of many other animals. In some parts of the world, insects are eaten by men; in South America large ants are prepared, packaged, and sold in supermarkets.

On the other hand, much illness is caused by insects. Mosquitoes, for example, transmit malaria, yellow fever, and filariasis, diseases that afflict at least a sixth of the world's population. Insects are a constant threat to man's food supply, both while it is growing and after it is harvested. Insects destroy clothes, wooden houses, and books. Many insects are parasitic on livestock (botflies, screwworms) and on man (lice, fleas). Even when they do little direct damage, biting and bloodsucking insects, such as gnats, sandflies, and mosquitoes, can be a tremendous nuisance.

Poisons introduced into an area to control obnoxious insects are equally destructive to beneficial ones. Spraying around a lake to reduce the nuisance of mosquitoes may kill off the food supply of the fishes. Furthermore, the poisons tend to accumulate in the bodies of animals that eat the insects until lethal levels are reached. Many songbird deaths have been attributed to such poisoning. Dangerous and destructive insects must be controlled but the methods of control should be more selective than those most widely used today. Selective controls are possible. In a campaign against screwworms in Florida, large numbers of the insects were reared in the laboratory, the males were sterilized by X-ray and released. So many males were released that most of the wild females mated with sterile males rather than with normal, wild males, and consequently produced no offspring. The screwworm population was eradicated, but other forms of life were unharmed.

Numerically there are more kinds of insects than of all other species of animals put together. No one knows precisely how many insect species there are, but the number probably exceeds 800,000. The number of individual insects is beyond ordinary comprehension. Insects live in all kinds of terrestrial habitats, from arctic regions to deserts to tropical jungles, and many have secondarily returned to fresh water, and a few to salt water. Figure 22–7 shows a small number of the many kinds of insects.

The body of an insect is divided into three main parts: a head with special sense organs and mouth parts; a thorax with three pairs of walking legs; and an abdomen, which lacks appendages except for special copulatory or egg-laying structures at the tip (see Figure 22–8). The jointed skeleton and the modification of some of the appendages into specialized mouth parts are common to all arthropods. The insects owe their success on land to a number of additional traits.

Locomotion. In addition to the walking legs, many insects have two pairs of wings on the thorax. The wings are not modifications of the original arthropod appendages but are entirely new structures. The development of wings opened up a new environment—the air—to the insects and undoubtedly contributed to their widespread distribution. The only other animals to invade this environment are vertebrates—birds, bats, and some extinct flying reptiles.

Nutrition. Chewing mouthparts seem to be the primitive form, but the mouthparts of many insects are modified into piercing and sucking structures so that they can feed on the fluids of an animal or plant without having to chew it up. This allows tiny mosquitoes to prey on large animals, such as man or horse. Almost all available sources of food are used by one insect or another.

Respiration. All adult insects depend on oxygen from the air and have a special respiratory system. Figure 22–8 shows a series of openings (*spiracles*) along the sides of the thorax and abdomen that open into an inner complex of tubes (*tracheae*) that branch to reach all parts of the body. Through them oxygen can pass into the body fluid and carbon dioxide can be expelled.

Circulation. Insects have an open circulatory system, that is, one in which much of the blood flows through a series of open sinuses (the *hemocoel*) rather

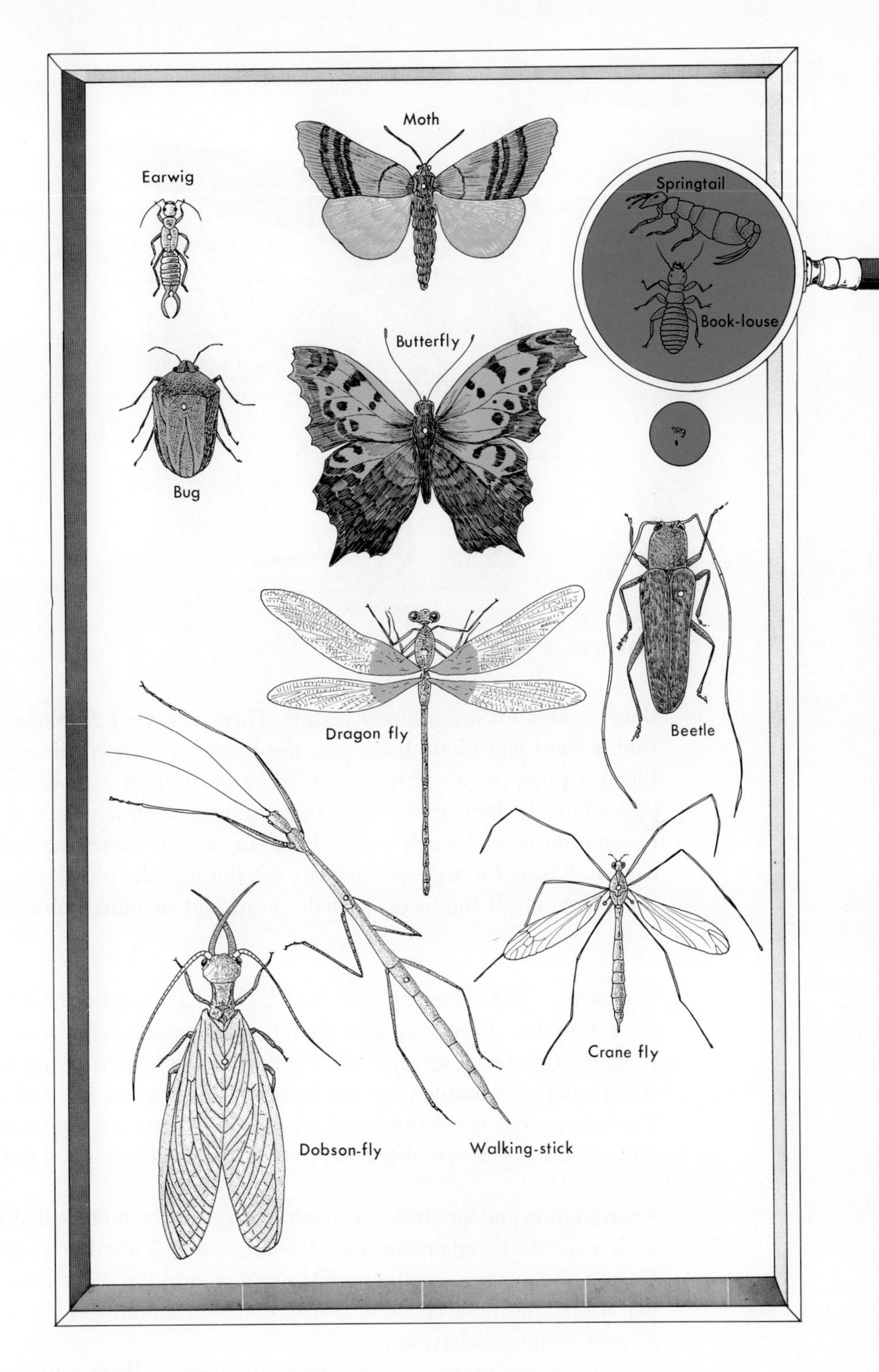

Figure 22–7. A few of the many kinds of insects.

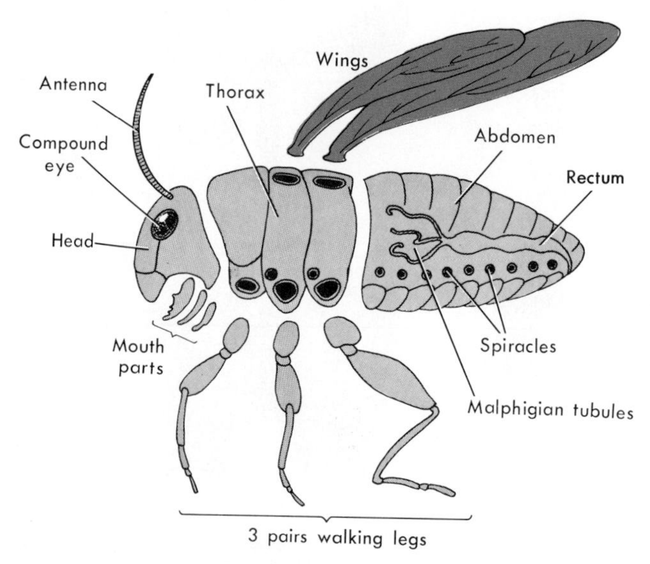

Figure 22–8. Structure of insect.

than through enclosed blood vessels. There is a well-developed heart lying in such a sinus and blood flows into the heart through openings in its wall. The blood is pumped out of the heart into a vessel that extends forward but soon opens into the hemocoel. Circulation through such a system is neither so fast nor so orderly as through a closed system, but the insect transports oxygen and carbon dioxide through its tracheae, not through the blood. Oxygen is the most immediate of all the insect's needs. Transport of other substances through its body can be slower.

Excretion. The excretory system of the insects is quite different from any we have thus far discussed. The excretory structures are blind tubules called *Malpighian tubules,* which lie in the part of the hemocoel surrounding the gut. Fluid enters the tubule from the hemocoel, and water and salt are reabsorbed. The nitrogenous wastes are discharged into the gut and are eliminated together with the feces in a very dry state, so that little water is lost through excretion.

Coordination and Control. The nervous system resembles that of the annelids, with a dorsal ganglion forming the brain and a double ventral nerve cord. Despite the relative simplicity of the system and the comparatively small number of neurons involved, insects are capable of a high level of coordination and of very complex behavior.

The sense organs are well developed, especially the eyes, which are composed of many closely packed lenses, each associated with a few sensory receptor cells. Insects can distinguish both the shape and the color of objects. They also have tactile and chemoreceptors, the latter often amazingly sensitive. A male moth can detect an odor emitted by a female nearly a kilometer (half a mile) away. And insects can hear, that is, detect vibrations in the air.

Insects also have well-developed hormonal systems. As in man, the hormones in insects are associated with processes of growth and development and also with homeostatic balance. Also as in man there is a close association between the insect's nervous and endocrine systems.

Reproduction. Most insects are bisexual, although some parthenogenetic forms of insects are known.

THE TERRESTRIAL EGG. All insects practice internal fertilization. Development of the insect embryo usually takes place within a hard-shelled egg, which protects the embryo from desiccation. The embryo could not afford to lose the water needed to excrete ammonia or urea, nor could it withstand the accumulation of large amounts of these poisonous substances. Nitrogeneous wastes are precipitated as harmless *uric acid.*

EMBRYONIC DEVELOPMENT. Like the terrestrial eggs of the vertebrates, insect eggs contain a relatively large amount of yolk. But the yolk, instead of being concentrated in the lower part of the egg, forms a central mass around the nucleus. In cleavage, the nucleus divides a number of times and the daughter nuclei migrate to the perimeter. Cell membranes form around the nuclei so that the blastula consists of an outer layer of cells surrounding an inner, undivided yolk mass.

LIFE HISTORY. In some insects, the young that hatches from the egg resembles the adult and development is simply by a series of molts as the insect grows with little change in form. Such development is said to be *ametamorphic.* Many insects, such as the flies with their maggots and the butterflies with their caterpillars (see Figure 22–9), have larval stages that are markedly different from the adults; often the larvae are aquatic as in the mayflies and dragonflies. When the young differ from the adults, there may be a gradual or incomplete *metamorphosis,* in which the young come to resemble the adult more closely at each successive molt. In complete metamorphosis, there are three distinct and very different life history stages: larva, pupa, and adult. These patterns of development are shown in Figure 22–10.

MOLLUSKS (MOLLUSCA)

Other than the insects and the vertebrates themselves, the mollusks are probably better known to most people than any of the other animals. Mollusks are widespread; many have colorful and attractive shells; there are about 110,000 kinds; some are very abundant, and many are good to eat. Mollusks

Figure 22–9. The larval stage of a moth. Tent caterpillars living on a community nest they have constructed on Carnegie Museum's Powdermill Nature Reserve. (Carnegie Museum—M. Graham Netting photo.)

are primarily a marine group, but some occur in fresh water and a few have made moderately successful invasions of land. The best known of the mollusks are the gastropods (snails, slugs), pelecypods (clams, oysters), and cephalopods (squids, nautilus, and octopods). Figure 22–11 shows a number of different kinds of mollusks.

Although some forms are so modified as to obscure the pattern, all mollusks are built on the same basic plan. There is a muscular foot; a covering for the body called the *mantle;* a shell that is secreted by the mantle; and a mass of internal organs, the *visceral mass.*

ECHINODERMS (ECHINODERMATA)

The echinoderms include the starfishes, sea urchins, and sand dollars. There are only about 6,000 species, all of which are restricted to the sea.

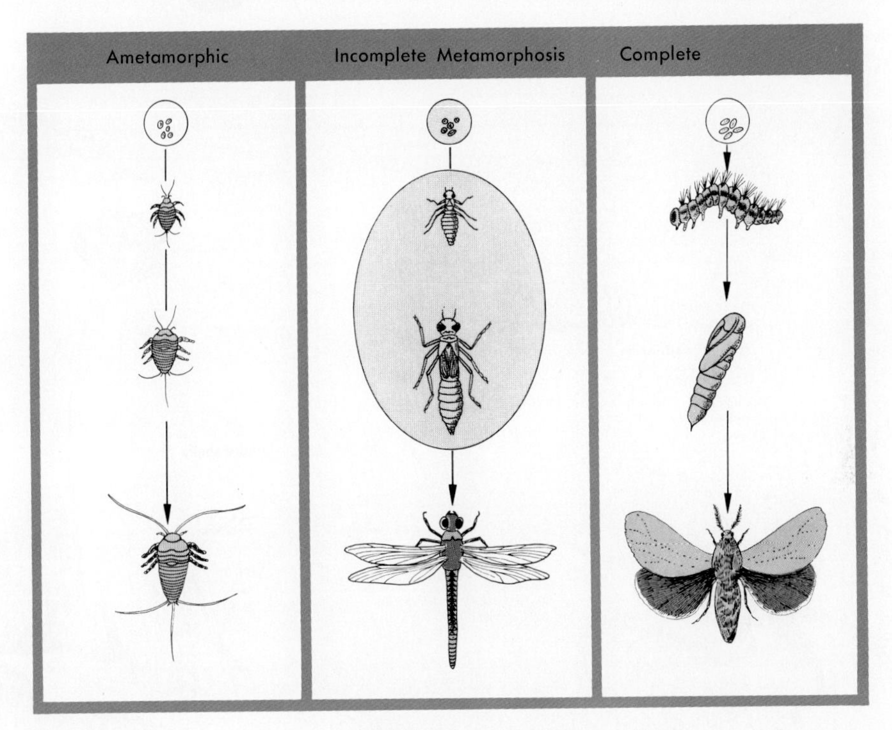

Figure 22–10. Patterns of insect development

Certain details of their structure and development indicate that they share a direct common ancestor with the vertebrates. They are deuterostomes, and their skeleton differs from that of other invertebrates in that it is internal and is formed from mesoderm rather than ectoderm. The chordates share this feature. Figure 22–12 shows some characteristic echinoderms.

This is the second phylum of multicellular animals that have radial symmetry; the first was the cnidarians. In the echinoderms, this symmetry is certainly a secondarily derived condition, for the larval stages of all echinoderms are bilaterally symmetrical and traces of this bilateral symmetry can still be seen in some adult starfishes.

CHORDATES (CHORDATA)

All members of this group have at some time in their life span (1) a notochord, (2) a hollow, dorsal nerve cord, and (3) pouches or slits in the wall of the pharynx. The embryonic development of the first two of these was discussed

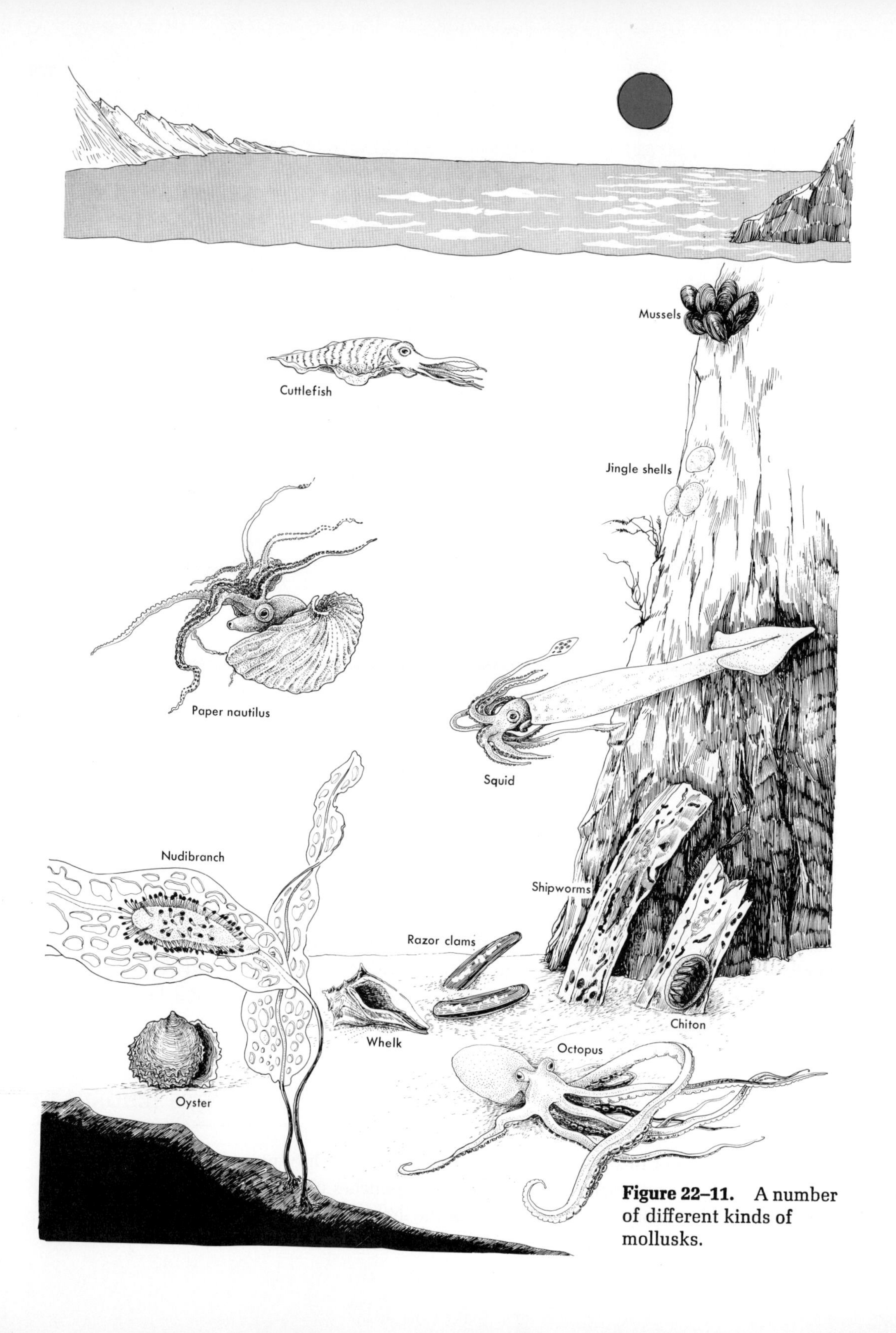

Figure 22–11. A number of different kinds of mollusks.

Figure 22–12. Some representative echinoderms. A: Starfish. B: Sand dollar. C: Sea urchin. D: Sea cucumber. E: Brittle star. F: Sea lily. G: Sun star.

in Chapter 12. The pharyngeal slits probably developed in the ancestors of the chordates as a feeding mechanism. Water was drawn in through the mouth and as it passed out through the pharyngeal slits, food particles were filtered out of it. Some primitive chordates living today are filter feeders (see Figure 22–13). Most fishes have jaws and teeth for grasping their prey; they do not need to strain it out of the water. They have developed gills on the walls separating the pharyngeal slits, now called *gill slits.* The fishes still pump a current of water out through the slits and over the gills and it is here that gas exchange takes place. The terrestrial vertebrates no longer need the open slits, either for feeding or respiration. The only trace of these slits that remains is a series of pouches that develop in the wall of the pharynx in the embryo. The first such pouch becomes the Eustachian tube, and the others normally disappear before birth.

In addition to the vertebrates, this phylum includes some small marine animals that lack vertebrae. One is amphioxus, whose embryonic development was discussed in Chapter 12 (see Figure 22–14).

A vertebrate has the characters of the chordates plus a vertebral column of separate units of cartilage or bone that surround the hollow dorsal nerve cord

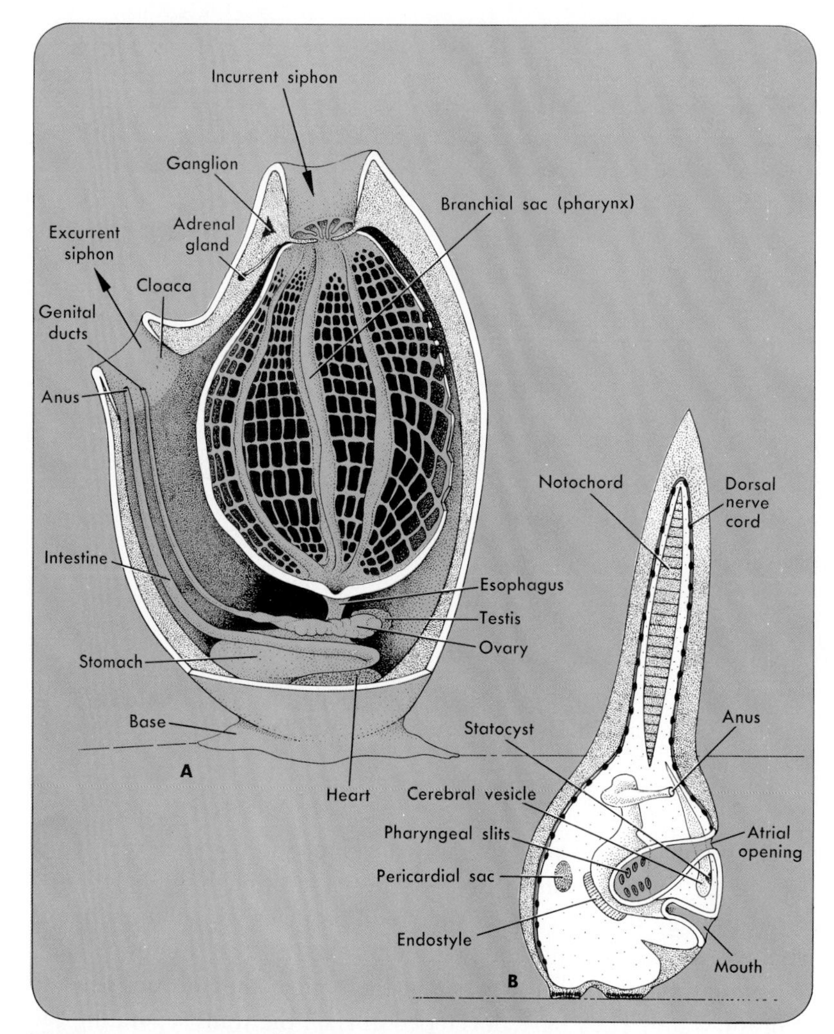

Figure 22–13. Internal structure of a tunicate, a filter-feeding chordate. The adult (A) is sessile, the larva (B) is free-swimming for a short time. The larva has a notochord and dorsal nerve cord as well as pharyngeal slits. The pharyngeal slits are the only traces of chordate structure that remain in the adult.

and, in the living vertebrates, more or less replace the notochord. All vertebrates are built on the same body plan. This basic pattern is shown in Figure 22–15.

Like the arthropods, some vertebrates are aquatic and others are terrestrial. Fishlike aquatic vertebrates are the cyclostomes (lamprey and hagfish);

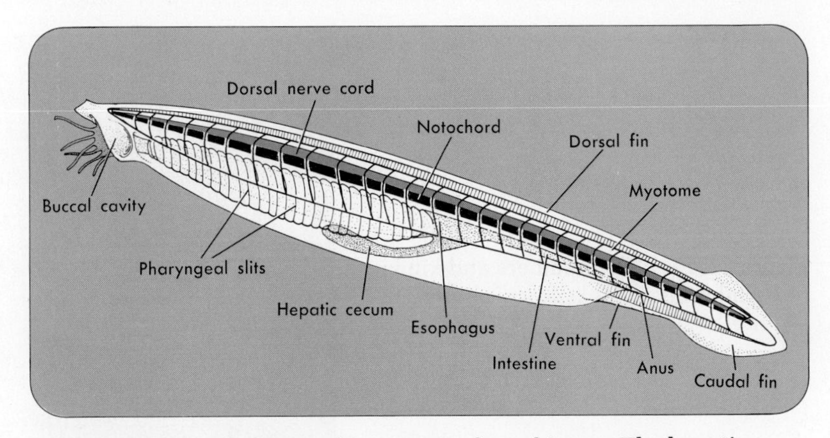

Figure 22–14. The internal structure of amphioxus. The hepatic cecum produces digestive enzymes. The body musculature is divided into separate segments called myotomes, an indication that metamerism is found in the chordates as well as in the annelids and arthropods.

the elasmobranchs (sharks, skates, and rays); and the bony fishes (trout, herring, and catfish). Terrestrial vertebrates are the amphibians (frogs and salamanders); reptiles (turtles, alligators, snakes, and lizards); birds; and mammals. See Figure 22–16.

The first section of this book is devoted to the description of a single vertebrate and in the section that follows, we describe the evolution of the other vertebrates.

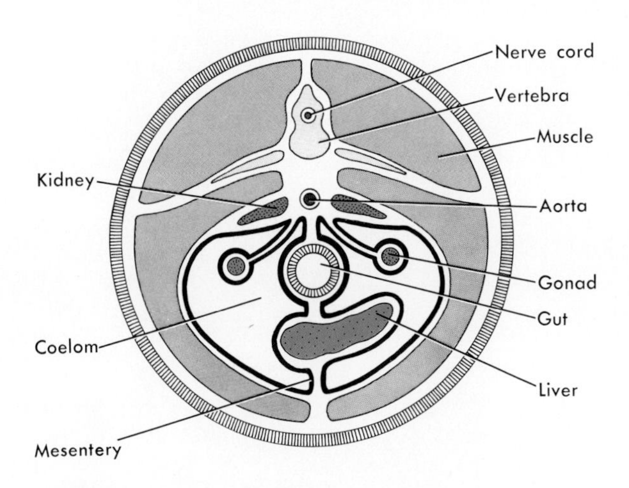

Figure 22–15. Vertebrate body plan.

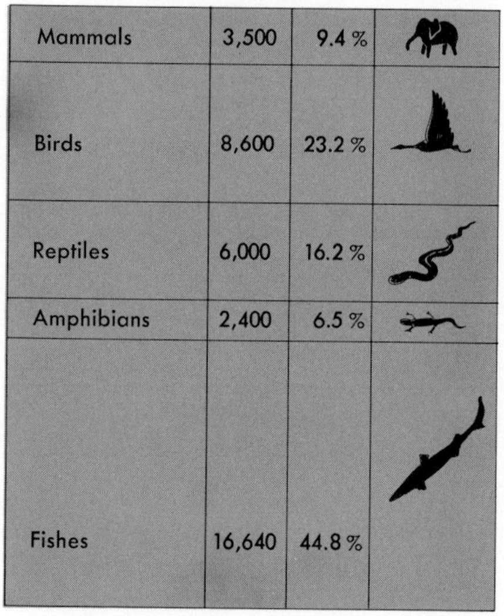

Mammals	3,500	9.4 %
Birds	8,600	23.2 %
Reptiles	6,000	16.2 %
Amphibians	2,400	6.5 %
Fishes	16,640	44.8 %

Figure 22–16. The numbers and kinds of vertebrates.

SUGGESTED READINGS

Barnes, R. D.: *Invertebrate Zoology,* W. B. Saunders Company, Philadelphia, 1968.

Blair, W. F., et al.: *Vertebrates of the United States,* McGraw-Hill Book Company, New York, 1957.

Hyman, L.: *The Invertebrates,* Vols. IV–VI, McGraw-Hill Book Company, New York, 1955–1967.

Orr, R. T.: *Vertebrate Biology,* 3rd ed., W. B. Saunders Company, Philadelphia, 1971.

Romer, A. S.: *The Vertebrate Story,* University of Chicago Press, Chicago, 1959.

The History of Life

How about cleaning up de whole mess of 'em and sta'tin all over again wid some new kind of animal?
MARC CONNELLY, *The Green Pastures*

CHAPTER 23

The Beginnings of Life, Evolution, and Geologic Time

In the last section we discussed some of the products of organic evolution, the 2 million or more different kinds of living things that inhabit the earth today. The fossil record makes it clear that many other kinds of living things existed in the past. How did life first arise, and how did it evolve into such a multitude of different kinds of animals and plants?

SPONTANEOUS GENERATION

To discuss the origin of life is to try to account for how something that does not happen now on earth once did happen.

History of the Theory of Spontaneous Generation

Just a few centuries ago, almost all men, scientists and laymen alike, believed in spontaneous generation. Everyone knew, of course, that men and domestic animals had parents, but they thought that mice and cockroaches developed from refuse, frogs and insects from mud, worms from horsehairs, and flies from decaying meat. This belief was held even by many theologians, for does not Genesis relate that God ordered the waters in the earth to bring forth living creatures, and this order, they thought, had never been rescinded.

About 1680 an Italian, Francesco Redi, cast the first doubt on the concept of spontaneous generation when he showed that if a container with raw meat in it is covered by a piece of gauze so that flies cannot enter and lay their eggs on the meat, maggots never develop in the meat.

About the same time, Leeuwenhoek discovered microscopic organisms. Although Leeuwenhoek himself did not believe that his "animalcules" were generated spontaneously, many other biologists did. In the next century, another Italian, Lazzaro Spallanzani, showed that nutritive broth sealed off from the air while boiling does not develop microorganisms and hence does not spoil. Some biologists objected to this evidence with the argument that the boiling made the broth and the air above it incompatible with life. It remained for Louis Pasteur (1822–1895) to answer this criticism with a simple modification of Spallanzani's work. Instead of sealing off the neck of the flask containing the broth, he drew it out in an S-shaped curve but left the mouth of the neck unsealed. With this arrangement, air molecules were able to pass from the atmosphere into the broth, but heavier dust particles and bacterial spores could not settle on it. In flasks prepared in this way, the broth generally stays clear and fresh. If the neck is broken so that dust can enter, the broth spoils.

Thus, step by step the idea of spontaneous generation was gradually whittled away until nothing was left of it. But today biologists recognize that it was not really the possibility of spontaneous generation that was disproved but merely that it does not occur so far as is known under the conditions that prevail on earth today. At one time, life did originate on earth.

Conditions on the Primitive Earth

A few billion years ago, conditions on the earth were quite different from what they are today. The atmosphere now consists of about 78 per cent nitrogen, 21 per cent oxygen, and 0.03 per cent carbon dioxide. But the primitive atmosphere must have been very different. Evidence indicates that the first atmosphere of the forming earth contained hydrogen, methane (CH_4), water vapor, and ammonia (NH_3). This atmosphere was probably soon replaced by one derived from volcanic gases that were emitted from vents in the condensing crust of the earth. This secondary atmosphere contained carbon dioxide, water vapor, and nitrogen. Molecular oxygen (O_2) was conspicuously absent or present in only minimal amounts. This is believed to have been the atmosphere present when life arose on earth. Remember that the chief components of protoplasm are carbon, hydrogen, oxygen, and nitrogen. All were present in the early atmosphere.

On the other hand, the primitive oceans are thought to have had about the same composition they do today, although they, of course, lacked dissolved oxygen.

Synthesis of Organic Molecules

In 1953, Stanley L. Miller devised an experiment in which he kept a mixture of water vapor, methane, ammonia, and hydrogen circulating over an electric spark. He maintained circulation by boiling the water in one arm of the apparatus and condensing it in the other. At the end of a week, he found that the water contained a number of organic molecules, including some of the amino acids that are most important in the formation of proteins and some simple fatty acids.

Since then, various mixtures of gases and various energy sources, such as ultraviolet light and heat, have been tried and a wide variety of organic molecules, including ribose and nitrogenous bases, have been synthesized. These molecules form only in a reducing atmosphere, such as the one postulated for the primitive earth. Figure 23–1 shows an experimental setup for the synthesis of amino acids.

Conditions for Polymerization

The atmospheric gases, along with these simple organic compounds, would have dissolved in the surface water with its salts and minerals. The ingredients for life were there. The next step must have been their combination into larger and larger aggregates. The nitrogenous bases must have joined with sugar and phosphate molecules to form nucleotides and these then must have

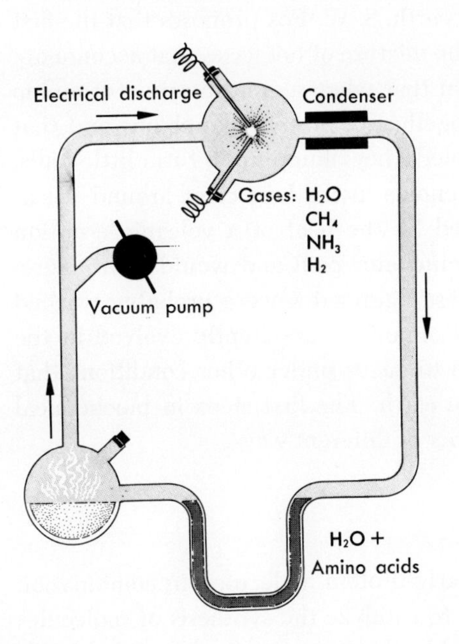

Figure 23–1. An experimental set-up for the formation of amino acids. Water vapor and gases are kept circulating past an electrical discharge by boiling water in the flask at the lower left. The water that accumulates in the lower part of the tube contains a variety of amino acids.

linked up into nucleic acid chains. Similarly, the amino acids must have joined to form polypeptides, and carbohydrates and fats must also have formed. In order for molecules to combine, they must come together. Were they concentrated enough in the early seas to have done so? Mathematically, the answer is that with enough time it was almost inevitable.

In a single toss of a coin, the chance it will not fall heads is one half. The chance that it will not fall heads in 10 consecutive tosses is $\frac{1}{2} \times \frac{1}{2} \times \frac{1}{2} \times \frac{1}{2} \times \frac{1}{2} \times \frac{1}{2} \times \frac{1}{2} \times \frac{1}{2} \times \frac{1}{2} \times \frac{1}{2}$ or 1/1,024. Thus, the chance that a head will appear in 10 tosses is about 999 out of 1,000. Ten chances have changed what was originally a fifty-fifty chance into a near certainty. With an astronomically large number of molecules in constant motion in the seas and with billions of years of time available, the spontaneous generation of life would also be inevitable.

Many biologists, however, believe that this is not a sufficient answer. There is no way of knowing how many simple organic molecules were present in the early sea, but scientists do have some idea of the time involved. Geologists believe that the earth was first formed about 4.5 billion years ago. The oldest rocks known today seem to be more than 3.5 billion years old and the oldest rocks containing evidence of life are about 3.5 billion years old. The evolution of organisms from the nonorganic world must have taken place in not more than a billion years.

Various ingenious suggestions have been made as to how the primitive constituents of life could have been concentrated—by evaporation of water from shallow pools, by a tendency of the molecules to adhere to wet clay surfaces, or by the impact of a comet on the earth. S. W. Fox proposes that the first simple biological compounds formed in the mixture of hot gases that accumulate before a volcanic eruption. He has shown that when a concentration of amino acids is heated, the acids tend to join together in protein-like chains and that when a solution of such protenoids is cooled, they clump up to form little balls. Fox and his co-workers have found protenoids in the lava crust around volcanoes. He suggests that protenoids formed by the heat of a volcanic eruption could have been washed together in a rainwater pool and would have aggregated as the water evaporated. Later these aggregates were probably washed down into the sea, because the first real organisms apparently evolved in the sea. Polymerization has also been shown to occur under other conditions that must have been present on the primitive earth. The first steps in biochemical evolution may have taken place in a number of different ways.

The Development of Enzyme Systems

It seems likely that some of these early protein molecules, in combination with nucleic acids, developed the ability to catalyze the synthesis of molecules like themselves. The first step was probably the incorporation of amino acids and other small organic molecules similar to those of which the synthesizing complex was composed. Meanwhile, scientists believe, the atmosphere was changing. Ultraviolet radiation from the sun dissociated some of the molecules of water vapor. The light hydrogen atoms escaped from the hold of the earth's gravitational force while some of the oxygen formed ozone (O_3). An ozone layer developed in the atmosphere and served as a screen to cut out the ultraviolet radiation. The production of oxygen from the dissociation of water stopped. Also, since ultraviolet radiation was the most abundant and constant source of energy for the synthesis of the first biochemical molecules, the production of these molecules must have been markedly reduced even if it was not halted entirely. When the supply of these molecules was used up, synthesis would have ended. N. H. Horowitz in 1945 suggested an explanation of why this did not happen.

Suppose there was in the sea a series of substances, A,B,C, . . . U,V,W,X, Y,Z, and further suppose that substance Z was the substance used by the proteins to synthesize additional protein. If there had been no change, protein synthesis would have halted when all of substance Z had been used up. Prior to this, however, some proteins had mutated and gained the ability to synthesize substance Z from substance Y, thus opening up a new source of supply. If later mutations made it possible to synthesize substance Y from substance X, then substance X from substance W, substance W from substance V, and so forth, it would make possible the widespread use of many compounds.

At some point these prebiotic synthesizing systems must have switched to some form of anaerobic respiration as a source of energy. The energy sources postulated for the synthesis of the first biochemical building blocks—heat, radiation, electricity—all tend to disrupt the tertiary structure of complex proteins, while at the time not enough oxygen was available for aerobic respiration. The living proteins of the sea did not use up all the available organic molecules because of other mutational shifts. The transitional atmosphere formed of volcanic gases contained carbon dioxide. With the accumulation of carbon dioxide the proteins could shift from the capture of organic substances (a heterotrophic existence), to the manufacture of such substances by photosynthesis (an autotrophic existence). Once autotrophs had developed, free oxygen, which is a by-product of photosynthesis

$$6\,CO_2 + 12\,H_2O + \text{energy} \rightarrow C_6H_{12}O_6 + 6\,O_2 + 6\,H_2O$$

began to accumulate. This made possible other mutational shifts from the inefficient means of gaining energy by fermentation to the much more efficient method of aerobic respiration:

$$C_6H_{12}O_6 + 6\,O_2 \rightarrow 6\,H_2O + 6\,CO_2 + \text{energy}$$

The development of photosynthesis and of aerobic respiration made abundant sources of energy available for the synthesis of the extremely complex molecules that are characteristic of life today. There are chemosynthetic and fermentation organisms living today, but compared to the other organisms they are few in number. It is noteworthy that among the earliest fossils are forms that are similar to modern blue-green algae. They also may have been able to carry on photosynthesis. The accumulation of oxygen that resulted led to the formation of the present atmosphere.

The Unity of Life

There may have been, and probably were, a number of different kinds of biochemical molecular aggregations formed in the early sea. But it seems probable that only one kind survived to give rise to all the living things known today. For one thing, there are a number of different kinds of nitrogenous bases, but only five, and always the same five, are, so far as we know, incorporated by all organisms into nucleic acids. The DNA code is apparently the same in all organisms, from bacteria to man. All organisms use ATP for the transfer of energy, and they all use the same kinds of amino acids to build proteins. The demonstration of the chemical unity of life is one of the great triumphs of modern biochemistry.

Further Steps

At some stage, of course, these molecular complexes reached a level of internal organization that would be recognized today as cellular. And they developed the ability to divide precisely enough so that each daughter cell had the requisites to carry on independently. The first step toward sexual reproduction, either by fusion or by transfer of parts, must also have occurred very early, but there is no evidence of how it took place.

Extraterrestrial Life

Because the origin of life on earth seems to have been more or less inevitable, it is most likely that life has also originated elsewhere in the universe many times. Some meteorites have been found to contain hydrocarbons and amino acids. These evidently were not produced by living organisms but they do show that extraterrestrial formation of precursors of biochemicals has occurred. With at least 100,000 planets like the earth in our galaxy alone, and with 100 million galaxies within telescopic range, many scientists now believe that life must be a cosmic event, not just a terrestrial one.

ORGANIC EVOLUTION

Why, once life had reached the stage of being able to use the energy of sunlight to synthesize its own substances, didn't it just continue in the form of simple, unicellular plants floating on the surface of the sea? What is the origin of the diversity of form and function that is as characteristic of life as its unity?

Darwinism

The idea that evolution had occurred was not original to Charles Darwin (1809–1882). It had been held by other naturalists, among them his grandfather, Erasmus Darwin, and the Frenchman Lamarck. But none of these earlier evolutionists was able to suggest a mechanism of evolution that was satisfactory to most other biologists. This was Darwin's great contribution. He not only accumulated a vast number of facts that could best be explained as the result of organic evolution but he also provided for the first time a reasonable explanation of how evolution could come about.

Natural Selection. Of course, Darwin knew nothing about genetics. But he did know that variations occur, even among the offspring of the same parents, and that much of this variation is inherited. And he knew that each generation of an organism normally produces many more offspring than can possibly survive. He reasoned that some of these offspring would be better adapted than others; they would be stronger, better able to withstand the hostile forces of the environment and to get the nutrients they needed from it. They would be the ones with the best chance of surviving and of passing on their traits to their offspring. Darwin called the power of the environment to select the characters of the next generation *natural selection*. From what we know of population genetics, it is clear that this selection would lead to a change in the gene pool in the direction of greater fitness, of better adaptation to the environment. Because conditions in the environment are constantly changing, organisms must also change to adapt to the new conditions. This will account for the evolution of a single population through time, but it does not account for an increase in the number of kinds of organisms.

Geographic Speciation. As the first organisms in the sea increased in numbers, they would necessarily spread out from the place of origin. Some of them would reach areas in which the environment was quite different. They might, for instance, be carried up a tidal estuary and perhaps become landlocked by the development of a sandbar across the mouth of the estuary. Conditions here would be different than in the open ocean. The organisms that could best survive would be the ones that had, or that soon developed, traits that enabled them to adapt to the new situation. Over the years, through mutation and recombination, their descendants would develop genetic complexes that differed significantly from the genetic complexes of their relatives in the sea. Suppose that the level of the water rose and the landlocked area was once again connected to the sea. The two populations would intermingle, and if they carried on sexual reproduction, cross-matings might occur. But the offspring of such matings, receiving part of their genetic material from one parent and part from the other, would lack the well-integrated genetic complex of either. They might be less fit, less likely to survive to reproductive age, than the offspring of intragroup matings in either of the parental stocks. This would give a selective advantage to intragroup matings and lead to the development of reproductive barriers, that is, of traits to prevent cross-matings. If one form liberated its gametes in the intertidal zone and the other did so in the open sea, intragroup matings would occur more frequently than intergroup ones. If one form reproduced in the spring and the other in the summer, the possibility of cross-mating would be reduced. As the genetic differences between the two stocks accumulated, the hybrid zygotes would become less and less capable of normal development. Eventually, even when cross-breeding took place, no viable offspring would be produced and the reproductive isolation of the two stocks would be complete. When two populations derived from a single parent population are still able to interbreed and produce fertile offspring, they are considered subspecies or geographic races. When they become reproductively isolated so that the exchange of genetic material is no longer possible, speciation has occurred. Two species have evolved from the original single species.

From this discussion it might seem that speciation is a process that took place only in the past. It is, however, still going on today. The factors involved in the hypothetical instance of speciation just discussed can all be observed in living populations.

Geographic speciation, then, involves four sets of factors: separation of populations; mutation and recombination; natural selection and, in small populations, genetic drift; and the development of genetic isolation.

Evolution Beyond the Species Level

Are these evolutionary processes sufficient to account for the development of much more widely divergent forms, for the evolution, say, of the mammals from primitive fish? Many biologists believe that they are. Once reproductive isolation is established, genetic differences will continue to accumulate and each daughter species may give rise in turn to other species. And so more and more divergent forms will appear. This is now the most widely accepted evolutionary dogma.

Recent studies of the amount of DNA per nucleus in a wide variety of organisms suggest that both increases and decreases in the quantity of genetic material have also had major roles in evolution. The big increases, probably caused by polyploidization, seem to have occurred in the lineages that made major adaptive shifts in their pattern of living. This is best documented for the groups that moved from an aquatic to a terrestrial existence. Primitive land plants have more DNA per nucleus than the green algae, and amphibians have more than fishes. When extra copies of chromosomes are added, the organism has more genes coding for a specific protein than it needs. The extra copies are suppressed, perhaps by the process known as *heterochromatinization*, in which the chromatin becomes tightly coiled. Genes in heterochromatic regions are not transcribed into RNA.

Almost all mutations are either neutral or harmful. Either the functioning of the enzyme is not affected or the enzyme is rendered less effective. It may no longer function at all. There is very little chance that a single mutation would produce a new enzyme with a new function that was of immediate benefit to the organism and that could replace the function lost. But if the animal or plant had a number of extra copies of a given gene, then those that were not being expressed could go through a series of mutations without in any way affecting the organism. New genes would accumulate. Some of them might code for enzymes that would be useful in a new environment, that would allow the organism to take on a new way of life. Excess DNA gives an organism the genetic flexibility that allows it to make a major adaptive shift. We believe that increase in amount of DNA is an important factor in the evolution of classes and phyla, which represent the major different patterns of existence.

There is also a tendency toward a reduction in amount of DNA. It is disadvantageous for an organism to have to replicate excessive amounts of inoperative DNA every time a cell divides. Once a major adaptive shift has been made, the more highly modified descendant species usually have less DNA than the primitive forms that are more similar to the ancestral stock. Figure 23–2 is a diagrammatic scheme of the evolution of the vertebrates showing where marked changes in amount of DNA occurred.

THE HISTORY OF THE EARTH

Because evolutionary change is so closely bound up with changes in the environment, the history of life on earth is closely tied in with the history of the earth itself.

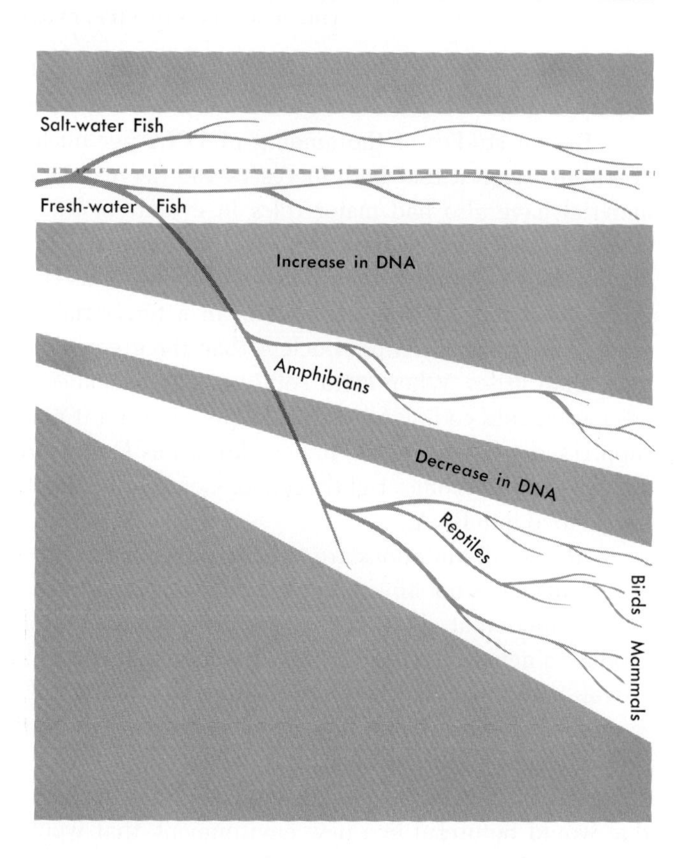

Figure 23–2. Diagram of the origin of the vertebrate groups. The ancestors of the fishes are thought to have lived in fresh water. They gave rise to the salt water fishes and later to the amphibians. There was a large increase in amount of DNA per nucleus in the line that led to the amphibians. The amphibians gave rise to the truly terrestrial reptiles, with a loss of DNA. The reptiles gave rise to both birds and mammals.

Continental Drift

Instead of being uniformly dense throughout, the earth seems to have a very dense fluid center (the *core*) surrounded by a less dense and less fluid covering (the *mantle*), which in turn is surrounded by an outer crust of even less dense, more or less solid material that forms the continental masses and ocean floors (see Figure 23–3).

Geologists believe that at one time all the continents and major islands of the world formed a single supercontinent that has been named *Pangaea.* The northern part, *Laurasia,* was more or less separated from the southern part, *Gondwanaland,* by a deep embayment called *Tethys Sea.* Laurasia comprised what are now North America, Greenland, Europe, and parts of Asia; Gondwanaland comprised what are now South America, Africa, Antarctica, Australia, India, and perhaps parts of China (see Figure 23–4). Then rifts developed in Pangaea, and the continents began to drift apart. South America separated from Africa, and North America separated from Europe. India, Australia, Antarctica, and Africa broke apart.

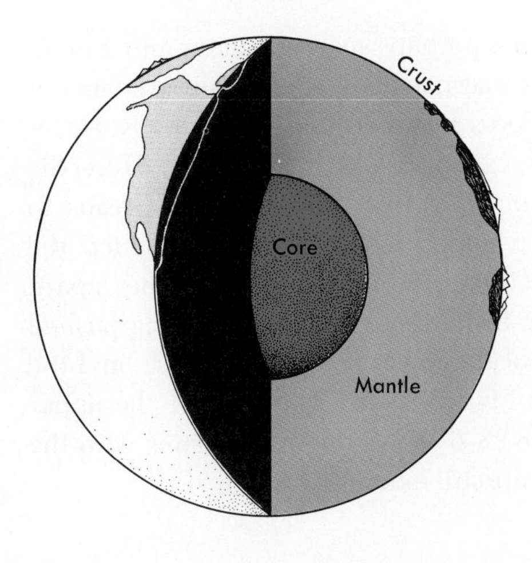

Figure 23–3. The earth with a section cut out to show the inner structure. The crust is a thin layer floating on the fluid mantle.

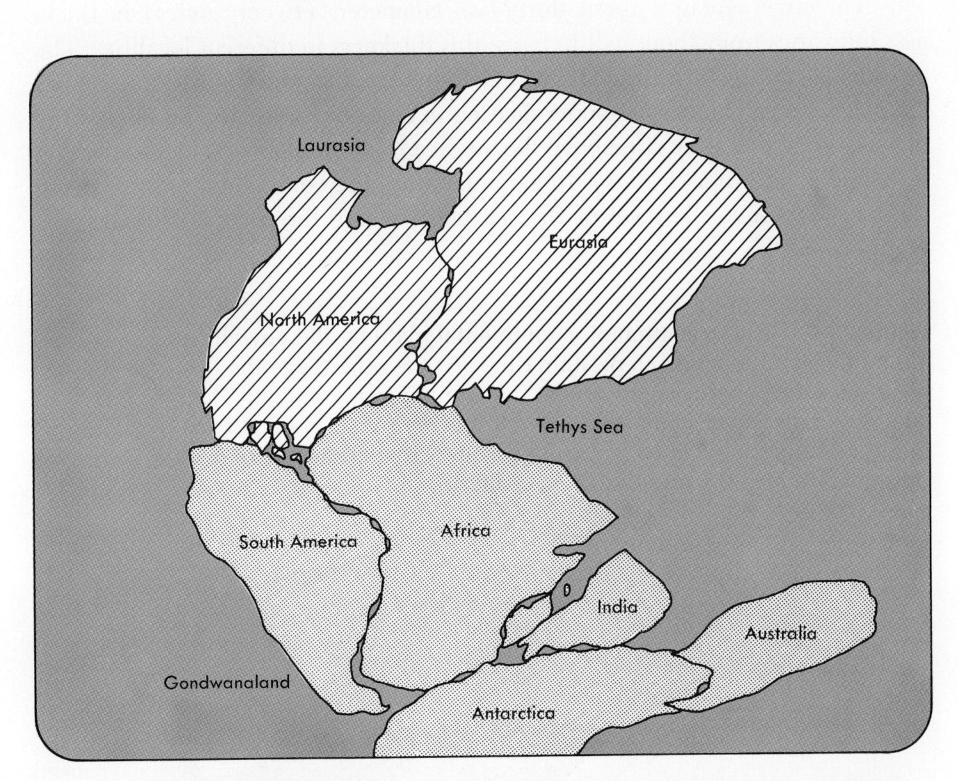

Figure 23–4. The supercontinent, Pangaea. (Modified from Colbert, *Wandering Lands and Animals*, 1973. By permission of E. P. Dutton and Company and Edwin H. Colbert.)

The crust of the earth is separated into perhaps six large plates and a number of minor ones. The continental masses that rise above the oceans are parts of these plates. Running along the ocean floors between the continents is a connected series of ridges, which are really great cracks in the crust. Here the molten material of the mantle underlying the crust wells up and spreads to either side. As it cools and solidifies, it is added to the plates that bound the ridge and so pushes the continents on either side of the ridge further apart. There are also deep trenches in the ocean floor where one plate is being pushed against and overrides another. The edge of the lower plate bends down and the material in it returns to the mantle (see Figure 23–5). The edge of the upper plate is thrown up into mountains. Figure 23–6 shows the major plates with the continents in their present position (they are still moving).

Isostasy

The crust averages about thirty-two kilometers (twenty miles) in thickness but varies from about half to twice this thickness in different localities. The continental masses are formed by regions where the crust is thicker. Figure 23–7 shows some blocks of copper floating in a sea of mercury. The thicker the blocks, the higher they rise above the surface. This principle of blocks floating in a fluid is known as *isostasy*. In a similar way, the continental masses, being thicker, rise above the level of the sea basin where the crust is thinner (see Figure 23–8).

Figure 23–5. The formation of a continental plate by the upwelling of mantle material along an oceanic ridge. At the other edge the plate rides over another plate and is folded up into mountains. The material of the lower plate is returned to the mantle. (Modified from Colbert, *Wandering Lands and Animals*, 1973. By permission of E. P. Dutton and Company and Edwin H. Colbert.)

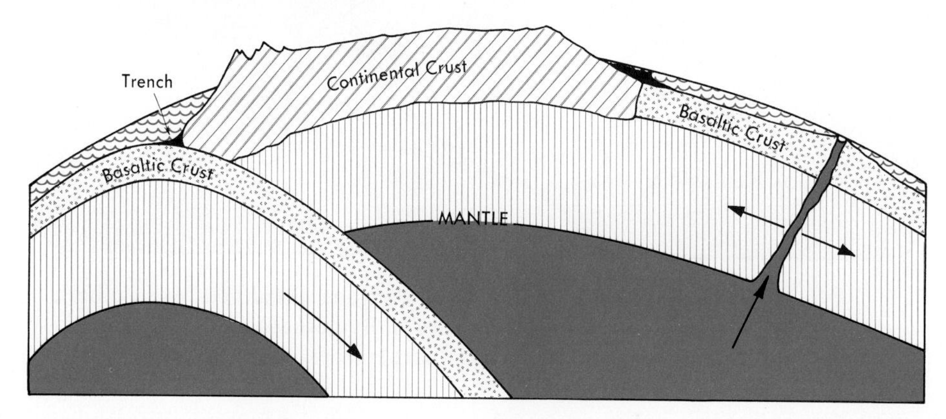

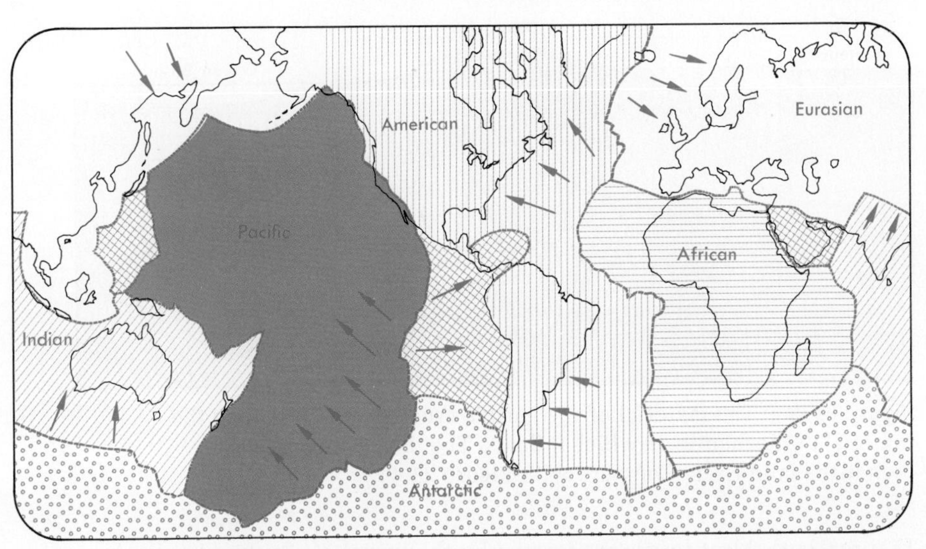

Figure 23–6. The major continental plates today. From Colbert, *Wandering Lands and Animals*, 1973. By permission of E. P. Dutton and Company and Edwin H. Colbert.)

The Geologic Cycle

The continental masses that rise up above the sea level are subject to erosion; the higher they rise, the more rapidly they are eroded. As their exposed surfaces are worn away, the crust becomes thinner. Eventually it tends to rise to keep the proper isostatic balance between the exposed and submerged portions.

A geological cycle covers the span from the time the continental masses are high through the time they are eroded down to relatively low plains, perhaps covered by shallow inland seas, and the readjustment period in which they once again rise high.

During the periods when the lands are being worn away, sediments tend to settle in low, shallow basins and slowly solidify into more or less compact sedimentary rocks. Because they are laid down in horizontal beds, they form a series of strata with the oldest on the bottom and the youngest on the top (see

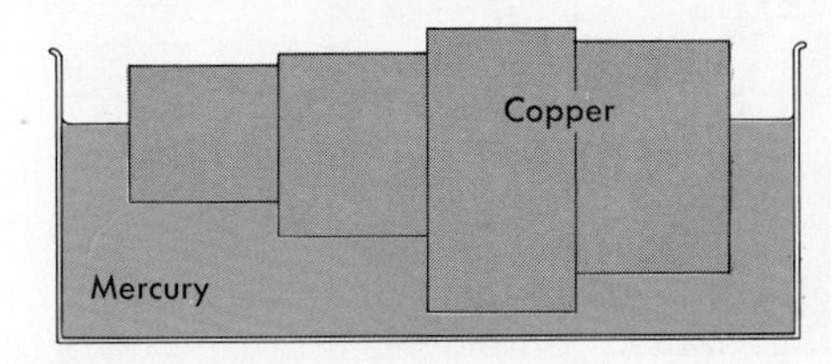

Figure 23–7. Blocks of copper floating in mercury. Copper weighs less than mercury. The thicker the block, the higher it projects above the surface of the mercury.

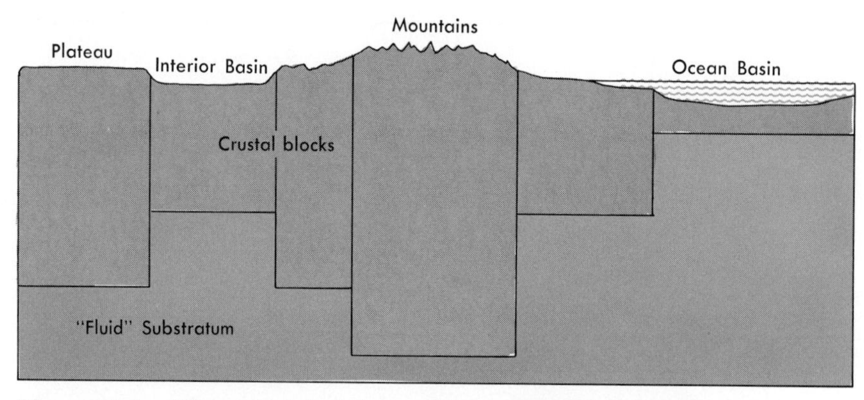

Figure 23–8. The material of the earth's crust weighs less than the material of the mantle on which it floats. The thicker the crust, the higher the land rises. Compare with Figure 23–7.

Figure 23–9). Most fossils are formed by the burial of plant and animal remains in these beds of sediments (see Figure 23–10). When the continents rise to regain their isostatic balance, there is widespread compression, folding, and erosion of the fossiliferous beds. These *orogenic* (mountain building) episodes of rapid elevation with fossil destruction are known as *geologic revolutions*. They

Figure 23–9. Stratified sedimentary rock formation near Sedona, Arizona. (Courtesy of Dr. James Menzies.)

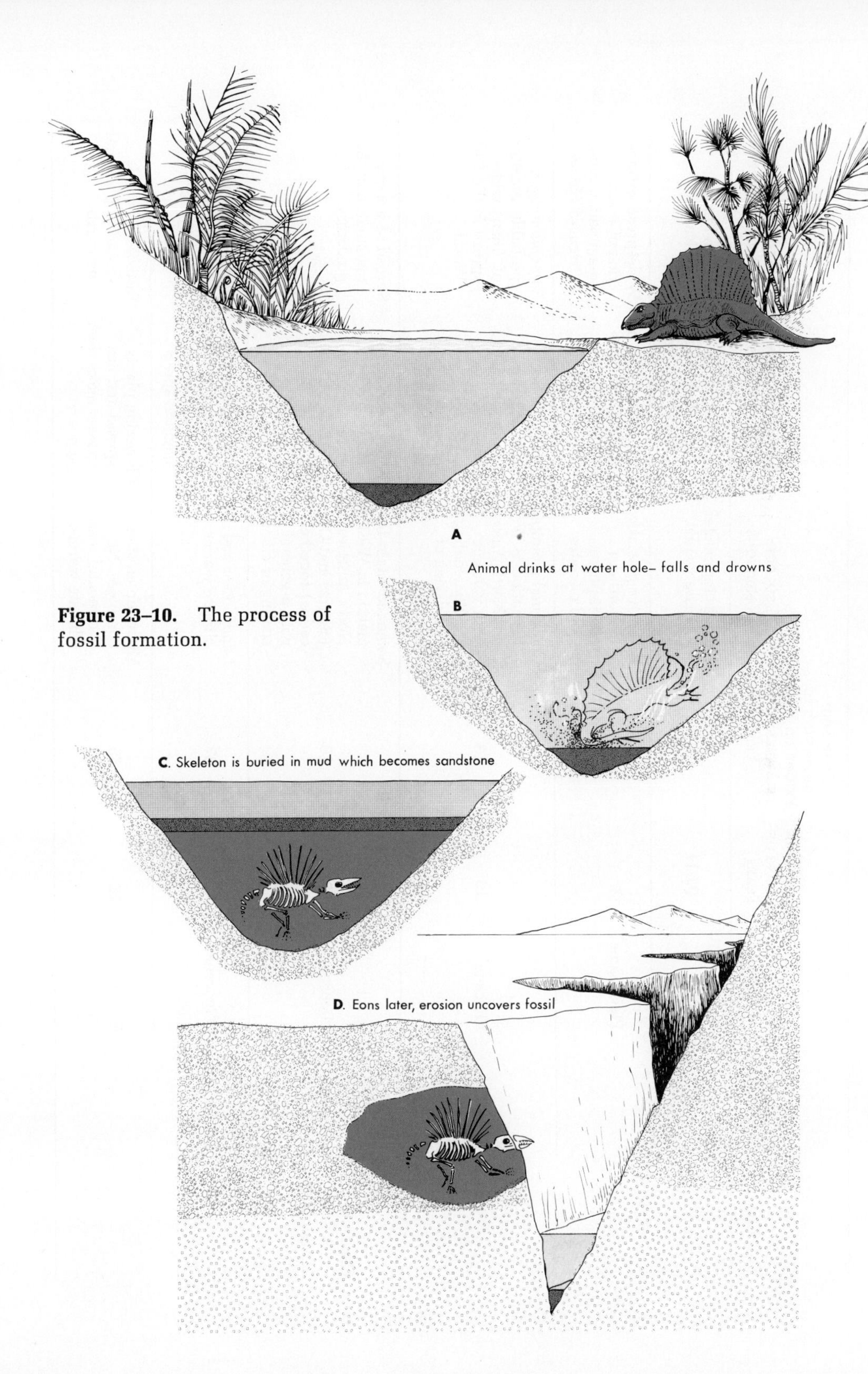

Figure 23–10. The process of fossil formation.

Table 23–1. Geologic Time Scale

Era	Period	Epoch	Duration in Millions of Years	Time from Beginning of Period or Epoch in Millions of Years	Geologic Conditions	Major Events in Plant Life	Major Events in Animal Life
Cenozoic (Age of Mammals)	Quaternary	Recent	0.011	0.011	Ice ages terminate? Climate warms.	Spread of herbaceous flowering plants.	Modern human civilization.
		Pleistocene	3	3	Extensive glaciation in Northern Hemisphere. Cold climates.	Widespread shift in ranges and some extinction with cooling climates.	Widespread extinction of many mammalian types. Origin of *Homo*.
	Tertiary	Pliocene	10	13	Continued mountain formation in Western North America with volcanic activity, cooling, and drying of climates.	Tropical floras retreat before cooling climates.	*Ramapithecus*, a probable ancestor of "man" widespread in the Old World.
		Miocene	13	26	Some mountain formation in Western North America (Sierras and Cascades) together with a general cooling of climates.	Widespread grasslands.	Spread of grazing mammals and their predators.
		Oligocene	11	37	Lands lower and climate warms.	Modern genera of flowering plants evolve. Grasses begin to spread.	Hominoid primate lines appear.
		Eocene	22	59	Mountain erosion continues. No continental seas. Climate warms.	Flowering plants spread and first grasses make their appearance.	Rise of birds and spread of placental mammals.

		Paleocene	5	64	Initial erosion of mountain systems. Climate begins to warm.	Rise of modern forests.	Archaic mammals spread.
Rocky Mountain Revolution							
Mesozoic (Age of Reptiles)	Cretaceous		72	136	Inland seas early. Later continents rise with formation of Andes, Alps, Himalayas, and Rocky mountains.	Anthophytes gradually become dominant as conifers wane.	Climax and extinction of most archosaur reptile stocks.
	Jurassic		46	182	Continents moderately high. Some continental seas in Europe and Western North America.	Earliest flowering plants. Cycads and conifers dominant.	Early birds.
	Triassic		49	231	Continents mostly above the seas. Widespread desert conditions with extensive terrestrial fossil formation. Volcanic eruptions at end mark beginning of breakup of Pangaea.	Increase in conifers, cycads, and ginkgoes. Seed ferns disappear.	Radiation and spread of reptiles. Climax of mammal-like reptiles.
Appalachian Revolution							
	Permian		50	280	Extensive elevation of lands. Appalachians formed; increasing aridity and glaciation.	Early cycads and conifers spreading. Waning of club mosses, horsetails, and ferns.	Modern insects appear. Early evolution of reptiles. Decline of amphibians.

Table 23–1. Geologic Time Scale—(continued)

Era	Period	Epoch	Duration in Millions of Years	Time from Beginning of Period or Epoch in Millions of Years	Geologic Conditions	Major Events in Plant Life	Major Events in Animal Life
	Pennsylvanian		40	320	Lands low at first with extensive coal swamps. Lands rising and cooling toward end.	Extensive coal forests of giant club mosses, horsetails, seed ferns, and primitive conifers.	First reptiles. Climax of amphibians.
	Mississippian		25	345	Climate warm and humid. Coal swamps develop.	Coal forest of club mosses, horsetails, ferns, and seed ferns. Primitive conifers.	Spread of amphibians. Rise of air-breathing arthropods, particularly insects.
Paleozoic (Age of Ancient Life)	Devonian		60	405	Continental seas reduced, lands higher and more arid. Some glaciation.	First horsetails, primitive club mosses, and ferns. First forests.	Origin of Amphibia. First terrestrial insects, snails, diplopods, etc.
	Silurian		20	425	Continental seas still extensive but lands beginning to rise with increasing aridity of land masses.	First land plants.	Ostracoderm proliferation. Origin of primitive groups of fishes, Eurypterids predominant.
	Ordovician		75	500	Extensive submergence of lands and continued very mild climates.	Marine algae.	Origin of the chordates.

	Cambrian	100	600	Lands low and climates mild.	Marine algae.	Trilobites particularly abundant. All major animal phyla save the Chordates present.
Second Great Revolution						
Proterozoic (Ancient Age)		1000±	1600±	Extensive sedimentation early. Later volcanic activity and extensive erosion.	Marine algae.	Jellyfishes abundant; worms, soft corals, protozoa, and at least two extinct phyla present.
First Great Revolution						
Archeozoic (Ancient Age)		2000±	3600±	Extensive volcanic activity and erosion. Some sedimentary deposition.	Beginnings of life.	Beginnings of life.

are rapid in terms of geologic time, although in relation to man they are slow. The elevation of major continental blocks is probably going forward today about as rapidly as it ever did.

Geologic Time Scale

These repeated cycles permit geologists to construct a calendar of geologic time in which the events of life on earth can be placed in their proper sequence.

Since life first left a record in the rocks, there have been four major geologic revolutions. The times between the revolutions, which are called *eras,* are: Archeozoic, Proterozoic, Paleozoic, Mesozoic, and Cenozoic. The eras are divided into *periods* based on lesser fluctuations in the major cycles, and the later periods are subdivided into still shorter *epochs.* Table 23–1 shows the geologic time scale.

The closer the epoch to the present time, the better is the fossil record for that epoch. This is because there has been less time and less opportunity for the fossils to be buried deep in the earth or to be destroyed by erosion or crushing, and also because many of the forms of life that make good fossils (like the mammals with their hard bones and teeth) were late in arriving.

With this history of the earth in mind, you should be better able to understand the history of life on earth, with its fluctuating periods of evolution and extinction, from the first record of life up to the present time.

SUGGESTED READINGS

Colbert, E. H.: *Wandering Lands and Animals,* E. P. Dutton & Co., Inc., New York, 1973.

Kenyon, D. H., and G. Steinman: *Biochemical Predestination,* McGraw-Hill Company, New York, 1969.

Laporte, L.: *Ancient Environments,* Prentice-Hall, Inc., Englewood Cliffs, N.J., 1968.

McAlester, A. L.: *The History of Life,* Prentice-Hall, Inc., Englewood Cliffs, N.J., 1968.

Ponnamperuma, C.: *The Origins of Life,* E. P. Dutton & Co., Inc., New York, 1972.

Sawkins, F. J., C. G. Chase, D. G. Darby, and G. Rapp, Jr.: *The Evolving Earth: A Text in Physical Geology,* Macmillan Publishing Co., Inc., New York, 1974.

Smith, H. H., ed.: *Evolution of Genetic Systems,* Brookhaven Symposia in Biology No. 23, Gordon and Breach, New York, 1972.

CHAPTER **24**

Journey onto Land

There is no fossil record of the origin of life, and very little record of the first four fifths of the history of life. If we seem to skimp this long span of evolutionary history, it is not because we think it unimportant, but because we know almost nothing about it.

THE PRE-CAMBRIAN

Only in North America is there good evidence of a major revolution separating the Archeozoic and Proterozoic eras. Throughout most of the world it is virtually impossible to date accurately any Pre-Cambrian formation and most of those that bear fossils are simply referred to as "late Pre-Cambrian" (= Proterozoic?).

The Pre-Cambrian represents perhaps five sixths of the time since the formation of the oldest known rocks. Until recently, knowledge of life in this great span of time was based on some deposits of carbon compounds that appear to be of organic origin and some microscopic fossils resembling present-day bacteria, blue-green algae, and star-shaped and parachute-shaped forms like nothing known today. These remains indicate that organisms like the Monera were present perhaps as much as 3 to 3.5 billion years ago. Green algae were possibly present 2 billion years ago. The earliest known fossil animals are perhaps 1 billion years younger than the first green plants. Recent findings in late Pre-Cambrian formations in Australia have revealed a surprisingly well-developed fauna in what was then the margin of a shallow sea. Over six hundred specimens taken from these beds include soft-bodied corals that are related to some modern forms, well-developed segmented worms with shieldlike heads, flat-ovate wormlike animals, at least six genera of jellyfishes, and two creatures that do not fit into any known phylum. One is small and hemispherical with three radiating, hooked arms; it has been named *Tribrachidium* and looks like a design for a modern pendant. The other is a small, circular organism about 25 millimeters (an inch) in diameter with a shield-shaped body bearing an anchor-shaped ridge. Figure 24–1 shows some of these earliest known animals. The remarkable diversity of this fauna suggests that animal life must have originated much earlier and had already undergone considerable evolution.

It is customary in historical geology to speak of times when certain types of animals or plants predominate as ages. Because of the number of jellyfishes in this Pre-Cambrian sea, it has been suggested that the Pre-Cambrian should be called the Age of Jellyfishes.

AQUATIC LIFE IN THE PALEOZOIC

The Paleozoic era began about 600 million years ago. The name means "ancient life," ancient, that is, as one looks back at it from the present, but in

Figure 24–1. Pre-Cambrian animal life. A and C: Sea pens (cnidarians). B: Sponge. D: Jellyfish-like animal. E: Segmented worm. F: Worm-like animal. G: Worm in burrow. H: Worm trail. (From "Precambrian Animals," by M. F. Glaessner, © 1961 by *Scientific American*, Inc. All rights reserved.)

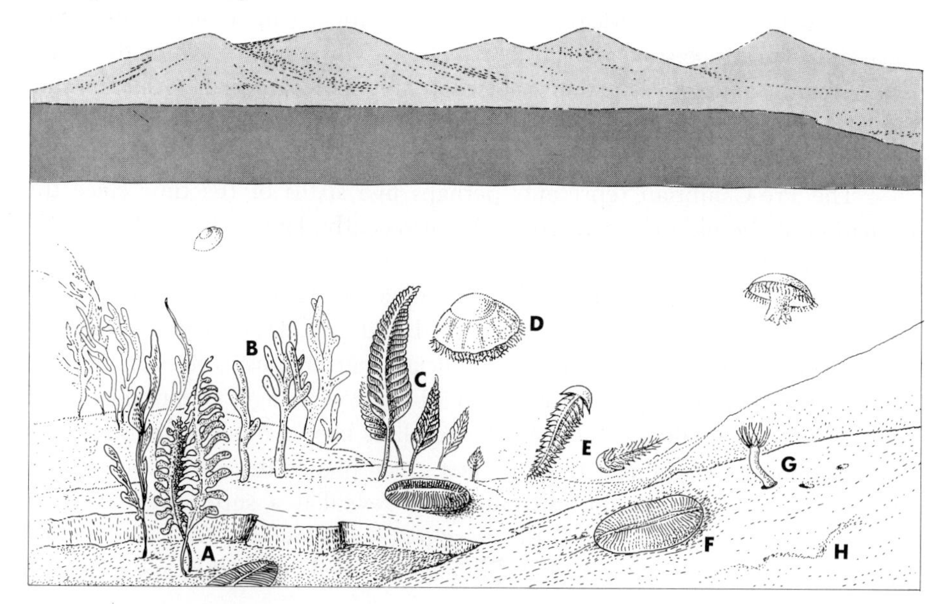

view of the length of time life had already been evolving, perhaps it would be better to consider the Paleozoic as the beginning of modern life.

Cambrian

In the Cambrian, at the beginning of the Paleozoic, the seas seem suddenly to be swarming with life. Except for chordates, all the major phyla of animals known today are represented among the Cambrian fossils. Many of the species were present in prodigious numbers.

The Cambrian Abundance. Several hypotheses have been proposed to explain this apparent sudden increase in the abundance and diversity of animal life following the second great revolution. Some believe that it is an actual record of the fact that many phyla came into existence then. Others feel that these types were probably present in Pre-Cambrian seas but that subsequent sedimentation and orogenic events have covered or destroyed their fossil remains. It has been suggested that it was during this time that the animals made the biochemical evolutionary steps that enabled them to synthesize hard parts from elements available in the sea. The presence of calcareous and chitinous hard parts made possible the extensive fossilization of forms that heretofore readily disintegrated after death into the compounds and elements from which they were formed and rarely indeed left any record of their existence in the rocks.

Another proposal is that the oxygen produced by the first photosynthetic plants combined immediately with iron and other elements dissolved in the sea to form oxides. It was only after these free elements had been oxidized that free oxygen could begin to accumulate in the air and water. Only when oxygen became widely available could the animals, which depend on cellular respiration for their energy, become at all common. They probably had a long evolutionary history before the Cambrian, but it may have been restricted to a few localized areas where oxygen was beginning to accumulate.

Cambrian Animals. The Cambrian seas (see Figure 24–2) were dominated by representatives of the sponges, cnidarians, mollusks, arthropods, echinoderms, and brachiopods. The latter resemble oysters and clams in having a mantle and calcareous shell consisting of two parts, but the resemblance is only superficial. Brachiopods have an entirely different internal structure, and the shell, instead of having right and left halves, as in a clam, has a smaller dorsal part fitting into a larger ventral part. Because the shells of some brachiopods are similar in shape to Roman lamps, they are often called *lampshells*. Only a few species of brachiopods are living today, but they were among the commonest animals of the early seas. In embryonic development they seem to stand between the protostomes and deuterostomes. Cleavage is radial and the coelom

Figure 24–2. Some Cambrian animals. A: Brachiopod. B: Trilobite.

develops by pouching from the archenteron, but the mouth forms in the region of the blastopore. Perhaps the brachiopods are an offshoot of the line that divided to give rise to both the protostomes and deuterostomes, but because this split occurred during the long Pre-Cambrian blank, we can only guess.

The most active and most abundant animals were primitive arthropods called *trilobites* (see Figure 24–2). Most were bottom-dwelling scavengers, but some were active swimmers. It has been estimated that 60 per cent of Cambrian fossils are trilobites.

Ordovician

During the Ordovician, the nautiloids, a group of cepalopods that first appeared in the Upper Cambrian, became very numerous and diversified. They were large, active predators, and their abundance may have been related to the abundance of edible trilobites; their subsequent decline to near extinction (only one form, *Nautilus,* remains today) is not so readily explained.

Also coming into prominence during the Ordovician were the *crinoids*, the sea lilies. These are echinoderms shaped like an inverted starfish attached to the sea floor by a stem.

The major event of the Ordovician was the appearance of the chordates. A series of tantalizing fragments of small bony scales and toothlike structures that are probably remains of a dermal armor have been found in North America and Russia. Unfortunately, they tell us little except that a bony chordate had appeared.

Silurian

By the late Silurian, early vertebrates known as *ostracoderms* (meaning "shell-skinned") were abundant. There were many different kinds of ostracoderms and they varied greatly in structure, indicating that the ostracoderms had had a long previous history of which we know next to nothing. They were fishlike in shape and were typically covered with an armor of bony plates or scales, but they lacked true jaws. A bony skull was present in some forms, but if they had other skeletal structures these must have been of cartilage, which does not fossilize as readily as bone does. Ostracoderms were small forms, ranging from about sixty to six hundred millimeters (a few inches to several feet) in length. Some were flattened bottom dwellers, others were fusiform in shape and were probably active swimmers. Figure 24–3 shows a typical ostracoderm. It had a large bony head shield bearing on the dorsal surface the eyes and single nostril, which were directed upward as is characteristic of bottom-living animals. Behind the head shield the body was covered with rows of bony

Figure 24–3. An ostracoderm, an early, fish-like vertebrate.

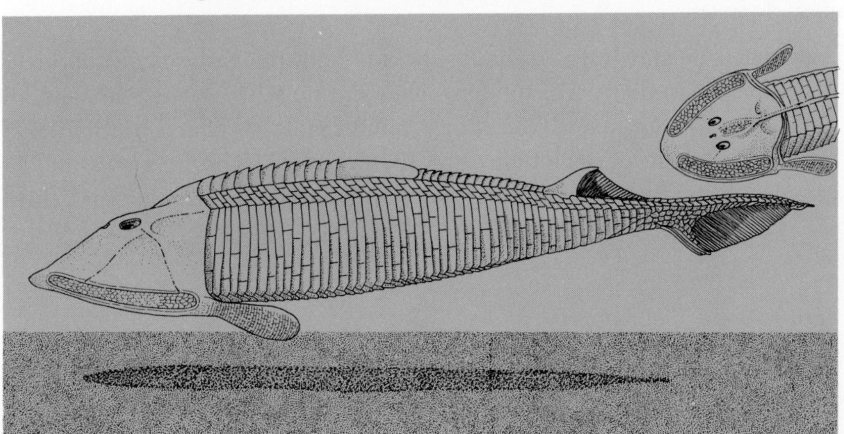

plates. The jawless mouth was on the underside of the head, suggesting that this ostracoderm lived by straining its food out of the bottom mud.

The ostracoderms are classed with the vertebrates, although there is no evidence that they had vertebrae, because only the vertebrates form true bone. This suggests that bone first evolved as a protective coat of dermal plates. What was it protective against? One hypothesis is that the ostracoderms first evolved in fresh water and the plates developed to help preserve osmotic balance. Because their body fluids were hyperosmotic to fresh water, osmosis would have tended to pull water into their bodies. The bony plates may have evolved as a waterproof covering, or they may have been for protection against the *eurypterids*, a group of large, predatory arachnids common in Silurian waters. These relatives of the modern spiders had a relatively small head, a series of body segments, and a short tail segment. Six pairs of appendages were variously modified for crawling, swimming, defense, and grasping prey. The eurypterids include the largest arthropods known, some of them reaching a length of three meters (ten feet). Although they became extinct by the end of the Paleozoic, they left as descendants the present-day scorpions and spiders.

Another major evolutionary event of the Silurian was the appearance of the first terrestrial plants, a few simple vascular forms. Before this time the early Paleozoic lands would have been a strange sight to human eyes, for there was no sign of anything other than rocks, sand, and dust.

Devonian

The Devonian is appropriately called the Age of Fishes. The vertebrate body plan is a successful one. Once it had appeared, the vertebrates evolved rapidly so that in the Devonian all of the major groups of fishlike animals were present.

Agnatha. The jawless vertebrates include both the early ostracoderms, which flourished up to the end of the Devonian and then became extinct, and another group, the *Cyclostomata* (meaning "round mouth"), which has persisted as a relic stock to the present time (see Figure 24–4). The lampreys and hagfish have lost the early ostracoderm armor plating but do have an internal, cartilagin-

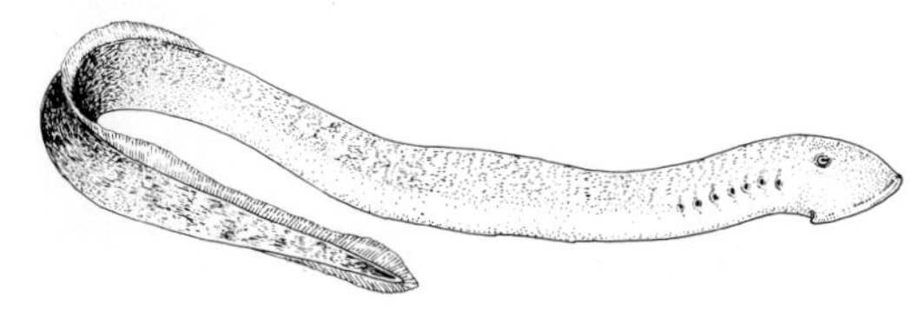

Figure 24–4. The lamprey, a modern cyclostome thought to be related to the extinct ostracoderms.

ous skeleton. The round, jawless, sucking mouth is armed with a rasplike "tongue." Cyclostomes are the only vertebrates that are parasitic. They attach themselves to fishes and rasp an opening in the body wall. They may either remain as external parasites, sucking the body fluids of the host, or burrow in and devour the tissues of the host.

Placoderms. The *placoderms* retained some of the bony armor of their ostracoderm ancestors. They also had bony jaws, with which they were able to grasp their prey instead of having to filter their food from the water or bottom mud or sand. The placoderms were active predators in both salt and fresh water. Some were tiny forms nearly completely covered with bony plates (see Figure 24–5). On the other hand, *Dinichthys* (see Figure 24–6) was an awesome creature reaching a length of about ten meters (over thirty feet) and having a gape of well over a meter (nearly four feet). Its jaws, instead of being armed with teeth, were margined with sharp, jagged edges of the jawbone proper. For it, a three-meter (ten-foot) eurypterid would have been easy prey.

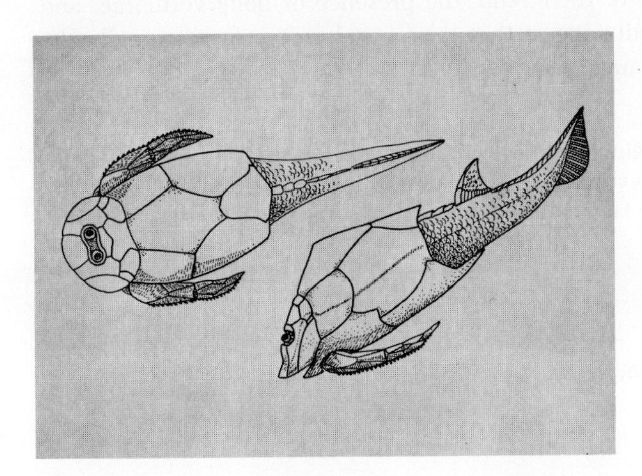

Figure 24–5. Placoderms, early fishes that had true jaws and paired fins.

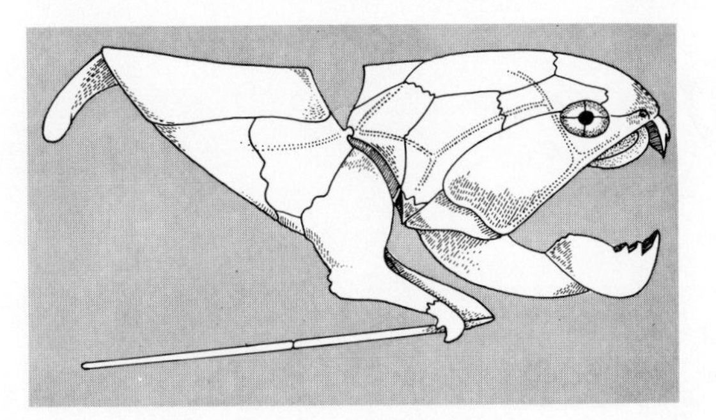

Figure 24–6. The skull of *Dinichthyes*.

Placoderms dominated the Devonian seas. Perhaps they could not withstand the competition of the higher fishes that appeared during this period, for by the end of the Devonian they had disappeared.

Chondrichthyes. The cartilaginous fishes include the sharks, rays, and weird creatures of the deep seas called *chimeras* (see Figure 24–7). They have lost the bony internal skeleton and heavy armor of the earlier fishes. Instead, their skeleton is composed of cartilage and their skin is covered with small, bony scales, each bearing a toothlike projection of dentine. Along the margins of the jaws, these scales developed into teeth.

Osteichthyes. The true bony fishes comprise the major group of fishlike vertebrates. They have a bony skeleton, true jaws, and paired fins. The group includes the spiny sharks (acanthodians), the ray-finned fishes, and the fleshy-finned fishes.

ACANTHODIANS. The so-called spiny sharks are not really sharks at all but belong to a primitive, extinct group of bony fishes. They first appeared in the Silurian. The structure of their skull, the presence of bony vertebrae, and the nature of their scales all suggest that they are close to, if not actually the ancestors of, all the other bony fishes.

Figure 24–7. Some cartilaginous fishes. A: *Cladoselache*, a shark-like fish of the late Devonian. B: A chimera, *Callorhinchus*. C: Electric ray, *Torpedo*. D: Hammerhead shark, *Sphyrna* (ventral view).

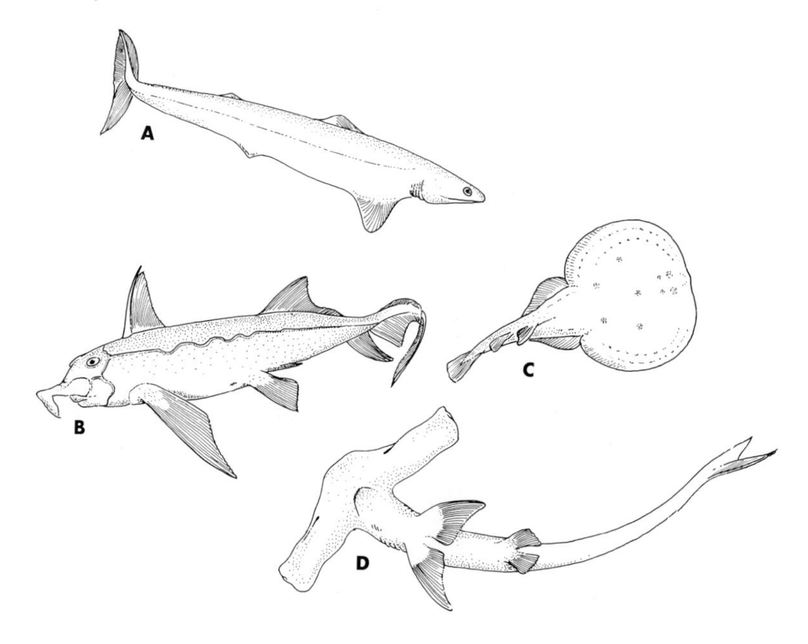

Small in size, the acanthodians were early residents of fresh water. At some time during the early evolution of the bony fishes, perhaps in the acanthodian stock from which the higher fishes evolved, an accessory breathing structure developed. This was a primitive lung in the form of a double sac branching off from the pharynx. Some paleontologists speculate that the climate of the Devonian was one of alternating rainy seasons and drought so that the streams in which these fishes lived tended to become stagnant pools that were low in oxygen during the dry season. When the oxygen content in the water dropped too low for respiration through the gills to supply their needs, these fishes could come to the surface and gulp air down into their primitive lungs. Most of the fishes that descended from the early lung-breathers now live in the ocean or in fresh waters that are less subject to seasonal drought. In these fishes, the lung has lost its connection to the pharynx and has become a hydrostatic organ. Only the lung fishes and brachyopterygians still use it for respiration, but its presence was a major factor in the ability of the vertebrates to invade the land.

RAY-FINNED FISHES (ACTINOPTERYGII). These are the fishes most people see in their mind's eye when they think of fish. Included here are the majority of the food and game fishes of the world—cod, haddock, bluefish, flounder, perch, trout, and many more (see Figure 24–8). They have bony

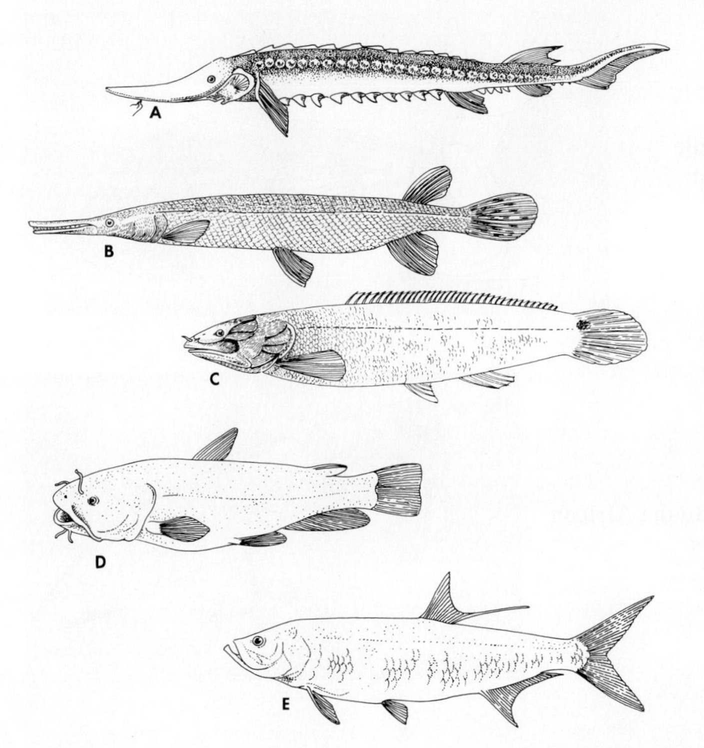

Figure 24–8. Living ray-finned fishes. A: A sturgeon, *Acipenser*, a primitive form. B: A garpike, *Lepidosteus*. C: A bowfin, *Amia*. These forms are slightly more advanced. D: A catfish, *Ameiurus*. E: A tarpon, *Tarpon*. The last two forms are advanced ray-finned fishes.

skeletons but their fins are thin membranes supported by special, nonbony rays. The nostrils lead into a nasal chamber that is purely sensory and lacks any connection with the mouth. The ray-finned fishes are an especially successful group of aquatic animals, numbering about twenty thousand species and occurring in all seas and in most of the fresh waters of the world.

FLESHY-FINNED FISHES (SARCOPTERYGII). In many ways the fleshy-finned fishes are similar to the other bony fishes, but their fins have a bony axis and a muscular lobe at the base (see Figure 24–9). These fins are stronger and more maneuverable than the fins of the actinopterygians. There are three groups of fleshy-finned fishes: the brachyopterygians; the lung fishes (dipnoans, see Figure 24–10); and the lobe-fins (crossopterygians). The latter are of special interest because they have passages that run from the nasal cavity to the mouth. They can thus pass air from the external nostrils through the internal nostrils into the oral cavity while keeping the mouth closed—a trait found in all the terrestrial vertebrates. The sarcopterygians flourished from the Devonian through the Mississippian; as fishes they were less successful than the ray-fins and only a few species survive today. But the Devonian lobe-finned fishes were the ones that gave rise to the terrestrial vertebrates. Figure 24–11 summarizes the evolution of the fishlike vertebrates.

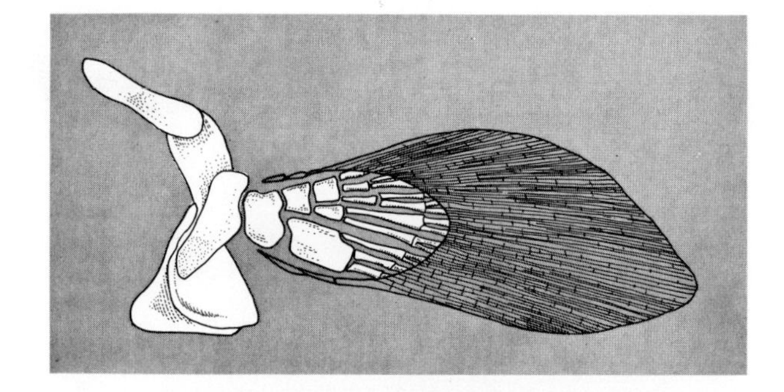

Figure 24–9. The fin of a fleshy-finned fish showing the girdle and bones that extend out into the fin.

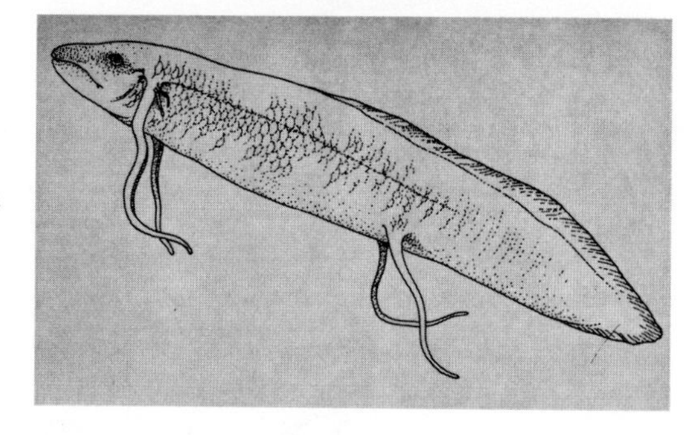

Figure 24–10. The living African lungfish, *Protopterus*.

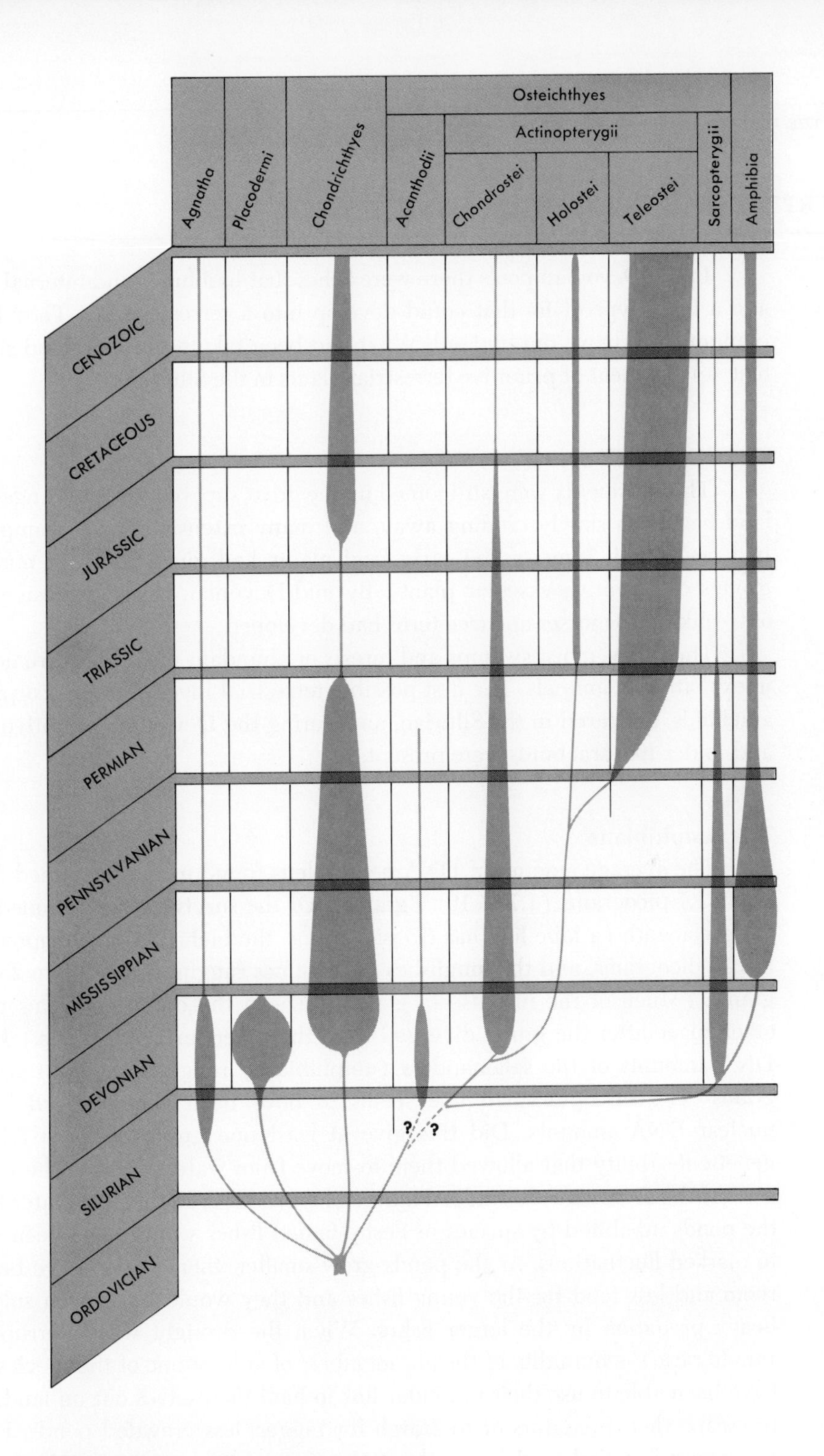

Figure 24–11. Chart summarizing the evolution of the fishes.

THE VERTEBRATES INVADE THE LAND

In the Devonian pools there were fishes that had lungs and internal nostrils and a bony type of fin that could develop into a terrestrial leg. They had the equipment to move out on land. What had been taking place on land since the first development of primitive terrestrial plants in the Silurian?

The Devonian Forests

The continents were still joined in one great supercontinent, Pangaea. The land had been slowly eroding away, and many extensive, low, swampy areas had developed. Some of the early land plants had given rise to a number of groups of primitive vascular plants. By mid-Devonian times, extensive forests of treelike club mosses and tree ferns had developed.

These low, moist swamps and forests of abundant plant life were not to go unexploited by animals. The first possible terrestrial invertebrates, scorpion-like arachnids, appeared in the Silurian, and during the Devonian, primitive insects and spider-like arachnids were present.

The Amphibians

The average amount of DNA per nucleus found in the ray-finned fishes is about 1.7 picograms (1.7×10^{-12} grams). Of the surviving fleshy-finned fishes, the coelacanth (a lobe-fin) has 6.5 picograms, the bichir (a brachyopterygian) has 12 picograms, and the lungfishes have values ranging from 100 to 284 picograms. (Much of the increase in genome size in the dipnoans seems to have taken place after the group diverged from the other sarcopterygians.) Nuclear DNA amounts of the salamanders (amphibians) range from 36 to 200 picograms. The sarcopterygian line seems to have been characterized by high nuclear DNA amounts. Did this give at least one group of these fishes the genetic flexibility that allowed them to move from water out onto the land?

If the Devonian was indeed a time of seasonal droughts, the water level in the ponds inhabited by species of fleshy-finned fishes would have been subject to marked fluctuations. As the ponds grew smaller, there would have been less room and less food for the young fishes and they would have been subject to heavy predation by the larger fishes. When the drought was interrupted by rain to raise the humidity of the air, members of at least one of the species might have been able to use their muscular fins to haul themselves out on land, either to escape their predators or to search for bigger, less crowded ponds. Perhaps the young first fled to shore in the rush to escape a pursuing adult. On land, they might have discovered an unexploited food supply in the primitive terrestrial arthropods and, perhaps, in dead animals left on the bank by the receding waters. If being able to move out on land gave these fish an advantage over their relatives that remained in the water, the former would leave more descendants. Natural selection would have favored the development of stronger limbs and other adaptations to terrestrial life. Eventually the fins evolved into legs and the animals had become *amphibians,* the first four-footed, terrestrial vertebrates.

The Coal Forests

The land continued low during the next two periods and extensive swamps and forests of succulent and primitive woody vegetation developed. The fossilized remains of these plants formed widespread beds of bituminous coal. For this reason the Mississippian and Pennsylvanian are usually grouped together as the Carboniferous. Not only club mosses, horsetails, ferns, and early seed-bearing plants called *seed ferns* went into the formation of these forests but primitive conifers had now appeared. Insects were abundant and diversified. See Figure 24–12.

In these forests, the amphibians, with all the land theirs to claim and without competition from any other vertebrates, evolved rapidly into a diversified group (see Figure 24–13). The Carboniferous is known as the Age of Amphibians.

The Waning Paleozoic

The Carboniferous closed with a gradual re-elevation of Pangaea. This tended not only to drain and dry the land but also to cool it. The bryophytes

Figure 24–12. Insect life in the Carboniferous.

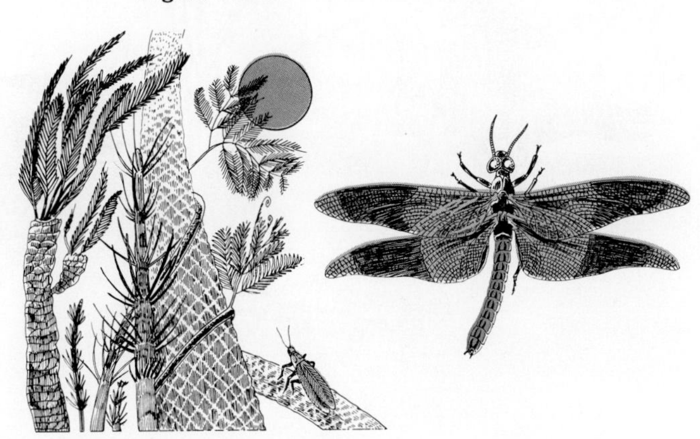

and more primitive vascular plants became less abundant, but the recently evolved conifers were better able to withstand drought and cold, and began to replace the primitive forms.

Origin of the Reptiles. Most amphibians never completely broke their ties to to the water. Principally this was because most of them failed to develop adequate reproductive adaptations to life on land. The reproductive habits of the amphibians are an example of a very usual evolutionary pattern. When a group enters a new environment and begins to diversify, the different lines evolve a variety of methods to meet a common problem. Frequently one complex of adaptive traits proves more successful than the others, and the possessors of this complex become the dominant members of the next age. Modern amphibians show far more diversity in reproductive pattern than the reptiles, birds, or mammals. Some have external fertilization, others internal fertilization. Many of them deposit their eggs away from water, in specially constructed foam nests, or in burrows, or under logs or rocks. Sometimes the eggs are carried about by either the male or female parent. Usually the egg has a relatively small amount of yolk, so that the young hatches in an immature state and passes through an aquatic, feeding, tadpole stage before metamorphosing into the adult body form. But in some amphibian eggs there is enough yolk to provide nutrients and the feeding tadpole stage is eliminated. A few species show intrauterine development. One line of early amphibians combined internal fertilization and a large-yoked egg with the development of a protective amnion and a calcareous shell to surround the egg. These animals became the reptiles, and this reproduc-

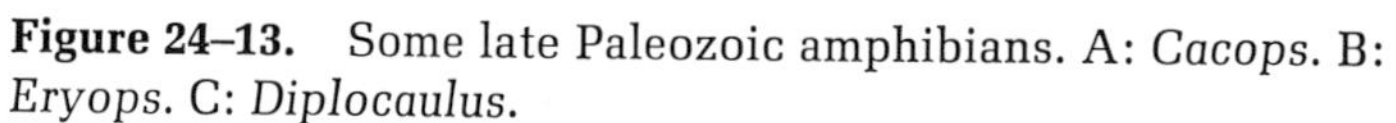

Figure 24–13. Some late Paleozoic amphibians. A: *Cacops*. B: *Eryops*. C: *Diplocaulus*.

tive pattern (or the derived one of intrauterine development with placenta formation) is found in all reptiles, birds, and mammals.

The shelled amniote egg was not the only advantage the reptile had over the amphibians. The bony fish armor was greatly reduced in the ancient amphibians (it is gone in modern forms). Probably this reduction was necessary because their lungs were still not very well developed and the early amphibians respired largely through their skins even as amphibians do today. But the reptiles had more efficient lungs and could also develop a new protective coating of scales. These scales are very different from the bony scales of the fishes, which are formed from the mesoderm. Reptile scales are cornified areas of the epidermis. The thin-skinned amphibians were susceptible to desiccation, but the reptiles, with both adult and egg protected from drying out, were free to move away from the vicinity of water.

Permian. The Paleozoic ended with the Permian. Figure 24–14 shows an artist's conception of an Early Permian scene. During this period the land was rising and the climate became more varied. Indeed, some of the most extensive

Figure 24–14. Life as it may have been in an early Permian swamp. The small animals in the water are amphibians. The two large animals on the bank are primitive reptiles (pelycosaurs). (From an original painting by Margaret Colbert, courtesy of E. H. and Margaret Colbert.)

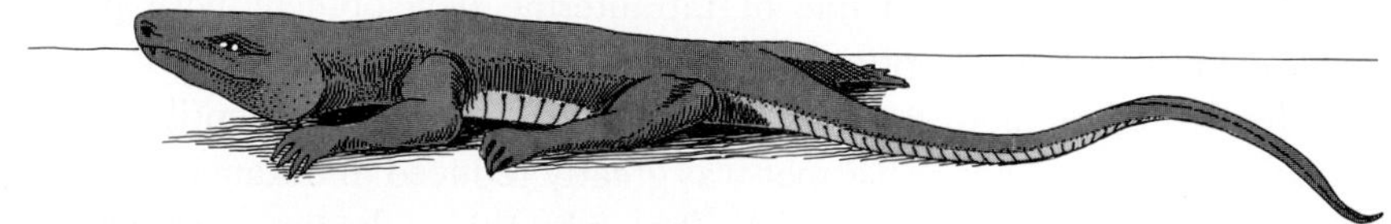

Figure 24–15. A cotylosaur, an early reptile.

glaciation the world has ever known occurred during this period, but the glaciers did not advance from the Arctic region as they did in the most recent glacial times; they were confined to parts of Gondwanaland. The Permian saw a great reduction in the seedless, primitive land plants and in the amphibians. The conifers with their seeds, the land-living arthropods, which had independently developed terrestrial eggs, and the reptiles with their scales and shelled amniote eggs were able to take advantage of the shifting environment and all became widespread. The early reptiles, a group known as the *cotylosaurs* or stem reptiles (see Figure 24–15), gave rise to the dominant animals of the Mesozoic.

SUGGESTED READINGS

Colbert, E. H.: *Evolution of the Vertebrates,* 2nd ed., John Wiley & Sons, Inc., New York, 1969.

———: *Wandering Lands and Animals,* E. P. Dutton & Co., Inc., New York, 1973.

Glaessner, M. F.: "Precambrian Animals," *Sci. Amer.,* Vol. 204, No. 3, 1961.

Kummel, B.: *History of the Earth,* W. H. Freeman and Co., San Francisco, 1961.

McAlester, A. L.: *The History of Life,* Prentice-Hall, Inc., Englewood Cliffs, N.J., 1968.

Moore, R.: *Evolution,* Life Science Library, Time Inc., New York, 1962.

Romer, A. S.: *Vertebrate Paleontology,* 3rd ed., University of Chicago Press, Chicago, 1966.

CHAPTER 25

Adaptive Radiation and Extinction

The thread of evolution leading from the origin of life up to man is a fascinating one to follow, but it is a very small part of the history of life on earth. Time and time again, some group of organisms has gone through an extensive evolution in which its members have come to occupy most if not all of the major habitats available to them. This pattern of evolution in which a single stock gives rise to many species that are variously adapted to different modes of life is called *adaptive radiation.* Unless you see the broad span of evolution as a series of repeated adaptive radiations, you will have only a meager and distorted view of the history of life on earth.

In the last chapter, the evolutionary line that led to the primitive terrestrial reptiles was emphasized. Later, we follow this line to the genesis of man, but first we discuss in some detail one of the most spectacular adaptive radiations, that of the reptiles during the Mesozoic. You should keep in mind, though, that this is not an isolated instance, but an example of a very common evolutionary pattern.

REPTILE RADIATION

The Paleozoic culminated with the rising and cooling of Pangaea and the consequent formation of environments that were inhospitable for many of the animals and plants of the Carboniferous. As the mountains began to wear away after this orogenic period (the Appalachian revolution), the land again became lower and warmer. It was occupied by expanding populations of cycads, ginkgoes, and conifers, all nonflowering seed plants. This was the world in which the reptiles formed their dynasty.

Of Names

Discussions of the evolutionary history of a group often seem overloaded with a cumbersome terminology. Most of the animals are now extinct and have no common names, only "book names" made up by scientists. These names are usually formed from Latin or Greek roots, and a knowledge of the meaning of some of these roots will help you to remember the names. *Saur* means simply "reptile" (more strictly, "lizard"). Names formed from this root include *dinosaur,* "terrible reptile;" *archosaur,* "ruling reptile;" *ichthyosaur,* "fish reptile;" *pterosaur,* "winged reptile;" *lepidosaur,* "scaly reptile." *Brontosaurus* is the "thundering reptile," and *Tyrannosaurus* is the "tyrant reptile." *Saurischia* means "reptile hips," and *ornithischia* means "bird hips."

Major Groups of Reptiles

The reptiles appeared first in the late Carboniferous, and before the end of the Permian they had diverged into five major stocks. These evolutionary lines can be distinguished by the formation of the skull. The brain of the reptile is small, and the part of the skull actually surrounding the brain is also small. But the skull case is overhung on each side behind the eye by a flange of bone, like the eaves of a roof. In the primitive stem reptiles the flanges were solid, and heavy muscles that closed the jaw were attached to the undersides of these flanges. In the various lines that diverged from this group one or two holes developed in the flanges. This allows the jaw muscles to pass through and attach to the outer surface of the skull, which is a more efficient arrangement mechanically. Figure 25–1 shows the arrangements of these openings in the major groups of reptiles. The bars of bone around the openings are called *arches,* and the names of the reptile groups are based on the Greek word *apsis,* which means "arch."

Anapsida. In the anapsids the skull flanges are solid, just as in the amphibians from which they arose. This group includes the stem reptiles (cotylosaurs) and the present-day turtles.

Synapsida. Here a single opening lies somewhat behind and below the eye. Common during the Permian, the synapsids were only a minor element in the great reptilian fauna of the Mesozoic. But during this time they gave rise to

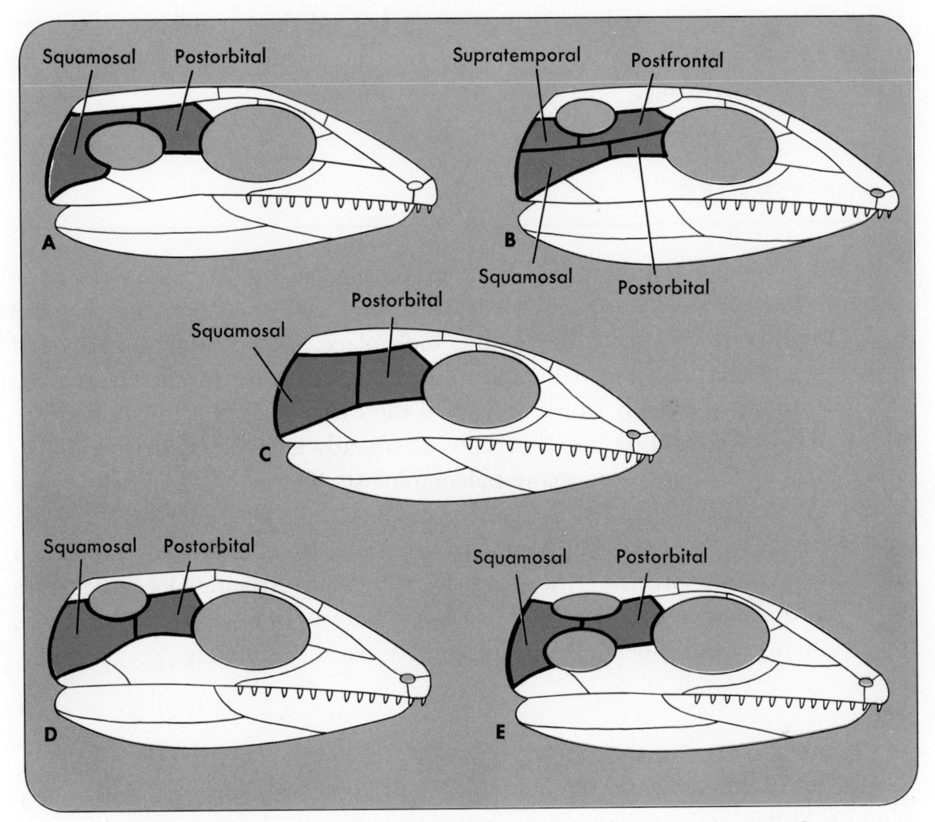

Figure 25–1. Reptilian skull types. A: Synapsid. B: Parapsid. C: Anapsid. D: Euryapsid. E: Diapsid (Redrawn from E. H. Colbert, *The Dinosaur Book*, 2nd ed., 1951. By permission of McGraw-Hill Book Company and E. H. Colbert.)

the mammals. The bony arch in your upper cheek region (the cheek bone) is the remnant of a synapsid arch.

Parapsida. The parapsids also have a single opening, but it is behind and somewhat above the level of the eye. These reptiles, the ichthyosaurs and their relatives, are all now extinct.

Euryapsida. Like the parapsids, euryapsids have a single opening behind and above the eye, but is bordered by different bones and apparently marks a distinct evolutionary line. These are the plesiosaurs and their relatives, and like the ichthyosaurs they are all extinct.

Diapsida. The diapsida have two openings behind the eye. This group gave rise to the modern lizards and snakes and to the archosaurs, the great ruling reptiles of the Mesozoic.

HISTORY OF THE REPTILES

All the major reptilian lines were established in the Permian, and some already showed considerable diversity. But it was in the Mesozoic that the great adaptive radiation of the reptiles took place. The continents were still broadly connected as parts of Pangaea during the Triassic, and newly evolved forms could spread widely from one part of the earth to another. The discovery in Antarctica of a fossil reptile, *Lystrosaurus,* already known from Africa, India, and China, finally convinced many paleontologists that continental drift was a reality. At the end of the Triassic, widespread volcanic eruptions heralded the end of Pangaea. The continents began to break apart and drift away from each other. It was a slow process. Thus, long after the North Atlantic Ocean began to form, North America remained tied to Europe by way of Greenland. But it was the beginning of the appearance of the world as we know it today.

Stem Reptiles and Turtles

The cotylosaurs were the first reptiles to arrive on the scene. Some were clumsy, heavy-bodied plant-eaters, others were more slender and agile, and probably fed on insects. The largest cotylosaurs reached three meters (ten feet) or more in length. The cotylosaurs gave rise to four other major reptilian stocks and their own immediate descendants, the turtles, still live today, but the stem reptiles themselves were extinct by the end of the Triassic. In addition to the reptilian epidermal scales, turtles have bony dermal plates on the back and belly region. These plates are fused to form a protective shell. Never a dominant group, these most primitive surviving reptiles are represented today by only a few hundred living species.

Marine Reptiles of the Mesozoic

In addition to marine turtles and lizards, two other reptile stocks took to the seas with marked success. These were the ichthyosaurs (Parapsida) and plesiosaurs (Euryapsida).

Ichthyosaurs. The ichthyosaurs were large, fish-shaped, predatory animals that undoubtedly lived much as the large sharks and porpoises (mammals) do today (see Figure 25–2). The reptile egg is terrestrial, and even today the marine turtles come to shore to lay their eggs. But the ichthyosaurs would have been as helpless on land as today's whales. Paleontologists long believed that they must have given birth to living young, and fossil ichthyosaurs have been found with the skeleton of the embryo in place in the body of the mother.

Plesiosaurs. These animals were unlike anything else in body form. They have been described as looking like a snake strung on the body of a turtle. A plesiosaur had a short, broad body, four heavy, paddle-like limbs, and an exceedingly long neck. Instead of actively pursuing its prey like an ichthyosaur, it may have simply floated at the surface, darting out its long neck to capture any unfortunate creature that chanced to swim by (see Figure 25–2).

Mammal-like Reptiles

The synapsids were present in the late Carboniferous and were the most common Permian reptiles. They declined to extinction during the Triassic, yet

Figure 25–2. Mesozoic seascape with an ichthyosaur and plesiosaurs.

they were the ones that developed the innovations that led to the rise of the mammals. It was not a simple, straight-line evolution, and several abortive side-lines are known. Among the more spectacular of these is that of the pelycosaurs. Some of these animals developed large sail-like fins that were supported by a series of mastlike spines down the back, thus earning these animals the name of ship-lizards (see Figure 25–3). The function of these fins remains an enigma. One paleontologist has suggested that they probably served to keep the body surface-volume ratio more or less constant as the pelycosaurs increased in size —some of them were over three meters (ten feet) long. The surface of the fin may have acted as a heat-absorbing or heat-radiating structure so that the animal could, in a limited way, control its body temperature. They may have been the first animals able to do this. Another suggestion is that the fins were used in behavioral displays. The group was not a successful one and died out at the end of the Permian. It was some of their relatives, the therapsids, that went on to give rise to the mammals. The therapsids were themselves a varied lot; some were large, heavy-bodied herbivores, others were active predators, wolf-like in their body build. Some of the Triassic therapsids had skeletons that were much more mammal-like than reptile-like, and by late Triassic times true mammals were present. The Mesozoic mammals were small, inconspicuous animals and apparently were not at all common until the Cretaceous. It was only with the onset of the Cenozoic that they became a conspicuous part of the fauna.

Figure 25–3. Skeleton of an early fin-backed reptile, *Edaphosaurus boanerges*. (Carnegie Museum—Leo T. Sarnaki photo.)

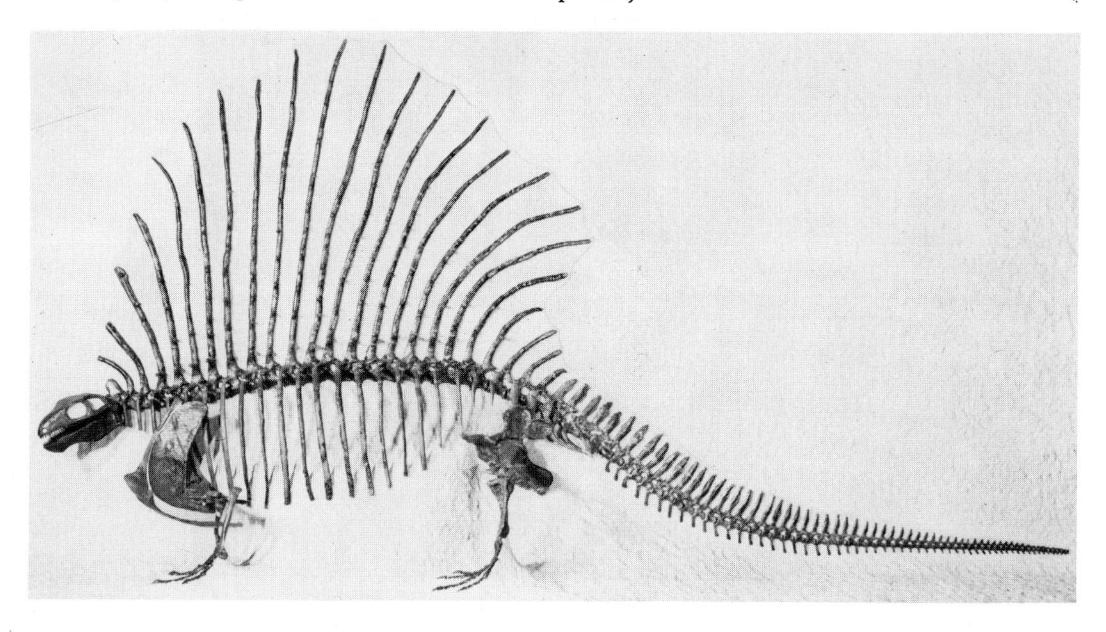

Diapsida

By the late Permian, at least, the diapsids had split into two distinct groups, the lepidosaurs and the archosaurs.

Lepidosaurs. This group includes primitive, lizard-like types and their living descendants, the tuatara (*Sphenodon*) of New Zealand (see Figure 25–4) and the lizards and snakes (Squamata). Only one group of lepidosaurs was at all prominent among the Mesozoic reptiles. The mosasaurs were huge marine lizards, some reaching nine meters (thirty feet) in length, that replaced the ichthyosaurs as the active predators of the Cretaceous seas. Of all the varied reptilian groups, only the Squamata can be considered successful today. There are about six thousand living species of lizards and snakes.

Archosaurs. The archosaurs included the largest and most spectacular of the Mesozoic reptiles and because of this are known as the "ruling reptiles." There are six evolutionary lines of archosaurs: pterosaurs, birdlike archosaurs, phytosaurs, crocodilians, and two groups of dinosaurs. Figure 25–5 shows some Triassic archosaurs.

PTEROSAURS. This is one of the two groups of archosaurs that took to the air, but they did it in an entirely different way than the birds did. Instead of feathered wings, a pterosaur had a flight membrane formed by a fold of skin between the front and hind limbs, much like that of a modern bat (a mammal, not a reptile). The pterosaurs seem to have been gliders rather than active fliers. They made their living near the seacoasts where physical features induced rising currents of air and where there was an abundance of marine fishes on which they fed (see Figure 25–6). Successful for a time, this evolutionary path ultimately proved to be a cul-de-sac, and by the end of the Cretaceous the pterosaurs had been replaced by the birds.

BIRDLIKE ARCHOSAURS. An entirely different line of archosaurs led to the birds. Birds have wings, feathers, lightweight hollow bones, and strong flight muscles. They are superbly adapted for flight. The birds are a highly successful group and, among the vertebrates living today, are second only to the bony

Figure 25–4. The tuatara, *Sphenodon punctatum*, of New Zealand.

Figure 25–5. Life in Late Triassic times. The animal in the water is an amphibian. The large crocodile-like animal on the bank is a phytosaur. Behind the phytosaur and in the left center are two kinds of thecodonts, representatives of the primitive archosaur stock. The two animals at the top left are early dinosaurs. (From an original painting by Margaret Colbert, courtesy of E. H. and Margaret Colbert.)

fishes in number of species. They did not develop all birdlike characteristics at once. As soon as the fossil evidence indicates that hollow bones and feathers were present (in the Jurassic), they are called birds, but the early birds still had teeth and long, reptilian tails (see Figure 25–7).

PHYTOSAURS AND CROCODILIANS. The phytosaurs were superficially crocodilian in aspect and probably lived much as the modern crocodiles do (see Figure 25–5). The skull of a crocodile or alligator has a solid bony plate, the secondary palate, that separates the nasal chamber from the oral chamber much as it does in man. A phytosaur had no such structure. This meant of course, that it could not breathe while it had food or drink in its mouth without danger of strangling. The crocodilians, with their bony palates, can eat and breathe at the same time. This may have been an important advantage, for the once numerous phytosaurs had been replaced by the crocodilians by the end of the Triassic. The surviving crocodilians are the only living archosaurs (see Figure 25–8).

Figure 25–6. Seascape with pterosaurs.

DINOSAURS, THE MAIN SHOW. Probably no groups of extinct animals are as familiar to the average man as are the dinosaurs. They are featured in comic books, advertisements, and science-fiction stories. They are ideal examples of rapid evolution, adaptive radiation, and extinction. But the rapidity of their evolution is relative. Man has been on the earth for only three million years but the dinosaurs held forth for forty times as long. Figure 25–9 shows some of these fantastic creatures.

Dinosaurs represent two main lines that evolved separately from primitive archosaurs in the Triassic—thus the word *dinosaur* does not indicate a single reptilian stock.

The two major lineages of dinosaurs differed fundamentally in the structure of the pelvic girdle. In one line (the Saurischia) the three bones formed a triradiate structure as they did in the primitive archosaurs. In the other line, a bar from the pubis had swung posteriorly to rest against the ischium as it does

Figure 25–7. *Archeopteryx* of the Jurassic, the oldest known bird.

in the birds, and another bar projected forward under the belly. This group is known as the Ornithischia. Figure 25–10 shows this difference.

Most people think of dinosaurs as large, clumsy brutes, but a few of them were small, agile animals, some no larger than a chicken. The saurischians soon divided into a carnivorous stock that developed *bipedal locomotion*, that is, they walked on their hind legs (see Figure 25–11) and an herbivorous stock that

Figure 25–8. A crocodile.

Figure 25–9. Dinosaurs of the Late Jurassic. The two animals in the foreground are carnivorous, bipedal saurischians. *Allosaurus*, on the left, was nearly ten meters (thirty feet) long. *Ornitholestes*, on the right, was about one and a half meters (five feet) long. The animal in the upper left is an herbivorous, quadrupedal saurischian, *Apatosaurus*, about twenty meters (seventy feet) long. In the middle right is an armored, quadrupedal ornithischian, *Stegosaurus*, about six meters (twenty feet) long. (From an original painting by Margaret Colbert, courtesy of E. H. and Margaret Colbert.)

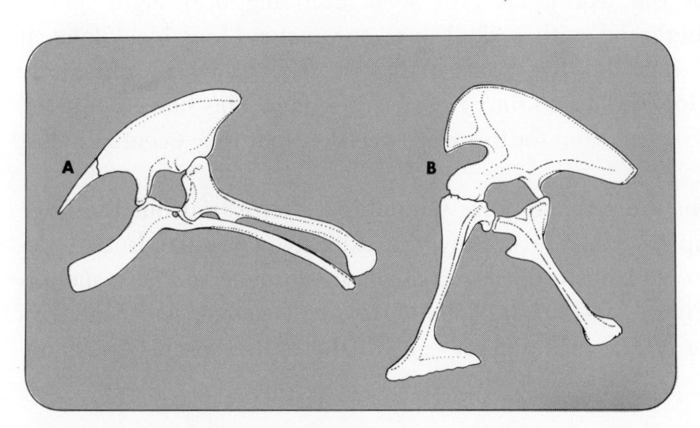

Figure 25–10. Dinosaur pelvic girdles. A: Ornithischian. B: Saurischian.

Figure 25–11. Skeleton of a bipedal dinosaur, *Dryosaurus altus*. (Carnegie Museum—Sydney Prentice photo.)

lumbered about on all fours (*quadrupedal locomotion*). The bipedal carnivores included *Tyrannosaurus rex* ("king of tyrant reptiles"), undoubtedly the most spectacular predator the world has ever seen. Standing near eight meters (twenty-five feet) high and with a skull over one meter (four feet) long, it must have been a formidable foe. Some of the smaller carnivores apparently became specialized for feeding on the eggs of larger ones. This line culminated in forms that were so specialized for this restricted diet that they eventually lost all sign of teeth.

The carnivorous saurischians were undoubtedly terrestrial but the huge herbivorous quadrupeds came to occupy the extensive marshlands that developed as the Mesozoic lands eroded away. These enormous animals, such as *Apatosaurus* (see Figure 25–12) and *Diplodocus*, half-waded, half-floated in the marshes, feeding on the lush aquatic growth there; they probably seldom

Figure 25–12. One of the largest of the dinosaurs, *Apatosaurus louisae*. From an original painting by Andrey Avinoff. (Carnegie Museum—Leo T. Sarnaki photo.)

came to shore except to seek other marshes or to lay their eggs. *Diplodocus* reached a length of nearly thirty meters (ninety feet).

The ornithischians were presumably all plant-eaters, but they had little else in common. Some were quadrupeds and some were bipeds. Some of the quadrupeds developed extensive body armor, others had elaborate head processes including large, projecting horns (see Figure 25–13). Most of the armored and horned dinosaurs seem to have been terrestrial. One of the weirdest was *Stegosaurus* (see Figure 25–14). A creature some six meters (twenty feet) long, it had a double row of huge, horny plates down its back that probably protected its spinal cord from the bites of *Tyrannosaurus*. Its tiny head was more than balanced by a heavy tail that was armed with a series of massive spines. Apparently more nervous tissue was needed to control the movement of the heavy hips, hind legs, and tail than to control the head and front end of the body, for an enlargement of the spinal cord in the region of the pelvic girdle was twenty times larger than the brain (see Figure 25–15). A newspaper writer, Bert L. Taylor of Chicago, commemorated this unusual state of affairs:

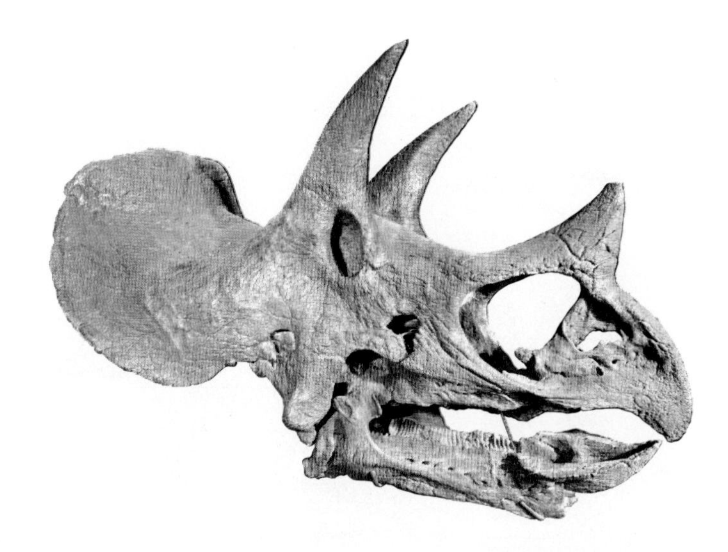

Figure 25–13. Skull of *Triceratops.* (Carnegie Museum —Hess photo.)

Figure 25–14. Skeleton of *Stegosaurus* in Carnegie Museum, Pittsburgh. (Carnegie Museum—Hess photo.)

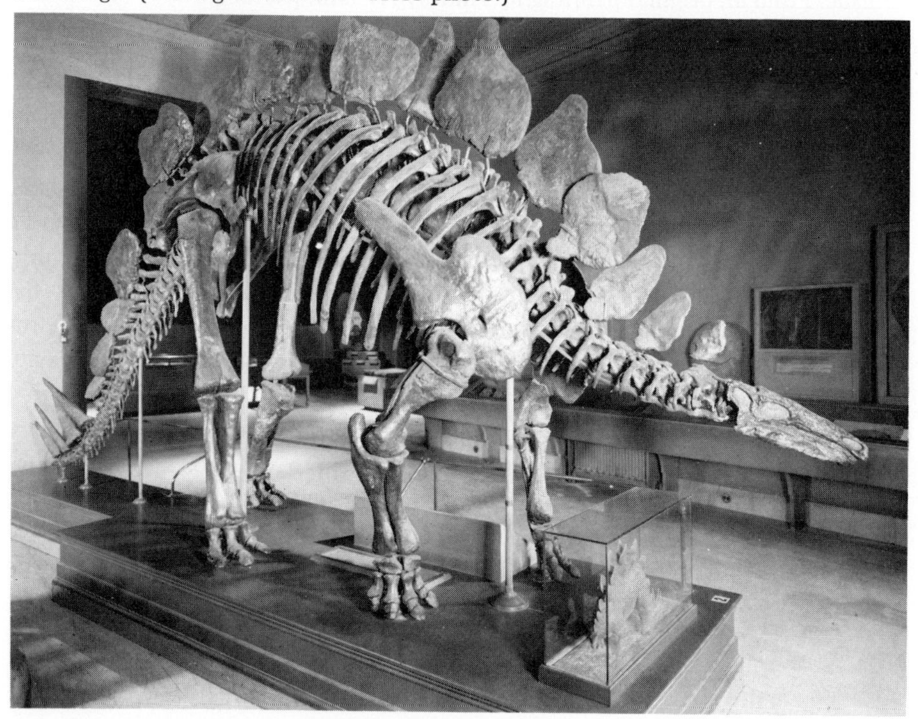

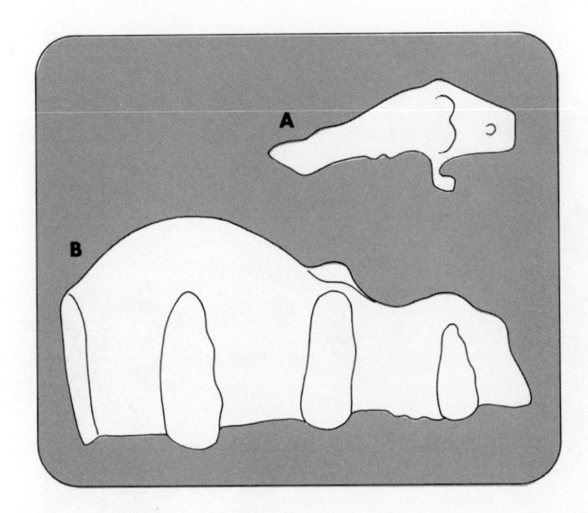

Figure 25–15. A: Brain. B: Sacral enlargement. In *Stegosaurus*

The Dinosaur

Behold the mighty dinosaur,
Famous in prehistoric lore,
Not only for his power and strength
But for his intellectual length.
You will observe by these remains
The creature had two sets of brains—
One in his head (the usual place),
The other at his spinal base.
Thus he could reason "*A priori*"
As well as "*A posteriori.*"
No problem bothered him a bit
He made both head and tail of it.
So wise was he, so wise and solemn,
Each thought filled just a spinal column.
If one brain found the pressure strong
It passed a few ideas along.
If something slipped his forward mind
'Twas rescued by the one behind.
And if in error he was caught
He had a saving afterthought.
As he thought twice before he spoke
He had no judgment to revoke.
Thus he could think without congestion
Upon both sides of every question.
Oh, gaze upon this model beast,
Defunct ten million years at least.

Many of the bipedal ornithischians followed *Diplodocus* and its allies to the marshes, and some of them became extremely specialized for this environment (see Figure 25–16). They were apparently waders that could go out in rather deep water and browse on the bottom. They developed fantastic head crests of a great variety of shapes. Some resembled the crest on the helmet of an

Figure 25–16. Late Cretaceous dinosaurs. In the left foreground is a ten-meter (thirty-foot) long carnivorous saurischian (*Gorgosaurus*). The crested ornithischian (*Corythosaurus*) in the water is about the same length. The two armored ornithischians (*Styracosaurus*) on the hill at the left are six meters (twenty feet) long and the one at the water's edge (*Ankylosaurus*) is about five meters (fifteen feet). Flowering plants are now present. (From an original painting by Margaret Colbert, courtesy of E. H. and Margaret Colbert.)

ancient Greek warrior. These crests were bony extensions of the skull and contained air passages that were connected to the nasal chambers. Undoubtedly these crests had some function. One suggestion (among several) is that they served as hollow air chambers in which a reserve supply of air could be stored to enable the animal to keep its head under water for a considerable period of time.

The evolution of the reptiles is diagramed in Figure 25–17.

DEATH AND EXTINCTION

The rise and spread of the great Mesozoic reptiles was dramatic indeed. They became the dominant animals of dry land, the swamps and marshes, the open seas, and the air. Equally dramatic was their disappearance. By the end of the Cretaceous, all were gone, not only the dinosaurs but the pterosaurs, ichthyosaurs, and plesiosaurs as well. They were not all wiped out in a day or a year. If you had been living then, you would probably have noticed little change during your lifetime. But by the scale of geologic time, the Great Extinction, as it is called, was very rapid.

Without the death and decay of individuals, the available carbon would long ago have been locked up in living organisms and there would be no more for the formation of new organisms. The extinction of entire species and groups of species is also necessary for evolutionary progress; the old must be removed to make place for the new. Sometimes a better adapted form appears and simply outcompetes the one it replaces. This may help explain the disappearance of many amphibian groups when the reptiles began their radiation, and the replacement of the pterosaurs by the birds. But it cannot explain the disappearance of the dinosaurs and the great marine reptiles. They had no competitors. The mammals of the time were small and comparatively few in number. They did not begin their great adaptive radiation until the land was left open by the removal of the dinosaurs.

In point of fact, biologists know much less about the processes of extinction than they do about the processes of speciation, although extinction is an equally important evolutionary fact.

Glenn L. Jepsen wrote:

> Authors with varying competence have suggested that dinosaurs disappeared because the climate deteriorated (became suddenly or slowly too hot or cold or dry or wet), or that the diet did (with too much food or not enough of such substances as fern oil, from poisons in water or plants or ingested minerals, by bankruptcy of calcium or other necessary elements). Other writers have put the blame on disease, parasites, wars, anatomical or metabolic disorders (slipped vertebral discs, malfunction or imbalance of hormone and endocrine systems, dwindling brain and consequent stupidity, heat sterilization), racial old age, evolutionary drift into senescent overspecialization, changes in the pressure or composition of the atmosphere, poison gases, volcanic dust, excessive oxygen from plants, meteorites, comets, gene pool drainage by little mammalian egg-eaters, overkill capacity by predators, fluctuation of gravitational constants, development of psychotic suicidal factors, entropy, cosmic radiation, shift of Earth's rotational poles, floods, extraction of the moon from the Pacific Basin, drainage of swamp and lake environments, sunspots, God's will, mountain building, raids by little green hunters in flying saucers, lack of even standing room in Noah's Ark, and paleoweltschmerz.[1]

We believe that the dinosaurs simply ran out of excess DNA. Modern reptiles have about three to five picograms of DNA, less than the amphibians or

[1] Glenn L. Jepsen, in "Terrible Lizards Revisited," *Princeton Alumni Weekly*, vol. 69, no. 10 (1963), p. 7.

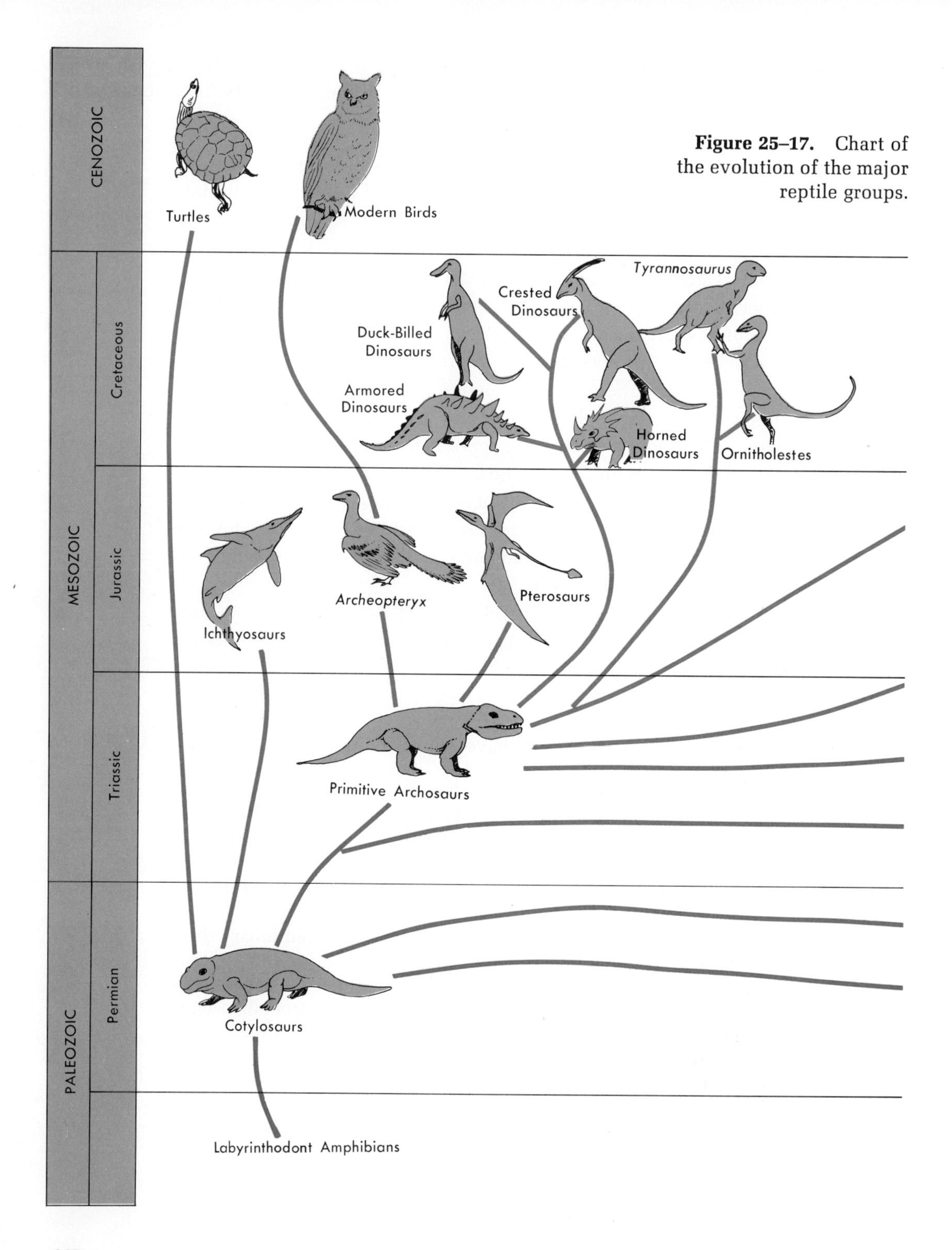

Figure 25–17. Chart of the evolution of the major reptile groups.

Crocodiles and Alligators
Snakes
Lizards
Rhynchocephalia
Mammals
Brontosaurs
Plesiosaurs
Phytosaurs
Pelycosaurs

fleshy-finned fishes. This suggests that the reptilian radiation was accompanied by a loss of DNA. The two groups of vertebrates that have the lowest amounts of DNA per nucleus are the ray-finned fishes and the birds. They are also the ones that have speciated most widely. It may be that in groups with much DNA, many mutations are masked by the presence of duplicate copies of the genes, whereas in those groups with less DNA, mutations are more readily expressed and can be acted on by natural selection. Such groups could speciate readily. But they would lack the genetic flexibility to adapt to major environmental changes. Dinosaur speciation reached its zenith in the Cretaceous. But the Mesozoic closed with the great Rocky Mountain Revolution, when the Rockies, the Andes, the Alps, and the Himalayas were thrust up. The climate grew colder. Did the same loss of DNA that allowed the dinosaurs to speciate so readily deprive them of the ability to adjust?

SUGGESTED READINGS

Colbert, E. H.: *The Dinosaur Book*, McGraw-Hill Book Company, New York, 1951.

———: *The Age of Reptiles*, Weidenfeld and Nicolson, London, 1965.

———: *Evolution of the Vertebrates*, 2nd. ed., John Wiley & Sons, Inc., New York, 1969.

———: *Wandering Lands and Animals*, E. P. Dutton & Co., Inc., New York, 1973.

Moore, R.: *Evolution*, Life Science Library, Time Inc., 1962.

Romer, A. S.: *Vertebrate Paleontology*, 3rd ed., University of Chicago Press, Chicago, 1966.

Sawkins, F. J., C. G. Chase, D. G. Darby, and G. Rapp, Jr.: *The Evolving Earth: A Text in Physical Geology*, Macmillan Publishing Co., Inc., New York, 1974.

CHAPTER 26

The Rise of the Mammals

The sweeping extinction of the majority of the reptiles cleared the way for the radiation of the mammals. Mammals had been present since late Triassic times; their long subordination during the Mesozoic suggests that their rise and spread would have been far less rapid without the disappearance of the dinosaurs.

Because the development of the mammals is closely correlated with the development of the modern flora, we should bring the history of plants up to date.

EVOLUTION OF MODERN PLANT LIFE

The early and middle Mesozoic forests were made up of cycads, ginkgoes, and conifers. By the mid-Mesozoic the conifers were much like modern forms and in places the forests were made up of dense stands of tall pines.

One of the dramatic events of the Cretaceous was the sudden rise of the flowering plants. The anthophytes are the most successful of the land plants. The flowers of many attract insects that convey pollen from one plant to another, a more effective means of ensuring fertilization than the wind dispersal of pollen on which the flowerless conifers depend.

The tissues surrounding the seeds can develop into edible fruits. These are eaten by animals, and the seeds may be swallowed and pass unharmed through the animal digestive tract to be deposited with the feces away from the parent plant. The naked seeds of the conifers usually simply fall to the ground below the parent tree. The flowering plants are thus able to spread far more rapidly. They soon displaced the conifers over much of the land.

Rather early in the Cretaceous the anthophyte line had already divided into dicots and monocots, and such modern kinds of trees as poplars, magnolias, figs, sassafras, (dicots), and palms (monocots) were present. By the end of the era, the forests looked decidedly modern, with trees such as maple, oak, beech, sycamore, walnut, and breadfruit, although the herbaceous anthophytes were not yet present.

As the Cenozoic opened, the plant life of the world was pretty much as it was at the end of the Cretaceous. Figure 26–1 shows what an early Cenozoic landscape might have looked like. The grasses and herbs appeared at this time. During the mid-Cenozoic, the lands began to rise and large areas received less rainfall. The forests dwindled but the grasslands spread and the grasses attained the significance they still have today. They are the primary source of food for many mammals, both the grass- and the seed-eaters and the carnivores that prey on them, even as they are for man and his domestic animals.

During the Cenozoic the final breakup of Laurasia and Gondwanaland occurred. North America broke away from Europe, although it was joined to Asia from time to time by a land bridge across the Bering Sea similar to the

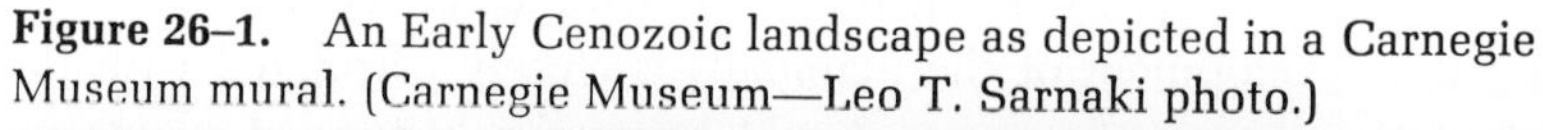

Figure 26–1. An Early Cenozoic landscape as depicted in a Carnegie Museum mural. (Carnegie Museum—Leo T. Sarnaki photo.)

bridge (the Isthmus of Panama) by which it is now connected to South America. Australia early became an island continent and South America too was isolated for much of the Cenozoic. These changes resulted in changes in the pattern of distribution of the plants and animals. A species that evolved on one continent could no longer spread freely to others. Marked differences developed in the kinds of plants and animals present, especially in South America and Australia, the most isolated of the continents.

MAMMALIAN BEGINNINGS

Mammals have more DNA per nucleus (about five to twelve picograms) than do the surviving reptiles. The line that gave rise to the mammals diverged early from the primitive reptilian stock and may have retained more of the DNA inherited from the amphibians.

Definition of a Mammal

Three characters combine to make the mammals the successful group they are. For one thing, like the birds they are able to maintain a high and constant body temperature (see Chapter 2). Hair, which is found in all mammals and only in mammals, probably first evolved as an adaptation to cut down on the loss of body heat. Reptiles have some behavioral control over their internal temperatures—they absorb heat by basking in the sun and move into the shade or into burrows when they get too hot—but by and large their temperature is more likely to fluctuate with the temperature of their surroundings and they grow sluggish in cold weather. Mammals are able to remain more constantly active.

Another great advantage the mammals have is the presence of mammary glands. Reptiles usually abandon their eggs as soon as they lay them, although a few reptiles guard their eggs until they hatch and a few have intrauterine development. Both the eggs and young reptiles are very susceptible to predation. In contrast, all mammals suckle their young. This means that the mother stays with them and is able to protect them from predators. What is probably even more important is that there is a period during which the young are able to learn from the parent.

Correlated with this learning period is an enormous increase, during the evolution of the mammals, in the size of the cerebrum, the part of the brain concerned with learning.

Hair and mammary glands are seldom preserved in fossils. When parts of the skull surrounding the brain are available, it is possible to deduce a great deal about the size and structure of the brain. But these remains are very rare for the earliest mammals and, such as they are, they indicate that the brains of the first mammals differed little from those of the advanced, mammal-like reptiles. The great expansion of the cerebrum came later. So other characters must be used to define the mammals if one wishes to decide when they first appeared. Two of the most important characters are found in the bones of the lower jaw and the ear. Mammals have only one bone on each side of the lower jaw (in man and some others the two halves fuse in the chin region to form a single bone). Reptiles have several bones on each side of the lower jaw. Mammals have a chain of three little bones in the middle ear, but reptiles have only a single bone. The teeth of mammals are usually differentiated into incisors, canines, premolars, and molars, whereas the teeth of reptiles are generally all very much the same.

The typical mammal characters did not develop all at the same time or in a single step and it will probably never be known when some of them first appeared. Because the evolution of the mammals forms a continuum, it really doesn't matter much where the line is drawn between the most advanced fossil reptile and the most primitive fossil mammal. For convenience, paleontologists draw it at the stage where the jaw consists of a single bone and there are three bones in the inner ear.

Early Evolution of Mammals

The therapsids flourished from the mid-Permian to the mid-Triassic. Although still considered reptiles, the carnivorous therapsids had begun to develop mammal-like characters. The lower jaw still had several bones, but one of them, the dentary, was much enlarged and, in the most advanced forms, was very similar to the mammal lower jaw. The teeth had begun to differentiate into a series of small, nipping teeth comparable to mammalian incisors, a single, tusk-like canine on each side, and a series of cheek teeth. In primitive reptiles the legs were ill-adapted to support the weight of the body. They spread out to the sides so that the animals must have waddled along in a clumsy, sprawling way. In the advanced therapsids the elbow had moved backward and the knee forward so that the legs had swung under the body and were able to support it (see Figure 26–2). This must have made it possible for the animals to move about much more rapidly.

THE TRUE MAMMALS

Most of the Mesozoic mammals were small, mouse- or rat-sized animals, and few fossils of them have been found. Because the documentary record of

Figure 26–2. Change in position of limbs during reptile evolution. A: A primitive, plant-eating reptile in which the limbs offer little support to the body. B: An advanced carnivorous therapsid.

the early evolution of mammals is so inadequate, and because the most primitive group alive today is essentially without a fossil record, ideas about the relationship of the major groups of mammals are based largely on the structure of the living forms.

Major Groups of Mammals

The mammals are divided into two major groups, Prototheria and Theria. Some paleontologists believe that the two lines evolved separately from the therapsids.

Prototheria. The Prototheria are represented today by the spiny anteaters and the duck-billed platypus (Figure 26–3) of the Australian region. They differ from all other living mammals in that the female lays eggs. It may well be that the early stocks of primitive mammals really belong here, but the fossil record gives no information about reproductive habits.

Theria. Two groups, the Metatheria (marsupials) and the Eutheria (placental mammals) are included in the Theria. In the marsupials, the female has a pouch or *marsupium* surrounding the nipples. The young form no placental attachment, or only a very transistory one, with the uterine wall. They are born in a very immature condition, make their way into the pouch, and attach to the nipples. There they live until they are able to become at least partially independent. Familiar marsupials are the opossums, kangaroos, and koalas (see Figure 26–4). All other living mammals are eutherians. They differ from the Prototheria and from most Metatheria in having an *allantoic placenta* to provide nourish-

Figure 26–3. Prototherians. A: Duck-billed platypus. B: Spiny anteater.

ment for the developing embryo. Because of this, they are commonly spoken of as placental mammals. Figure 26–5 shows the three different patterns of mammalian life history.

History of Marsupials

The separation of the metatherian and eutherian lines had occurred by late Mesozoic. The oldest known marsupial fossils have been found in the United

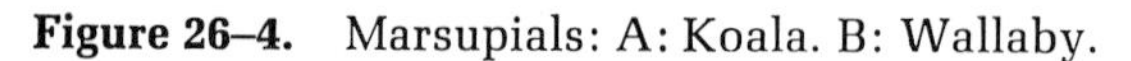

Figure 26–4. Marsupials: A: Koala. B: Wallaby.

Figure 26–5. Mammalian life histories. A: Young platypuses hatch from eggs but are suckled by the mother. B: Opossum embryos are not nourished by a placenta. Instead they make their way into the mother's pouch and attach to the nipples. C: In man as in all placental mammals the fetus receives nourishment through the placenta. The baby is suckled after birth.

States. Today marsupials occur only in Australia, New Guinea, Tasmania, and South America, except for the opossum, a recent immigrant from South America that ranges well up into North America. The marsupials may have reached Australia from South America while the two were still connected by way of Antarctica, but the connection was broken before the placentals had passed. The marsupials had the continent entirely to themselves (except for monotremes and bats) until mice, men, and dogs arrived in the late Cenozoic. In Australia, the marsupials underwent an extensive adaptive radiation. Some became cursorial carnivores much like modern dogs. Some went underground and became mole-like. Others developed the habits and general appearance of muskrats, rabbits, or wolverines. Many, like the koalas, do not resemble any placental mammals (see Figure 26–6).

History of Placental Mammals

The first known placental mammals were small, shrewlike insect eaters of the late Cretaceous. Like modern shrews they were long-snouted with many fine, sharp teeth. They had tiny brains and their feet were slender and flexible, the toes terminating in pointed claws. The structure of the feet suggests that the early placentals may have been at least partly arboreal. It was from such as these that all the modern placental mammals evolved.

Spread of Placental Mammals. The exact place of origin of the placental mammals is unknown, but certainly it was not Australia. They must have first appeared at a time and place that did not allow them to cross into Australia before it became isolated. This meant that while the marsupials had Australia to them-

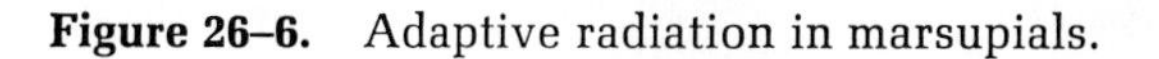

Figure 26–6. Adaptive radiation in marsupials.

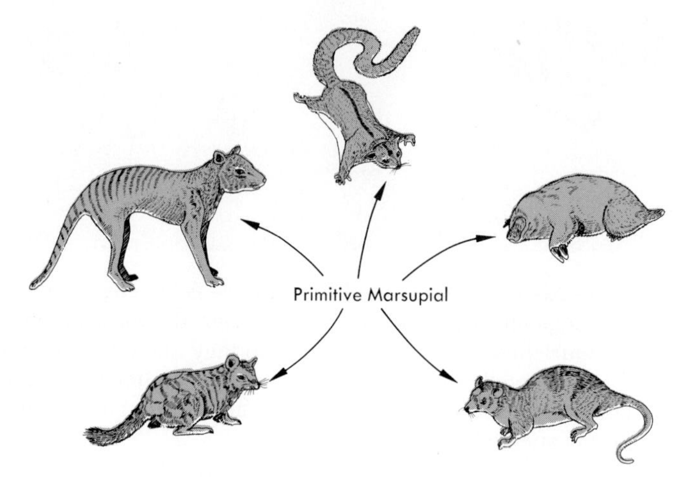

selves the placentals and marsupials shared much of the rest of the world. The placentals are apparently better adapted, for they have replaced the marsupials except in South America where a remnant of the stock still lingers.

ADAPTIVE RADIATION OF PLACENTALS. Like the marsupials in Australia, the placentals underwent a rapid adaptive radiation and now occupy every major habitat on the earth. Rabbits, antelopes, foxes, and bears live on the surface of the ground, squirrels and monkeys in the trees, moles underground, bats in the air, otters, beavers, and muskrats in fresh water, and seals and whales in the sea.

CONVERGENT EVOLUTION. When two different stocks both undergo adaptive radiation, members of those stocks may, through evolutionary processes, come to look very much alike. This convergence of superficial characters in two separate stocks, brought about by adaptation to a common habitat, is called *convergent evolution.* It is well illustrated by the two major mammalian stocks; both the marsupials and placentals produced molelike forms, rabbitlike forms, doglike forms, and so on. This is shown in Figure 26–7.

Figure 26–7. Convergence between marsupials and placentals.

Primitive Marsupials

Primitive Placentals

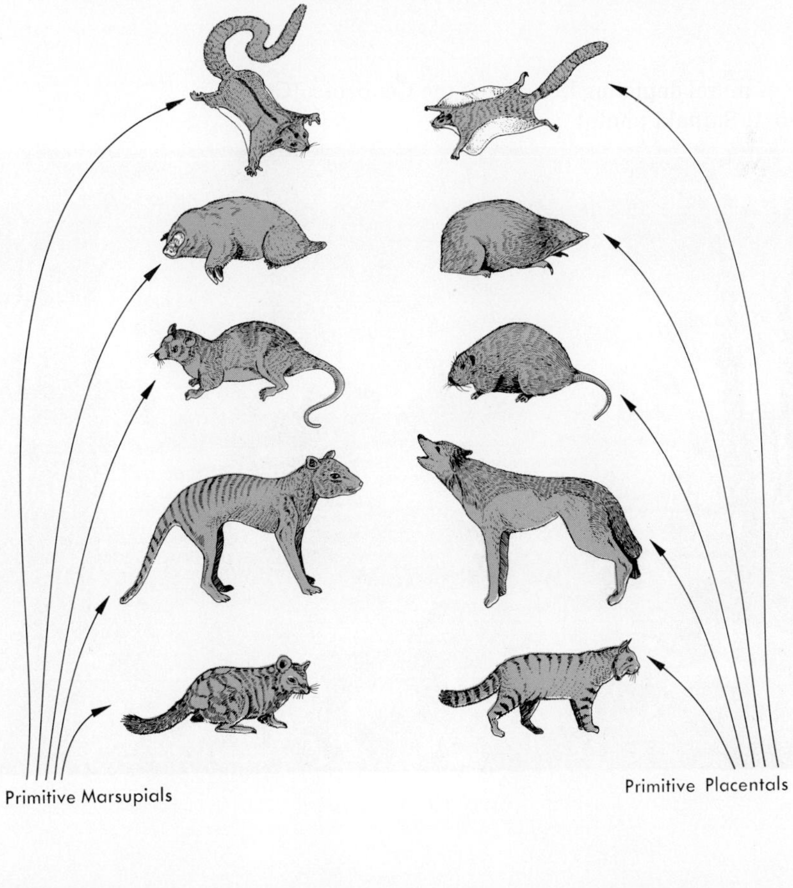

Plants, Herbivores, and Carnivores. The first small mammals apparently fed on insects and other invertebrates. As the mammals increased in size, some of them added other vertebrates to their diet. The rise of the flowering plants in the late Mesozoic made an abundance of plant food available and led to the development from the primitive carnivores of herbivores that browsed on the leaves and tender twigs of the trees and shrubs. The spread of the grasslands during the Cenozoic was followed by a great expansion of the herbivore lines. Some, like the mice and ground squirrels, were seed-eaters. Others developed into grazers, cropping the grass as cattle and sheep do today. And the carnivores followed the herbivores into the grasslands. The selection pressure exerted by predation produced a number of differentiated stocks among the herbivores. The horses became specialized for swift running. The cattle and antelopes developed sharp horns that could be used as defensive weapons. The elephants came to rely mainly on their size for protection. Figure 26–8 shows some of the varied Cenozoic herbivores. But natural selection worked not only on the herbivores but also on the carnivores. Some specialized for the long chase, such as the wolves, others for pouncing and killing such as the cats. Figure 26–9 shows big carnivores attacking a big herbivore. As with the dinosaurs, there was an evolutionary race between offense and defense. The first insectivorous mam-

Figure 26–8. A mural depicting life during the Cenozoic. (Carnegie Museum—Leo T. Sarnaki photo)

Figure 26–9. Sabertooth tigers attack a mammoth mired in a tar pit in what is now Los Angeles, California. Mural by Ottmar F. Von Fuehrer. (Carnegie Museum—Leo T. Sarnaki photo.)

mals were probably largely solitary, but the more highly evolved mammals developed complex behavior patterns that involved cooperation between individuals. The herbivores organized into family groups and herds that allowed mutual defense while at the same time some of the carnivores formed family groups and packs that engaged in cooperative hunting. Such groups are unknown among the reptiles.

Decline of the Mammals. As the Pleistocene waned, many of the large herbivores (see Figure 26–10), and with them the large carnivores that preyed on them, became extinct. Predation pressure by man was probably a major, but not the only, cause. The wave of extinction is still going on. Only in the African game preserves is it possible today to glimpse something of the wealth of the late Cenozoic mammal fauna (see Figure 26–11), and much of what remains may be doomed. The human population explosion requires that more and more land be given over to man's domestic crops and animals, his cities and highways. The relentless pressure on the waning herds continues. This is indeed something new. Never before has a single species been the dominant form of life over the whole earth.

Figure 26–10. Two species of glyptodonts, large, armored South American herbivores that became extinct at the close of the Pleistocene. A pointer dog is shown in the insert to indicate the size of these relatives of the modern armadillos.

MAN'S RELATIVES, THE PRIMATES

It is unfortunate that, of all the mammals living today, the primates have one of the poorest fossil records, but it is easy to understand why. Primates as a group are tree dwellers and they tend to live in tropical regions. There is hardly any place less conducive to fossil formation than a forest floor in the tropics. With relatively high temperatures, a rich source of organic matter, and

Figure 26–11. Zebra, gazelles, and gnu at Ngorongoro Crater, Tanganyika. (Carnegie Museum—M. Graham Netting photo.)

plenty of moisture, the forest floor is a virtual culture of decay bacteria, and the remains of any tree dweller that fall there are soon rotted away with little chance of preservation. Fossiliferous beds bearing a few early primate remains are known from what are today temperate zones, but these beds are mostly early Cenozoic in age, when these regions had essentially tropical conditions. Poor as the record is, it is sufficient, when combined with data on the structure of living forms, to give a fairly adequate idea of primate evolution.

Primate Structure

In all but the most primitive primates the hands and feet are either *prehensile* (adapted for grasping) or derived from a prehensile type. The thumb is typically *opposable*, that is, it is set at an angle to the digits and can be swung around to touch their tips. A well-developed collarbone is present in primates, though absent in many mammals. The digits usually have flattened nails instead of claws; the mammary glands are in the pectoral region; and the cerebrum is enlarged and organized with a degree of complexity found in no other group. Most of these characteristics are adaptations for an arboreal existence. The grasping hands and feet are functional in climbing and the clavicle takes the load off the shoulder muscles when the animal swings by its arms. The senses are also organized for life in the trees. One of them, smell, has proved to be of little value to tree dwellers and with no selection pressure to sustain it, it has deteriorated far below the level found in many terrestrial animals. On the other hand, sight, including the ability to judge the distance to a nearby limb, is of the utmost importance. In primates, not only are the parts of the brain concerned with seeing well developed but the eyes both focus forward so that the fields of vision can overlap, giving rise to stereoscopic vision. In all but the most primitive primates, the bony bar that separates the eye socket from the original synapsid skull opening supports a bony plate that completely closes off the socket from the rear. No movement of the jaw muscles can in any way be transmitted to the eyeball and interfere with vision.

Modern Primates

There are two major groups of living primates: prosimians and anthropoids.

Prosimians. The lemurs, lorises, and tarsiers are the most primitive of the living primates; they are all Old World. The lemurs of Madagascar are mostly small animals, somewhat squirrellike in appearance and habits, and quite arboreal. In many ways they are little advanced beyond the shrewlike ancestors of the primates. The muzzle is long and pointed, the eyes are directed outward

Figure 26–12. A young bushbaby, *Galago senegalensis*. The young male is about five weeks old and has a body length of about ten centimeters (four inches) and a tail length of fifteen centimeters (six inches). (Photographed in South West Africa by E. G. Franz Sauer.)

Figure 26–13. The tarsier, *Tarsius* (Chicago Zoological Society, Brookfield, Ill.— Martin Deutsch photo.)

more than in the higher primates, and the bony eye socket is still incomplete. Although most of the digits have flattened nails, some bear nearly typical claws. Lorises and their allies are more widely distributed in tropical Africa and southeastern Asia and are somewhat more advanced than the lemurs. Figure 26–12 shows a young bushbaby, a member of this group. The tarsiers are restricted to the islands of the East Indies. Small and with ratlike tails, they are both nocturnal and arboreal. The large eyes are directed forward (see Figure 26–13)

Figure 26–14. A New World monkey. (Courtesy of David J. Chivers.)

and each is surrounded by a complete bony socket. The brain is larger than in the lemurs. They are specialised in many ways, but on the whole, the tarsiers seem to be somewhat intermediate between the lemuroids and the New World monkeys, the platyrrhines. Fossils from the lower Cenozoic of North America and Europe indicate that the present-day stock is only a remnant of a much more widespread earlier one.

Anthropoids. The anthropoids are divided into two groups: the New World monkeys, and the Old World monkeys, apes, and men. The New World monkeys (including the marmosets) are called *platyrrhines* (meaning "flat-nosed") because the nostrils are directed forward and are separated by a low, wide septum. All New World monkeys have three premolar teeth on each side of each jaw. The marmosets have flat nails on only the first digits, which are not opposable. The platyrrhine monkeys have nails on all the digits and have an opposable thumb. Most of them also have prehensile tails (see Figure 26–14). This characteristic is restricted to the New World monkeys; some of the Old World forms have tails, but only monkeys from South and Central America can swing by them.

The last group of primates, the one to which man belongs, is discussed in the next chapter.

SUGGESTED READINGS

Carrington, R.: *The Mammals,* Life Science Library, Time Inc., New York, 1963.

Colbert, E. H.: *Evolution of the Vertebrates,* 2nd ed., John Wiley & Sons, Inc., New York, 1969.

———: *Wandering Lands and Animals,* E. P. Dutton & Co., Inc., New York, 1973.

Kummel, B.: *History of the Earth,* W. H. Freeman and Co., San Francisco, 1961.

McAlester, A. L.: *The History of Life,* Prentice-Hall, Inc., Englewood Cliffs, N.J., 1968.

Moore, R.: *Evolution,* Life Science Library, Time Inc., New York, 1962.

Romer, A. S.: *Vertebrate Paleontology,* 3rd ed., University of Chicago Press, Chicago, 1966.

CHAPTER **27**

Man Emerges

Watching the antics of New World spider monkeys in the zoo is a pleasant way to spend a Sunday afternoon. But a person interested in man's closer relatives should turn to the Old World forms. The catarrhines include the African and Asian monkeys, the great apes, and man. They all have the same dental formula: two incisors, one canine, two premolars, and three molars in each row. All have nails on all the digits, and their nostrils are close together and open downward (this is what *catarrhine* means). The tail, be it long, short, or rudimentary, is never prehensile. Catarrhines tend to be larger than the New World platyrrhines; the size increase is most marked in the gorilla and man.

OLD WORLD PRIMATES

There are three main groups of Old World primates: the monkeys and baboons of the family Cercopithecidae (from Greek *kerkos*, meaning "tail" plus *pithecos*, meaning "ape"); the gibbons and great apes of the families Hylobatidae and Pongidae; and man and his close relatives of the family Hominidae.[1]

Cercopithecids

An Old World monkey (see Figure 27–1) can be easily distinguished from a New World monkey by the shape of the nose, the lack of prehensile tail, and

[1] Before going further, the student should read the discussion of classification in the introduction to Appendix C.

Figure 27–1. Spectacled langur and baby, an Old World monkey. (Chicago Zoological Society, Brookfield, Ill.—Martin Schmidt photo.)

the presence of bare calloused areas on the buttocks, which tend to be brightly colored in the males.

The cercopithecids include a group of arboreal forms with smaller thumbs and complicated stomachs for the temporary storage of the plants on which they feed. Members of a second group, which contains the majority of Old World monkeys, have well-developed thumbs, but have rather simple stomachs and special pouches in the cheeks for food storage. This group contains not only typical arboreal monkeys but also the ground-dwelling baboons and mandrills. In the latter, the muzzles are long and doglike and the canine teeth are like tusks, although the tooth formula is the same as that of man. Like man, the baboons and mandrills are truly terrestrial, but they walk about on all fours rather than on their hind legs.

The Manlike Apes

The four kinds of manlike apes are the gibbon of the family Hylobatidae, and the orangutan, chimpanzee, and gorilla, the great apes of the family Pongidae. They all lack external tails and have larger brains than the monkeys. Figure 27–2 shows the skulls of the manlike apes.

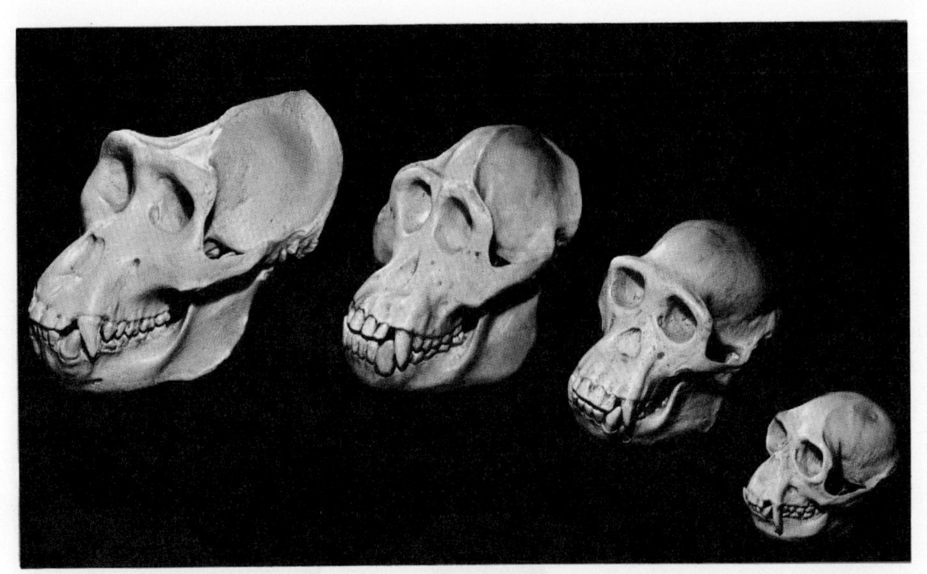

Figure 27–2. The skulls of the manlike apes. Left to right, gibbon, chimpanzee, orangutan, and gorilla. (Courtesy of Ward's Natural Science Establishment, Inc., Rochester, N.Y.)

Gibbon. The gibbon (see Figure 27–3) is the smallest of the manlike apes and is also the most different. Gibbons are the only apes that customarily walk erect but they are seldom seen on the ground. Less than a meter (three feet) in height, and with extraordinarily long arms that reach to the ground when they walk, the gibbons are most adept at *brachiating*, that is, swinging from branch to branch by the arms. With great agility, they hurl themselves about through the trees, making swings of from six to twelve meters (twenty, thirty, or even forty feet). This activity demands precise coordination, accurate stereoscopic vision for judging distance, and excellent judgment of the strength of tree branches. Gibbons have all of these traits, and they can move through the trees at speeds that make it difficult to follow them with a movie camera. There are several living species of gibbons, all inhabitants of southeast Asia.

Orangutan. The "Old Man of the Woods," as the orangutan is sometimes called, leads a different kind of life. It is a bulky animal, only 120 to 150 centimeters (four or five feet) in height but weighing up to 72 kilograms (160 pounds). The arms are long and the body is covered with long, reddish-brown hair. The brain is about a third as large as that of man. Although arboreal, orangutans move slowly and methodically through the trees, testing each branch of doubtful strength before trusting their weight to it. They build simple

Figure 27–3. The Lar Gibbon. (Chicago Zoological Society, Brookfield, Ill.—Sam Kipnis photo.)

nests in which they sleep. Entirely herbivorous, orangutans feed on both fruits and foliage. Like humans, they normally bear only one offspring at a time. They are at present found only in Borneo and Sumatra (see Figure 27–4).

Chimpanzee and Gorilla. The other two great apes, the chimpanzee and gorilla, are the most manlike of all. They are very similar to each other and seem to be closely related. Both are inhabitants of Africa.

The chimpanzee (see Figure 27–5) is more terrestrial than the gibbon or orangutan, although it is still much more arboreal than man. A large male chimpanzee may be 150 centimeters (5 feet) tall and reach a weight of 80 kilograms (175 pounds). Although an excellent climber, it spends much time on the ground, where it runs around on all fours. It is also capable of bipedal walking. Chimpanzees are primarily herbivorous but do occasionally eat animals.

The largest of the manlike apes is the gorilla (see Figure 27–6). Reaching a height of nearly 2 meters (6 feet) and a weight of 270 kilograms (600 pounds), it is pound for pound several times stronger than man. The females and young sleep in trees and climb them in the daytime to pick fruit, but otherwise spend most of their time on the ground. Large males are usually too heavy to climb. The gorilla's more terrestrial habits are reflected in its structural adaptations for

Figure 27–4. The Orangutan. (Chicago Zoological Society, Brookfield, Ill.—Ted Farrington photo.)

Figure 27–5. The Chimpanzee. (Chicago Zoological Society, Brookfield, Ill.—J. Musser Miller photo.)

Figure 27–6. The Gorilla. (Chicago Zoological Society, Brookfield, Ill.—William Vokoun photo.)

terrestrial life. The gorilla's arm and hand are shorter, and the thumb is more opposable than in the other great apes. When they walk, gorillas carry the weight of the body almost entirely on the sole of the foot, whereas the other apes walk more on the side of the foot. Gorillas can stand in a semi-erect position, but they still use their hands for walking. Although the gorilla is larger than man in body bulk, its brain is only about half the size of that of the most primitive living man.

HOMINOID EVOLUTION

Interesting as they are, the gibbon, orangutan, chimpanzee, and gorilla are not man's direct ancestors, but rather sister species, living on earth at the same time. To find man's ancestors we must go back in geologic time and examine the fossil record.

The interpretation we give to this record is by no means the final one. Many more forms have been discovered and named than are mentioned here, but so much of the fossil material is fragmentary that any attempt at this time to trace out the evolutionary history of the manlike apes and man in precise detail would be nothing more than an exercise in futility. What follows seems to be the most likely interpretation of the available evidence.

All the manlike primates assigned to the families Hylobatidae, Pongidae, and Hominidae are called *hominoids*. The latter should not be confused with *hominid*, which means a member of the family Hominidae.

The Early Hominoids

The earliest material that sheds some light on the evolution of the hominoids has been found in fossil beds of the Oligocene age in Egypt. This region was then apparently covered with a dense tropical forest through which meandering rivers flowed. Fossils found in the lower parts of these beds include the left half of the lower jaw of a small primate that has been named *Oligopithecus*. The structure of the teeth suggests that this primate might have been ancestral to the entire hominoid stock. Numerous other fossil jaws of about the same age belong to two other genera of small primates, *Apidium* and *Parapithecus*, which seem to represent the line that gave rise to the Cercopithecidae. It thus appears that the hominoid line had separated from that of the monkeys by the Lower Oligocene.

Somewhat more recent in age than *Oligopithecus*, and also somewhat larger, is *Propliopithecus*, which is known only from a few jaws and teeth. This form may well have given rise to the great apes and man and also to the gibbon. From the Upper Oligocene comes an almost complete skull of an animal named *Aegyptopithecus*. It was about the size of a gibbon, but its teeth were more like those of a gorilla, and it may have represented the stock that was ancestral to both the Pongidae and Hominidae. Another, smaller animal of the same age (*Aeolopithecus*) is more gibbon-like in its teeth and the shape of its jaws, suggesting that the gibbon line had separated from the line leading to the great apes and man by the Upper Oligocene.

Widespread in the Miocene of Africa, Europe, and Asia were a number of species of *Dryopithecus*. Enough material has been found to show that the dryopithecines had apelike skulls and limbs. Like the great apes, they were probably capable of both climbing and of "knuckle-walking," that is, of walking on all fours with the forelimbs contacting the ground on the middle joints of the flexed fingers. Figure 27–7 shows what *Dryopithecus* may have looked like. Various species of *Dryopithecus* are thought to have been ancestral to the chimpanzee and gorilla and also to man.

Toward the Hominids

One of the ways in which man differs from the great apes is that in man the rows of molar teeth are curved rather than parallel to one another. The canine teeth are small and like the incisors in shape, and are not separated from

Figure 27–7. Reconstruction of the Miocene ape *Dryopithecus* (*Proconsul*). Animals like this may have been ancestral to the great apes and also to man.

the incisors by a gap. See Figure 27–8. Pieces of jaws, both upper and lower, from India and Africa indicate that in the Upper Miocene and Lower Pliocene there lived a creature that may have been on the direct line leading to modern man. This hominoid, named *Ramapithecus,* probably descended from *Dryopithecus* and may represent the last stage in the evolutionary sequence leading to the first genus we call man, *Australopithecus,* from the Pliocene and Pleistocene of Africa. Figure 27–9 is a suggested family tree of the hominoids.

The evidence is scanty but it is possible that *Ramapithecus* lived at the forest edge and ate mostly vegetable food, which he gathered in open areas

Figure 27–8. The shape of the dental arch. A: in the gorilla; B: in *Dryopithecus*; C: in modern man.

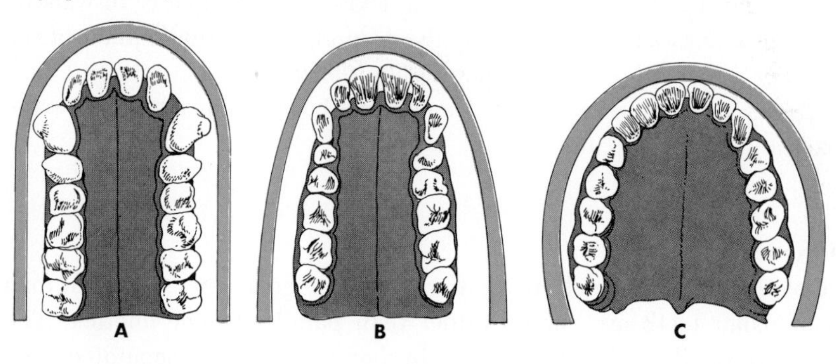

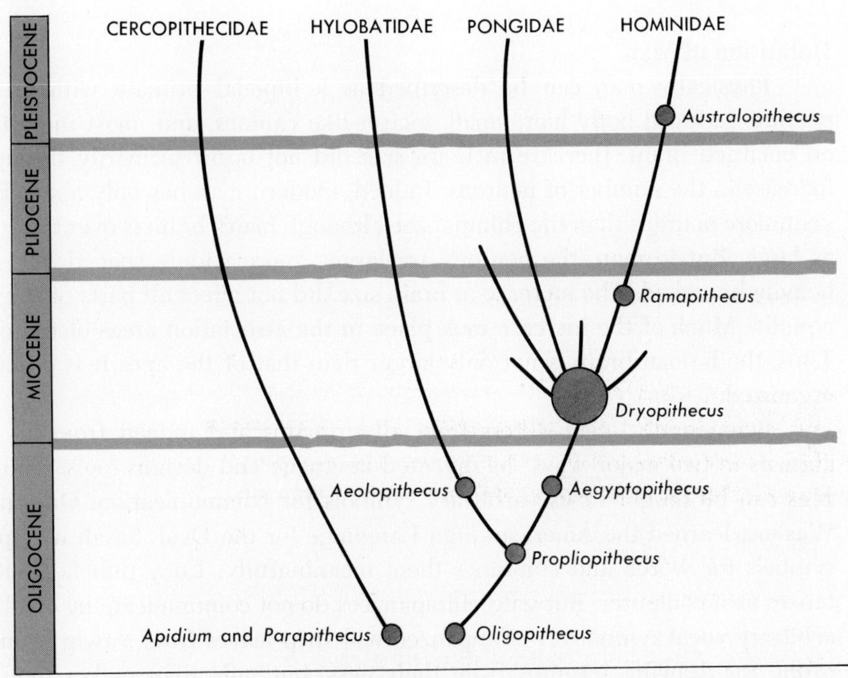

Figure 27–9. The family tree of the hominoids.

around rivers and lakes. When he was not foraging he may have spent his time in the trees, using both his arms and legs in climbing as the orangutan does. On the ground *Ramapithecus* may have been bipedal, although he could not have been as efficient a walker as man. *Dryopithecus* is thought to have eaten fruit, which is soft and does not require much chewing. The food gathered by *Ramapithecus* on the ground would have included seeds and tubers, items that must be well chewed. The change in the dental apparatus was probably correlated with the shift in diet. In particular, the reduction in the size of the canines allowed a sidewise grinding movement of the jaws.

EARLY EVOLUTION OF MAN

Between *Ramapithecus,* who lived perhaps 12 million years ago, and *Australopithecus,* who may go back as much as 5 million years, there is a gap in the fossil record. All we have is a single tantalizing tooth from a deposit that is perhaps 9 million years old. This tooth seems to be intermediate in shape between the molars of *Ramapithecus* and those of *Australopithecus.*

Definition of Man

Physically man can be described as a bipedal primate with an erect posture, reduced body hair, small, incisor-like canines, and, most importantly, an enlarged brain. Increase in brain size did not occur primarily through an increase in the number of neurons. Indeed, modern man has only about 25 per cent more neurons than the chimpanzee although man's brain is over three times as large. But in man, the neurons are larger, more widely spaced, and more heavily branched. The increase in brain size did not affect all parts of the brain equally. Much of the increase took place in the association areas of the cortex. Thus, the human brain is not only larger than that of the apes it is differently organized.

Behaviorally, man differs from all primates and indeed from all other animals in two major ways; he invented language and designs tools. Chimpanzees can be taught to use arbitrary symbols for communication. One, named Washoe, learned the American Sign Language for the Deaf. Sarah uses plastic symbols for words and combines them meaningfully; Lucy punches out sentences on a computer. But wild chimpanzees do not communicate by combining arbitrary vocal symbols. A chimpanzee will strip leaves from a twig to make a probe for drawing termites from their nest, but only man makes tools to a predetermined pattern in advance of his need for them. In a way, language and tool-making are comparable activities. Both impose an arbitrary form on external reality, and both probably evolved concomitantly with thc restructuring of the brain.

Australopithecus, The Tool Maker

The first tool-making man was apparently a member of the genus *Australopithecus*. Remains of these hominids have been found in southeastern and eastern Africa, and they may have occupied other regions as well. *Australopithecus* was a true biped, although still not as good a walker as modern man. Although his lower pelvis was apelike, his ilium had definitely developed to the stage of bipedal man (see Figure 27–10). His brain was somewhat larger than that of an ape of comparable size but less than half the size of the brain of modern man. There is evidence that the restructuring of the brain had begun. The associative areas of the cortex in *Australopithecus* were larger than in the ape brain.

The australopithecines lived from perhaps 5 million to 1.5 million years ago, from the Middle Pliocene to the Basal Pleistocene. There were two main kinds, a small, slender one weighing about 18 to 32 kilograms (40 to 70 pounds) and having small premolars and molars, and a larger one weighing about 36 to

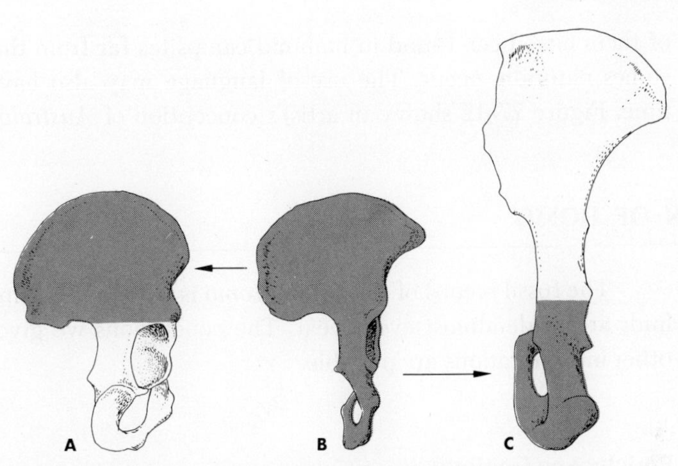

Figure 27–10. Pelvic girdles. A: of modern man; B: of *Australopithecus*; C: of an ape. The upper part of the girdle of *Australopithecus* is more like that of bipedal man, the lower part more like that of an ape.

64 kilograms (80 to 140 pounds), with massive grinding teeth that suggest a primarily vegetarian diet. The larger australopithecine line probably became extinct, but the small species, *Australopithecus africanus,* apparently gave rise to another species, *Australopithecus habilis,* with a somewhat larger brain. (*Australopithecus habilis* is considered by some anthropologists to be the earliest member of the genus *Homo.*)

Australopithecus africanus lived in open savanna areas, probably in small, multimale troops of hunter-gatherers, with the males cooperating in hunting and the females remaining at the campsite to care for the children and gather plant food. Between 2 and 3 million years ago the australopithecines were making tools. The first primitive stone tools are known as pebble-tools, for they were formed by the simplest modifications of large pebbles like those found in streams (see Figure 27–11). There is no doubt that they were real tools; clusters

Figure 27–11. Man's earliest stone tools were crudely chipped pebbles, shown here in side and cutting-edge views.

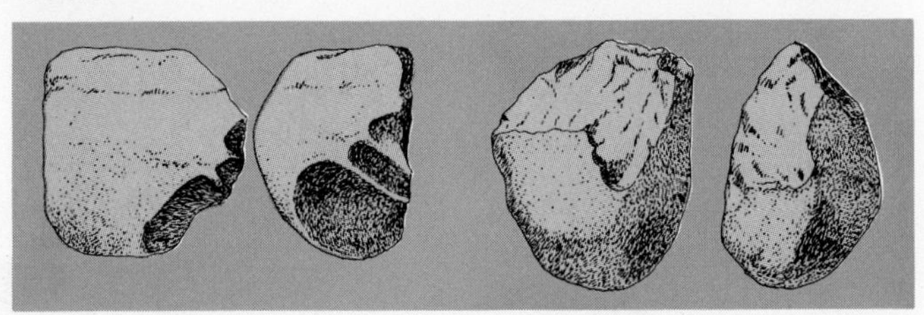

of them have been found in hominid campsites far from the places where such stones naturally occur. The use of language may also have developed at this time. Figure 27–12 shows an artist's conception of *Australopithecus*.

EVOLUTION OF HOMO

The fossil record of the genus *Homo* is still very incomplete, although new finds are made almost every year. The conclusions we give here are tentative; other interpretations are possible.

Pleistocene Geology

The evolution of man is correlated with a series of unusual geological events. The elevation of the continents and cooling of the climate that started in mid-Cenozoic times culminated in the Pleistocene. In the 3 million years since the end of the Pliocene, the world has suffered a series of marked fluctuations in climate. During cold periods, glaciers formed in the higher latitudes and moved south to cover much of Eurasia and North America. The glaciers retreated during the warmer interglacial periods. There were minor fluctuations in climate as well and there is disagreement among Pleistocene geologists as to how many

Figure 27–12. Artist's conception of *Australopithecus.*

major glacial ages should be recognized. (There may have been as few as four or as many as seven.) Glaciers are still present in Greenland and Antarctica and in many mountain regions, although we are witnessing the retreat of the latest of the extensive continental glaciers. Whether it was the last, whether we are entering an interglacial period, or whether this is a minor fluctuation is still unknown.

While the glaciers advanced and withdrew the climate of Africa fluctuated between dry and more humid conditions, but whether and how the fluctuations in Africa were correlated with glacial and interglacial periods has not been determined.

The Pleistocene began about 3 million years ago. Climatic fluctuations occurred during the Basal Pleistocene (Villafranchian), but glaciation was limited to mountain regions. The first extensive continental glaciation marked the beginning of the Lower Pleistocene, but the time when this occurred has not yet been determined. The Lower Pleistocene comprises at least four cold and three warm periods. It ended about 300,000 years ago. The Middle Pleistocene, with two glacial and two interglacial episodes, lasted until about 125,000 years ago and the Upper Pleistocene, with one interglacial and one glacial, lasted until 10,000 years ago. These divisions of the Pleistocene are based on marked changes in the distribution of plants and animals, with the extinction of many species. Table 27–1 shows the divisions of the Pleistocene.

Fossil Men

During the breakup of the continents, Africa retained a connection with Eurasia, as it does today. *Australopithecus* evolved in Africa, but by the Basal Pleistocene had spread into Asia and had reached Java in the East Indies. There were probably a number of semi-isolated stocks in Africa and Asia, subspecies really, that were subjected to similar selection pressures and evolved in parallel. The various populations differed somewhat from one another, but were all basically similar. They became larger, with better developed brains, smaller faces, and more erect postures. They had heavy brow ridges, no chins, and their skulls were very thick, thicker than those of either *Australopithecus* or modern man, *Homo sapiens*. Their skulls were broad and rather flat, not narrow and high-domed as in modern man. They are called collectively *Homo erectus*.

During the Lower Pleistocene, various races of *Homo erectus* were widely spread in Europe, Asia, and Africa. A variety of names have been given to the fossils representing these races. Thus, the first to be found, in Java, was named *Pithecanthropus erectus*, specimens from China were called *Sinanthropus pekinensis*, and some from Africa were called *Atlanthropus mauritanicus*. These were probably all representatives of a single widespread species. Their tools

Table 27–1. The Divisions of the Pleistocene*

Age	Climatic Conditions	Time from Beginning in Years
Postglacial (Recent)	Warm	10,000
Upper Pleistocene	Cold	75,000
	Warm	125,000
Middle Pleistocene	Cold	170,000
	Warm	225,000 (?)
	Cold	265,000 (?)
	Warm	300,000 (?)
Lower Pleistocene	Cold	340,000 (?)
	Warm	380,000 (?)
	Cold	430,000 (?)
	Warm	(?)
	Cold	(?)
	Warm	(?)
	Cold	(?)
Basal Pleistocene (Villafranchian)	Warm	(?)
	Cold	3,000,000

*Data from K. W. Butzner, *Environment and Archeology*, Aldine-Atherton, Chicago, 1971.

were more sophisticated than those of *Australopithecus,* they hunted larger game, and in China and Europe, at least, they had discovered the use of fire. The first *Homo erectus* may have lived 3 million years ago and the species survived until perhaps half a million years ago.

The trend toward increase in brain size continued during the Pleistocene; when brain size reached that of modern man the fossils are classified as *Homo sapiens.* The development of *Homo sapiens* from *Homo erectus* probably involved both the parallel evolution of the various populations and migratory movements that brought the groups into contact with one another and allowed a certain amount of gene flow between populations. These early representatives of *Homo sapiens* are referred to collectively as *Archaic man*. They were present in Europe during the Middle Pleistocene and evolved there into the best known of the fossil men, *Neandertal man*, who lived during the first part of the last glaciation. He was stockily built, with a long, low skull, heavy brow ridges, and a receding chin (see Figure 27–13). A skilled tool-maker and hunter, he was able to kill even the huge cave bear, which was as big as a grizzly bear. He buried his dead. Populations of Archaic man, which were similar in many ways to Neandertal man, have been also found in Africa and Asia.

Figure 27–13. Skeletons. A: of Sapiens Man; B: of Erectus Man. It used to be thought that the latter walked in a crouched position, as shown, but it is now believed that Erectus was quite as erect as Sapiens.

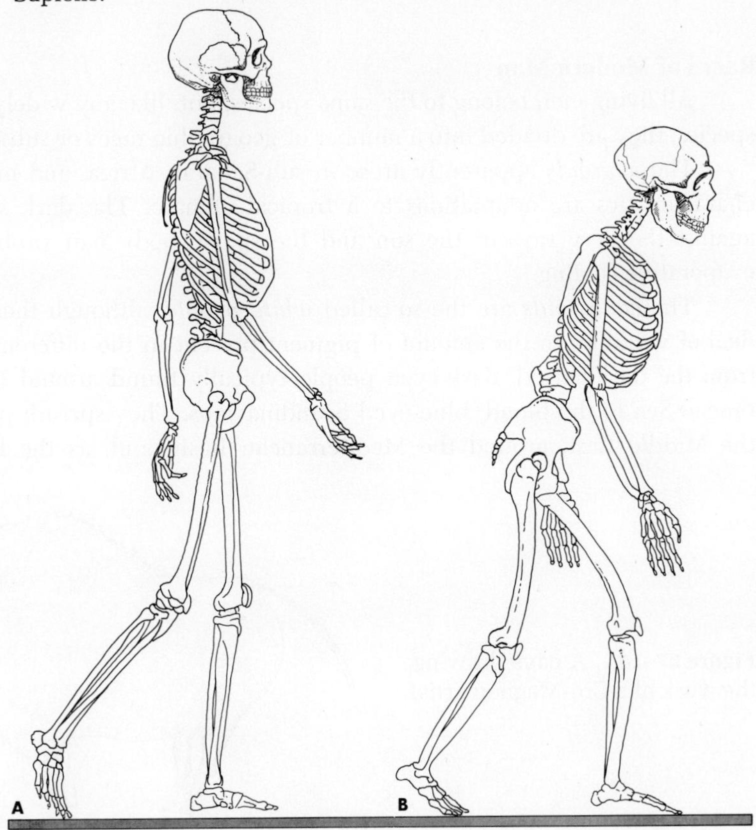

Between about 100,000 and 50,000 years ago, Archaic man began to be replaced by Modern man, who differs from his predecessor mainly in the shape of his skull. It is high-domed and thin-walled, with reduced brow ridges and a well-developed chin. In Europe, the first representatives of Modern man are known as *Cro-Magnon man,* the skilled artist who left such vivid impressions of late Pleistocene animals on the walls of caves (see Figure 27–14). Just where Cro-Magnon man came from is not known—perhaps from eastern Africa, or Arabia, or western Asia. He did not evolve from Neandertal man of western Europe.

Once Modern man had appeared, he spread rapidly through Europe, Africa, and Asia. Some time toward the end of the last Ice Age, wandering bands of hunters began making their way from Asia to North America, probably by a land bridge across what is now the Bering Strait. Although much of the northern part of the continent was still covered by glaciers, there were ice-free corridors that gave access to the rich game lands to the south, and these first Americans spread rapidly.

Races of Modern Man

All living men belong to the same species, but, like any widely distributed species, they are divided into a number of geographic races or subspecies.

The *negroids* apparently arose in sub-Saharan Africa, and many of their characteristics are adaptations to a tropical climate. The dark skin protects against the fiery rays of the sun and the sparse body hair probably speeds evaporative cooling.

The *caucasoids* are the so-called *white people,* although there is a great deal of variation in the amount of pigment present in the different subgroups, from the dark-haired, dark-eyed people typically found around the Mediterranean Sea to the blond, blue-eyed Scandinavians. They spread, perhaps from the Middle East, around the Mediterranean Basin and, as the latest of the

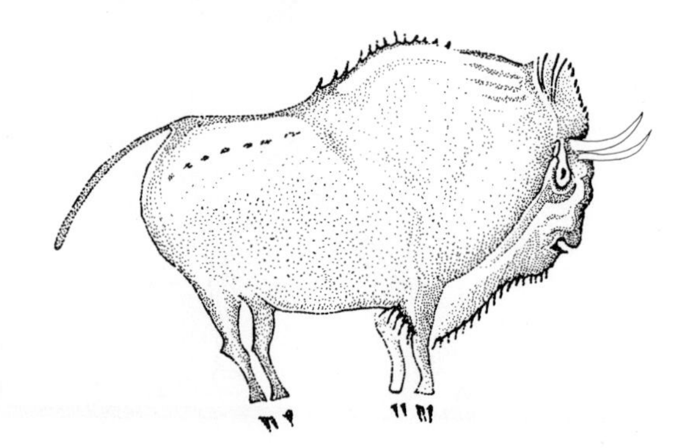

Figure 27–14. A cave drawing, the work of a Cro-Magnon artist.

glaciers retreated, northward through Europe. (Cro-Magnon man was caucasoid.) The tendency of the more northern groups to lose pigmentation is probably correlated with the reduced amount of sunlight available in northern latitudes.

Meanwhile the *mongoloids* were evolving in northeastern Asia. It has been suggested that the original mongoloids may have been like the American Indians, in whom the "typical mongoloid face" is not highly developed. The most exaggerated form of this face is found in the Eskimo and people of eastern Siberia, and seems to be an adaptation to the extremely cold climate in which they live. The eyes are protected both from the cold and from the glare of the sun reflected from the snow by folds of skin at the inner corners and by pads of fat. The small nose and flat face present little surface for freezing. The beard is sparse, reducing the danger that water vapor in the breath will condense on it and freeze. The American Indians probably branched off from the mongoloids before this face was fully evolved.

There are many other racial groups, such as the Asiatic Indians; the Australoids; and the Polynesians of Hawaii, New Zealand, and the Pacific Islands. Each was originally adapted by natural selection to the part of the globe in which it lived. The extreme mobility of modern man has been too recently achieved to have yet wiped out the racial differences that are simply a reflection of the evolutionary process.

SUGGESTED READINGS

Buettner-Janusch, J.: *Origins of Man,* John Wiley & Sons, Inc., New York, 1966.
Butzner, K. W.: *Environment and Archeology,* Aldine-Atherton, Chicago, 1971.
Dobzhansky, T.: *Mankind Evolving,* Yale University Press, New Haven, 1962.
Editors of Life: *The Epic of Man,* Time Inc., New York, 1961.
Eimerl, S., and I. DeVore: *The Primates,* Life Science Library, Time Inc., New York, 1965.
Howell, F. Clark: *Early Man,* Life Science Library, Time Inc., New York, 1965.
Jolly, A.: *The Evolution of Primate Behavior,* Macmillan Publishing Co., Inc., New York, 1972.
Pilbeam, D.: *The Ascent of Man,* Macmillan Publishing Co., Inc., New York, 1972.
Readings from *Scientific American:* "Human Variations and Origins," W. H. Freeman and Co., San Francisco, 1967.
Simons, E. L.: "The Earliest Apes," *Sci. Amer.,* Vol. 217, No. 6, 1967.
Weckler, J. E.: "Neanderthal Man," *Sci. Amer.,* Vol. 197, No. 6, 1957.

The World Today

No man is an island. . . .
JOHN DONNE, *Devotions, xvii*

CHAPTER **28**

The Physical Environment

No organism is sufficient unto itself. From the most minute bacterium to the largest tree or whale, everything on earth lives in and is dependent on an external environment. This environment includes all the external physical and biological factors that affect the organism. The relationship between an organism and its environment is mutual, for just as the environment affects the organism, so the organism, in turn, produces changes in the environment. Remember how the presence of green plants must have changed the composition of the primitive atmosphere. The study of the relationships between organisms and their environments is known as *ecology*.

POPULATIONS, COMMUNITIES, AND ECOSYSTEMS

The individual members of a species found in the same general area form a *population* (see Figure 28–1). How inclusive a population is depends on the limits of the area being studied. It is possible to discuss the human population of a single neighborhood, New York City, the United States, or the world. In practice, though, the more inclusive the population the more difficult it is to study it in depth. The populations studied by ecologists usually represent only a small part of a species. The black bass found in a single lake, the goldenrod plants in an old field, and the fungi of a given species on a rotting log are all populations in the ecological sense.

All the populations of plants, animals, fungi, and monerans in an area form a *community*. The populations in a community fall into three broad classes:

Figure 28–1. Part of a population of Elephant Seals on Guadalupe Island, Mexico. (Carnegie Museum—C. J. McCoy photo.)

producers, consumers, and decomposers. The producers are the autotrophs, chiefly the green plants that convert the energy of sunlight into chemical energy and build simple substances into biocompounds. The consumers include the animals that eat the plants (herbivores) and the ones that eat other animals (carnivores). The decomposers are the bacteria and fungi that break down the remains of plants and animals into simple substances again (see Figure 28–2).

The community plus the nonliving environment in which it lives form an *ecosystem*. This is the basic unit of study of ecology. As with a population, an ecosystem may be more or less inclusive. A rotting log in the forest floor can be studied as an ecosystem, or the whole forest can be studied. An ecosystem may be more or less sharply set off from surrounding ecosystems, as a lake is set off from the surrounding dry land, or one ecosystem may grade gradually into another.

The nonliving factors in a given ecosystem determine very largely the kinds of organisms found there.

Figure 28–2. A decomposer, a coral fungus growing on a fallen log in Carnegie Museum's Powdermill Nature Reserve. (Carnegie Museum—M. Graham Netting photo.)

WATER, AIR, AND SUBSTRATE

The physical environment of living organisms can be broadly divided into water, air, and substrate.

Water

Some water must be available in any area that supports life.

Aquatic Environments. Life probably originated in the ocean, and many organisms still live in water. As a medium, water differs in several important respects from air. For one thing, it does not transmit light so readily. In shallow waters, the producers may include rooted aquatic plants, such as water lilies and pickerel weed and algae with holdfasts, but in deep water the producers are floating plants. Microscopic algae are the major producers of the oceans. They are limited in their downward distribution by the penetration of light. Where there is not enough light for photosynthesis, plant life is absent. In the depths of the oceans and deepwater lakes, animal life is dependent on the detritus that drifts down from above.

Oxygen from the atmosphere dissolves in water, and oxygen is added to water by the photosynthetic activity of plants, but it is not so abundant in water as it is in air. When there is a large accumulation of organic detritus, or when a body of water is heavily polluted with organic wastes, the number of decomposers may increase to the point where they remove so much oxygen by their respiratory activities that fishes and other aquatic animals die through oxygen starvation.

Water and Terrestrial Environments. Water is constantly being evaporated from the surface of the oceans and other bodies of water, and it is also passed into the atmosphere by the transpiration of plants. This atmospheric water is precipitated and returned to earth, usually as rain or snow. Some of the water that falls on the ground percolates down through minute pores between the particles of soil. At a greater or lesser depth below the surface is a zone permanently saturated with water called *ground water.* The upper limit of the ground water is the water table. The amount of rainfall that a given area receives, its distribution throughout the year, and the height of the water table determine in large part whether the ecosystem will be forest, grassland, or desert.

Air

Air, the medium for most terrestrial organisms, is a harsher environment than water. It is subject to more extreme fluctuations in temperature, it provides less support for the body, and it draws water from living bodies. We have already discussed many of the adaptations that allow plants and animals to live in this medium. These adaptations include supporting tissues, protective coverings, internal respiratory surfaces, reproductive mechanisms to allow fertilization away from water, and protection for the developing embryo.

The percentage of the various gases in the atmosphere is usually relatively constant under natural conditions. It can be so altered by the burning of coal, oil, and gas to provide heat and transportation for man, and by industrial air pollution, that some forms of life may be unable to survive in the vicinity of cities. Smog from Los Angeles is apparently killing pine trees in the San Bernardino National Forest ninety-six kilometers (sixty miles) away.

Substrate

Ecologically, substrates are the surfaces in the environment on which organisms live or move or the solid substances through which they burrow. Soil, the most common substrate on land, is formed by the weathering of rock, which is broken down into finer and finer particles. Soil contains more or less organic detritus from plants and animals. Soil particles are classified according to size as sand, silt, or clay. Most soils contain a mixture of particles. Coarse-textured sandy soils allow water and air to move down through them more easily than the fine textured clay soils. Sandy soils are better aerated and easier to burrow through and consequently support a larger population of burrowing animals than the denser clay soils. On the other hand, the rapid movement of water through coarse soils increases *leaching*, the washing away of organic materials and chemical compounds. Such soils are likely to be dry and impoverished. Rock that is only slightly weathered provides a very barren substrate on which only lichens may be able to find a foothold.

VARIABLE FACTORS IN THE ENVIRONMENT

The physical factors in the environment may vary extensively. Temperature, light, humidity, and air currents change not only from time to time but also from place to place. An organism has a range of tolerance for certain factors in the environment. It is able to live in a given environment if the fluctuations of these factors do not exceed the limits it is able to tolerate. The range of tolerance to a given environmental factor may be wide for one organism and narrow for another. The crocodile has a narrow range of tolerance for temperature. It is unable to live in areas much beyond the tropical parts of the world because it cannot survive cold weather. The Norway rat has a much wider range of thermal tolerance and, like man, can be found in most parts of the globe.

An animal's range of tolerance to variation of a given physical factor is an inherited adaptation to the conditions it is likely to encounter in its natural environment. Within this range it is able to adjust to changes by short-term, reversible responses, such as the increase in the rate of sweat production as a man moves from an air-conditioned room to a hot summer street. An animal may also be able to shift the range of variation to which it is able to adjust by a slower process called *acclimatization*. A goldfish that has been living in water at a temperature of 4.5°C (about 40°F) dies if the temperature is suddenly raised to 30°C (86°F). One that has been living for a while at 25°C (77°F) can withstand temperatures as high as 37.2°C (99°F) but no lower than 6°C (43°F). Figure 28–3 shows the ranges of thermal tolerance of the goldfish. Acclimatization is advantageous for animals that live in regions where there are marked seasonal fluctuations in climatic factors.

An organism must be adapted to its physical environment in all respects. For the sake of brevity, the following discussion is limited to only two basic types of adaptations in animals, those concerned with temperature and those concerned with maintaining osmotic balance.

Temperature

Most animals are restricted to temperatures within a range of −1°C (30°F) to 50°C (122°F). Some protozoans and insects live in hot springs where

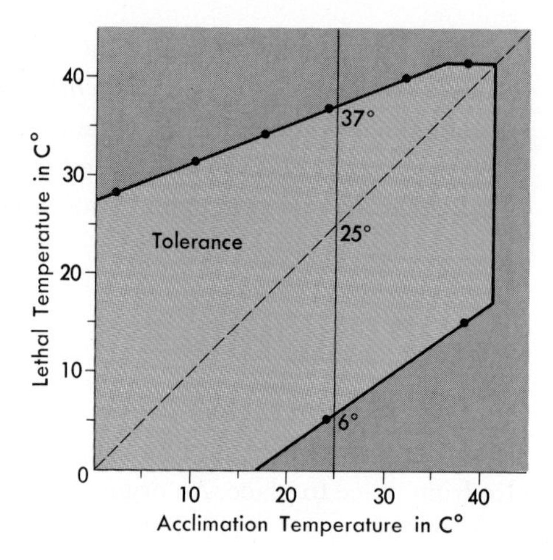

Figure 28–3. Thermal acclimation in the goldfish. If a fish is kept for a time at a temperature indicated on the dashed line, the high and low temperatures that will be fatal to it can be read on the upper and lower solid lines. Thus a fish kept at 25°C will die if the temperature is suddenly raised to 37° or dropped to 6°. (From F. E. J. Fry et al., Revue Canadienne de Biologie, 1, 1942.)

the temperatures range around 50°C, and a number of Arctic fishes live in water that is just a shade above freezing.

Ectothermic Animals. As long as an animal is carrying on metabolic activity, it is producing heat. For most animals, the rate of heat production is so low and the rate at which heat is dissipated to the environment is so great that their body temperatures are determined largely by heat acquired from the environment. Such animals are said to be *ectothermic*. Many ectothermic animals exert behavioral control over their internal temperature. A lizard moves to the shade or sun to adjust its body temperature according to its needs. A lizard that lives in the high mountains of Peru spends the night in a burrow where it is protected against freezing. In the morning it crawls slowly out, climbs a tuft of grass, and turns its back to the sun. By absorbing solar radiation it can raise its internal temperature as high as 30°C (86°F) although the air temperature may be at freezing.

Endothermic Animals. Mammals and birds differ from the majority of animals in having metabolic rates that are high enough so that their internal body heat is produced by their own oxidative activities. They are said to be *endothermic*. They have mechanisms by which they are able to maintain their internal temperature at a relatively constant level in spite of fluctuations in the external environment. In man, and apparently in the other endothermic animals as well, there is a center in the hypothalamus that exerts a precise control over body temperature. (Premature infants are placed in incubators in which the temperature can be carefully controlled because the thermal control center in the hypothalamus does not become completely functional until just prior to the normal time of birth.)

In a room in which the temperature ranges between about 26°C (80°F) and 31°C (88°F), a naked man at rest would feel tolerably comfortable. This range is known as *thermal neutrality*. In such a room, the appropriate body temperature of about 37.2°C (99°F) is maintained simply by the control of blood flow through the skin. Chapter 2 contains a discussion of the various mechanisms by which man controls the rate at which heat is dissipated from the body surface when the temperature of the environment varies from thermal neutrality. Modern man has changed his range of thermal neutrality by wearing clothes to retain body heat.

The different races of man show adaptations to the different temperatures of the regions in which they evolved. The negroids, who adapted initially to tropical regions, in general have a rangy build in which the surface to volume ratio is high so that body heat can be quickly dispersed. The mongoloids evolved in a cold climate, and their short, stocky bodies and rounded faces present a minimal amount of surface for the dispersion of heat. Figure 28–4 shows the difference in body build of a Nilotic Negro and an Eskimo.

All mammals are not adapted for temperature control in exactly the same way that man is. The Eskimo husky, like other Arctic land mammals, has a thick coat of fur to insulate it from extreme frigid conditions. Indeed, it is so well insulated that its metabolism does not increase to produce heat for warming the

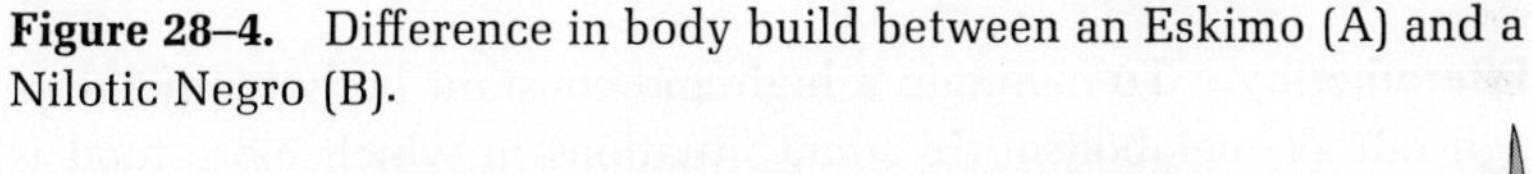

Figure 28–4. Difference in body build between an Eskimo (A) and a Nilotic Negro (B).

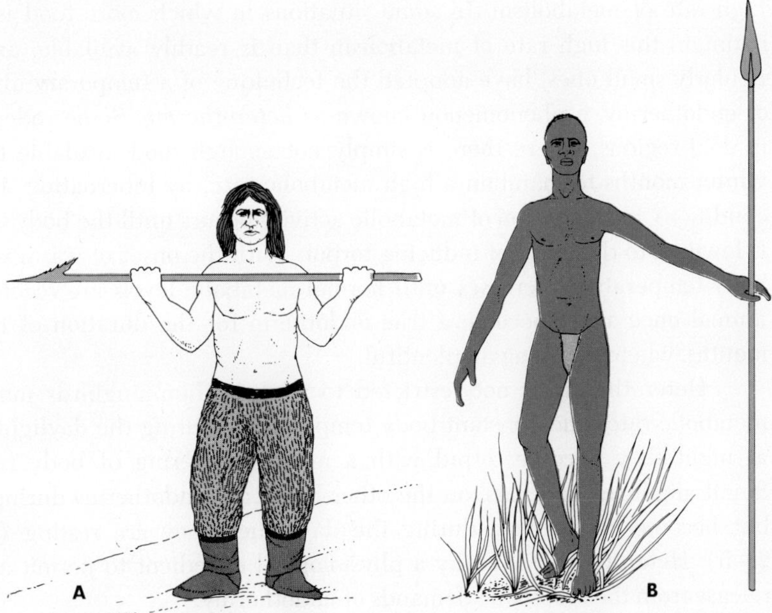

body until the external temperature drops to about − 29°C (− 22°F). Marine mammals, such as the seals, whales, and walruses, are also able to tolerate low temperatures. Hair is of little use as insulation when it is wet; these animals have thick layers of dermal fat (blubber) for insulation. The whales, which always stay in the water, lack body hair, but seals and walruses, which come out to rest on surface ice, have both blubber and hair. The hair prevents the skin from coming into direct contact with the ice so that it is not damaged by the freezing temperature.

For temperatures above the range of thermal neutrality, some mammals induce evaporation in other ways than by sweating. One method of providing surface moisture for evaporative cooling is practiced by the kangaroos and to a lesser extent by rabbits, cats, and cows. They moisten their body hair by licking it with their tongues, but it seems likely that this is not as efficient a method as either sweating or panting. An overheated dog pants with his mouth open. As inhaled air moves through the narrow passages of the nasal chambers on its way to the lungs, it absorbs heat from the walls of the chambers so that when it reaches the lungs its temperature is about the same as the internal body temperature. If the air is exhaled through the nose, heat passes back from the air to the walls of the chambers so there is little net heat loss. But if the air is exhaled through the wider passages of the mouth, most of the air does not come in contact with membranes lining the passage, little heat is resorbed and there is a marked increase in the amount of heat removed from the body.

Heterothermy. To maintain a high and constant body temperature requires a high rate of metabolism. In some situations in which more food is needed to maintain this high rate of metabolism than is readily available, animals, particularly small ones, have adopted the technique of a temporary abandonment of endothermy, a phenomenon known as *heterothermy.* Some rodents can live in cold regions, where there is simply not enough food available through the winter months to maintain a high metabolic rate, by hibernating. Under these conditions, a diminution of metabolic activity occurs until the body temperature is lowered to the point of inducing torpor. With the onset of warm weather, the body temperature increases until normal metabolic levels are reached and the animal once again become a true endotherm for the duration of the summer months, when food is again plentiful.

Heterothermy is not restricted to winter. Hummingbirds maintain high metabolic rates and constant body temperatures during the daylight hours, but at night they become torpid with a marked lowering of body temperature. Small, night-feeding bats, on the other hand, are endotherms during the night, but become ectothermic during the day when they are resting (see Figure 28–5). Heterothermy is really a physiological expedient to permit a temporary release from the metabolic demands of endothermy.

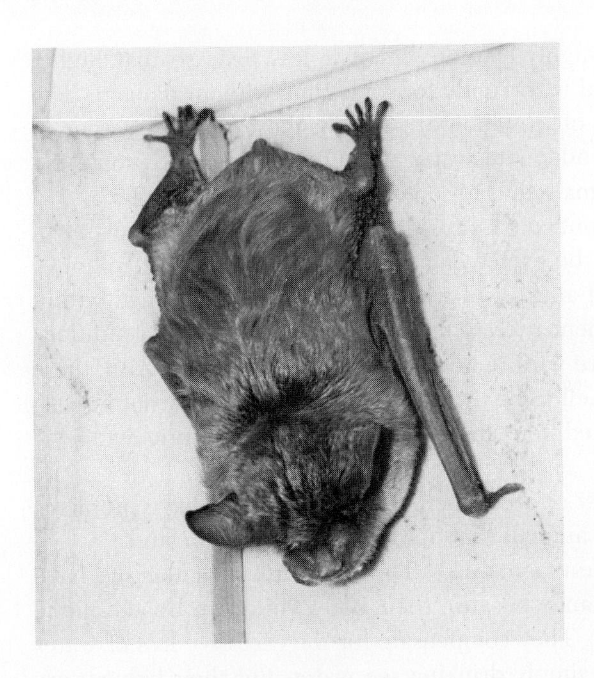

Figure 28–5. A bat, *Eptesicus*, a heterothermic mammal. When it is active at night its temperature remains high. During the day it rests quietly and its body temperature drops. (Carnegie Museum—M. Graham Netting photo.)

Water Balance

Animals must be able to control the water balance in the body in relation to the medium in which they live. Remember that osmosis is the movement of water through a membrane from a region of a relatively low concentration of salts or other solutes into a region of a higher concentration of solutes. Because all animals must have at least part of the body lined or covered by membranes through which food and oxygen can enter, they all have a gateway through which water can enter or leave the body. This is of major importance to animals that live in media that are either hyper- or hypo-osmotic to their own body fluids and to animals that have only salt water to drink. The maintenance of osmotic equilibrium involves not only salt balance but also nitrogenous wastes, for these usually must be dissolved in water to be removed from the body.

Nitrogenous Wastes. Animals can convert nitrogenous wastes into three different forms for elimination: ammonia, urea, and uric acid.

Ammonia is highly soluble and very toxic. Animals that excrete their wastes mainly in this form are usually small aquatic forms—small, so that they have a large surface relative to their bulk to enable the ammonia to pass rapidly out of the body, and aquatic, so that there is ample water available to dissolve the ammonia and carry it away. Aquatic invertebrates and larval amphibians excrete ammonia. Fishes also excrete large amounts of ammonia, mostly through their gills.

Urea, like ammonia, is highly soluble, but it is less toxic so that higher concentrations of it can be held in the body for some time without damage. The animals that use this method of excretion are mainly those that are larger in size and have a more or less adequate water supply in which to dissolve the urea. Adult amphibians, mammals, and to some extent, fishes, excrete urea.

Finally, uric acid is a nontoxic and essentially nonsoluble form of nitrogenous waste. Because it can be excreted in an almost solid state, it requires little water for its removal. All animals in which the young are confined within an egg shell during development excrete uric acid, for under these conditions the wastes, confined as they are with the embryo, must be nontoxic. Also, there is no spare water within the shell to be used as a solvent. The terrestrial insects, reptiles, and birds all lay shelled, terrestrial eggs, and all excrete uric acid.

Salt. Animals that spend most or all of their time in salt water have problems that are different from those of animals that lvie in fresh water or on land.

WATER BALANCE IN MARINE ANIMALS. The bony fishes in the sea have special problems of water balance because their body fluids are hypo-osmotic to the water in which they live. They compensate for the excessive loss of water to the sea through osmosis by simply drinking sea water. But their kidneys are not adapted for the removal of the excess salt they gain in drinking. The nephric tubules are short and straight and the glomeruli are reduced (absent in some species). Not much water filters through the kidneys, and not much urine is produced. This conserves water, but leaves the animal with an excess of salt. This salt is removed from the body by special secretory cells in the gills that actively excrete salt. The sharks and rays lack the salt removal cells of the bony fishes and are able to tolerate life in the sea by their ability to retain enough urea and certain other compounds in the blood to bring the body fluids up to the same osmotic concentration as sea water.

Marine turtles have kidneys that are unable to excrete the large amounts of salt taken in when they eat and drink. They have specialized salt-excreting glands close to the eyes, with ducts that empty at the corners of the eyes. Drippings from these ducts are seen as tears when the female turtles come ashore to lay their eggs. As the Gryphon said to Alice when she asked about the Mock Turtle's tears: "It's all his fancy, that: he hasn't got no sorrow, you know." (See Figure 28–6.)

Marine birds that feed far from land have problems similar to those of the sea turtles. They have special salt-secreting glands that empty into the nasal cavity. The salt glands work only when there is an excess of salt, rather than constantly, as do the kidneys. When the glands are functioning, the salt solution drips from the tip of the beak, giving the bird a runny nose.

Marine mammals such as the seals have solved their water problems

Figure 28–6. The Mock Turtle's tears serve to eliminate excess salt from the body.

differently than have the turtles and the birds. They also lose water to the sea because their body fluids are hypo-osmotic to the water in which they live, and unlike the turtles and birds, they excrete urea and thus need extra water to eliminate their nitrogenous wastes. They have no glands for salt excretion. They get their water not by drinking but from the bodies of the fishes on which they feed. Because the osmotic concentration in the bodies of the fishes is essentially the same as in the body of the seal, the problem is not to get rid of excess salt but to conserve water. A fish diet is rich in protein and produces large amounts of nitrogenous wastes. Seals have evolved kidneys that are capable of forming urine that is so concentrated that the animal can gain as much water from the bodies of the fishes as it loses through respiration and urination. Female seals, being mammals, have an additional problem of water loss through the formation of milk for the young. Their mammary glands produce milk with about ten times the concentration of butterfat that cow's milk has. This rich milk is presumably formed simply as a water conservation measure, for the baby seals seem to grow no faster on it than when they are fed cow's milk.

WATER BALANCE IN FRESHWATER ANIMALS. Animals inhabiting fresh water, like the freshwater fishes and amphibians, have the reverse problem of that of the marine forms. Their body tissues are hyperosmotic to the water in which they live. They tend to gain water from the environment and lose salt to it. They have evolved kidneys that produce large amounts of a very dilute urine for the removal of water from the body. To keep the body salt at its proper level, they depend on getting some salt from their food and a small amount from the fresh water in which they live. Freshwater fishes have special glands associated with the gills that actively absorb salt, and the skin surface itself absorbs salt in the amphibians. Some fishes that migrate between fresh and salt water, such as the salmon, are apparently able to reverse the direction of salt transport by the gills. One amphibian, the crab-eating frog of the Old World, is able to live in tide-water pools and thus must conserve water, rather than get rid of it. It does so in the same way the sharks do, by concentrating the urea in the blood. Figure 28–7 shows the osmotic concentration of the body fluids of various groups of animals as compared to sea water.

WATER BALANCE IN TERRESTRIAL ANIMALS. Terrestrial animals are also hyperosmotic to their environment. They maintain water balance by taking in as much water as they lose during the average day. A terrestrial mammal loses water in four ways and gains it in three. Over a period of time, these gains and losses must balance each other out. Table 28–1 shows how water is gained and lost by a terrestrial mammal.

Remember that when foods are oxidized, water is released, e.g.,

$$C_6H_{12}O_6 + 6\,O_2 \rightarrow 6\,CO_2 + 6\,H_2O.$$

Figure 28–7. The osmotic relationships between various kinds of vertebrates and the environment. The body fluids of the elasmobranchs and the crab-eating frog are isosmotic to the sea water in which they live. The body fluids of other vertebrates are hyperosmotic to fresh water but hypo-osmotic to sea water. (Redrawn from Knut Schmidt-Nielsen, *Animal Physiology*, 2nd ed., © 1964. By permission Prentice-Hall, Inc., Englewood Cliffs, New Jersey)

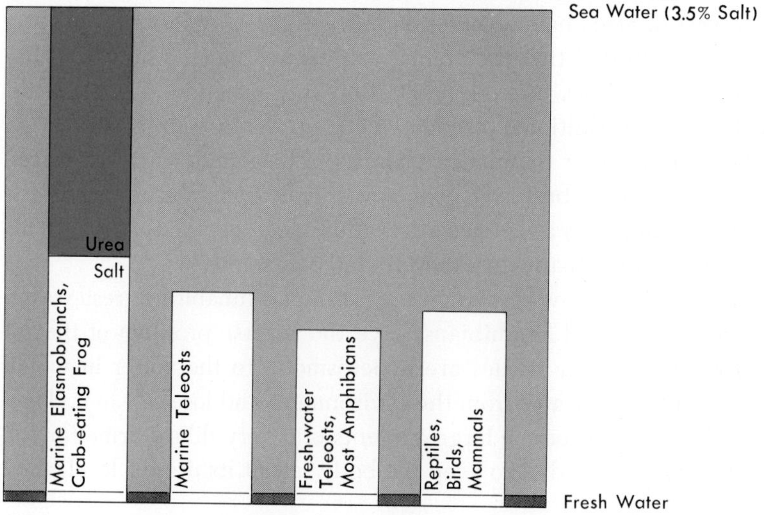

Table 28–1. Water Gain and Water Loss by a Terrestrial Mammal

Gain	Loss
A. Water of oxidation	A. Evaporation from lungs
B. Water in food	B. Evaporation from skin
C. Drinking water	C. Water in feces
	D. Water in urine

This is called the *water of oxidation.* If you ingest 350 grams (a gram equals 0.035 ounces) of carbohydrates, 100 grams of fat, and 100 grams of protein a day you will gain about 340 grams of water of oxidation. In addition, all foods, even the driest crackers, contain more or less water. Lettuce is 95 per cent water, meat about 60 per cent, crackers 5 per cent. Men vary in their eating habits so that it is impossible to give an average figure for the intake of water with food, but even the driest diet is an important source of water. A center in the hypothalamus that controls drinking apparently regulates the osmotic concentration of body fluids rather than the actual amount of water in the body, for although the stimulus to drink is increased by the depletion of water in the body, it is stimulated just as readily by an addition of salt.

Loss of water from the body involves loss through defecation, urination, evaporation from the lungs, and evaporation from the skin. Probably the average human adult loses about 100 grams of water a day in the feces, although this amount varies with the diet and with the general condition of the body. The water loss in urine may run as high as 2,400 grams a day. The amount lost is under the control of the antidiuretic hormone of the pituitary, which determines the amount of water in the body by controlling the amount reabsorbed by the kidney tubules (see Chapter 7). The water lost by the lungs also varies. An average man living at ordinary room temperature and at 50 per cent humidity loses about 300 grams of water a day by this route. Finally, man loses water constantly through sweat. In the range of thermal neutrality, this loss averages only about 500 grams a day, but under the stress of high temperatures, the loss may reach 15,000 grams a day.

Some mammals that live in desert regions have developed special adaptations that permit them to live under conditions of high temperatures and low water availability.

The camel of the Sahara has two physiological adaptations that allow it to make extended journeys into the desert away from water. A camel's temperature drops overnight to about 34°C (93°F). As it travels across the desert during the day, instead of getting rid of all the heat produced by sweating, it retains some heat to warm the body tissues, which may reach a temperature of

41°C (106°F) before the day is over. Heat produced above this level is dispersed through sweating, but here again the camel has an advantage for it can let its body water loss continue until only about 60 per cent of the normal body water is present. A loss of 10 per cent of body water causes serious illness in man and a loss of 20 per cent is fatal.

In the deserts of the American Southwest, there are small rodents called kangaroo rats whose kidneys have become so efficient for the conservation of water that these animals need never take a drink. But if they are brought into the laboratory and fed a diet of soy beans (a very high protein diet) they will even drink salt water to procure the water needed to carry away the excessive amounts of urea. They are able to do this because their kidneys can produce urine that is three times as rich as man's in salt and four times as rich in urea. Table 28–2 compares the ability to concentrate urine in man, the laboratory rat, and the kangaroo rat. Man is less efficient than either of the rodents.

Figure 28–8 compares kidney structure and function in a marine, a fresh-water, and a terrestrial vertebrate.

Figure 28–8. The relationship between the environment and the structure and functioning of the kidney. Fresh-water fish gain water from the environment and lose salt to it. The nephric unit has a well-developed glomerulus but lacks Henle's loop. Large amounts of dilute urine are produced. Marine fish lose water to and gain salt from the environment. The nephric unit is reduced and little urine is passed. Terrestrial mammals lose both salt and water to the environment, the nephric unit is well developed, and both salt and water are reabsorbed through the tubule.

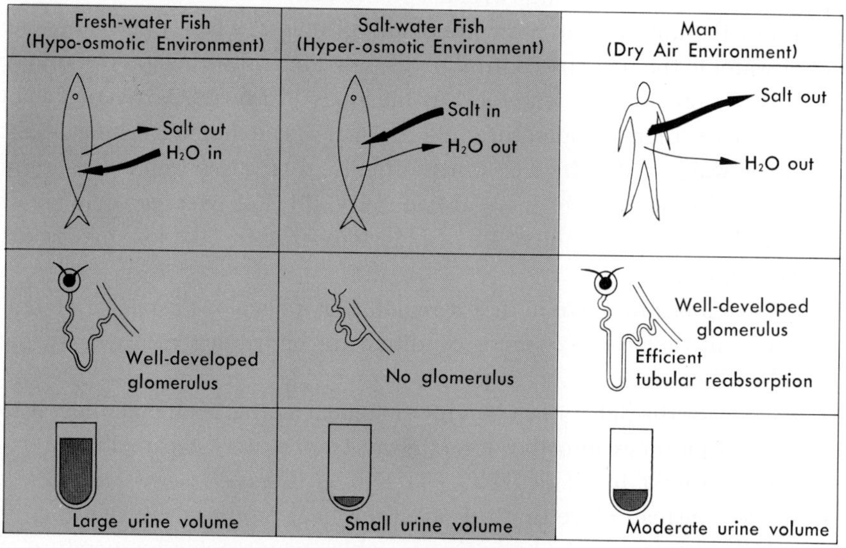

Table 28–2. Maximum Concentrations of Salt and Urea in Urine

	SALT	UREA
Man	2.2%	6%
White Rat	3.5%	15%
Kangaroo Rat	7.0%	23%

Only a few of the variables in the physical environment have been discussed. The nature of the soil, humidity, air currents, barometric pressure, tides, and a host of others are all major environmental factors for some organisms. One feature of the physical environment, though, is of utmost importance to all living things, for it is the ultimate source of organic energy. This is light. In the next chapter we discuss how the energy of light is captured through photosynthesis, and how materials are cycled between organisms and the environment.

SUGGESTED READINGS

Colinvaux, P. A.: *Introduction to Ecology,* John Wiley & Sons, Inc., New York, 1973.

Gordon, M. S., et al.: *Animal Physiology: Principles and Adaptations,* 2nd ed., Macmillan Publishing Co. Inc., New York, 1972.

Schmidt-Nielsen, B.: "Comparative Morphology and Physiology of Excretion," in *Ideas in Modern Biology,* J. A. Moore, ed., Natural History Press, Garden City, New York, 1965.

Schmidt-Nielsen, K.: *How Animals Work,* Cambridge University Press, London, 1972.

Smith, H. W.: *From Fish to Philosopher,* CIBA edition, revised and enlarged, CIBA Pharmaceutical Products, Inc., Summit, N. J., 1959.

CHAPTER 29

Energy Flow and Chemical Cycles

The sun is the ultimate source of energy for most living things on earth. Life as we know it is possible here only because the radiant energy of the sun can be captured and converted into chemical energy. This involves a series of complicated chemical processes known as *photosynthesis.*

PHOTOSYNTHESIS

Photosynthesis is the process in which the energy of light is used to convert carbon dioxide and water into sugar.

The weight of the biochemical substances produced each year by photosynthesis is many times as great as that of the combined products of all the farms, mines, and factories in the world. Plants synthesize nearly 87 trillion kilograms (about 87 billion tons) of biocompounds above those they use in their own respiration. In this synthesis, they use nearly 128 trillion kilograms (about 128 billion tons) of carbon dioxide and nearly 52 trillion kilograms (52 billion tons) of water, and they release nearly 52 trillion kilograms (52 billion tons) of oxygen into the atmosphere.

Efficiency of Photosynthesis

When energy is converted from one form to another, some of it is always lost, mostly in the form of heat. Indeed, the efficiency of most conversion processes is only about 25 per cent or less. Photosynthesis is relatively efficient; about 30 per cent of the light energy absorbed by the chlorophyll is converted to chemical bond energy. But because very little of the solar energy that reaches the earth is absorbed by chlorophyll, the amount of solar energy actually captured and converted to chemical energy is very small indeed. It has been estimated that only about 1.6 per cent of the light reaching a cornfield during the growing season is utilized in photosynthesis. Because the growing season is only about one third of the year, the annual capture of the light that reaches the field is in the neighborhood of 0.5 per cent. The amount of energy that reaches the man who eats the corn is even less. About one fourth of the photosynthetically produced food is used by the plant in its own respiration, leaving 0.375 per cent of the energy locked in the tissues of the plant. Because only about one third of the photosynthetic products go into the formation of corn grains, in the final analysis only about 0.125 per cent of the light energy falling on the field is actually converted into stored food available to man. And corn is one of the most efficient of all plants in its ability to convert light energy into the energy of food.

Process of Photosynthesis

The green pigment, chlorophyll, is essential for photosynthesis because it is the only compound that can absorb solar energy and convert it to chemical energy. Structurally the pigment is similar to the heme group of hemoglobin and the prosthetic groups of the cytochromes. Instead of a central atom of iron, chlorophyll has one of magnesium. Chlorophyll itself has to be synthesized by the plant, and this synthesis usually does not occur until light is available. If a bean seedling is germinated in the dark, it does not start photosynthesis as soon as it is placed in the light, for it takes a while to synthesize enough chlorophyll. In plant cells, chlorophyll is situated in special organelles called *chloroplasts;* in higher plants they are usually round or saucer-shaped. Like a mitochondrion, a chloroplast is surrounded by a double membrane. Within is a ground substance, the *stroma,* and some localized bodies, the *grana,* formed of stacks of membranes. Just as the enzymes and electron transport molecules required for cellular respiration are arranged in an orderly way in the mitochondria, so apparently the chlorophyll molecules and the enzymes and electron transport systems needed for photosynthesis are organized within the chloroplasts.

Initially the equation for photosynthesis was written as simply the reverse of that of cellular respiration:

$$6\,CO_2 + 6\,H_2O \rightarrow C_6H_{12}O_6 + 6\,O_2$$

This equation was interpreted to mean that carbon dioxide was split, the carbon was added to the water, and the oxygen released. Then it was found that all of the oxygen released came from the water and not from the carbon dioxide. To produce this much oxygen, twelve molecules of water must be split. If the equation is to balance, it is necessary to postulate that water is also produced as an end product of photosynthesis. Because of this, the equation is now written:

$$6\,CO_2 + 12\,H_2O \rightarrow C_6H_{12}O_6 + 6\,O_2 + 6\,H_2O$$

This led to a much better understanding of photosynthesis. Like cellular respiration, it is a complex series of reactions. Some of these reactions depend on light, whereas others can take place in the dark.

Light Reactions. The products of the light reactions are known, but many of the intermediate steps have still to be clarified. These steps are mediated by enzymes. What follows is an abbreviated and much simplified account of a complex process.

As light falls on the leaves, about 4 per cent of it is absorbed by chlorophyll. Of the rest, about 16 per cent is either reflected or transmitted through the leaves and the remaining 80 per cent is converted into heat. The green appearance of the plant results from the fact that more red and violet wavelengths are absorbed than green ones. The light that is reflected consists of more green wavelengths. See Figure 29–1.

The energy of the light absorbed by chlorophyll is used for activating chlorophyll molecules:

$$\text{chlorophyll} + \text{light} \rightarrow \text{activated chlorophyll}$$

Activated chlorophyll is chlorophyll in which electrons have been raised to a high energy level. They are released, and as they pass through the steps of the photosynthetic process they fall back to lower energy levels and eventually return to the cholorphyll.

Much of the energy released as the electrons return to lower energy levels is used to divide water molecules into hydrogen (H) and hydroxyl (OH):

$$\text{activated chlorophyll} + 4\,H_2O \rightarrow \text{chlorophyll} + 4\,H + 4\,OH$$

The free hydrogen is temporarily bound to the acceptor NADP:

$$NADP + H \rightarrow NADPH$$

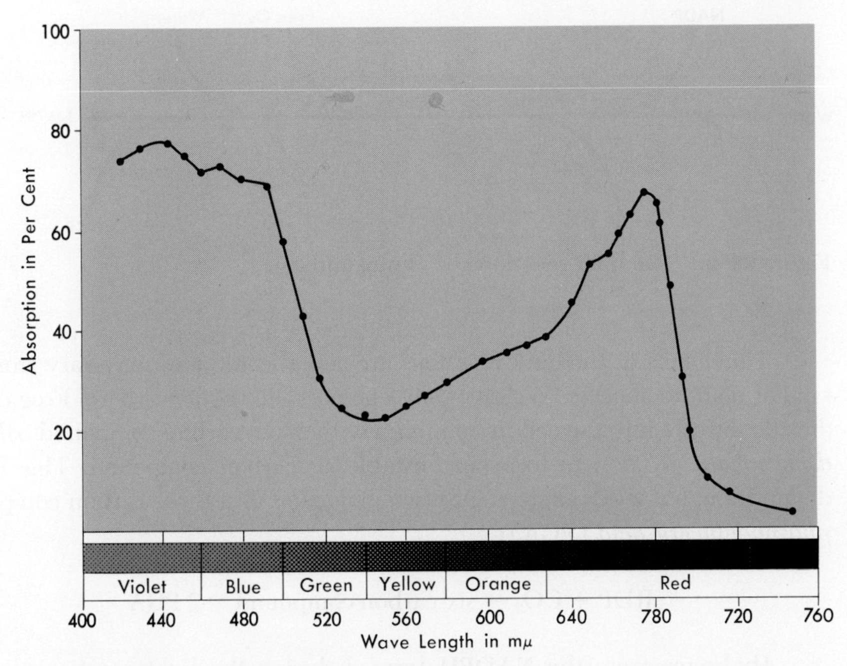

Figure 29–1. Differential absorption of light rays by the sea lettuce, *Ulva.*

The hydroxyl groups are recombined into water with the release of oxygen:

$$4\,OH \rightarrow 2\,H_2O + O_2$$

In the meantime, the remainder of the energy of the activated chlorophyll is used to combine ADP with inorganic phosphate for form ATP:

$$\text{activated chlorophyll} + \text{ADP} + \text{Ⓟ} \rightarrow \text{ATP} + \text{chlorophyll}$$

The series of reactions that involve the use of light can be summarized by a flow diagram (see Figure 29–2).

The light reactions are the unique part of photosynthesis. The other reactions do not require the presence of chlorophyll and can take place in many cells, including animal ones.

Dark Reactions. The remaining reactions involved in photosynthesis can take place in the dark, although they do not have to have darkness to be carried to completion. This is where the energy of sunlight is actually stored in food.

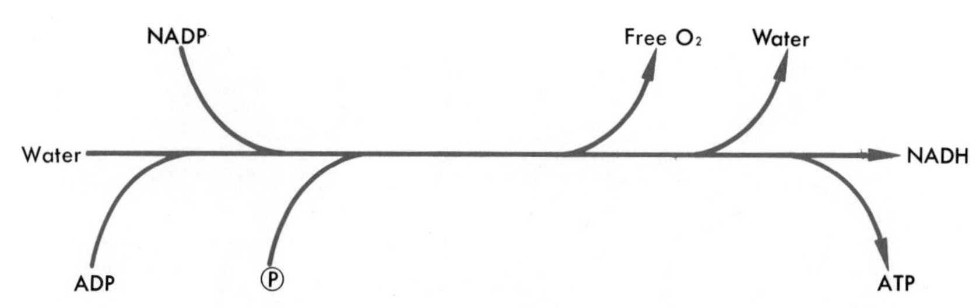

Figure 29–2. The light reactions of photosynthesis.

The details of the dark reactions are complicated and may vary from one kind of plant to another. We give only one possible sequence here. Free carbon dioxide diffuses into the cell. It combines with a five-carbon compound, *ribulose diphosphate* (RDP), to form an unstable six-carbon compound. This breaks down through a series of steps into two molecules of a three-carbon compound, *phosphoglyceric acid* (PGA):

$$\text{RDP} + CO_2 \rightarrow \text{six-carbon compound} \rightarrow 2\ \text{PGA}$$

Hydrogen from the NADPH formed during the light reactions reduces the PGA by removing oxygen from it to form a three-carbon sugar called *phosphoglyceraldehyde* (PGAL) and water:

$$\text{PGA} + \text{NADPH} \rightarrow \text{PGAL} + H_2O + \text{NADP}$$

Energy for these reactions is provided by the ATP produced during the light reactions.

The PGAL produced during photosynthesis can be used to resynthesize RDP, or two molecules can combine to form a six-carbon sugar (see Figure 29–3).

Combining the flow charts shown in Figures 29–2 and 29–3 gives the composite shown in Figure 29–4. The items on the left are those that enter into the

Figure 29–3. The dark reactions of photosynthesis.

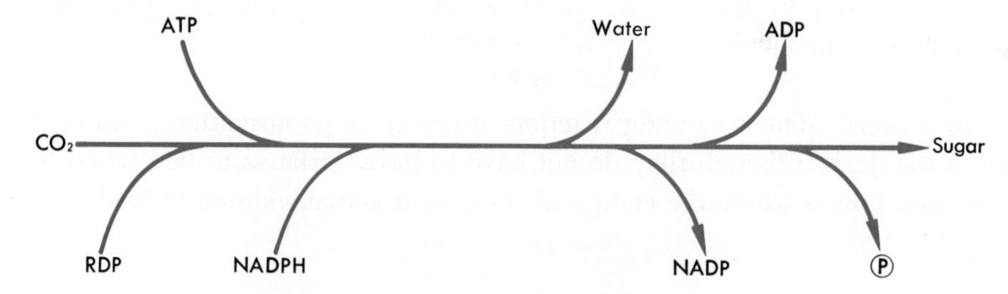

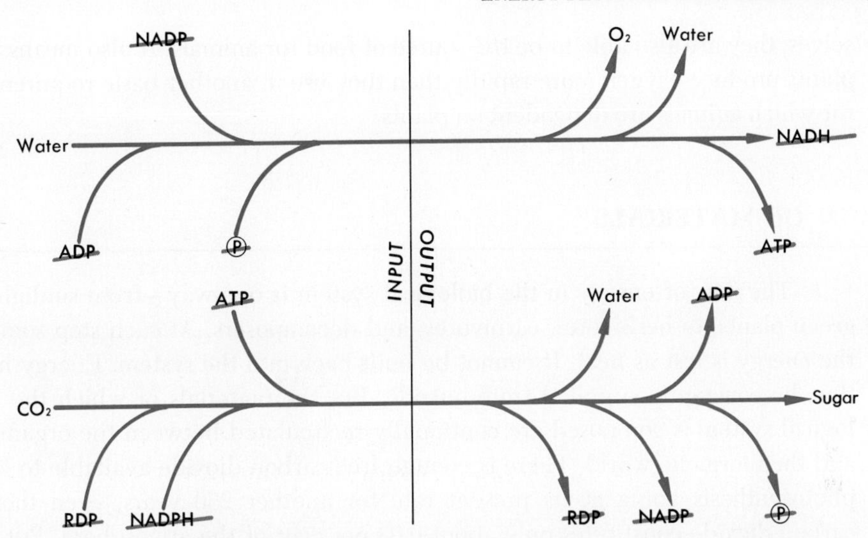

Figure 29–4. Input and output of the combined reactions of photosynthesis.

light and dark reactions (input), and those on the right represent the products of both series of reactions (output). NADP, NADPH, ADP, ATP, and Ⓟ and RDP are cyclical, that is, they can be used over and over again. Canceling them out from the charts leaves only water and carbon dioxide entering into the process as input and water, oxygen, and sugar appearing as output. This was expressed more formally earlier in the chapter as:

$$6\,CO_2 + 12\,H_2O \rightarrow C_6H_{12}O_6 + 6\,O_2 + 6\,H_2O$$

Photosynthesis and Respiration

Plants are able to produce ATP in the chloroplasts by photosynthesis, and the energy of this ATP is used to build carbohydrates. Carbon dioxide is incorporated and oxygen is released. But plants also have mitochondria and carry on cellular respiration, producing ATP, using oxygen, and releasing carbon dioxide. Photosynthesis, or at least the initial light reactions, can be carried on only in the light, but respiration goes on both day and night. It is obvious that if the two processes should proceed at equivalent rates, with respiration breaking down biocompounds as rapidly as photosynthesis builds them, none would be left for the repair and growth of the plant and the plant would soon die. Under natural conditions, photosynthesis proceeds at eight to twelve times the rate of respiration. It is this excess of production over consumption that allows plants to grow and store food. Because they make more than enough food for themselves, they are also able to be the source of food for animals. It also means that plants produce oxygen more rapidly than they use it, another basic requirement for which animals are dependent on plants.

CIRCULATION OF MATERIALS

The flow of energy in the biological system is one-way—from sunlight to green plants, to herbivores, carnivores, and decomposers. At each step some of the energy is lost as heat. It cannot be built back into the system. Energy must thus be constantly supplied from outside. But the materials of which the biological system is composed are continually recirculated between the organisms and the inorganic world. There is enough free carbon dioxide available to keep photosynthesis going at its present rate for another 250 years, even though carbon dioxide constitutes only about 0.03 per cent of the atmosphere. But 250 years is only the briefest moment when the span of life on earth is considered. Consequently, there must be a constant cycling of carbon dioxide so that it can be used by successive generations of organisms. It seems likely that any particular carbon atom has been through a carbon cycle millions of times since life has been on earth. Similar cyclic use is known for many other elements, for example, nitrogen and sulfur. These cycles are known as *biogeochemical cycles.* Figure 29–5 represents in chart form how such elements as carbon, oxygen, and nitrogen may be used over and over again.

The Oxygen Cycle

During photosynthesis, water is split, the hydrogen is combined with carbon dioxide to form organic compounds, and the oxygen is released to the environment. This oxygen is used by both plants and animals for respiration, in which it serves as the ultimate acceptor of the hydrogen released when organic compounds are oxidized. And so water is formed again.

The Carbon Cycle

The source of the carbon that enters the biological system is the carbon dioxide in the air and dissolved in the waters of the earth. This carbon dioxide is used by plants to build the three food compounds: carbohydrates, fats, and proteins. The foods may be broken down by cellular respiration by the plants themselves or by the animals that eat the plants. Either way, free carbon dioxide is returned to the environment. But the carbohydrates, fats, and proteins also go into building the bodies of plants and animals. The carbon dioxide is then locked

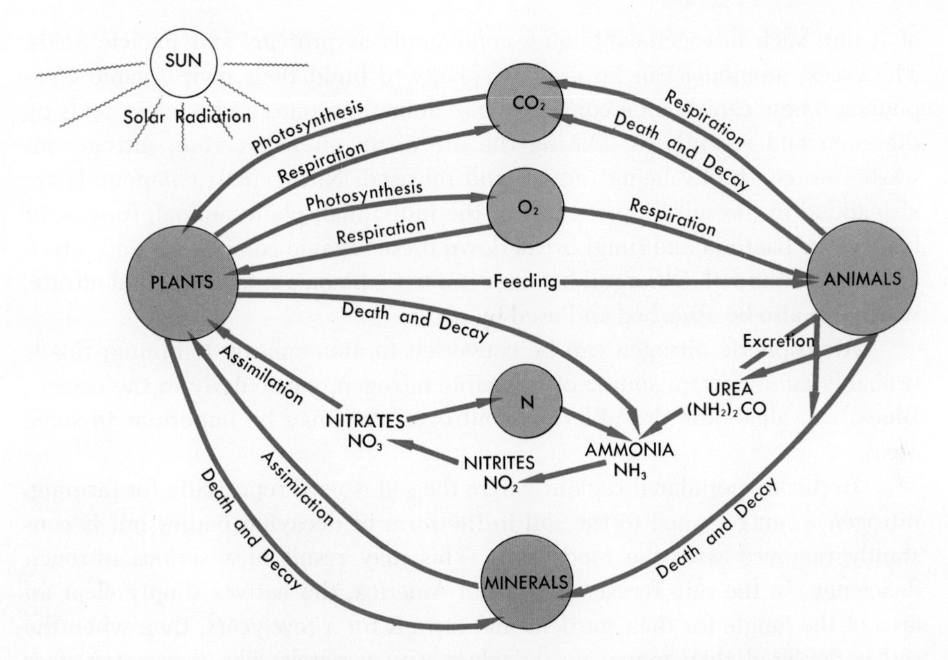

Figure 29–5. Some of the major cycles of materials between organisms and the environment.

up until the organisms die. If decay ensues, the carbon dioxide is released to the environment. If the dead bodies are protected from decay bacteria, they may accumulate and become covered by protective layers of earth to form deposits of stored carbon compounds in the form of coal, peat, oil, and gas, the so-called fossil fuels. These may subsequently be recovered by man and burned to produce heat, with a consequent release of carbon dioxide back into the atmosphere. Most industrial plants are not efficient enough to completely oxidize all of the carbon compounds in the fossil fuels, and usually there is an unburned residue of intermediate compounds, the soot and smoke produced along with the heat and carbon dioxide. Some of these residues are poisonous and some are mutagenic.

The Nitrogen Cycle

Nitrogen is constantly being lost to the soil by erosion, by leaching, by the removal of crops, and by the activity of denitrifying bacteria. As we pointed out in Chapter 18, the free nitrogen of the atmosphere cannot be used by plants for organic synthesis. It must first be combined with hydrogen. This is the role of the nitrogen-fixing bacteria. They reduce nitrogen to ammonia and build some of it into such nitrogen-containing compounds as proteins and nucleic acids. The excess ammonia can be used by plants to build their own organic compounds. These can then be converted into animal proteins and nucleic acids by digestion and resynthesis. During the life of an animal, certain nitrogenous wastes are constantly being formed and released. Nitrogenous compounds are also added to the soil by the death of the individual, plant, animal, fungus, or bacterium. Bacteria and fungi break down these organic compounds and return ammonia to the soil. Other soil bacteria convert ammonia into nitrite and nitrate, which can also be absorbed and used by plants.

Atmospheric nitrogen can be converted to ammonia by lightning; this is probably an important source of utilizable nitrogen, particularly in the oceans. Blue-green algae are also able to fix nitrogen and may be important in some areas.

In thickly populated regions where the soil is used repeatedly for farming, nitrogen is not returned to the soil in the form of decaying bodies but is constantly removed with the food crops. This may result in a serious nitrogen deficiency. In the rain forests of tropical America, the natives simply clear an area of the jungle for their gardens and farm it for a few years; then when the soil is depleted they move on and clear a new patch. The first patch goes back to jungle, and its fertility is gradually restored. The North American Indians learned to use the soil repeatedly by placing a dead fish in each hill of corn as they planted. Decay bacteria made the nitrogen of the fish available for the corn. Since the time of the Greeks, men have known that fertility can be restored to a field by planting a crop of legumes, such as beans or clover. Some of the nitrogen-fixing bacteria do not live free in the soil but in nodules on the roots of such plants. Roots containing such nodules are very rich in nitrogen, and when they are allowed to decay in the soil they replenish its stores of utilizable nitrogen.

The recent widespread use of insecticides, herbicides, and fungicides raises a very important question. What effect do these poisons have on the soil bacteria?

SUGGESTED READINGS

Colinvaux, P. A.: *Introduction to Ecology,* John Wiley & Sons, New York, 1973.

Haynes, R. H., and P. C. Hanawalt: *The Molecular Basis of Life,* W. H. Freeman and Company, San Francisco, 1968.

Kucera, C. L., and J. J. Rochow: *The Challenge of Ecology,* C. V. Mosby, St. Louis, Mo., 1973.

Nobel, P. S.: *Plant Cell Physiology: A Physiochemical Approach,* W. H. Freeman and Company, San Francisco, 1970.

CHAPTER **30**

The Biotic Environment

The transfers of energy and materials discussed in the last chapter involve interactions not only between organisms and the physical environment but also between one kind of organism and another. These energy transfers are a very important aspect, although not the only one, of the relationship between a plant or animal and its biotic environment.

FEEDING

Transfer of energy takes place when one organism feeds on another. There are limits to the feeding ability of all animals. In the first place, only certain species are adapted to feed on the photosynthesizing plants; the rest have to eat other animals. In the second place, animal food items must be of the right size and be present in the appropriate quantities. Obviously, the prey cannot be too large; a garter snake can neither catch nor swallow a cow. But neither can the prey be too small. A predator cannot afford to feed on any creatures that are so tiny and so scattered that it has to expend more energy catching, eating, and digesting than it gains from oxidizing its prey.

Not all animals are so restricted in diet as the dichotomy herbivore-carnivore suggests. Squirrels and primates eat bird eggs and insect grubs as well as plant food; foxes and bears eat berries. Some species of birds and mammals, such as robins, raccoons (see Figure 30–1), and men, should really be thought of as *omnivores* (meaning "eat everything"). Food transfer between organisms can be thought of as food pyramids or food webs.

Figure 30–1. An omnivorous mammal. A family of raccoons (*Procyon lotor*) at Carnegie Museum's Powdermill Nature Reserve. (Carnegie Museum—M. Graham Netting photo.)

Food Pyramids

At the base of every food pyramid is some type of plant life that is capable of transforming solar energy into chemical energy. Because one plant does not feed upon another (with the exception of certain parasitic species, see Figure 30–2) the next step is a plant-eating animal (a herbivore) that can convert plant energy into animal energy. Next is a primary carnivore capable of making its

Figure 30–2. The Indian Pipe is a flowering plant that lacks chlorophyll and lives as a parasite on a fungus. (Carnegie Museum—M. Graham Netting photo.)

living by feeding on the herbivores. The next step might be the final one, in which a secondary carnivore preys on the primary one. Figure 30–3 illustrates a food pyramid that might be found in the Antarctic. At the base are microscopic marine algae known collectively as *phytoplankton* that drift around in the seas and convert light into chemical energy. Feeding on the phytoplankton are minute animals (*zooplankton*) that also drift with the sea. They are present in such enormous numbers that despite their small size, large primary carnivores, such as the whales, can feed on them. Finally man, as a secondary carnivore, may capture and eat the whale. In any food pyramid, there is a tremendous loss in the amount of energy transferred. In this idealized case, only about one tenth of the energy of the phytoplankton is converted into zooplankton, and one tenth of the energy of the zooplankton is converted into primary carnivore. If man eats the whale, only one tenth of the energy of the whale is transferred to man, so that he actually gains only 0.1 per cent of the original energy of the

Figure 30–3. A hypothetical food pyramid in the Antarctic. (From "Whales, Plankton and Man," by W. E. Pequegnat, © 1958 by *Scientific American*, Inc. All rights reserved.)

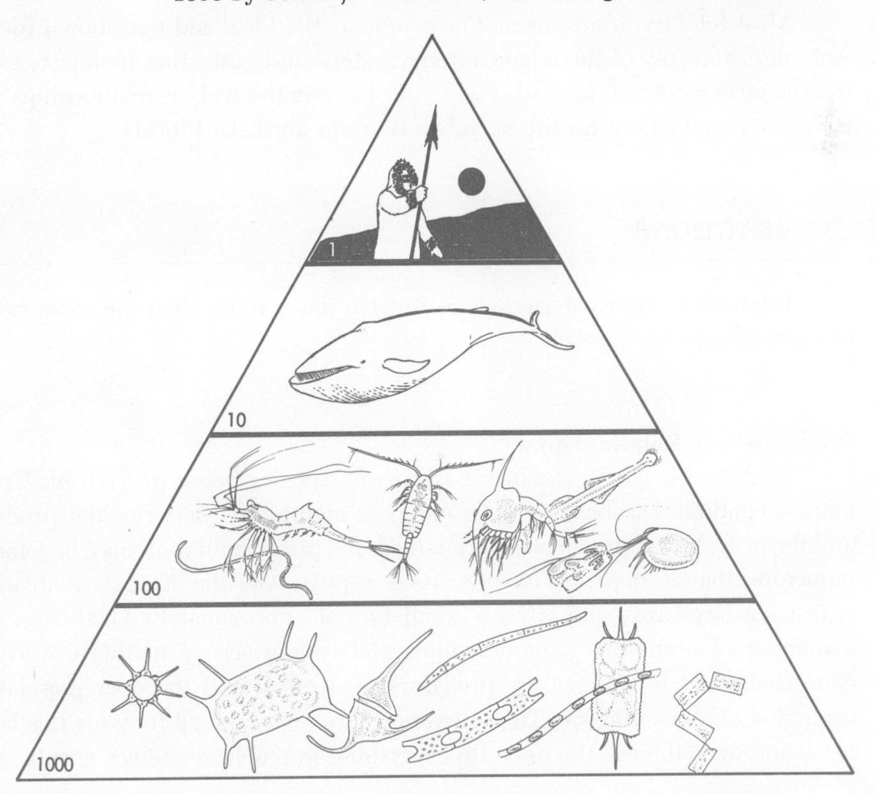

phytoplankton. This figure of a 90 per cent loss of energy from one step to another in a food pyramid is probably a conservative estimate for most food pyramids.

Food pyramids can also be measured simply in mass or weight. It has been calculated that it takes over 8,000 kilograms (nearly 18,000 pounds) of alfalfa to produce 1,000 kilograms (2,250 pounds) of beef, which will in turn produce 48 kilograms (105 pounds) of boy. This also shows why food pyramids usually have only two or three steps. They are limited to five or six steps at most for the very simple reason that so much mass and energy are lost at each step that one with more steps could hardly sustain the terminal population.

In regions where the human population is very dense, the diet usually contains little animal food because the primary food resources are insufficient to provide for the loss of energy resulting from the introduction of an intermediate step between plants, the producers, and men, the consumers. Because most plant proteins are deficient in some of the essential amino acids, protein-deficiency diseases are common in these areas.

Food Webs

Most food pyramids are not as simple as the idealized one shown for the Antarctic. So many of them have alternate steps and paths that biologists today use the term *food web* instead. Figure 30–4 shows the web of relationships that have been worked out for the St. Johns River in northern Florida.

INTERSPECIFIC REACTIONS

Interactions between organisms involve much more than the mere eating of one organism by another.

Predator-Prey Relationships

Predation is not necessarily bad for the species being preyed on. Under natural conditions, a balance of numbers is maintained between the predator and the prey. When predators are removed, the prey population may become so numerous that it depletes its own food supply. On the Kaibab Plateau in Arizona, a large area supported a population of approximately 4,000 deer, and a number of predatory mountain lions and wolves. Early in this century, a concerted effort to remove the predators succeeded, and the deer population expanded to about 100,000. There was nowhere near enough food for this large herd, and even though the deer ate everything in reach, seedlings, shrubs, and

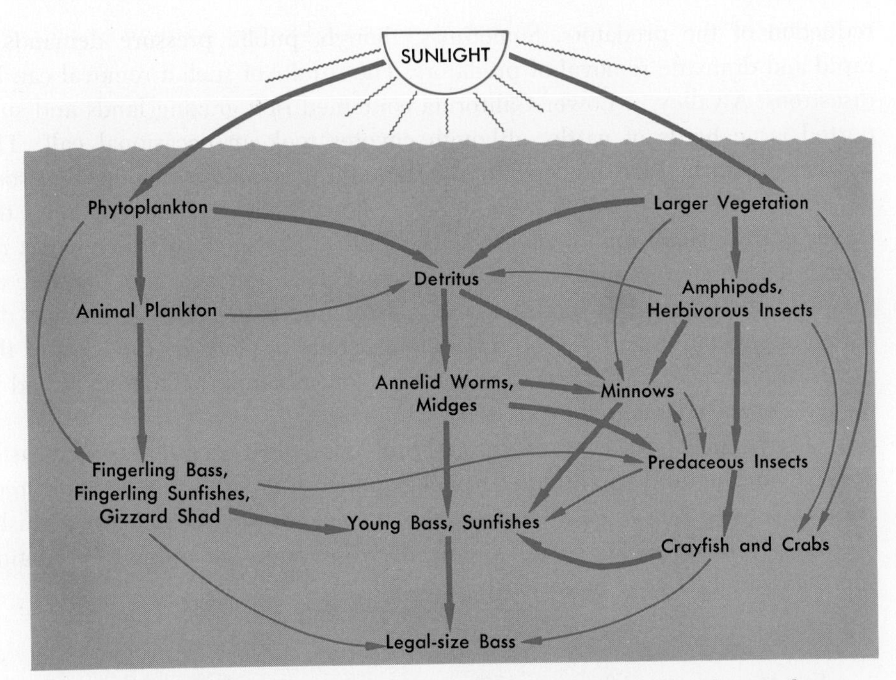

Figure 30–4. A food web in the St. Johns River, Florida. (Redrawn from J. S. Rogers, T. H. Hubbell, and C. F. Byers, *Man and the Biological World*, 2nd ed., 1952. By permission of McGraw-Hill Book Company.)

all, within two years, over half of them starved. The population finally declined to below where it was before.

Preventing a population from outrunning its resources is not the only benefit conferred by predators on the prey species. Because predators tend to capture the individual that is most easily caught, they remove the diseased and unfit from a population and thus may help control the spread of parasites and promote genetic fitness by eliminating individuals with deleterious genes. In a classic study of the ecologic relationships of the wolves of Mount McKinley National Park, Adolph Murie found:

> The caribou is the main food of the wolf, and a heavy toll of the calves is taken. Yet the park herd of between 20,000 and 30,000 animals is apparently maintaining its numbers. After the first few days in the life of the calves the hunting of them by wolves necessitates a chase which usually eliminates the slowest and weakest.

Fortunately, most conservationists today realize that the expansion of the habitat and food supply is a better way to maintain a large population than the reduction of the predators. Sometimes, though, public pressure demands a rapid and dramatic removal of predators. The results of such a removal can be disastrous. A valley in Lower California contained rich grazing lands and supported large herds of cattle, although coyotes took an occasional calf. The ranchers, disturbed by the predation, initiated a poisoning campaign that soon wiped out the coyotes. But the chief prey of the coyotes had not been the calves but the small rodents living in the valley. The death of the coyotes resulted in a population explosion of the rodents. This put too much pressure on the plants that served as food for both rodents and cattle. The valley was denuded of vegetation and the topsoil, no longer held in place by the roots of the plants, was blown away. The valley could no longer support the herds it had in the days of the coyotes.

The balance between predator and prey is a delicate one that can be easily upset. Under natural conditions, a quail population is able to withstand predation by hawks. But when a farmer clears his fields and does not leave brushy borders to provide shelter for the quails, they are more vulnerable to predation and their numbers may be seriously reduced.

Symbiosis

Symbiosis means the living together of two species. The conditions under which two species live symbiotically range from those in which both species profit by the arrangement to those in which one derives the benefit and the other suffers.

Mutualism. The symbiotic relationship in which both species profit is called *mutualism.* Many flowering plants depend on insects for pollination, and the insects in turn depend on the pollen and/or nectar of the plants for food (see Figure 30–5). This mutualistic relationship is of utmost importance to man since many of his most important crops are pollinated by insects, usually bees. Indiscriminate spraying for insect control in Arizona sharply reduced the bee population, which in turn markedly reduced the cotton yield.

Domestication can be defined as the establishment of a mutualistic relationship between man and a plant or animal species. Man may kill some of the individual animals for food, but he insures the survival of the species by providing food, shelter, and medical care for others.

Commensalism. In *commensalism,* one number of the pair benefits while the other neither gains nor loses. One the island of Jamaica, there is very little standing water on the ground, for the topography is extremely rugged. But bromeliads, large air plants that grow on the trees, have water stored in the

Figure 30–5. A checkerspot butterfly (*Euphydryas anicia*) pollinates Colorado sunflowers during its visits for nectar, an example of mutualism. (From *An Introduction to Ecology and Population Biology* by Thomas C. Emmel. By permission of W. W. Norton & Company, Inc. Copyright © 1973 by W. W. Norton & Company, Inc.)

bases of their leaves. There creatures such as mosquitoes and frogs find water in which to reproduce to their own advantage, but this in no way harms the bromeliads.

Parasitism. The final type of symbiosis is that in which one member benefits and the other suffers. The most usual kind of *parasitism* is a special kind of predation, differing only in the closeness of association between the host and parasite. Much that has been said about predation applies also to parasitism. If a parasite is so destructive that it brings about the host's death, the parasite itself dies, for it has killed off its own food supply. As a consequence, there is selection pressure toward the parasite doing less and less harm to the host. Parasitism tends to evolve toward commensalism. The rabbits and myxomatosis in Australia, discussed in Chapter 17, are a case in point.

Some birds, such as the European cuckoo and the American cowbird, practice social parasitism. The female bird lays her eggs in the nest of another kind of bird so that the host mother supplies the foster young with the food that should have gone to her own offspring.

Habitat Formation

Plants have a profound effect on the lives of animals that do not feed on them simply by their contribution to animal habitats. They provide protection from extremes of heat and cold in forested regions as compared to deserts; they hold moisture in the soils; their transpiration activity raises the humidity of the air; they provide shade; and they contribute appreciably to the formation of soil. Rotten logs serve as homes for many creatures, and the leaf mold provides a living place for innumerable tiny, and not-so-tiny, species of the forest floor. Many birds, squirrels, and other animals live in trees and use plant material to construct their nests.

Buffering

Buffer species are those that intercede between predator and prey. In the north central United States, otters eat 25.9 per cent forage fishes, 25.3 per cent amphibians, and 22.7 per cent game fishes. If the number of forage fishes is reduced, the otters will prey more extensively on the game fishes and amphibians. The forage fish population serves as a buffer to protect the amphibians and game fishes from extensive predation by the otters.

Because most populations tend to alternate between periods of abundance and scarcity, and because predators are inclined to feed most heavily on the most abundant species, the buffering interaction tends to be fluctuating and reciprocal. The cotton rats and quail of the southeastern United States show such a relationship. The rats are the normal prey of a number of species of hawks, but if the rat population declines, the hawks begin feeding on the quail. This reduces predation pressure on the rats and they can then increase in numbers until it becomes more profitable for the hawks to hunt them rather than the quail. Thus, the rats may be buffered by the quail one year and serve as buffers for the quail population the next year.

Protective Coloration, Warning Coloration, and Mimicry

An animal frequently gains protection from predators because its color and shape allow it to blend into the background. Sometimes an animal bears a remarkable resemblance to an inedible object. A walkingstick insect looks like a twig and a dead-leaf butterfly at rest looks like a dead leaf. Animals that are poisonous or unsavory are often brightly colored and boldly patterned. This warning coloration wards off attacks by predators, which soon learn by experience that such a species is dangerous to tackle or not good to eat. Sometimes a harmless, edible species evolves through convergence the appearance of a harmful species found in the same region and thereby gains the protection of the warning coloration. This is called *mimicry*. Several species of harmless kingsnakes have developed patterns similar to those of the poisonous coral snakes. This type of mimicry is known as *Batesian mimicry* after the naturalist who first put forward a reasonable explanation of the fact that many edible species of butterflies have patterns remarkably like those of inedible species inhabiting the same area. The one restricting feature on Batesian mimicry is that the species being copied, the model, must never be greatly outnumbered by the copier, the mimic, for then the predators would be more likely to learn that the pattern signaled something palatable. Only when the percentage of dangerous or distasteful individuals is high is there any selective advantage to gaudy recognition characters.

In *Müllerian mimicry,* two or more species, all of which are harmful or unsavory, come to look alike through convergence. If a number of species adopt the same warning hallmarks, the number of victims will be no greater as each new generation of predators learns to recognize them, but the sacrifice of individuals will be spread over several species rather than falling entirely on one. Many species of Latin American coral snakes show Müllerian mimicry.

Competition

An animal's niche is the role that it plays in the economy of nature, not just where it lives but how it lives, what it feeds on, and what it provides food for. Two species cannot occupy exactly the same ecological niche at the same time. This was well demonstrated by an experiment using two species of the protozoan *Paramecium.* Two equal-sized populations of *Paramecium caudatum* and *Paramecium aurelia* were introduced into a culture flask in which the available food supply was the only limiting factor. Within about two weeks, *P. aurelia* had completely eliminated *P. caudatum* and was the sole survivor. Experiments such as this with other forms of life have repeatedly shown that when two species are in direct competition in a single niche, one will survive and the other will be eliminated. However, the survivor under one set of conditions may not be the survivor under different conditions. If two species of flour beetle, *Tribolium castaneum* and *T. confusum,* are put into direct competition, when the temperature and humidity are high, *T. castaneum* is the survivor, but when the temperature and humidity are lower, *T. confusum* turns out to be the survivor. The axiom that no two species can occupy the same niche at the same time seems to be the explanation of why there have been, through geologic time, many more species on earth than have ever occupied it at any one moment. The evolution of new species is mostly preceded by the opening up of niches through the extinction of previously existing species.

POPULATIONS

The individuals of a single species in a given area make up a population.

Characteristics of Populations

Populations have certain definite characteristics that are inherent in the group, as distinct from the individuals composing the group. Among the most important of these characteristics are population density, birth rate, death rate, and age distribution.

Population Density. The density of a population is measured by the number of units per area. For example, the human population is measured by the number of individuals per square kilometer. The density and changes in density of a population are indicative of its success or lack of success.

A population whose density stays the same is not an unchanging population, for in any population there is a difference in the ability of individuals to leave offspring. One generation is descended from only part of a previous generation. Figure 30–6 shows a bacterial population in which the entire third generation is derived from only half the individuals in the first generation. In the United States and in most western European countries today, any one generation is estimated to have been derived from about 50 per cent of the previous generation. In certain Asiatic countries, far less than half of the population contributes to the succeeding generation.

Birth Rate. The birth rate, or natality, of a population is the number of new individuals produced in a given unit of time. There is a theoretical maximum rate based on the largest number of individuals that can be produced in a unit of time under ideal conditions. These conditions are never more than momentarily met, though, and a more realistic figure is the actual birth rate.

Death Rate. The death rate has a theoretical minimal figure and also a more realistic actual figure. The theoretical minimum is based on each individual living out the full life span characteristic of the species. Figure 30–7 gives a

Figure 30–6. A constant population size is maintained in a colony of bacteria though not all individuals of one generation contribute to the next generation. (Bruce Wallace and Adrian M. Srb, *Adaptation*, 2nd ed., © 1964. By permission Prentice-Hall, Inc., Englewood Cliffs, New Jersey)

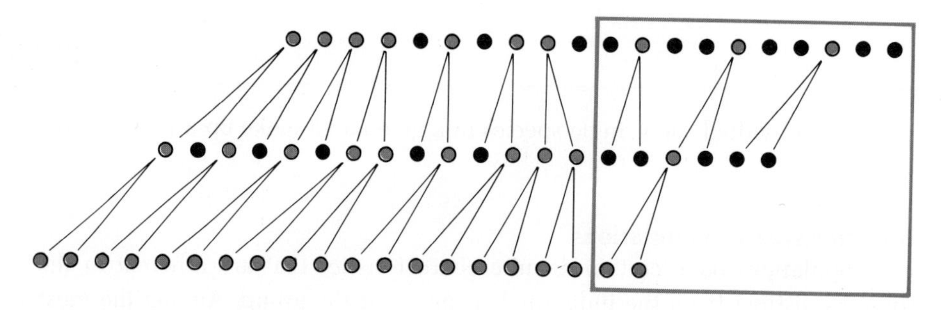

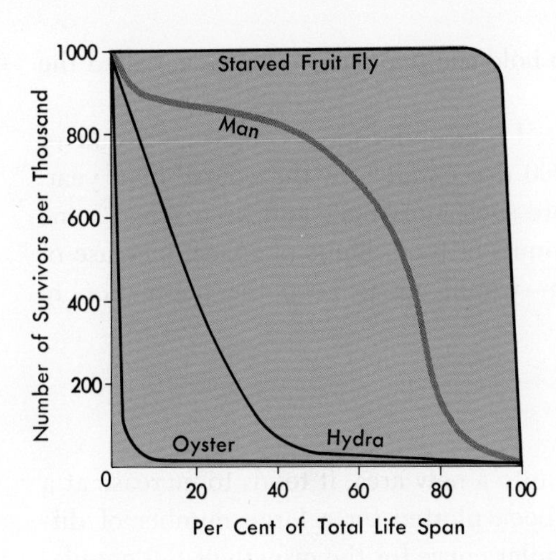

Figure 30–7. Survival curves for four species in which the number of surviving individuals is plotted against the percentage of the total life span characteristic for each species. For starved fruit flies, the span is five days, for man about a hundred years. Survival curves for most plants and animals resemble the curve for hydra. (From Claude A. Villee, *Biology*, 4th ed., 1962. By permission of W. B. Saunders Company and Claude A. Villee.)

comparison of the survival curves of four species. Unfed fruit flies nearly all live five days, the allotted span for flies under these conditions, and then all die at once. The death rate thus approaches the theoretical minimum. Most organisms tend to have a very large percentage of their offspring die long before they have reached the average life span of the species. Man has modified his death rate curve both by reducing infant mortality and by increasing the longevity of the average individual.

Age Distribution. Another characteristic of a population is the average age of the individuals that make it up. There are three ages in the life span of the individual: prereproductive, reproductive, and postreproductive. When information on the proportions of the different age groups in the population is available, it is possible to predict something about its future. Populations with a high proportion of young individuals are rapidly growing ones, whereas those made up of a higher percentage of postreproductive individuals grow at a much slower rate. Theoretically every population should reach a point of equilibrium in which the proportions of the different age groups remain constant, but other factors are constantly modifying these proportions.

Structure of Population

The existing population is simply the result of the birth rate being held in check by the factors that keep it from reaching its theoretical maximum, and the actual death rate that determines the average age of the survivors. In other words, the existing population is produced by the interaction of the theoretical potential reproductive rate of the species, called its *biotic potential,* and all the factors in the environment that tend to hold the population in check, called the *environmental resistance.*

For example, it has been calculated that under ideal conditions a pair of houseflies would leave 5,598,860,000,000 descendants in the course of a year. This never happens because long before such numbers could be reached some factor or factors in the environment—limits of food, limits of space, increase of predators or parasites, or competition—would act to keep the population in check.

Population Growth Curve

When a population is introduced into a new area, it tends to increase at a definite, predictable rate. Rates have been plotted for a large number of different species; many of them form a similar curve for the growth of the population. Such a growth curve is shown in Figure 30–8. At first, the population increases at an accelerated rate, then at a steady but rapid rate, then at a decelerated rate, until finally it reaches equilibrium, from which point on it remains more or less constant as long as conditions remain unchanged. In some mammals, at least, the negative acceleration phase that occurs in populations, even when there is an abundance of food and water, is thought to be brought about by the stress of overcrowding. The increased competition between individuals that results from overcrowding may stimulate the hypothalamus to bring about the release of adrenocorticotrophic hormones. Upsets in the hormonal balance of the body can result in a reduction of the reproductive rate and in a lessened resistance to disease.

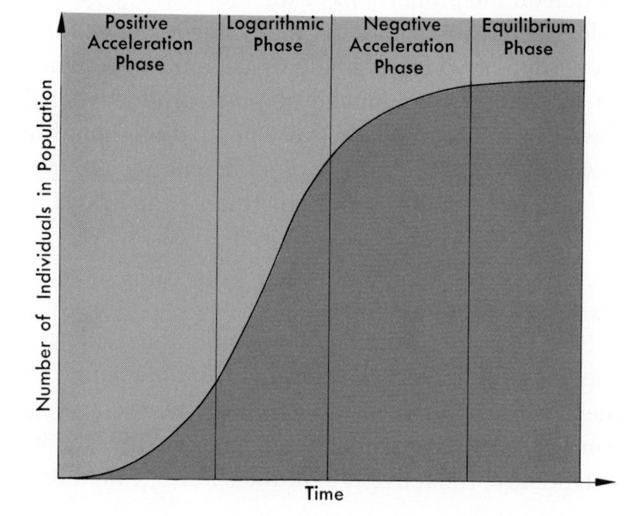

Figure 30–8. A typical population growth curve.

A plot of the growth of the human population from 500 A.D. to the present makes a growth curve that fits the accelerated and logarithmic phases of the generalized growth curve (see Figure 30–9). The twelve years following World War II saw a marked increase in birth rate in the United States, the "Baby Boom." The average number of children per woman reached 3.7 in 1957. Then it began to decline; it has now fallen below 2.0, less than the number needed to offset the death rate. This does not mean that the population of the United States has stopped growing. Aside from the increase in population added by immigration, the women now of reproductive age and those approaching reproductive age were born during the baby boom. They are having fewer children, but there are more of them. Still there is a chance that the population of the United States will have stabilized in another generation or so. Unfortunately, these figures do not apply to the developing nations. In these countries, the introduction of modern medical and sanitary practices has sharply reduced the death rate, although there has been little reduction in the birth rate. The threat of famine still hangs over many parts of the world.

A society in which only enough children are born to replace the individuals removed by death, and in which, because of medical advances, most people live out their full life span, will have problems of its own. What will

Figure 30–9. The growth curve of the human population.

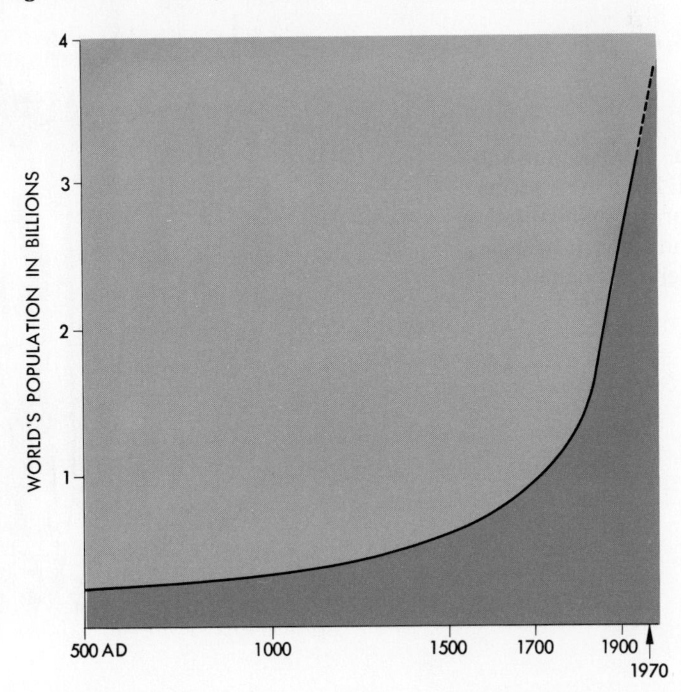

it be like to live in a society in which there are twice as many people between the ages of fifty and eighty as there are between the ages of fifteen and thirty?

COMMUNITIES

A community is made up of a number of different populations of plant and animal species that inhabit the same general area.

Kinds of Communities

The nature of the community inhabiting a given land area is largely determined by climatic conditions. When the amount of rainfall is low, the area supports a desert community, with succulent plants such as the cacti that can store water in their stems (see Figure 30–10), and low, branchy shrubs with

Figure 30–10. Organ-pipe and saguaro cacti have thick stems for water storage, an adaptation to desert conditions. (Carnegie Museum—Neil D. Richmond photo.)

small leaves to cut down on the surface area from which water can be lost. Figure 30–11 shows a typical desert community. Animals that live in deserts are frequently nocturnal to avoid the high daytime temperatures and often have special adaptations for water conservation such as those of the kangaroo rat and camel (Chapter 28). When water is somewhat more abundant, grasses are the dominant plants, as in the prairies and plains of the United States. Typical grassland animals include the large grazers and the carnivores that prey on them. More humid regions support forest communities (see Figure 30–12). Animals that live in forests are often either arboreal or adapted to living in the leaf litter on the forest floor. Temperature also plays a role in determining the type of community. In Arctic regions, and on high mountains above the timber line, are found tundras, with lichens, low-growing grasses, and quick-blooming annuals adapted to a short growing season. Figure 30–13 shows a mountain tundra, also called an *alpine meadow*. Mammals that live in such regions are cold adapted, with thick coats and short legs and ears to cut down on the surface area from which heat can be radiated. Large mammals are usually migratory, moving up and down the mountain or north and south as the snow line retreats or advances.

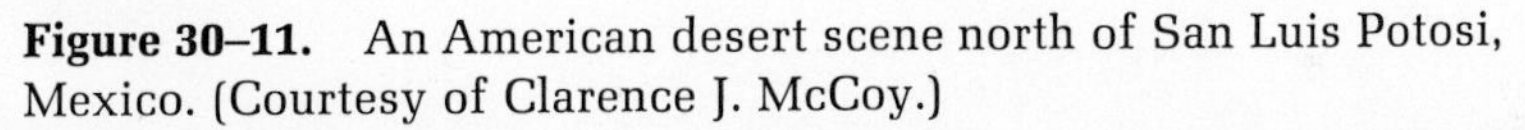

Figure 30–11. An American desert scene north of San Luis Potosi, Mexico. (Courtesy of Clarence J. McCoy.)

Figure 30–12. A lowland, tropical forest in Honduras. (Courtesy of A. F. Carr.)

Succession

If any new physical environment becomes available, such as a sandy dune blown up by the wind or a rocky slope left bare by a retreating glacier, it soon comes to support a community. The original community, though, is not the final one, for one community will modify the environment in such a way as to make it habitable for a different set of populations, so that the community gradually changes from time to time. This succession of communities is known as a *sere*. When any area first becomes available to life, it presents a rigorous environment that can be tolerated by only a few species. It lacks the soil elements needed by many plants, and usually undergoes extreme fluctuations of temperature and moisture content. Some hardy species are able to tolerate these rigorous conditions and get a foothold to form a pioneer community (see Figure 30–14). As time goes on, this pioneer community adds humus to the soil, its roots and foliage tend to maintain the moisture content at a more equitable level, and the plants provide food and cover for animals that can now enter into the community life. All these changes tend to alter the conditions in such a way that other, less hardy species can move in. These species are better adapted to the changed conditions than the pioneers and gradually replace them. The

Figure 30–13. Alpine tundra, Mt. Evans, Colorado. (Courtesy of Ward's Natural Science Establishment, Inc., Rochester, N.Y.)

Figure 30–14. A pioneer community. Ice plants (*Mesembryanthemum*) growing on a gravel plain. (Carnegie Museum—Clarence J. McCoy photo.)

newcomers will, in turn, modify the area and ultimately be supplanted by another complex of organisms, still better fitted to live and reproduce in the ameliorated environment.

The succession of one community by another is not unending, though, for finally a *climax community* will tend to establish itself and become self-perpetuating so long as it is left undisturbed and the climatic conditions remain the same. In the southeastern United States hardwood forests of oak, hickory, and magnolia seem to be the climax (see Figure 30–15).

The successive stages of a sere have various characteristics. Early communities tend to be made up of relatively few species but are extremely productive. More energy is entering the system through the conversion of solar energy into chemical energy than is being lost through respiration. As a sere approaches a climax, it comes to consist of more and more diverse populations; energy loss now approaches energy gain, resulting in a decrease in overall community production and an increase in the rate of transfer of energy and materials within the community. Seres can be interrupted by such natural phe-

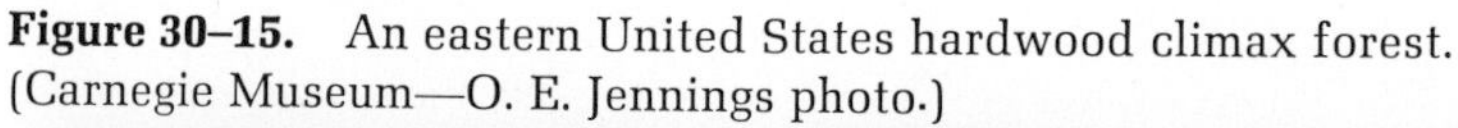

Figure 30–15. An eastern United States hardwood climax forest. (Carnegie Museum—O. E. Jennings photo.)

nomena as floods, droughts, or fire, or by extensive grazing or browsing by herbivores. Such interruptions can prevent the sere from proceeding to the climax or can return it to an earlier stage. In the southeastern states, pinewoods are naturally succeeded by the climax hardwood community. But the pines are more valuable commercially than are the hardwoods. Controlled burning prevents the young hardwoods from establishing themselves so that the community is maintained in the subclimax (see Figure 30–16).

Lightning-caused fires are important in maintaining some climax communities. The redwood trees of California are highly resistant to fire, but their saplings are not competitive against the less fire-resistant plants that crowd in when fire is kept out of the redwood forests. Only periodic burning of the forest floor allows the redwoods to reproduce themselves.

The successive community stages are of greatest importance to mankind. To keep productivity at the level necessary to grow harvests large enough to feed him, man must maintain extensive areas in the early successional stages.

Figure 30–16. Controlled burning of pine woods, Pineland Plantation, Georgia. This prevents the establishment of the hardwood climax forest. (Carnegie Museum—M. Graham Netting photo.)

When a man removes the weeds from his corn patch, he is doing nothing more than interrupting succession and maintaining the community as an early, highly productive grassland community. Figure 30–17 shows a man-made early grassland community—a field of wheat.

Early communities are very fragile. Because only a few species are present, a disease that decimates a single species can have a drastic effect on the community as a whole. This is especially true of man's crops. Not only are all the plants in a corn field members of a single species but they are also highly inbred. They have much less genetic diversity than do the members of a natural species. If one plant is susceptible to a given disease or unfavorable environmental condition, then the chances are that all the others are too. In 1970 a corn leaf blight wiped out many fields of corn in the United States and there were no other plants around to fill the niche that was left vacant. But when a blight killed all the chestnuts in the eastern hardwood forests, the community

Figure 30–17. A wheat field, a man-maintained early successional stage. (Photograph courtesy of the United States Department of Agriculture.)

as a whole was hardly disturbed because other hardwood trees could take the place of the chestnuts.

Climax communities, especially forest ones, also have an ameliorating effect in buffering and controlling such physical forces as water and temperature, and thus contribute to the stability of the environment. To ensure both a productive and a stable environment, man must constantly maintain a good mixture of early successional stages for production and mature stages for stability in the communities over which he has control.

SUGGESTED READINGS

Boughey, A. S.: *Ecology of Populations,* Macmillan Publishing Co., Inc., New York, 1968.

Clapham, W. B., Jr.: *Natural Ecosystems,* Macmillan Publishing Co., Inc., New York, 1973.

Colinvaux, P. A.: *Introduction to Ecology,* John Wiley & Sons, Inc., New York, 1973.

Emmel, T. C.: *An Introduction to Ecology and Population Biology,* W. W. Norton & Company, Inc., New York, 1973.

Kucera, C. L., and J. J. Rochow: *The Challenge of Ecology,* C. V. Mosby, St. Louis, Mo., 1973.

Murie, A.: *The Wolves of Mount McKinley,* Fauna of the National Parks of the United States, Fauna Series No. 5, U.S. Government Printing Office, Washington, D.C., 1944.

Zinsser, H.: *Rats, Lice, and History,* Little Brown and Company, Boston, 1935.

CHAPTER 31

Behavior

Man has always been interested in animal behavior. The primitive hunter had to know what his quarry was likely to do—would it hide, flee, or charge? The herdsman needed similar knowledge about the animals in his keeping, and he also looked for ways to modify their behavior, to make them do what he wanted them to do. Most modern men are neither hunters nor herdsmen. But the doctrine of evolution has given man an even more cogent reason for studying animal behavior, for behavioral patterns, like physical characters, have a genetic basis. They are inherited and they are subject to the action of natural selection. If a man is to understand his own behavior and that of his fellow men, he needs to know what it is he has inherited.

STUDY OF BEHAVIOR

It is not easy to give a simple definition of behavior. It is sometimes described as a response to the environment. But a flower plant in a window box grows toward the light; this is a response to an environmental stimulus, although most people would hardly think of it as behavior. Nor are physiological responses, such as the increase in red blood cells when a man moves to a high altitude, usually considered behavioral, although the physiological state of an animal can modify its behavior. Perhaps the simplest definition is that behavior includes all the acts performed by an animal.

Biologists have approached the study of behavior in three ways. Neurophysiologists have been concerned with the functioning of the nervous system, with the mechanisms of sensory stimulation, impulse propagation, and synaptic transmission, and with the identification of functional areas in the brain. Behaviorists have concentrated on problems of learning, that is, the modification of behavior by environmental factors. They have usually worked with cage-reared animals of a few species (most often the white rat). The behaviorist's approach is experimental and he attempts to shield the animal from any outside stimuli except the one whose effect he wishes to measure (usually a schedule of rewards and punishments). Ethologists are interested in innate behavior patterns, their evolution, the way they adapt the animal to its environment, including the other animals with which it shares the environment. Their approach is comparative. They study as many different species as possible under as natural conditions as possible.

These three methods of studying behavior are moving closer together. Few behaviorists today maintain the extreme position that the mind of an animal is a complete *tabula rasa*, a blank page on which every behavioral pattern is written by the environment. They are studying more kinds of animals in more varied situations. Ethologists recognize that even innate patterns can be more or less modified by learning. Both must ultimately explain their findings in the terms of the neurophysiologist, who in turn must relate his findings to overt behavior patterns. When a unified science of behavior is achieved, it will encompass the findings of all three.

We have already discussed some of the results of neurophysiological studies in Chapter 9. Here we are mostly concerned with the ethological approach.

ELEMENTS OF BEHAVIOR

What you see when you watch an animal do something is usually a sequence of muscular movements by the animal. (It may be lack of movement or a change of color resulting from the migration of pigment cells.)

Fixed Action Patterns

A squirrel, scurrying along the forest floor, finds an acorn. If it is hungry, it sits up on its haunches, grasps the nut in its front paws, gnaws through the shell, extracts the nutmeat, and eats it. If the squirrel is not hungry, it holds the nut in its mouth, hunts around until it finds a place that suits it, digs a hole with its front paws, pushes the nut into the hole, and buries it. See Figure 31–1. Each of these behaviors consists of a sequence of fixed action patterns, a series of highly coordinated muscular movements that an animal can perform without

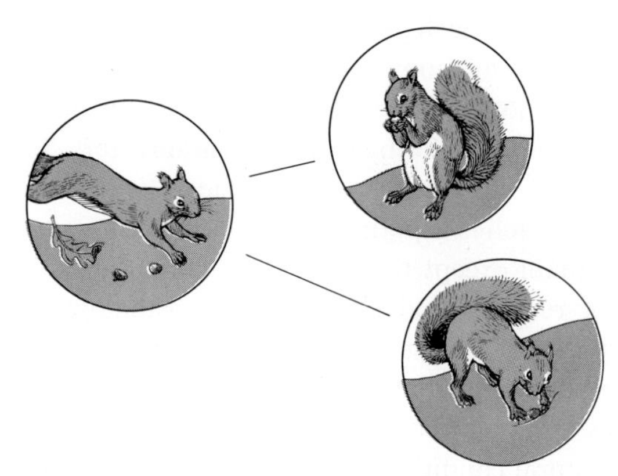

Figure 31–1. A squirrel finds a nut. If he is hungry, he sits up and eats it. If he is not hungry, he buries it. (Original drawing by Margaret Colbert.)

previous practice or without having learned them from observation of another animal. The sequence of motor elements in a fixed action pattern is constant. A squirrel always sits up or digs, or scrapes dirt into a hole in the same way.

These behavior patterns of the squirrel are innate. A young European squirrel was hand reared in a cage with a bare floor and fed only liquid food. When the squirrel reached the age at which it would normally have been eating and burying nuts, it was given some hazel nuts. The first few it gnawed open and ate in typical squirrel fashion, then it picked one up in its mouth, carried it to the corner of its cage, and went through the motions of burying it. It had had no previous experience with nuts or with digging, and it had never seen an adult squirrel eat or bury a nut.

A fixed action pattern is species specific, that is, all normal members of the species perform the action in the same way. But it is not necessarily, or even usually, confined to a single species. The studies on nut-burying behavior were made on the European red squirrel, but anyone who has watched an American gray squirrel has seen the same behavior.

Like other innate characters, fixed action patterns may be retained even though they have not had any obvious functional value for a long time. Varieties of hornless goats and cattle still carry on the fixed action sequences of battling with their nonexisting horns just as their horned relatives do.

A fixed action pattern is not the same as a reflex or a chain of reflexes. A reflex is a fixed response to a stimulus, but the response of a squirrel to a nut varies with the squirrel's state of hunger. Furthermore, if the squirrel is scolding another squirrel, or chasing a female, or running from a dog, it will pay no attention to the nut. The muscular movement of a reflex such as the knee jerk occurs once when it is triggered by a stimulus. If the movement is to occur again, the stimulus must be repeated. A fixed action sequence, although usually initiated by some external stimulus, will continue even when the stimulus is removed. A nesting graylag goose, like many other ground-nesting birds, will retrieve an egg that has rolled out of the nest. She stretches out her neck, hooks her lower bill over the egg, and rolls it back by drawing in her head. See Figure 31–2. Once she has started the sequence, she will continue it to completion even though the stimulus, the egg, is removed.

A fixed action sequence is frequently preceded by, or accompanied by, an orientation movement called a *taxis.* If a circling hawk spots a flock of birds, before it begins its dive it must position itself so that it is aiming at one particular bird. Unlike a fixed action sequence, a taxis is dependent on the continuation of the stimulus. When the graylag goose is rolling an egg into her nest, she constantly moves her bill from side to side to correct the "wobble" that results from trying to roll an object that is ovoid rather than round. This is a taxis component of the behavior. When the egg is removed before the task is completed, the sidewise movements stop. A taxis movement is more variable than a fixed action pattern because of its greater dependence on variable external stimuli.

Sign Stimuli

A fixed action sequence is frequently performed in response to an external stimulus and usually only one or a few of the many potential stimuli in the animal's environment will elicit the response. Such a stimulus is called a *sign stimulus.* It is typically very simple. The common tick climbs out on the end of a twig and responds, not to the sight of a mammalian host but to the odor of butyric acid, a component of sweat. When this odor reaches it from a mammal passing underneath, the tick drops from its perch to the host's back and proceeds to drink its blood. If a sweaty man has been sitting on a rock under the perch and then gets up, the tick will drop to the rock and try to suck blood from it. It is the odor, not the presence of the host, that is the sign stimulus causing the tick to drop. In the spring, an English robin will attack another male robin that enters its breeding territory. The sign stimulus is the red breast of the rival. The bird will attack a tuft of red feathers mounted on a stick but not a stuffed specimen of a male robin that lacks the red breast. Again, a hen responds to the sign stimulus of the distress call of a chick by running to the rescue, but if the

Figure 31–2. A graylag goose retrieves an egg that has rolled out of the nest.

chick is put under a glass jar so that the hen can see it but not hear it, she is utterly indifferent to its frantic struggle to escape. It is only one aspect of the situation, not the total situation, that serves as a sign stimulus.

A single object may present a number of different sign stimuli to an animal. Which stimulus the animal responds to varies with its motivational state. Like the graylag goose, a brooding herring gull will roll an egg back into the nest. A number of different egg models were presented to nesting gulls. In this way it was found that the sign stimulus for egg rolling is the speckling on the shell. The more heavily speckled the models were, and the more the specks contrasted with the background egg color, the stronger was the response. The shape of the egg made no difference. The birds rolled square or cone-shaped models into their nests and brooded them as long as they were speckled. Away from its own territory, a gull is not adverse to eating the eggs of another gull. Here the sign stimulus by which an egg is recognized is its unbroken shape. When the chicks hatch, adult gulls quickly remove the broken shells from the vicinity of the nest. (Broken eggs attract predators that prey on the chicks.) The sign stimulus for removal is the jagged white line that surrounds the hole through which the chick emerged. When a gull is presented with an empty shell with a smooth-edged hole, the sign stimulus for removal is lacking and the bird rolls the shell into the nest.

Additive Stimuli. A sign stimulus usually releases only one reaction, but sometimes several sign stimuli each evoke the same response. When these stimuli are present together, their releasing values may supplement each other so that the strength of the response increases. If a belligerent male cichlid fish erects its dorsal fin, takes up a position parallel to its opponent, and lashes its tail, it evokes three times as much fighting response as it would if it presented only one of the sign stimuli. The intensity of the reaction may depend, then, not only on what sign stimuli are present but also on how many there are.

Superoptimal Stimuli. Exaggerated stimuli sometimes produce a stronger response than normal ones. Gulls try to incubate huge painted wooden eggs in preference to ones of natural size, although they cannot even sit on the super-eggs without toppling off. Similarly, the ringed plover prefers eggs on which black spots have been painted on a white background to its own less-contrasting, light brown eggs with dark brown spots.

It seems likely that this phenomenon of strong response to supernormal stimuli is really more common than reports indicate. Hungry nestling songbirds respond to the sight of the parent, or to the shaking of the nest when the adult alights on the rim, by gaping broadly. The yellow lining of the mouth exposed by the gape is a sign stimulus to the parent to drop food in it. Perhaps a mother songbird actually prefers feeding a cuckoo young in her nest because its enormous inviting gape is a superoptimal stimulus compared to the smaller gape of her own young. Humans too may be susceptible to superoptimal stimuli. Think of the way artists accentuate certain aspects of the human body in painting advertisements. The innate fondness for infants may be the basis of the popularity of the widely loved cartoons of the late Walt Disney, who gave his characters supernormal baby faces (see Figure 31–3).

Releasers. Most of the sign stimuli we have discussed here belong to a special class called *releasers*. A releaser is a sign stimulus presented by one animal that evokes a response in another. Releasers are frequently means of communication between members of the same species. The begging of young birds releases feeding in the adults. Expressive movements, which indicate the emotional state or intentions of an animal, also serve a communicative function and may be classed as releasers. The difference in mood between a dog that is wagging its tail and one that has its tail tucked between its legs is clear not only to humans but also to other dogs, and their reactions may be modified accordingly.

An intraspecific releaser and its response evolve together. A clear-cut, albeit perplexing, case has recently been reported in crickets. The singing of the male cricket attracts the female. The rhythmic structure of the song differs

Figure 31–3. The rounded facial contours of young animals may serve as sign stimuli to arouse parental behavior in adults.

from one species of cricket to another, and the female moves to the calling male of her own species rather than to one of another species. A hybrid form was produced in the laboratory by crossing two species that do not normally interbreed. The songs of the male hybrids were intermediate between those of the two parent species, and the female hybrids turned toward the calls of the male hybrids in preference to the calls of either of the parent species. The genetic mechanism by which releaser (call) and response are linked is unknown.

Releasers are also found in interspecific interactions. Many small songbirds flee at the sight of spots that look like large eyes, probably because eyes may signal an approaching predator. The birds themselves prey on insects. Some moths have conspicuous eyespots on their hind wings. The spots are hidden when the moth is at rest but at the approach of a bird they suddenly flash forth as the moth unfolds its wings. See Figure 31–4. The startled bird halts in its attack and the moth escapes.

Motivational State

A squirrel may respond to a nut by eating it, burying it, or ignoring it. Furthermore, the strength of the stimulus needed to evoke a fixed action sequence may vary from time to time. Sometimes it takes a very strong stimulus to produce a response, sometimes a weak stimulus elicits a full response from the same animal. If an animal has had no opportunity to perform a given fixed

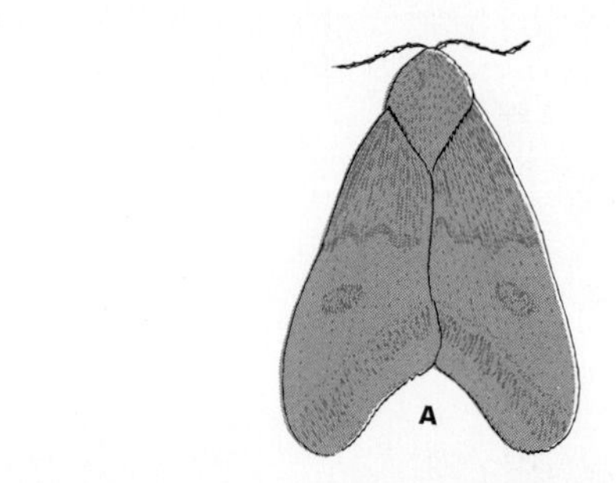

Figure 31–4. A startled moth spreads its wings to show eyespots.

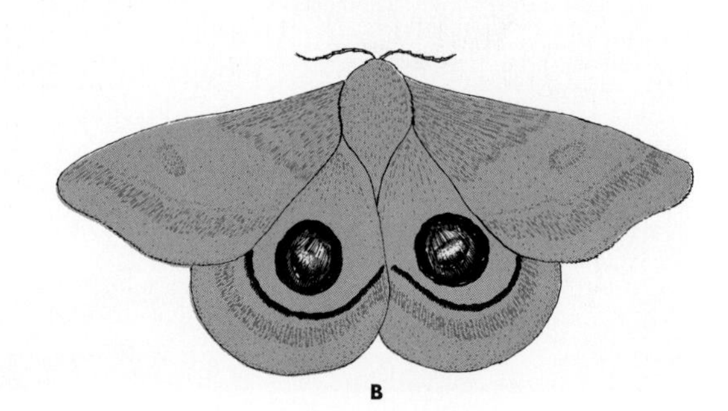

action sequence for some time, it may do so even in the absence of the stimulus that usually evokes it (a *vacuum activity*). A well-fed pet starling, which had never had an opportunity to catch insects, would sometimes fly from its perch, go through the motions of catching something, return to its perch, make killing movements, and finally swallow. Apparently there are intervening variables between the stimulus and the response. The nervous system of an animal is not just a relay station or telephone switchboard. It can modify the way an animal perceives and reacts to a stimulus, and it also seems able to initiate activity.

One intervening variable is the length of time since the action was last performed. Small birds will "mob" an owl they find resting during the day. They flit about and scold the owl. Gradually the birds' response wanes and then they will not mob the owl again for several minutes. The maturational state of an animal is also an intervening variable. A small puppy does not respond to a female in heat the way an adult dog does. Learning, which we discuss in a following section, can also modify behavior. Beyond these variables, the motivational state of an animal changes from time to time. The animal shows a tendency to eat or drink or mate. Many tendencies are clearly related to the physiological state of the animal.

The Hypothalamus. The hypothalamus in the brain monitors the condition of the blood. If the water content of the blood is too low, the animal searches for water. If the lateral part of the hypothalamus of a rat is injured, the animal cannot be forced to drink. Indeed, it will die of dehydration even when water is readily available. Other areas of the hypothalamus control the activities that are associated with feeding or with thermoregulation.

Endogenous Rhythms. The rhythmic changes in activity known as *endogenous rhythms* seem to have a physiological basis. These rhythms are timed so that they correspond to changes in the environment, to the seasons, to phases of the moon, to tides, or to day and night. The environmental rhythms act as entraining agents, but the animals are not continually dependent on them. If an animal is active at night and sleeps during the day, it will continue in this established pattern if it is kept in constant darkness or constant light. It is as if it had an internal clock to tell it when to wake up and when to go to sleep. Such a daily rhythm is called *circadian* (meaning about a day). The internal clock can be reset. If the periods of light and dark are reversed so that the animal is kept in the dark during the day and in the light at night, it will gradually shift its activity period to conform to the new time schedule (see Figure 31–5). Man is one of the many animals showing circadian rhythm. If you should fly, say, from the United States to Japan, you would feel off balance for the first few days because your circadian rhythm had not yet adjusted to the new time schedule.

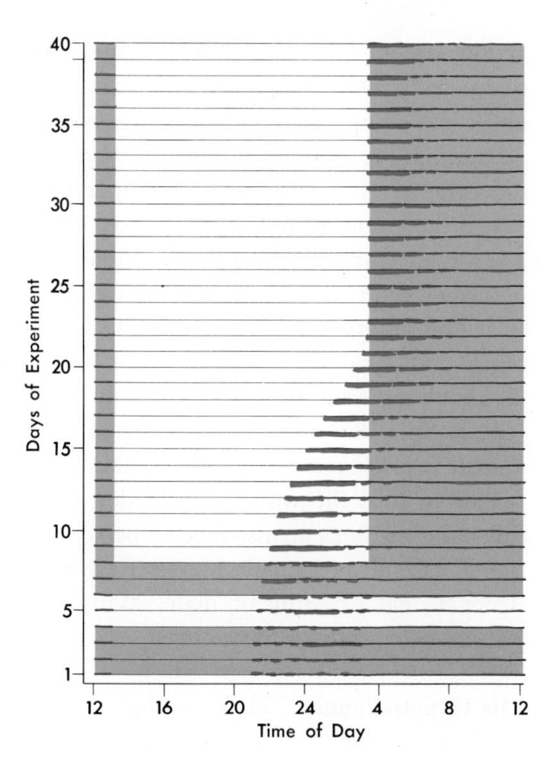

Figure 31–5. A flying squirrel is normally active at night. When one was kept in constant dark, or in constant light, it maintained its normal rhythm. When an artificial day-night pattern was established, the squirrel shifted its time of activity to correspond to the new period of darkness. From P. J., De Coursey, "*Phase Control of Activity in a Rodent,*" *Cold Spring Harbor Symposia,* **25:**49 1960.)

Hormones. The influence of the level of hormones on behavior is especially well shown in the various activities associated with reproduction. If a cock is castrated, it will cease crowing and attempting to mate, but if the appropriate sex hormones are injected into the castrate, it will again tread on the hens and announce the coming of dawn.

One of the best-known examples of the interaction of nervous and hormonal factors is that of nest building by canaries. Like many other songbirds, a canary builds her nest in two phases. First, she forms the basic cup of grass, then she lines it with soft feathers. An increase in estrogen initiates construction of the nest. As the canary builds the grass cup she tries it out, sits in it, and molds it into shape. As egg-laying time approaches, she begins to lose feathers, again under the influence of hormones, so that a bare brood patch develops on her breast. Other hormones make the skin of the brood patch sensitive to touch and heat. As she moves around the nest, the skin is stimulated by the scratchy grass and she responds to the stimulus by lining the nest cup with soft feathers. The entire nest building is an integrated interaction between external stimuli and hormonal and nervous mechanisms that results in specific sequences of fixed action patterns.

Hierarchical Organization

The behaviors associated with such tendencies as eating, drinking, sleeping, and mating are frequently organized into well-marked sequential phases (*hierarchies*). First there is a search—the hungry hawk flies up, soars and circles as it hunts for its prey, or more precisely hunts for a sign stimulus that will switch its behavior to the next phase. The search is called *appetitive behavior.* It may be very similar, whether the bird is searching for food, or nesting material, or a mate. It is succeeded by a taxis as the bird orients itself to the stimulus. The fixed action sequences of stooping, striking, and killing follow. Finally the hawk eats its prey. (See Figure 31–6). The action that terminates such a sequence, here eating the prey, is called a *consummatory act.* It is normally followed by a quiescent period during which the animal will not again perform that particular behavior sequence, although it may, of course, be doing something else.

CLASSIFICATION OF BEHAVIORS

Behavior patterns can often be classified by the tendencies of which they are the expression. The tendency to escape from danger usually takes precedence, although it may be suspended, or at least its expression may be modified, by the tendency to protect the young. Eating, drinking, and mating are obvious categories. Many animals have behavior patterns that are concerned with the care of offspring. These include nest building and den construction as well as feeding, cleaning, and guarding the young. Some animals have marked behavior

Figure 31–6. The hunting sequence of a hawk includes orientation on a specific prey (a taxis movement) and culminates in the consummatory act of eating. (Original drawing by Margaret Colbert.)

patterns associated with elimination. A dog will not defecate in its home area. It will not use a box as a cat will. This is why it has to be walked and may be so annoying to the neighbors. A cat, on the other hand, digs a hole, defecates in it, and buries its feces. Both patterns have obvious sanitary advantages for animals that have fixed home sites.

Thermoregulatory behavior includes not only panting and shivering but also such actions as moving from sun to shade and digging a cool hole in the dirt in which to rest. Some behavior patterns are concerned with the care or grooming of the body—a cat washing its face or a bird preening its feathers. Many animals have marked exploratory behavior. A rat will investigate and learn a maze even without being rewarded.

Usually the stimulation of one behavior pattern inhibits the appearance of another, but there are times when two or more tendencies are present at the same time. Several types of behavior are associated with the simultaneous arousal of conflicting tendencies.

Conflict Behavior

In conflict (*ambivalent*) behavior both tendencies are expressed. When two aggressive animals threaten each other, the tendencies to attack and to retreat are both aroused and the result may be that the animals will circle each other. Small boys squaring off for a fight also circle. Mark Twain describes an aggressive encounter between Tom and a strange boy: "Neither boy spoke. If one moved, the other moved—but only sidewise, in a circle; they kept face to face and eye to eye all the time." The ambivalence may be successive, as when two dogs engaged in a boundary dispute alternately attack and flee. See Figure 31–7.

Figure 31–7. Conflict behavior. Both the man and the gull have adopted a threatening posture but remain frozen in this position rather than attacking. Such behavior patterns result from the simultaneous arousal of the desire to attack and the desire to escape.

Displacement Activity

When two behavior patterns are aroused at the same time, the result may be an action that is entirely different from either of the ones normally shown (see Figure 31–8). If you are asked a question to which you do not know the answer, you may be in a conflict situation. You want to answer but you are afraid you will show your ignorance if you do. You may then scratch your head —a behavior unrelated to either of your conflicting tendencies. Behavior of this sort is called *displacement activity* and is very common. Fighting cocks in a pit may suddenly stop fighting and, without running, start pecking at imaginary grains on the ground. The conflict between aggression and flight shows up as a fixed action pattern that is part of feeding behavior.

Redirected Response

An individual may express his aggressive drive but direct it elsewhere than to his adversary (see Figure 31–9). A blackbird in a fight may peck aggressively at a leaf or twig instead of his foe, as an angry man may pound on a table or stamp on the floor.

LEARNING

The question of whether behavior is instinctive or learned, inherited or produced by the impact of environment is really meaningless. As with physical characteristics, a behavioral trait can be both genetically determined and dependent on the proper environment for its development.

Figure 31–8. Displacement activity. A starling, confronted with a rival, preens its wing. An irritated man who does not wish to display anger scratches his head.

Figure 31–9. Redirected response. A blackbird pecks a leaf rather than its rival. An angry man pounds his fist on a table rather than hitting his opponent.

Interaction of Instinct and Learning

Learning has been defined as "that process which manifests itself by adaptive changes in individual behavior as a result of experience." Consider again the squirrel and the hazel nut. An immature squirrel gnaws all over the shell of its first nut and may take fifteen minutes to open it. An experienced squirrel gnaws along a groove in the shell and cracks it in a couple of minutes. The behavior of gnawing nuts is innate, but the squirrel must learn the best way to crack each of the several different kinds of nuts it may encounter. On the other hand, a squirrel buries a nut as efficiently the first time it tries as the fiftieth. The best way does not differ with the kind of nut.

Each different species of bird has its own song. If a newly hatched white-crowned sparrow is isolated so that he never hears a male of his own species singing, when he is mature he will produce a simplified song that in its phrasing is still recognizably a white-crowned sparrow song. If he hears only males of other species as a juvenile, he will still produce this simple song, but he must hear a male of his own species singing during his first three months of life in order to sing the full song himself when he matures. Apparently, he inherits the basic phrasing and the tendency to copy only the song of his own species and then learns the precise details.

We cannot generalize from this to other species of birds. When a song sparrow reared by canaries starts to sing, he will produce the normal song sparrow song that he has never heard. But a bullfinch fostered by canaries will sing like a canary even though other bullfinches are singing in the same room. How much of a given behavior pattern is innate and how much is learned can vary even between closely related species.

Types of Learning

At the neurophysiological level, learning may well be a unitary process, but at the behavioral level, it is convenient to recognize different kinds.

Habituation. A scarecrow in a field may be effective for the first few days; thereafter the birds will ignore it. They have become habituated to it and no longer respond to the stimuli that first alarmed them. Newly hatched chicks peck at any small object on the ground but they soon ignore pebbles and other non-food items. Habituation means that an animal learns to ignore neutral stimuli, those that promise neither reward nor punishment. Almost all animals can become habituated.

Associative Learning. Associative learning includes both classical conditioning and trial and error learning. The study of classical conditioning stems from the work of I. P. Pavlov, who presented hungry dogs with food. Each time the food was offered, a bell rang. The dogs salivated when meat powder was blown into their mouths, which was a reflex action. Soon they began to salivate when the bell rang, even though they were given no food. The dogs had formed an association between the sound of the bell and food. The response that results when such an association is formed between unrelated stimuli is called a *conditioned reflex.* How frequently it occurs under natural conditions is not known.

Trial and error learning in the laboratory is often called *instrumental conditioning* or *operant conditioning.* A cat is shut in a cage that it can open from the inside by pulling a loop. As it moves about trying to get out, by chance it pulls on the loop, the door opens, and the cat escapes. After this has happened a number of times, the cat learns to pull the loop every time it wants to leave the box. By trial and error, it has learned to associate a specific action with a desired result. The behavior that results from trial and error learning is in no sense reflexive. A rat learns to respond to a flashing light by pressing a lever that releases a food pellet into its cage. If the rat is to the right of the lever when the light flashes, it moves to the left; if it is to the left, it moves to the right. Individual rats differ in the way they press the lever; some use the right paw, some use the left, and some press it down with the chin. The animal apparently forms a mental construct of the desired outcome and modifies its movements accordingly. Most of the learning animals do in the wild is probably by trial and error.

Latent Learning. An animal can learn without the reinforcement provided by rewards and punishment. It becomes familiar with its surroundings during exploratory behavior. Later this knowledge may help it escape a predator or find food, but at the time it is not associated with any material benefit. The song the bullfinch learns as a juvenile is not even expressed until some months later.

Insight Learning. Insight learning is the ability to achieve a desired end by combining previous experiences in a new way, without going through the steps of trial and error learning. A chimpanzee can pile up boxes and climb them to reach a banana suspended from the roof of its cage. Watching the animal as it looks at the banana, at the boxes, and at the spot on the floor beneath the banana leaves little doubt that it is mentally combining previous experiences to arrive at a solution of the problem.

HUMAN ETHOLOGY

The ethological study of human behavior is just beginning, yet already it is clear that much that has been learned about animal behavior is also applicable to man. We have indicated the similarity of human and animal reactions to the arousal of conflicting tendencies. Many expressive movements in man, such as crying, smiling, and laughing, seem to be innate. They are common to all races and are found even in children who are born blind and deaf. If a newborn baby is held upright with its feet pressed against a hard surface, it makes walking movements. Walking is not the only method of bipedal locomotion—kangaroos, many desert rats and mice, and many birds hop. All men can learn to hop, yet all men walk. Walking is not a learned behavior but a sequence of fixed action patterns that appears spontaneously at the appropriate stage in the development of the child. Man differs from other animals in the extent to which his innate behavior patterns can be modified by learning, but he does not lack these patterns.

SUGGESTED READINGS

Darwin, C.: *The Expression of the Emotions in Man and Animals,* John Murray, London, 1872.

Dethier, V. G., and E. Stellar: *Animal Behavior,* 3rd ed., Prentice-Hall, Inc., Englewood Cliffs, N. J., 1970.

Eibl-Eibesfeldt, I.: *Ethology: The Biology of Behavior,* Holt, Rinehart and Winston, Inc., New York, 1970.

Hinde, R. A., ed.: *Non-Verbal Communication*, Cambridge University Press, London, 1972.
Jolly, A.: *The Evolution of Primate Behavior*, Macmillan Publishing Co., Inc., New York, 1972.
Lorenz, K. Z.: *King Solomon's Ring*, Thomas Y. Crowell Company, New York, 1952.
Manning, A.: *An Introduction to Animal Behavior*, Addison-Wesley Publishing Company, Reading, Mass., 1967.
Tinbergen, N.: *Curious Naturalists*, Basic Books, Inc., Publishers, New York, 1958.

CHAPTER **32**

Social Behavior

Man is a social animal, living in groups with others of his kind. He is not alone in this, for many animals form social groups. A group is considered social when its members stay together as a result of the way in which they react to one another; if they merely happen to be in the same place because the individuals are responding to some other factor in the environment, as insects cluster around a light, they form an aggregation, not a social group. At the lowest level, social behavior may simply involve the coming together of two individuals for the purpose of reproduction, with the animals leading solitary lives the rest of the time. At the highest level are the very complex social organizations of insects such as ants, bees, and termites, in which each individual is completely dependent on the group and each contributes to the welfare of the group. Fascinating as insect societies are, though, they are less important for an understanding of human society than the more loosely organized associations of the vertebrates.

MATING

When reproduction simply involves the release of eggs and sperm in water, there may be no interaction between individuals at all. Both sexes simply release their gametes in response to a given set of environmental conditions so that the eggs and sperm are shed at about the same time and place. Usually though, and always when fertilization is internal, reproduction involves the cooperation of the two individuals. Sometimes the interaction is very brief. The male finds a female, discharges his sperm, and the two then separate. Often, though, pairs are formed that remain together for a time; it may be only for a season or it may be for life.

Finding Mates

Before mating can take place, the individuals have to come together. For animals that live in social groups outside the breeding season, this may not be much of a problem. The prospective mates are already members of the group. But sometimes animals are solitary or the males and females live in separate groups for most of the year. Then behavior patterns that serve to bring the two together are necessary.

Approach. Sometimes the male selects the breeding site and advertises to attract females. In the spring, male frogs move to ponds and each selects a calling station. The sound of the chorus draws the females to the pond. Males of migrating songbirds arrive in the spring before the females. Again each male songbird selects an area and sings vigorously from a conspicuous perch; when the females come they are attracted by the singing males. Some male birds, such as the grouse and prairie chicken, gather in a communal display area called a *lek*. They may jump up and down, spread ornamental tails, ruffs, or wings, and make loud booming noises to attract the females (see Figure 32–1). Among mammals, solitary males usually wander widely during the mating season searching for females.

Recognition. It is not enough for the male and female to be together in the same area, they must be able to recognize each other as potential mates, that

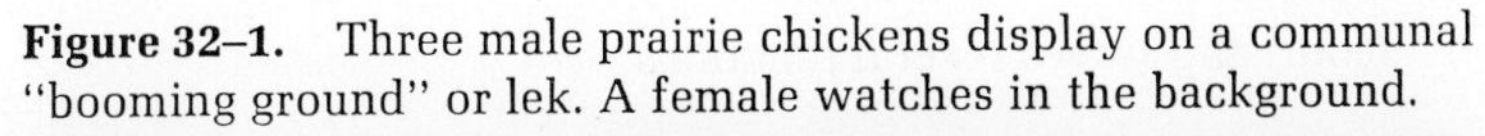

Figure 32–1. Three male prairie chickens display on a communal "booming ground" or lek. A female watches in the background.

is, not just as members of the same species but as individuals of the appropriate sex and reproductive state. Structural features, sounds, odors, and behavior can all act as recognition characters. Male and female flickers are colored much alike, but the male has a black moustache patch at the corner of the mouth (see Figure 32–2). When a similar patch was painted on the female of a pair, her mate attacked her as if she were another male. The male of the common green lizard (*Anolis*) of the southeastern states displays a brilliant red throat fan during the breeding season. Another male is stimulated by this either to flee or to spread his own throat fan, which may start a fight. A ripe female neither flees nor displays and so is recognized and accepted. Male dogs apparently recognize a bitch in heat by smell.

Courtship

Even after the male and female have found each other, copulation usually does not take place at once. There is a longer or shorter period of courtship, an engagement period, if you will. Some birds that mate for life form pairs during their first summer, although they will not be ready to reproduce until the next spring. When mating is only for a season, the courtship period may still last for several weeks.

Courtship activities are many and varied. A male salamander may creep under his mate and carry her for a piggyback ride. Snakes of a courting pair often glide along side by side while the male caresses the female with his chin and flickers his tongue over her body. Figure 32–3 shows a pair of courting snakes. A male bird may bow and display before his mate, he may present her with food or nesting material, or the pair may preen each other. Carnivorous mammals frequently engage in a rough-and-tumble that is hard to distinguish from real fighting.

Courtship serves several functions.

Timing. Females of some species of animals are able to store sperm, sometimes for months or years, but usually for successful mating the gametes must be produced at the same time. Most animals reproduce only at certain seasons

Figure 32–2. The prominent black mark at the corner of the mouth of the male flicker serves in sex recognition.

Figure 32–3. Courtship behavior of the King Cobra. (From C. J. and O. B. Goin, *Introduction to Herpetology*, W. H. Freeman and Company, copyright 1962. New York Zoological Society photo.)

of the year, not throughout the year as man does. The breeding season is timed so that the young are produced when conditions are most favorable. Thus, outside of the tropics most breeding takes place in the spring and summer when more food is available. When the gestation period is long, as with deer, copulation takes place in the fall and the young are born in the spring. During the breeding season, the male is usually ready to produce sperm at any time, but the female may only ovulate once or a few times. Insemination must take place when she is ready to ovulate. The presence of her mate and his courtship activities stimulate hormone production and bring her to a state of reproductive readiness. The females of some birds and mammals do not ovulate at all unless stimulated to do so by a courting male. The female repulses the male until she is ready and indicates her willingness by a characteristic posture or action.

Conflict, Appeasement and Ritualization. Because an animal that approaches too closely may be a predator, most animals tend to avoid bodily contact with others. Also, during the mating season animals are frequently belligerent toward other members of the species. This means that, during courtship, both members of the pair may have conflicting drives. The male may be stimulated to attack the female or to flee from her, as well as to court her, and the same is true for the female. Much courtship activity shows the characteristics of simultaneous arousal we discussed in the last chapter.

Frequently, a fixed action pattern that appeared first in a behavior sequence as a result of simultaneous arousal becomes incorporated as a normal part of the sequence. Thus, displacement preening of the wings is part of the normal courtship display of a drake and now serves as a signal indicating the bird's readiness to mate. When a fixed action pattern belonging to one behavior sequence is modified by evolution to serve as a signal in another context, it is said to be *ritualized.* It is usually performed in an exaggerated, stereotyped way and the animal has often evolved special colors and structures that emphasize the movement. The spectacular tail of the peacock evolved as part of a ritualized courtship display. It is sometimes possible, by comparing related species, to deduce the course of evolution of a ritualized display. Courtship preening in the shelduck resembles ordinary preening. The mallard displays a brightly colored wing patch, but still preens. The gargany no longer preens, but touches a blue wing patch, whereas the mandarin points to an enlarged, bright orange feather (Figure 32–4).

An animal that is belligerent must be persuaded not to attack its potential mate. A female bird frequently adopts the posture of the begging young to appease the male and he may respond by feeding her. Anyone who has watched cardinals at a feeding station during the breeding season probably has seen this. Sometimes the male appeases by food begging. Appeasement signals and responses are often ritualized.

Human dances are highly ritualized. They are frequently performed as courtship displays, or as threat (war dances), or for appeasement of a deity. These are all situations in which simultaneous arousal is common and dances contain many ritualized movements derived from conflicting drives. Consider the circling movements that are a common feature of the dances of many diverse cultural groups or the exaggerated stomping of the feet found in the dances of the American Indians and the Highland Scots.

Reproductive Isolation. If reproduction is to be successful, the animals must breed with members of their own species. Many of the behavior patterns that

Figure 32–4. Evolution of ritualized preening in ducks. A: Preening during courtship by the shelduck does not differ greatly from ordinary preening. B: The mallard lifts his wing to display a brightly colored patch and preens behind it. C: The garganey indicates a blue patch at the front of the wing. D: The mandarin touches an expanded orange feather in the wing.

we have discussed have the additional function of preventing crossbreeding. Usually the signals by which individuals attract their mates and the various maneuvers during courtship are distinct for each species. And the individuals of a species tend to react only to the appropriate signals. Courtship is usually sequential, and at each step the partner must give the appropriate response if it is to continue. The male of the three-spined stickleback (a fish) builds a nest of plant material in a pit that he digs in the sandy bottom. When another fish appears in the vicinity, he dashes toward it and starts a threatening, zigzag dance toward and away (*ambivalent behavior*). A ripe female adopts a head-up position, displaying her silvery belly swollen with eggs (*appeasement behavior*). Her color and shape serve as releasers for the next phase in the behavior of the male. He swims toward the nest. If she follows him, he shows her the nest by turning on his side and thrusting his nose into the entrance. The female than pushes into the nest. The male prods her tail base with a series of quick thrusts. She spawns and leaves the nest, and he then enters and fertilizes the eggs. Figure 32–5 shows courtship in the stickleback. If at any stage in this sequence one of the fish fails to make the appropriate response, the next step is blocked.

FAMILY LIFE

Among birds, mammals, and some fishes, care-giving behavior is highly developed. Eggs must be brooded until they hatch, and the young must be protected and, usually, fed. The male may leave after mating is accomplished so that the care of the young devolves solely on the female. In some birds like the penguin and fishes (stickleback) the male parent takes over. Often the pair remains together and both contribute to the welfare of the young. Family life may involve three sets of interactions: parent-young; parent-parent; and young-young.

Parent-Young Reactions

Sometimes the young are born or hatch from eggs in an advanced state. Their eyes are open, they are clothed with hair or feathers, and they are able to walk and follow their parents almost at once. Such young are said to be *pre-cocial* (see Figure 32–6A). They are found among herbivores that live in open places—cattle, sheep, chickens, and ducks. Other young are much more immature at birth. They are frequently naked and blind; their neuromuscular systems are poorly developed so that they are incapable of the motor coordination that is necessary for them to follow their parents. They remain for a longer or shorter time in the nest, den, or burrow in which they were born. They are said to be *altricial* (see Figure 32–6B). Carnivorous animals typically have

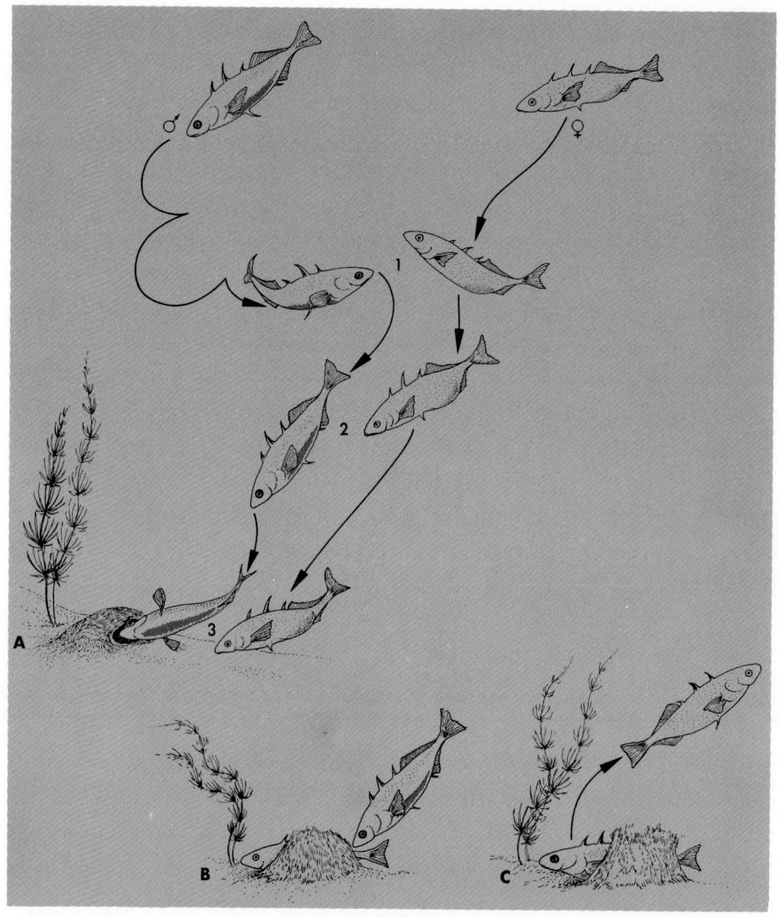

Figure 32–5. Breeding activities in the stickleback. A: Courtship. B: Spawning. C: Fertilization.

altricial young. These differences result in differences in the parent-young relationship.

Feeding. The young of precocial birds are not fed by their parents. Chickens hatch with the fixed action pattern of pecking at spots on the ground already developed, and they quickly learn to distinguish between edible and inedible objects. They respond to the clucking of the mother hen when she has found food by running to her. Gulls have altricial young. When the adult gull arrives back at the nest after a fishing expedition, the chicks peck at a colored spot at the base of its bill. The parent regurgitates some food, picks up a bit with its bill, and waits until one of the chicks succeeds in grabbing it. The colored spot

Figure 32–6. A: Precocial young piping plovers. (Philip Carpenter, from National Audubon Society.)

B: Altricial young chimney swifts. (Robert Knickmeyer, from National Audubon Society.)

serves to release the pecking response in the chicks, and the pecking itself is the sign stimulus for the parent to regurgitate. Feeding, then, depends on an interaction between parent and young. Baby mammals, of course, are suckled by the mother. A newborn pup attempts to suck anything warm and soft with which it comes in contact, but it soon learns to search for and suck a nipple. When wolf cubs are ready to eat meat, adult members of the pack return to the den and regurgitate food for them.

Elimination. When the young remain in the nest or den for an extended period, there must be some mechanism to prevent fecal material from accumulating. Baby barn swallows deposit their wastes in a neat packet in the mouth of the mother, who either eats it or carries it off and drops it elsewhere. A mother dog licks the anal region of the pup, which stimulates urination and defecation; the waste products are licked off by the mother.

Protection. When a parent bird sits on the eggs, it is doing more than protecting them from predators. Unless they are kept warm, the embryo birds die. Before their feathers grow out, very young birds also need to be guarded against changes in temperature, so the parents continue to sit when they are not hunting for food.

Young animals usually do not initially show fear. During this period they are closely guarded by the parent. Later the young respond to a warning signal by crouching down and remaining still, by running to a hiding place, or by running to the parent. In some ground-nesting birds, the parent leaves the nest at the approach of a predator and flutters off as if injured. When the enemy has been attracted away to a safe distance, the parent will fly up and may either attack the predator or return to the nest.

Learning. "It is a wise father that knows his own child." Many mammals and birds apparently do not recognize their offspring. The songbird feeding a cuckoo is a case in point. Apparently, the parent has learned only the location of the nest. Other species, though, do learn to distinguish their young from others of the same species. Parent herring gulls, which nest in closely packed colonies, will feed only their own chicks. Female fur seals may leave their young on shore in a "nursery" while they go off fishing, but when they return after many hours each mother seal goes straight to her own offspring. Female mammals clean their young by licking them just after they are born and this seems to be important in establishing the mother-young relationship. If a young goat is removed from its mother before she has cleaned it and then returned a few hours later, the mother will neither accept her offspring nor will she allow it to suckle.

A special kind of learning called *imprinting* takes place in the young of many precocial species. Shortly after birth, the young develop a tendency to follow a moving object. After they have followed it for a while, they apparently learn to recognize the object and thereafter regard it as their "mother," running to it when danger threatens and searching for it if they are separated from it.

They have become imprinted on it. Normally, of course, the first moving object a young animal sees is its own mother; it learns to follow her, and so remains close to its source of protection and, for young mammals, food (see Figure 32–7). The initial tendency to follow is unspecific. Young ducklings will follow a bird of another species, or a man, or a box pulled along the floor. Once imprinted on something, no matter how inappropriate, the animal will no longer follow its own mother. The time during which an animal can be imprinted is quite short. Mary's little lamb was apparently taken from its mother at an early age and cared for by Mary. It had become imprinted on her.

The effects of imprinting may last long beyond the period of infancy. If an animal has been imprinted on an inappropriate parent, it may subsequently not recognize members of its own species as potential mates but may make sexual advances to individuals of the foster parent species. A white peacock reared in the reptile house at a zoo ignored all peahens and persistently courted the Galapagos tortoises.

Parent-Parent Interactions

The members of a pair frequently remain together and both contribute to the care of the young. Parent birds may take turns incubating the eggs, or the female may incubate while the male stands guard nearby. The parent not incubating may bring food to its mate, and both parents usually feed the young. Sometimes a second clutch of eggs is laid before the young of the first brood are

Figure 32–7. A white-tailed deer fawn with its mother. (National Audubon Society Inc.—Leonard Lee Rue photo.)

independent. The female then may incubate while the male takes over the care of the young, or the parents may alternate the duties. Families of this sort are less common among mammals, probably because only the female can suckle. Such families occur in some carnivores, in which the female remains close to the den with the young and the male brings food to her. In general, as one would expect, family groups of father, mother, and offspring are more common among forms with altricial young. These family groups are advantageous, but they do pose some problems.

Recognition. A bird sitting on the nest tends to regard any approaching animal as a potential threat. It must learn to recognize its mate as an individual. Perhaps this is one of the functions of the courtship period. At any rate, for many species there is evidence that the members of the pair do recognize each other.

Timing. If both parents are to share in the care of the eggs and young, their activities must be synchronized. They cannot both incubate at the same time, nor can they both be away from the nest for long at the same time. Some birds have elaborate changeover ceremonies. When a herring gull has been off foraging for several hours, the drive to brood becomes stronger than the drive to hunt. It returns to the nest area, plucks some nesting material, and walks up to the nest, perhaps giving a certain call. The other partner then usually leaves the nest. If the latter's brood drive is still strong, it may refuse to depart and there may be a struggle. If one partner, for some reason, fails to relieve its mate, the sitting bird, lacking the stimulus to depart, may remain on the nest for days.

Young-Young Interactions

When more than one young are produced at a time, the interactions between them are also an important part of family life.

Competition. Young animals compete with each other for the food brought by their parents. The young animal that begs most vigorously presents the optimal stimulus to the parent and is one that is fed. Normally this results in an equitable division of food, because the hungriest of the young is the most ardent beggar and is likely to be the one that has not been fed recently. It sometimes happens, though, that one of the young is a weakling and begs less strongly than the others. It gets too little food, grows progressively weaker, and eventually dies. Here natural selection operates against the individual to maintain fitness in the species.

Play. Play is an important activity in the lives of many animals. It is most marked in juvenile mammals but is also present in young birds and in the adults of social species of both birds and mammals. Parents may play with their offspring and, when the animals live in social groups, playmates include not only siblings but other young of about the same age. Play is often easy to recognize but it is hard to define. It contains elements of many other behavior patterns, such as prey-catching, feeding, fighting, and sexual activity, often mixed together without any fixed sequence. The arousal of play does not seem to depend on the normal motivational states for such activities and it does not carry through to the completion of the consummatory act. A well-fed kitten will pounce on a sibling's tail, but it will not attempt to kill and eat the tail. During play a young animal perfects many of the skills it will need in later life. Play is an important part of socialization. In particular, for some animals, lack of the opportunity to play with other young animals during the juvenile stage seems to prevent successful sexual behavior later. Rhesus monkeys reared in isolation with dummy mothers are completely unable to mate with each other. If they are allowed to play with each other for a few hours as juveniles, they are more successful in their later mating activities.

Breaking the Bond

As the young reach the stage of being able to take care of themselves, they come more and more to resemble the parent in appearance. The parent's tendency toward care-giving behavior wanes, and an aggressive tendency normally aroused by competitors may take over. The adults begin to threaten the young (see Figure 32–8). For a while, the young may ward off attack by adopting appeasement attitudes, which often resemble the postures of the female during courtship. Eventually, though, the family may separate, with the individuals scattering or becoming part of a larger social group. Sometimes the family remains as the nexus of a social organization.

GROUP LIFE

Social groups beyond the level of the immediate family are often formed by birds and mammals and occasionally by other animals. Figure 32–9 shows part of a herd of ostriches. The size of the group depends largely on the position of the animals in the food chain. If a tenth of a hectare (quarter of an acre) provides sufficient food for one herbivore, then a hundred of them can live on a ten-hectare (twenty-five acre) plot. But the carnivore that feeds on the herbivore needs access to much more than a tenth of a hectare. Group feeding

Figure 32–8. South African Ostrich, *Struthio camelus australis*. An adult male, at the beginning of the reproductive cycle, breaks up the family flock by threatening and attacking the yearling birds, and driving them away before taking up territory with from one to three mates. (Photographed in South West Africa by E. G. Franz Sauer.)

Figure 32–9. South African Ostrich, *Struthio camelus australis*. A small segment of a large herd of ostriches. The (temporary) herd in the communal pasture and at the water hole is composed of a number of flocks that maintain their identities within these large temporary assemblies. (Photographed in South West Africa by E. G. Franz Sauer.)

is consequently more common in herbivores than it is in carnivores, except for the marine birds that feed on schooling fish. When carnivores do form social groups, the groups are usually smaller than those formed by herbivores.

Advantages of Group Life

There are other advantages to group life beyond the obvious one of facilitating the finding of mates.

Feeding. When one member of a social feeding group discovers a source of food, its actions attract the other members of the group. The sight of a gull diving at a school of fish draws other gulls to the scene. When there is a greater number of animals looking for food, it is more probable that one will find it, and the group as a whole benefits.

Several lions hunting together are twice as likely to catch their prey as is a solitary lion. The lions are able to kill larger animals, such as buffalo, and they utilize their kill more fully. They can consume a zebra in an hour and can prevent a takeover by scavengers. A single lion, however, can be driven from its kill by a pack of hyenas.

Protection. A group of animals gathered together is more likely to attract the attention of predators than is a single animal. The odor is stronger and the group presents a larger visual target. Many small, terrestrial animals are solitary and those that live in colonies are usually burrowing forms, such as the prairie dogs (rodents) of the central and western United States. They feed in the open but remain close to their burrows, to which they retreat at the slightest alarm. Prairie dogs gain protection from the group. If one prairie dog spots a hawk approaching, it gives an alarm call as it dives into its burrow, and all the prairie dogs in the vicinity duck down in their burrows. The alarm spreads rapidly through the whole prairie dog colony. Pronghorn antelopes have a white rump patch. When one is frightened, the hair on the patch stands erect and the conspicuous white rump serves as a sign stimulus to start other antelopes bounding away. See Figure 32–10. A predator does not dash headlong into a group but attempts to isolate a single victim from the group. It is the animal that lags behind, or panics and darts from the group, that is lost. Here it is obvious that selection pressure works to maintain the tendency of the animals to remain together. A group of animals can provide better protection for the young. A herd of musk oxen attacked by wolves gather in a circle with the cows and calves in the center and the bulls with their sharp horns facing outward.

Learning. Members of a social group are able to learn from each other. Jackdaws (relatives of the crows) will attack any living creature if it is carrying a

Figure 32–10. The conspicuous white rump patch of a startled antelope serves as a sign stimulus to elicit escape behavior in other members of the herd. (National Audubon Society, Inc.—Allan D. Cruickshank photo.)

black thing dangling or fluttering. (It could be a jackdaw caught by a predator, but "black thing dangling" is itself a sign stimulus for attack. Konrad Lorenz was once viciously set on by his flock of tame jackdaws when he pulled his black swim trunks from his pocket.) The attack is accompanied by a harsh, rattling cry. If an animal, or man, has aroused the rattling attack several times, the birds now recognize him and will begin rattling when he appears even if he is not holding the offending object. The rattling response to the sign stimulus is innate. However, unlike the young of many other species, a young jackdaw does not recognize a potential predator at first sight. He learns that a cat is dangerous by hearing the rattling response of adult members of the flock. Thereafter he will rattle at the sight of a cat although he has never seen one carrying a black object. Thus, knowledge of potential enemies is passed from generation to generation of jackdaws by tradition.

Social Adjustment

Animals that live together must have means of adjusting to each other. There must be some mechanism for keeping the group together and, if the animals are normally aggressive toward others of the same species as many birds and animals are, some way of suppressing intragroup fighting.

Recognition. Members of small groups are apparently able to recognize each other individually. The members of a wolf pack greet each other with friendly, tail-wagging ceremonies but will attack a stranger. See Figure 32–11. Birds in small flocks also seem to recognize the individual members of the flock. Large groups such as the migratory flocks of birds may be held together not by individual recognition but by the tendency of the members to respond to some species-specific signal. Birds may be attracted by a brightly colored rump patch that shows when one flies, or a wing patch, or by the characteristic call of their own species.

Social Facilitation. Social facilitation means the tendency of one member of a group to do what the others are doing. If one drake in a flock of ducks begins courtship display, the other drakes are also stimulated to such display. Social facilitation probably functions to keep the group together. When a flock of birds is feeding on the ground, one bird may signal its decision to fly by making intention movements, crouching, and then rearing up as if about to take off without actually doing so. The other members of the flock may be stimulated by this and all fly up together. Man too is subject to social facilitation. Did you ever yawn when somebody else near you yawned? Mob action is the result of the spread of emotion through a group by social facilitation.

Dominance. One means by which intragroup fighting is suppressed is the establishment of a *dominance hierarchy*. In a flock of chickens, one hen is dominant to all the others. If another bird approaches too closely when the dominant hen is feeding, she will peck it, but it never pecks back. A second hen may be pecked by the dominant bird, but she in turn pecks all the others, and

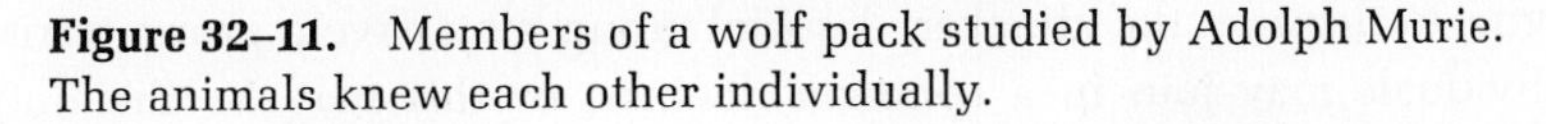

Figure 32–11. Members of a wolf pack studied by Adolph Murie. The animals knew each other individually.

so a definite hierarchy is formed down to the lowest bird who can be pecked by every other bird but can peck no one. Such dominance hierarchies are common in social animals, but not universal. They appear to be more rigid in herbivores than in carnivores that engage in cooperative hunting. Dominance hierarchies are lacking in a lion pride (a social group).

Social rank is frequently determined by tournaments rather than true fights. Oryx gazelles are equipped with long, sharp-pointed horns. They may use them to gore an attacking lion, but in dominance combats they fence with them rather than attempt to stab each other. Bighorn sheep engage in butting contests. Rarely is an animal injured in such tournaments. Social rank is often expressed by ritualized gestures. A subordinate chimpanzee who has annoyed a dominant chimpanzee wishes to avoid punishment and at the same time to stay with the group. He may "present" in the posture of a female soliciting copulation—a ritualized submission gesture. The dominant chimpanzee may respond by mounting the subordinate briefly.

Territoriality. Usually an animal does not roam at random but has a definite home range to which it confines its activities. A part of the home range that is defended against invasion by other members of the species is called a *territory*. Males of songbirds in the spring each establish a territory before the arrival of the females. The boundaries of the individual territories are set by aggressive encounters between the males. Once territories are established, their limits are recognized. Two birds may bluff or threaten each other at a boundary, and one may chase another for a short way, but as soon as the chased bird is safely within his own territory, he becomes the aggressor and drives the intruder back. A small dog driven back into his owner's yard immediately turns and begins to threaten the attacker. A social group may have a group territory. The individuals may join in a common defense, or this may be the duty of the dominant male. Animals recognize each others' territories, and an intruder will usually flee rather than fight. Territories are maintained largely by ritualized displays. Much of the singing of birds in the spring is a warning to other males that the singer is within his own territory and will, if necessary, fight to maintain it. An animal may make intention movements of fighting when faced by an intruder, and this is usually sufficient to warn the intruder away. If an intruder is attacked, it usually makes submissive movements, which may resemble the appeasement movements of the female during courtship. These movements normally inhibit the aggressive drive of the attacker until the intruder has a chance to escape.

When animals that are normally territorial are overcrowded, as in a laboratory cage, they may switch their pattern of social organization to the establishment of a dominance hierarchy. A member of a territorial species that fails to establish a territory in which he is dominant usually does not breed, so that the change in organization of the group sharply reduces its reproductive rate and acts to relieve the overcrowded condition.

Types of Social Organization

There are many different patterns of social organization. Pairs of migratory song birds have individual territories in the breeding season but may later join in large feeding flocks. The young of the year sometimes form a separate flock from that of the adults. The feeding flocks are nonterritorial, and, because food at that time of year is abundant and widespread, there is little aggression. Marine birds, such as the gulls, have extensive hunting ranges and usually feed in flocks, but, although they nest in closely packed colonies, each pair has a small individual territory within the immediate vicinity of the nest. Flock-living ground birds, such as pheasants, turkeys, and chickens, have precocial young and usually range over a large feeding area. Frequently the flock consists of a male with a harem of hens and the accompanying chicks. Territories are usually not established, although the female may defend the vicinity of her nest. Dominance hierarchies are well developed.

Although sociality among carnivorous mammals is frequently limited to mating and a close but transitory female-young relationship, social groups that show a high degree of cooperation, not only in hunting but in the care and training of the young, are sometimes found. Wolves tend to be monogamous; a pair of wolves may mate for life. The pack usually consists of relatively few members but may include more than the parents and young of the year. A pack that Murie studied closely in the Mount McKinley area contained five adults: a mated pair, two single males, and a single female. Later two more males joined the group. All the adults spent much time during the day resting in the vicinity of the den in which the mated female had her pups. See Figure 32–12. One male seemed to be dominant, and the others fawned as they approached him, but encounters were always friendly and there was no intragroup fighting. Surprisingly, the dominant male did not seem to be the father of the pups. The adults hunted in various combinations. When the pups were very small, the mother

Figure 32–12. Members of Murie's wolf pack relax near the den.

stayed with them and the others brought food to her and, later, to the pups. As the pups grew older, the mother sometimes joined a hunting group, and when she did the other female remained with the pups. Murie once saw the group attack a strange wolf that approached the den area. The animals had a hunting range about 85 kilometers in diameter (51 miles), and the seven adults and pups traveled together throughout the winter. The next spring, the pair again had pups in the same den; the other female had mated with one of the bachelors of the previous summer and had her pups in a den a few miles away. Later she moved them to the original den. One of the other bachelors was attached to the first pair, but the other two had disappeared, perhaps the victims of trappers. None of the young of the previous year were present. After the pups were old enough to leave the dens, the two families were often seen together. Murie says, "The strongest impression remaining with me after watching the wolves on numerous occasions was their friendliness."

Primate societies are very varied, with their pattern being at least partly determined by the ecological niche of the animals. Thus, diurnal, arboreal, leaf- and fruit-eaters such as the howling monkeys and langurs usually live in one-male harems or small groups. There is little aggression between the members, but the group territory is well marked and strongly defended by all, frequently by ritualized shouting matches with neighboring groups. Open country omnivores such as the savanna baboons live in large bands of up to a hundred members. Individual troops avoid each other, but there are no defended territories. Dominance hierarchies are well developed and only the dominant males can copulate with females.

Of the manlike apes, gibbons are diurnal, arboreal omnivores. They live as mated pairs with their subadult young in sharply defined territories. The arboreal, fruit-eating orangutans are solitary. Gorillas, semiterrestrial leaf-eaters, form small groups that may include one or more adult males. Groups have rather small and extremely overlapping home ranges but they do not defend territories. Dominance hierarchies are well developed, but subordinate males have ready access to females in heat. Chimpanzees are semiterrestrial omnivores that live both in forests and in open savannas. The chimpanzees of an area seem to form discrete regional populations of forty to eighty animals that know each other personally. Within the population, shifting small groups form and separate again. Groups may be all male, or mixed male and female, or a female with her young of various ages, or several females with young. Again, there are dominance hierarchies but no territorial defense and all males have access to females in heat.

Various attempts have been made to reconstruct the social structure of early hominids from a consideration of animal societies. Emphasis is sometimes placed on primitive man's hunting habits, sometimes on his open savanna environment, and sometimes on his anthropoid ancestry. Most omnivorous primates eat bird eggs, insects, and other small animals they come upon by chance. Cooperative hunting of larger game (monkeys, young baboons) has been reported for chimpanzees, but it always seems to be of animals discovered fortuitously in the vicinity of the group. Primitive man probably always relied largely on roots, seeds, and fruits, but meat-eating played a larger role in his life than in the lives of the other omnivorous primates. He was an active hunter and early learned to pursue large prey. He probably lived in fairly small groups that included several adult males who cooperated in hunting. Beyond this we can only speculate. Whether each group had a discrete territory, whether there was a dominance hierarchy, whether they formed mated pairs, or harems, or were completely polygamous, cannot be determined. The invention of agriculture shifted man's position in the food pyramid and allowed the growth of larger social groups, villages, towns, and cities.

SUGGESTED READINGS

Hinde, R. A., ed.: *Non-Verbal Communication,* Cambridge University Press, London, 1972.

Jolly, A.: *The Evolution of Primate Behavior,* Macmillan Publishing Co., Inc., New York, 1972.

Lawick-Goodall, J. van: *In the Shadow of Man,* Dell Publishing Co., Inc., New York, 1971.

Lorenz, K. Z.: *King Solomon's Ring,* Thomas Y. Crowell Company, New York, 1952.

Mech, L. D.: *The Wolf: The Ecology and Behavior of an Endangered Species,* The Natural History Press, Garden City, N. Y., 1970.

Murie, A.: *The Wolves of Mount McKinley,* Fauna of the National Parks of the United States, Fauna Series No. 5, U.S. Government Printing Office, Washington, D.C., 1944.

Schaller, G. B.: *The Year of the Gorilla,* The University of Chicago Press, Chicago, 1964.

———: *The Serengeti Lion,* The University of Chicago Press, Chicago, 1972.

Tinbergen, N.: *Social Behavior in Animals,* John Wiley & Sons, Inc., New York, 1953.

CHAPTER 33

Man in Nature

Many a man has dreamed of owning a square mile or two of land that was unmodified by human activity, but it is doubtful if such a place exists today outside of such inhospitable environments as Antarctica or some unscaled Himalayan peak. And if a man owned such an area, he could not know it, for as soon as he entered it to enjoy it, his visit would immediately remove it from the pristine. Even in the remotest Amazon jungles and most barren deserts, man's influences from hunting and the transmission of disease have been felt. Man has become a species whose environment includes essentially the entire world (see Figure 33–1). With his developed power of transportation, he has not only invaded all of the tolerable places on the face of the earth but has also come to depend on most of them to meet some of his needs. In practically any modern city, foods and beverages come from almost the entire world: coffee from South America, sugar from the West Indies, spices from the Orient, wheat from western Asia and North America, rice from the tropic lowlands, and wines from western Europe and western North America. Minerals such as iron, tin, and copper are shipped throughout the globe, and tobacco and medicines pass from one person to another around the world. The entire earth is rapidly becoming a single global human community.

We indicated in Chapter 30 that the population of man on earth is at present still in its logarithmic phase, but perhaps some actual figures will make the picture clearer. Although no exact count is possible, it is estimated that the world's present population is nearly 4 billion. The birth rate exceeds the death rate—for every fourteen people who die, thirty-four babies are born. As a result, the human population is now increasing at a rate of about 2 per cent a year. Nearly 80 million people are added to the world's population each year.

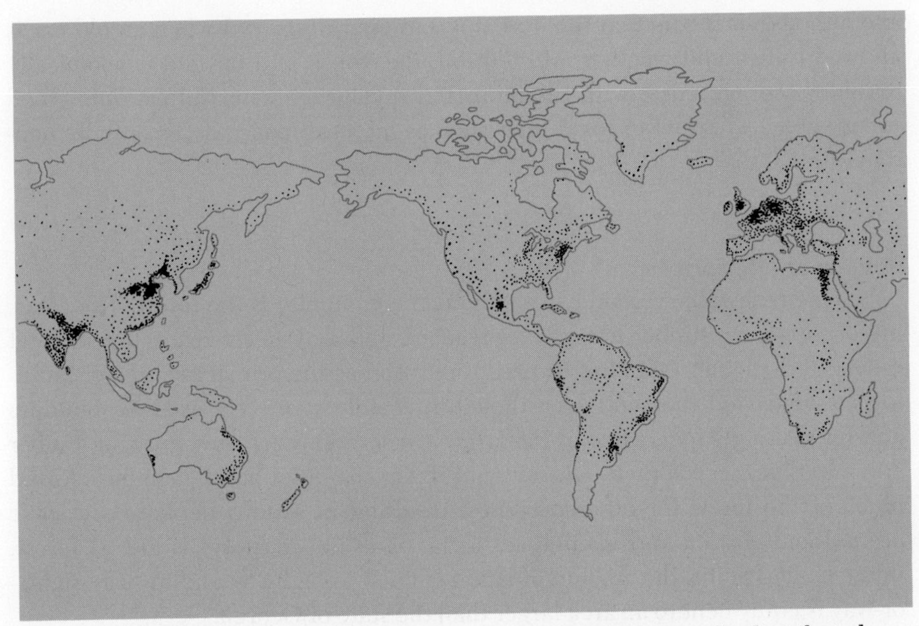

Figure 33–1. The distribution of man. He is unevenly distributed because of variations in climate and resources as well as of historical developments. (From Marston Bates, *Man in Nature,* 2nd ed., © 1964. By permission of Prentice-Hall, Inc., Englewood Cliffs, New Jersey)

Mankind is thus a rapidly expanding population occupying a region that is limited in size. As every organism does and must, he utilizes the resources of his environment, but he is able to exert more control over his environment than any other animal.

ENVIRONMENTAL RESOURCES

The resources of the environment are of two basic types: renewable and nonrenewable. The ordinary biological community is a self-maintaining system, dependent on the sun as an initial source of energy, but otherwise essentially self-sufficient. Such elements as carbon, oxygen, and nitrogen are constantly recycled. Almost nothing is lost but instead is passed through a series of cycles from one organism to another and between organisms and the inorganic environment. The types of resources available to such a community are said to be renewable, for they can be used over and over again.

When man developed the use of metals, he began to exploit nonrenewable resources. When iron is mined, there is less in the earth than there was before, and although it is true that the iron atoms are not destroyed when an old car is allowed to rust and weather into the soil, the iron is still beyond economically feasible recovery. Also, with the industrial revolution, man shifted from wood (renewable) as a fuel to fossil fuels such as oil, coal, and gas (essentially non-renewable).

Renewable Resources

One renewable resource of the human community is the forest. The clearing of forests for timber has not yet reached and may never reach the stage of universal depletion. Man has turned from wood as the primary source of building materials and fuel, and even though it is still extensively used for building and for making paper, some of the larger corporations are now growing timber at a rate that exceeds their harvesting of it. On the other hand, man has cleared much of the forest for other purposes, resulting in serious depletion in some areas. Land cleared and maintained for agricultural purposes is lost as forest; in the United States the amount of land assigned to highway and railway rights-of-way is equivalent to an area larger than the state of Georgia.

Because soil is continuously being formed in natural communities, it is generally considered a renewable resource. But man has, in places, tended to overcultivate and overgraze his lands to such an extent that soil is eroded away and carried ultimately to the sea faster than it is being formed (see Figure 33–2).

Resources of the sea, including the plankton, have hardly been utilized (except for the over-exploitation of the great whales). Man does not farm the sea. Although his agricultural practices have undergone extensive revolution, his fishing methods are still rather primitive—his tools are still essentially simple nets and hooks, although the hooks may be made of steel rather than bone.

Man's treatment of renewable resources differs most from that of other animals in the interruption of natural cycles and food webs. When a natural grassland is plowed under and replaced by wheat, the result is not simply that wheat now grows where native grasses grew before. Organisms that used the natural grasses as the first step in the food pyramid are forced to move on; other organisms that eat wheat move in and increase and new food webs are formed. When wheat is harvested, essential materials are removed from the community instead of remaining in the biogeochemical cycles.

Man is beginning to understand that wise use of his renewable resources must involve production processes that maintain the capacity of the environment at the level needed to meet human needs. Although the ideal would be the maintenance of those biological cycles that have kept life going on earth for over 2 billion years, it is doubtful that knowledge of these cycles is as yet sufficient to meet this ideal. Even so, much progress has been made. Whereas

Figure 33–2. The havoc of soil erosion on an Alabama farm prior to the introduction of modern farming practices. (Courtesy of the United States Department of Agriculture.)

man once let his soil erode away, he now may conserve it by sound agricultural practices (see Figure 33–3), and his improved techniques in foresting and animal husbandry make the future brighter.

When man removes something from a mine—iron, copper, silver, uranium —he is depleting a source of nonrenewable material, just as when he burns a fossil fuel. In using the nonrenewable as well as the renewable resources from his environment, man is unique among organisms.

Large-scale removal of nonrenewable resources from the earth has been going on for only a couple of hundred years, a short time in terms of human history. Even so, pessimists point out, it cannot continue very much longer. Oil and natural gas are already in short supply. Optimists believe that man's ingenuity will find substitutes that can be made from renewable resources before the nonrenewable ones are spent. Metals will be replaced by plastics that can be synthesized, oil will be made from alcohol produced by plants through fermentation, and perhaps energy will be tapped directly from the sun.

When we analyze pollution we find that it fits into three major categories —natural pollution; results of utilization of energy; and litter.

Natural pollution is that resulting from natural causes and not from any of man's cultural activities. Volcanoes and forest fires started by lightning bring particulate matter into the air as pollution. Natural seepage of oils from crevices in the bottom of the sea has added nearly a billion liters (a quarter of a billion gallons) of petroleum hydrocarbons to the oceans. As long ago as 1776, Fr. Pedro Font reported seeing balls of tar along the coast at Santa Barbara.

For as long as we use energy we will have some pollution. The Second Law of Thermodynamics as stated by Kelvin and Planck says "No cyclic process is possible whose result is the flow of heat from a single heat reservoir and the performance of an equivalent amount of work on a work reservoir." In other words, we can never get as much useful work out of a machine as we put into it. Some energy is always lost in the form of heat or converted into some other form of energy. Thus coal-fired furnaces give off smoke, nuclear plants give off heat, and animals must defecate to rid themselves of undigested remains. Chemical wastes from manufacturing plants can be very destructive. See Figure 33–4. With increased technology we may improve the situation, but we will not be able to eliminate pollution entirely.

Only the third form of pollution, litter, is under our complete control. This we can contain by educating the public and by legislative action.

IN RETROSPECT AND PROSPECT

Probably at no other time in the earth's history could man have reached his present stage of physical and cultural evolution. Not until late in the Paleozoic had enough fossil fuels been formed to permit an industrial revolution. Not until late in the Cenozoic had the grasslands been present for a long enough time to permit the development of great herds of large herbivores. In the Pleistocene, human societal behavior had evolved to the point where cultural development was possible. Grasses and herbivores and the cultural transmission of knowledge all contributed to the development of settled, agricultural communities in which enough food could be produced so that it was no longer necessary for every individual to expend most of his energy in the search for food. The settled communities could grow into cities, with a marked division of labor between individuals.

Figure 33–3. Aerial view of farmlands in Pennsylvania where modern farming practices are carried on. (Courtesy of the United States Department of Agriculture.)

Figure 33–4. The kill of fishes in an Ohio stream polluted by sugar beet wastes. (Courtesy of the United States Department of Agriculture —W. E. Seibel, photo.)

So here we are today with the Ice Ages behind us (we hope), with our reasoning powers developed to the point where seemingly nothing is impossible in our efforts to control and modify our environment.

Man has extended the range of his behavioral responses far beyond any other animal. With his greatly enlarged brain, he has moved from innate and simple learned behavior into the cultural transmission of knowledge and ultimately to the development of great civilizations. In doing so, he has altered his environment by the overexploitation of some of the renewable resources, and within the past several hundred years he has begun to make appreciable inroads into the store of nonrenewable resources. He is at present living well beyond his biological income, and any species that continues to live beyond its biological income eventually faces decimation or extinction. To complicate matters even further, man has, because of his large population and cultural activities, begun to produce more waste products than he has yet learned to dispose of.

The maintenance in a small pond or woodland of a stable community made up of organisms other than man remains a biological problem, and biologists can supply answers to such problems. But man has so molded and altered his environment that the answers to his major problems—overexploitation, population explosion, and pollution—cannot be given by biology alone but must take in as well such fields as economics, sociology, psychology, and political science. And now we must face the question most of us wish need never have been asked: "Where do we go from here?"

Will man let his population run rampant to his own destruction, like a parasite that overexploits the resources of its host and goes with it to extinction? Or will he voluntarily limit his population, get out of the resource cycles no more than he puts in, learn to use solar energy or some other source that is for the foreseeable future unlimited, and develop a civilization far beyond any known today?

SUGGESTED READINGS

Emmel, T. C.: *An Introduction to Ecology and Population Biology,* W. W. Norton & Company, Inc., New York, 1973.

Hardin, G.: *Nature and Man's Fate,* The New American Library, New York, 1959.

———— *Population, Evolution and Birth Control,* 2nd ed., W. H. Freeman and Co., San Francisco, 1969.

APPENDIX **A**

Some Basic Chemistry

This appendix is designed to provide students who have not had any chemistry with a background for understanding the chemical basis of living processes.

ATOMS AND ELEMENTS

All matter is composed of minute particles called *atoms*. They are almost inconceivably tiny—less than twenty-five hundred millionths of a millimeter (a hundred millionths of an inch) in diameter. In spite of the great diversity of substances found in the world, there are only about a hundred different kinds of atoms. When a substance is made up of only one kind of atom, it is called an *element*. The element oxygen contains only oxygen atoms and the element iron contains only iron atoms. The elements are designated by letters, usually the first one or two letters of the English name, but sometimes of the Latin name. For example, *H* stands for *hydrogen*, *Ca* for *calcium*, and *Fe* for *iron*, which is *ferrum* in Latin.

Structure of Atoms

All atoms are basically similar in structure. They have a compact central core, or *nucleus*, which contains most of the mass of the atom, and one or more *electrons*, minute, negatively charged particles orbiting around the nucleus.

Nucleus. The simplest atomic nucleus is that of hydrogen. It consists of one positively charged particle called a *proton*. Other elements have two, three, four

. . . protons in the nucleus. Thus, helium has two, carbon six, nitrogen seven, oxygen eight, iron twenty-six, uranium ninety-two. Other particles found in the nucleus are called *neutrons*. They are like protons in mass but they lack electric charges. Hydrogen usually does not have a neutron, but sometimes the nucleus of a hydrogen atom contains one or two neutrons, and sometimes atoms of other elements have extra neutrons. This makes the atoms heavier but does not alter the way they react chemically. Atoms of the same element that have different numbers of neutrons are called *isotopes*.

Electrons. The number of electrons orbiting an atomic nucleus is usually the same as the number of protons in the nucleus. Electrons are not distributed at random around the nucleus but are arranged in definite zones (orbital shells or energy levels). Each shell can contain a definite number of electrons. The first one, that is, the one nearest the nucleus, can have two electrons, the next eight, the next eighteen, and so on.

Electrons can be removed from the atom, or they may sometimes move from one energy level to another. Because they are held by the attraction between the positive charges in the nucleus and their own negative charges, the closer an electron is to the nucleus, the more tightly it is held and the more energy it takes to remove it. It also takes energy for an electron to move from a level close to the nucleus to one further out. If an electron moves from a higher orbit to a lower one, energy is released. If an atom is stripped of its electrons and then brought close to a source of electrons, it picks them up, filling the lower levels first.

Although the third orbital shell can hold more than eight electrons, it is not filled unless the fourth shell contains one or two electrons. Thus, potassium has eight electrons in the third shell and one in the fourth, calcium eight and two, titanium ten and two. The same is true for the higher energy levels. In other words, the outer shell of an atom never contains more than eight electrons.

Some Biologically Important Atoms

It is impossible to draw an atom to scale. If the nucleus were drawn the size of a pea, the nearest electron would be about eighteen meters (sixty feet) away. Also, the picture is clearer if the electrons are not shown as orbiting. Figure A–1 is a schematic representation of three biologically important atoms. Hydrogen has one proton in the nucleus and one electron in the first shell; carbon has six protons and six neutrons, and two plus four electrons; oxygen has eight protons and eight neutrons, and two plus six electrons. Because the protons carry positive charges and the electrons negative ones, and because there are the same number of protons as electrons, the atoms are electrically neutral.

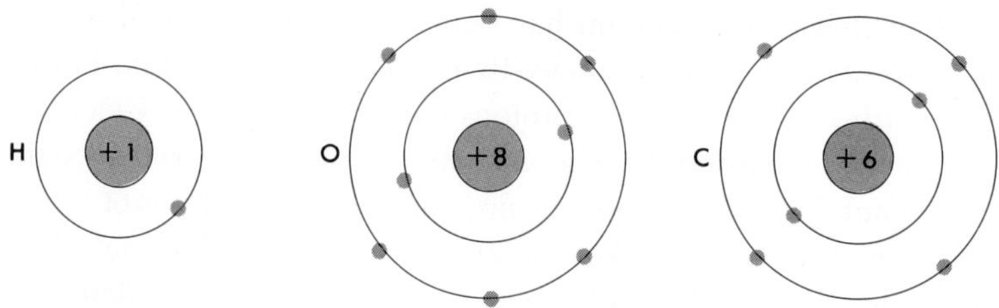

Figure A–1. Atoms of hydrogen, oxygen, and carbon.

MOLECULES

Atoms are able to join together to form *molecules*. Sometimes the molecules contain only a single kind of atom. Thus, oxygen atoms usually do not exist singly but each is joined to another oxygen atom. This is called *molecular oxygen*, expressed symbolically as O_2. Here the subscript indicates that the molecule consists of two oxygen atoms. Two separate atoms of oxygen would be expressed as 2O. When the molecules of a substance contain two or more different kinds of atoms, it is called a *compound*. Water is a compound, the molecules being composed of two atoms of hydrogen and one of oxygen, H_2O. See Figure A–2. The forces that hold atoms together are called *chemical bonds*.

The number of electrons an atom has in its outer ring determines the number and characteristics of the other atoms with which it can form bonds. The most stable configuration for an atom is to have eight electrons in its outer ring (two if only the inner ring contains electrons). Thus, oxygen has six electrons in its outer ring. It tends to pick up two electrons to bring the number up to eight. When two hydrogen atoms each share their single electron with an oxygen atom, this requirement is fulfilled and a water molecule is formed. Because oxygen needs two electrons to fill its outer ring, it can form two bonds.

On the other hand, sodium (Na) has eleven electrons, arranged two,

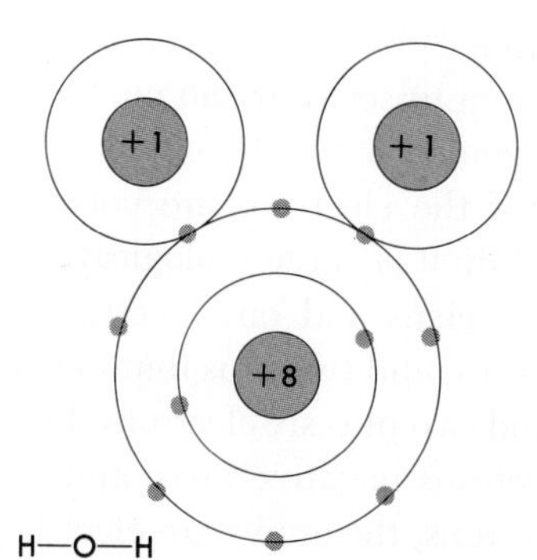

Figure A–2. Water molecule.

eight, and one. It tends to pass the outer electron to another atom in order to achieve an outer ring of eight. It can form a single bond. If an oxygen atom gets one extra electron from a hydrogen atom and one from a sodium atom, sodium hydroxide (NaOH) is formed. Whether an atom tends to be an electron donor or an electron acceptor depends on the number of electrons it has in its outer shell. If it has fewer than four, it loses them, if it has more than four, it picks them up. The number of electrons an atom can either give up or accept, and hence the number of bonds it can form, is known as its *valence*. Hydrogen has a valence of one, oxygen has a valence of two.

There are three types of chemical bonds.

Electrovalent or Ionic Bonds

When a sodium atom loses the outer of its eleven electrons, it is no longer electrically neutral. It has eleven positive charges in the nucleus, but only ten negative electrons; it is electrically positive. A chlorine atom has seventeen electrons, arranged two, eight, and seven. It can pick up the electron lost by a sodium atom. Now it has seventeen positive charges in the nucleus and eighteen electrons; it is electrically negative. Since positive and negative charges attract, atoms of sodium and chlorine are attracted to each other to form table salt, sodium chloride (NaCl). The force that holds the atoms together is called an *electrovalent* or *ionic bond*. The atoms do not form discrete molecules, but crystals, in which each sodium atom is surrounded by six chlorine atoms and each chlorine atom by six sodium atoms. See Figure A–3.

When a crystal of table salt is dissolved in water, the atoms separate, but

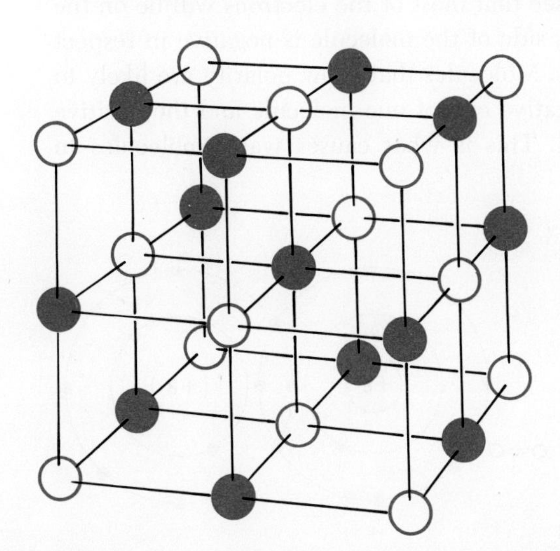

Figure A–3. Part of a salt crystal.

the chlorine retains the electron it accepted from the sodium. Atoms that have either given up or accepted electrons and so carry electric charges are called *ions*. They are indicated by either a plus or minus superscript. The sodium ion is written Na^+, indicating that it has a positive charge, and the chlorine ion is written Cl^-, indicating its negative charge.

Covalent Bonds

Atoms frequently form bonds not by the transfer of electrons but by the sharing of them. This is what happens when two oxygen atoms form a molecule. Each atom shares two of its electrons with the other (see Figure A–4). The shared electrons may orbit either nucleus, thus giving each atom eight electrons in its outer shell. A carbon atom, with four electrons in its outer shell, neither donates nor accepts electrons but can form *covalent bonds* with a great variety of other atoms, including other carbon atoms. This makes possible the formation of the long carbon chains that are so important in biochemical molecules. The ends of a carbon chain may join to form a ring.

Polar Bonds

When atoms combine to form a molecule, there is a definite, geometric relationship between them. A molecule is three-dimensional. Sometimes, because of its shape, a molecule has polarity. Although the molecule itself is electrically neutral, there are more positive charges on one side, more negative charges on the other. This is because the nuclei of some atoms hold more electrons close to them than do the nuclei of other atoms. Look again at the picture of the water molecule; it is easy to see that most of the electrons will be on the side where the oxygen atom is. This side of the molecule is negative in respect to the side with the hydrogen atoms. Molecules that show polarity are likely to be attracted to each other—the negative end of one molecule and the positive end of another form a weak bond. This is what causes water molecules to adhere to one another.

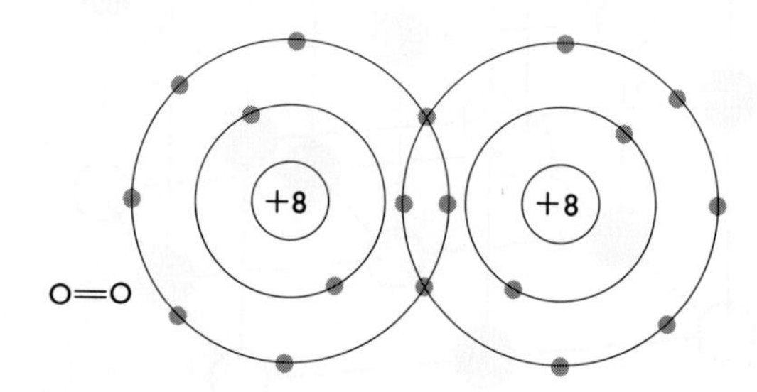

Figure A–4. Oxygen molecule.

In living organisms, polar bonds between hydrogen atoms on one part of a molecule and nitrogen or oxygen atoms on the same or another molecule are very common. They are called *hydrogen bonds*.

Chemical Formulas

The simplest way to describe a molecule is to indicate the kind and number of atoms it contains. Water is H_2O; the sugar glucose is $C_6H_{12}O_6$. Chemical formulas such as this, called *empirical formulas*, tell very little about the molecule. They do not show the relationship of the atoms. A water molecule is not just a string of atoms H—H—O. H—O—H is a better way of expressing it, and

  O
 / \
H   H is better still. A chemical formula set up so that it indicates the relationships of the atoms to each other is called a *structural formula*. The number of lines drawn between two atoms indicates the number of bonds between them. Because a molecule is three-dimensional and a structural formula is two-dimensional, such a formula still does not give a complete picture of the molecule. A structural formula of the molecule of a sugar, glucose, is shown in Figure A–5. For the sake of simplicity, when chemists draw structural formulas of carbon ring compounds, they usually do not show the carbons in the ring. To interpret these formulas, you should mentally insert carbon atoms at the places where the lines intersect.

Sometimes two substances have the same empirical formula but differ in the arrangement of the atoms in the molecule. A number of different sugars have the empirical formula of glucose but are quite different because their atoms are arranged differently. Such differences can only be shown by structural formulas. The structural formula of fructose, a sugar found in fruits, is shown in Figure A–6. Substances that have the same empirical formula but differ in structural formula are called *isomers*.

Sometimes a group of atoms behaves as a single unit in chemical reactions. Some important groups are the carboxyl group ($-COOH$), the amino group ($-NH_2$), the hydroxyl group ($-OH$), and the phosphate group ($-H_2PO_4$). A number of substances may be classed together because they have the same

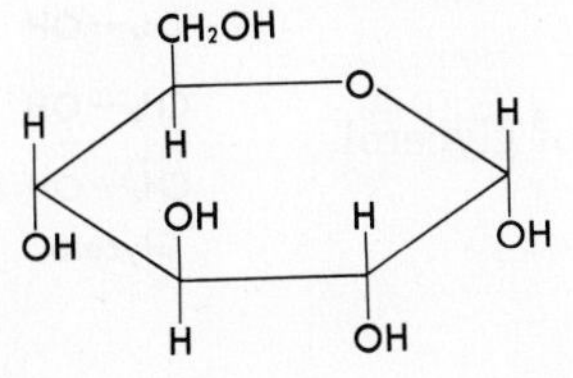

Figure A–5. Molecule of glucose.

Figure A–6. Molecule of fructose.

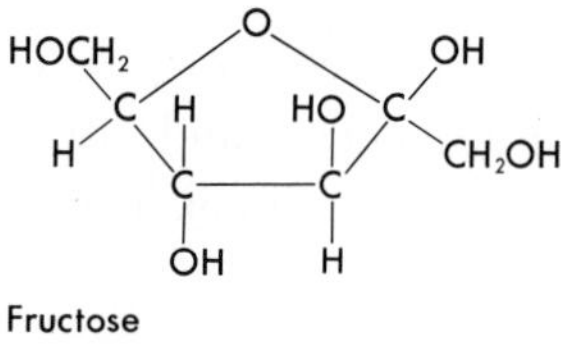

Fructose

group or groups. Alcohols are molecules of hydrogen and carbon in which one (or more) of the hydrogens is replaced by a hydroxyl group. Methane is CH_4, methyl alcohol is CH_3OH. The structure of glycerol, a three-carbon alcohol, is shown in Figure A–7. Fatty acids are hydrocarbon chains with a carboxyl group at one end:

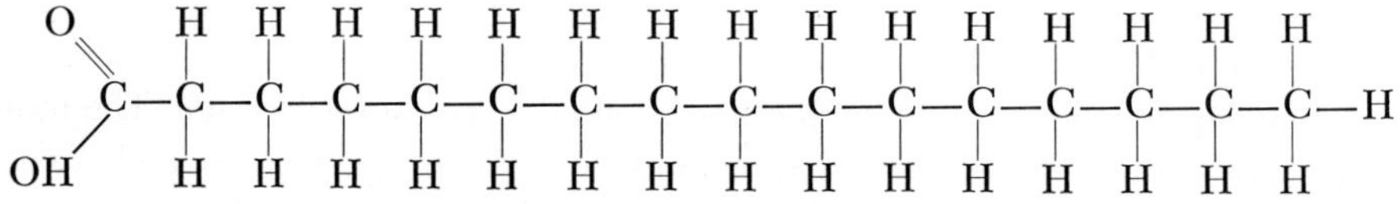

```
O    H  H  H  H  H  H  H  H  H  H  H  H  H  H  H
 \\  |  |  |  |  |  |  |  |  |  |  |  |  |  |  |
   C—C—C—C—C—C—C—C—C—C—C—C—C—C—C—C—H
  /  |  |  |  |  |  |  |  |  |  |  |  |  |  |  |
OH   H  H  H  H  H  H  H  H  H  H  H  H  H  H  H
```

Chemical Reactions

Various types of reactions take place between molecules. Reactions are written as chemical equations, a shorthand method of expressing what is happening.

A *synthetic reaction* is one in which two or more molecules combine to produce a larger molecule. Carbon dioxide and water combine to form carbonic acid:

$$CO_2 + H_2O \rightarrow H_2CO_3$$

Sometimes more than one product is formed in a synthetic reaction. Two molecules of glucose combine to form maltose and water:

$$C_6H_{12}O_6 + C_6H_{12}O_6 \rightarrow C_{12}H_{22}O_{11} + H_2O$$

What has happened is that one of the sugars has yielded a hydrogen and the other a hydroxyl group. This left each sugar with a free bond so that they could

Figure A–7. Molecule of glycerol.

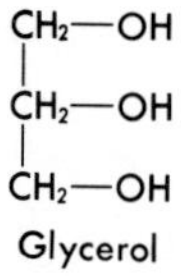

CH_2—OH
CH_2—OH
CH_2—OH
Glycerol

come together, and the hydrogen and hydroxyl united to form water (see Figure A–8). Such reactions, in which water is formed, are called *condensations*. Notice that there are the same number and same kind of atoms on each side of the equation, that is, it is balanced. Because atoms are neither created nor destroyed in chemical reactions, they must all be accounted for.

A *decomposition reaction* is the opposite of a synthetic one. The large molecule breaks down into smaller ones. When a molecule is split by the addition of water, the reaction is called *hydrolysis*. It is the opposite of condensation. The fact that a reaction can go either way is frequently expressed by drawing a double arrow in the equation.

$$CO_2 + H_2O \rightleftarrows H_2CO_3$$

An *exchange reaction* is one in which atoms trade places between two substances to produce two new substances. Hydrochloric acid and sodium hydroxide interact to form salt and water:

$$HCl + NaOH \rightarrow NaCl + H_2O$$

In a *rearrangement reaction*, one isomer is converted into another isomer of the same substance by a reorganization of the atoms. Fructose can be changed into glucose.

Figure A–8. The formation of maltose and water from two glucose molecules.

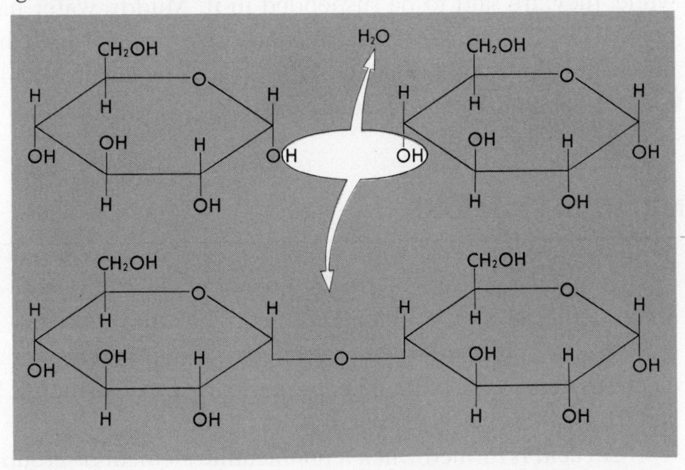

Oxidation-Reduction

The term *oxidation* was first introduced to describe the combining of oxygen with a substance. When oxygen combines with iron to form ferric oxide (rust), the iron has been oxidized. But because in such reactions the oxygen removes electrons from the substance being oxidized, the word is now used to indicate the removal of electrons, even when oxygen is not involved. When a substance has lost electrons, it is said to be oxidized. The opposite of oxidation is *reduction*, the gain of electrons. Because when one substance gains electrons, another must lose them, the two processes go hand in hand. When sodium and chlorine combine to form salt, the sodium loses an electron; it is oxidized. The chlorine gains an electron; it is reduced. Many oxidation-reductions involve the transfer of hydrogen. A substance from which hydrogen is removed is oxidized because it has lost the hydrogen electron. A substance to which hydrogen is added is reduced because it has gained the hydrogen electron.

MIXTURES

When two or more different substances are intermingled but are not combined chemically into compounds, they form a *mixture*. Air is a mixture of gases —oxygen, carbon dioxide, nitrogen, and so on. Biologically, most important mixtures are formed with water. If the individual molecules of a substance are equally dispersed between the water molecules and do not settle out, the substance is said to be dissolved in the water. Sugar dissolves in a cup of coffee. If the particles are larger, so they tend to settle out when the mixture stands for a while, they are said to be suspended in it. Muddy water has dirt particles suspended in it. Particles of intermediate size may be held in suspension indefinitely by the bombardment of the water molecules, which are in constant motion. Such mixtures are called *colloids*.

SOME CHEMICAL DEFINITIONS

A *metal* is an element that is a donor of electrons. Iron, copper, and sodium are examples of metals. Nonmetals act as electron receivers.

A *base* is formed when a metal, or a group that acts as a metal in donating electrons, combines with one or more hydroxyl groups. Sodium hydroxide, NaOH, is a base.

An *acid* is formed when a nonmetallic element or group combines with one or more hydrogens. Hydrochloric acid, HCl, and sulfuric acid, H_2SO_4, are examples.

When an acid or base is dissolved in water, some of the molecules dissociate to form ions. Hydrochloric acid dissociates into H^+ and Cl^- ions, sodium hydroxide into Na^+ and OH^-. If only a few of the molecules dissociate, the acid or base is said to be weak; if many dissociate, it is said to be strong. In pure water, a few of the water molecules are dissociated into ions. Because an acid is a substance that yields hydrogen ions and a base is a substance that yields hydroxyl ions, water can be considered as both an acid and a base. If a solution contains more hydrogen ions than hydroxyl ions, it is acidic; if it contains more of the hydroxyl ions than the hydrogen ions, it is basic; if it contains equal numbers of both, it is neutral. The degree of acidity is expressed by a scale of numbers called the pH scale. A neutral solution has a pH of 7. The lower the pH number, the more acidic the solution is. If the pH is 6, it is only weakly acidic, but if the pH is 2, it is very strongly acidic. If the pH is higher than 7, the solution is basic.

When an acid and base dissociate together in water, the metallic ions combine with the basic ions to form a salt and the hydrogen and hydroxyl ions form water:

$$NaOH + HCl \rightarrow NaCl + H_2O$$

SOME IMPORTANT BIOLOGICAL MOLECULES

Sugars

Two sugar molecules are shown in Figures A–5 and A–6. Notice that each molecule is composed of carbon, hydrogen, and oxygen, and that the hydrogen and oxygen are in the same proportions as in water, that is, two hydrogens to one oxygen. Sugars may have a varying number of carbons; the most important ones in biology have either five or six. Sugar molecules can combine by condensation to form long chains.

Fats

Fats are composed of three fatty acid molecules joined to a three-carbon alcohol (glycerol) in the same way that the sugar molecules join together; that is, by condensation.

Amino Acids

An amino acid consists of a carbon atom with an amino group attached to one bond, a carboxyl group to the opposite bond, and a hydrogen to a third bond. The fourth bond is occupied by one of a number of different groups.

Figure A–9 shows two amino acids. The carboxyl group of one amino acid bonds to the amino group of another by condensation. Proteins are long chains of amino acids.

Nucleotides

A nucleotide has three parts: one of two five-carbon sugars, a phosphate group, and any one of five nitrogenous bases, ring-shaped structures containing nitrogen. Figure A–10 shows the five bases. Nucleic acids are chains of nucleotides formed by the linkage of the phosphate group of one nucleotide to the sugar of another.

Figure A–9. Two amino acids.

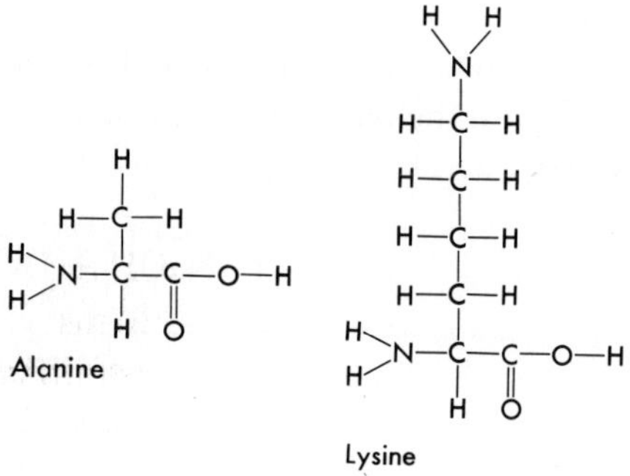

Alanine

Lysine

Figure A–10. Five nitrogenous bases found in nucleic acids.

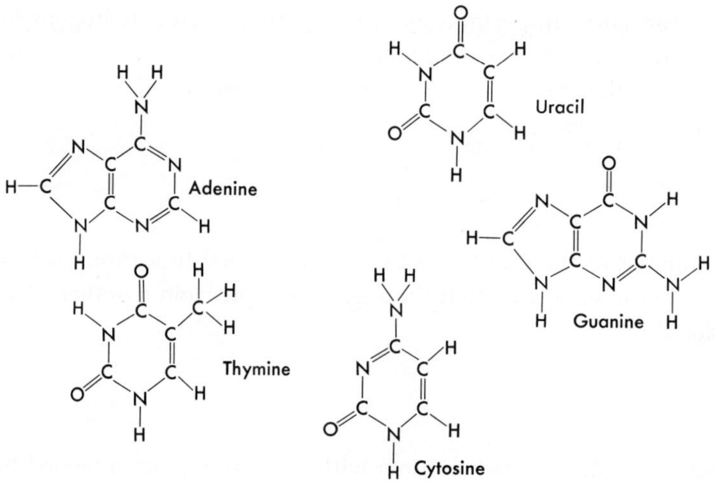

Adenine

Uracil

Guanine

Thymine

Cytosine

APPENDIX **B**

The Metric System

Until we arrive at the point where we think entirely in the metric system, we sometimes need tables for converting the English system to the metric.

The metric system has the advantage that its units are based on multiples of ten—thus 10 millimeters are 1 centimeter, 10 centimeters are 1 decimeter, and 10 decimeters are 1 meter.

LENGTH AND AREA

10 millimeters = 1 centimeter = 0.3937 inch
10 centimeters = 1 decimeter = 3.937 inches
10 decimeters = 1 meter = 3.28 feet
1000 meters = 1 kilometer = 0.62137 mile
1 square meter = 1550 square inches
10,000 square meters = 1 hectare = 2.471 acres

WEIGHT

1 milligram = .015432 grains
1 gram = 15.432 grains = .03527 ounces
1000 grams = 1 kilogram = 2.2046 pounds
1000 kilograms = 1 metric ton = 2204.6 pounds

VOLUME

1000 cubic centimeters = 1 liter = 1.0567 liquid quarts
10 liters = 1 decaliter = 0.284 bushels or 2.64 gallons

ENERGY

Energy is the capacity to do work. It may be measured in either heat or force times distance.

Heat. The basic unit of heat is the *calorie*, written with a small *c*, which is the amount of heat necessary to raise one gram (1 cubic centimeter) of water from 14.5° to 15.5° centigrade. More commonly used in biological sciences is the kilogram Calorie (written with a capital C). One Calorie = 1000 calories.

Work. Work is equal to force time distance, ($W = Fd$). The basic unit of work in the metric system is the *erg*. It is the amount of work done when a force of 1 *dyne* moves through a distance of 1 centimeter. The force of one dyne is that necessary to accelerate a mass of 1 gram 1 cm per second per second.

TEMPERATURE

Temperature is a measure of the average speed of molecules. In the metric system it is measured in degrees Celsius or centigrade. The freezing point of pure water at sea level is 0°C, whereas that of boiling water under the same conditions is 100°C. The respective figures for the Farenheit scale are 32°F and 212°F. To convert the Farenheit scale to centigrade take 5/9 of (°F − 32).

APPENDIX C

Classification of Organisms

A man's observations and experiments on plants and animals would be of little value if he could not communicate information about the organisms he had studied to others. Thus, the naming of animals and plants is a matter not only of convenience but of necessity. Up until the eighteenth century, most species really had no names; they were referred to by short descriptive phrases such as "the little gray mouse with big ears and long whiskers." With the spread of interest in zoology and botany during the eighteenth century, there came an increase in efforts to provide names for all known kinds of animals and plants. Men also felt the need to classify them, to arrange them in convenient groups in order to talk, not just about this monkey or that monkey, but about primates. One of the better attempts at naming and classifying was devised by a Swedish botanist, Carolus Linnaeus (von Linné). We still use the Linnaean system of naming plants and animals, and our method of classification is a modification of his.

SYSTEM OF CLASSIFICATION

Various schemes have been proposed for dividing organisms into major groups, called *Kingdoms*. Originally, every living thing was considered to be either a plant or an animal. Leeuwenhoek's discovery of the world of microscopic organisms led to the proposal of a third Kingdom, Protista, to include all single-celled forms. Realization of the fundamental differences between the procaryotes and eucaryotes resulted in the establishment of a fourth Kingdom, Monera, for the former. Recently a five-kingdom approach has been suggested, with the fungi and slime molds separated from the plants as the Kingdom Fungi. The distinction of the monerans is unquestioned. Plants, fungi, and animals differ from each other both structurally and in basic nutritional strategy. Plants make their own food, animals ingest it, and fungi absorb nutrients from the environment. The main problem lies with the Protista. This Kingdom includes a wildly heterogeneous group of organisms, some obviously related to algae, some to fungi, and some to animals. The Kingdom Protista does not even include all single-celled organisms, for the yeasts, which are believed to be descended from multicellular fungi, are classed with the latter. We discuss four Kingdoms here, separating the fungi from the plants and dividing the protists among the other groups. You should, however, be aware that this is only one of a number of possible arrangements.

Each Kingdom is divided into a relatively small number of basic groups called *divisions* in the Plant Kingdom and *phyla* (singular: *phylum*) in the Animal Kingdom. The phyla or divisions are divided into a series of classes, for example, class Mammalia (mammals) and class Aves (birds) in the phylum Chordata. Classes again are divided into orders, such as the orders Primates and Rodentia in the mammals. The orders are divided into families and families into *genera* (singular: *genus*). The genus is the smallest group that includes a number of different kinds of species. These are the main categories, but as our knowledge of animals and plants has increased, it has been found necessary to expand the system by the inclusion of sub-, infra-, and super- categories in some groups (see Figure C–1).

The basis of Linnaeus' classification was the grouping together of organisms that resemble each other structurally. The species included in one genus are more like each other than they are like the species of another genus, and the members of one class resemble each other more than they do the members of another class. Linnaeus did not believe in evolution but thought that every species was created as a separate and distinct entity from the beginning. But because he based his system on structural characters, and because the more closely animal or plant species resemble each other structurally, the more similar they usually are genetically, it has been possible to convert Linnaeus' system into an evolutionary classification. The species in one genus are more closely related to each other than they are to the members of another genus, and it is the same for all the categories.

NAMING ORGANISMS

The scientific name of a plant or animal is made by combining the generic and specific terms. Thus, the dog, which is in the genus *Canis,* is known as *Canis*

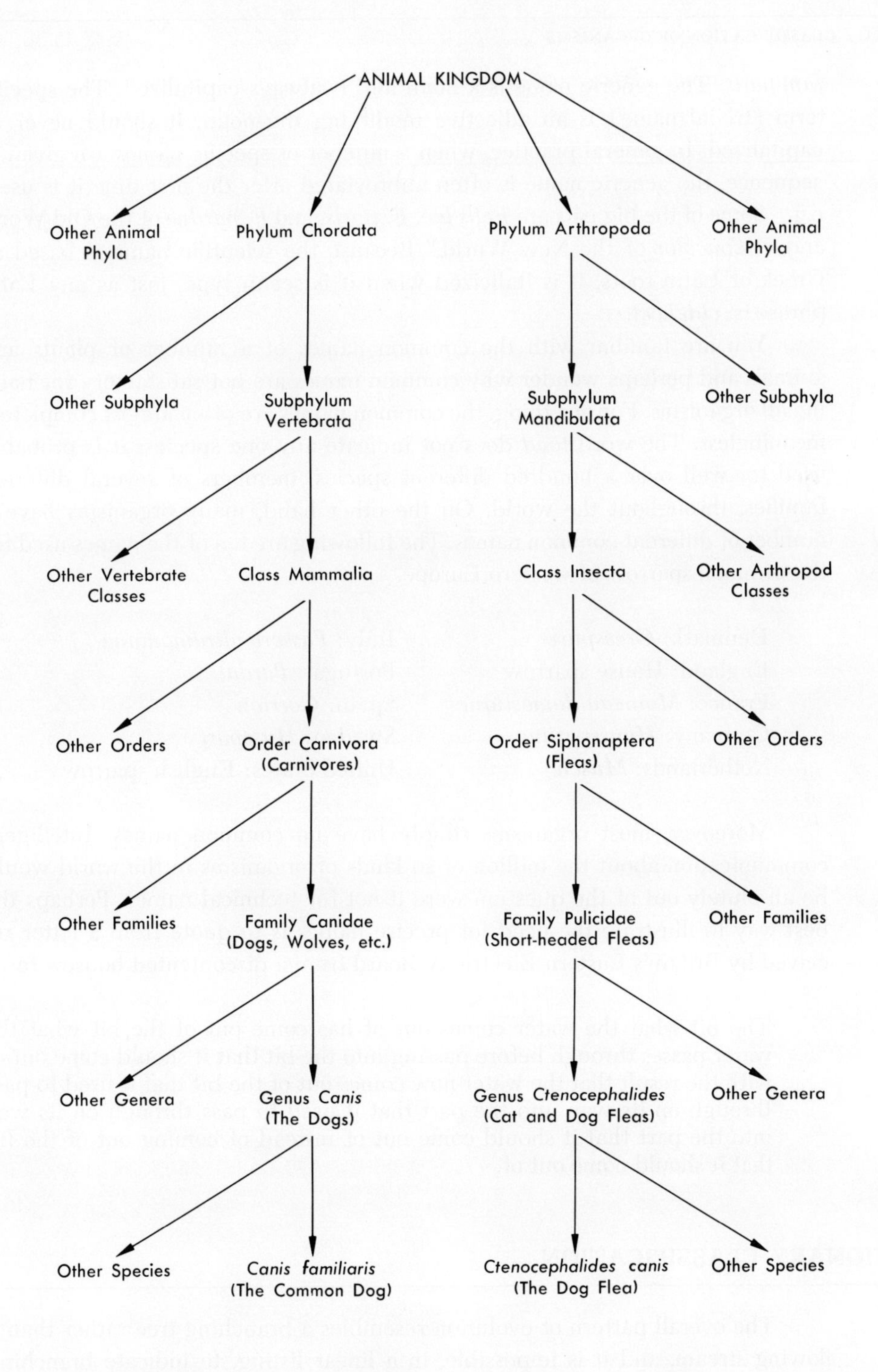

Figure C–1. Classification of a dog and his fleas.

familiaris. The generic name is a noun and is always capitalized. The specific term (trivial name) is an adjective modifying the noun; it should never be capitalized. In general practice, when a number of specific names are given in sequence, the generic name is often abbreviated after the first time it is used, e.g., "Some of the big cats are *Felis leo*, *F. tigris*, and *F. pardus* of the Old World and *F. concolor* of the New World." Because the scientific name is based on Greek or Latin roots, it is italicized when it is set in type, just as any Latin phrase is, *videlicet*.

You are familiar with the common names of a number of plants and animals and perhaps wonder why common names are not satisfactory for naming all organisms. For one thing, the common names are often almost completely meaningless. The word *toad* does not indicate any one species; it is probably used for well over a hundred different species, members of several different families, throughout the world. On the other hand, many organisms have a number of different common names. The following are ten of the names used for the common sparrow of western Europe.

Denmark: *Grasspurv*
England: House sparrow
France: *Moineau domestique*
Germany: *Haussperling*
Netherlands: *Musch*
Italy: *Passera altramontana*
Portugal: *Pardal*
Spain: *Gorrion*
Sweden: *Hussparf*
United States: English sparrow

Moreover, most organisms simply have no common names. Intelligent communication about the million or so kinds of organisms in the world would be absolutely out of the question were it not for technical names. Perhaps the best way to illustrate the need for precise names is to quote from a letter received by Britain's Eastern Electricity Board from a discontented housewife:

> The bit what the water comes out of has come out of the bit what the water passes through before passing into the bit that it should come out of with the result that the water now comes out of the bit that it used to pass through on its way into the part that it used to pass through on its way into the part that it should come out of instead of coming out of the bit that it should come out of.

EVOLUTIONARY CLASSIFICATION

The overall pattern of evolution resembles a branching tree rather than a flowing stream, and it is impossible, in a linear listing, to indicate branching relationships. The classifications given here are arranged with the more primitive forms first, followed by the more advanced ones, and with related phyla grouped together. To that extent it is *phylogenetic,* that is, it indicates the evolutionary history of the groups.

It is not a final classification. The best evidence of evolutionary relationship is that provided by the fossil record, and for many large groups the fossil record is woefully inadequate or completely lacking. Here recourse must be had to a comparison of the characters of the living forms. From time to time new discoveries are made, that is, new fossils are found or new characters are investigated, that make it necessary to shift the position of some group. It is also true that by no means all of the living organisms have been studied and described. Some biologists estimate that as many more species are yet to be discovered as have already been named. This is especially true for microscopic forms, yet even the large terrestrial vertebrates are still incompletely known. The numbers given are purely tentative. Fossil forms are not included.

Kingdom Monera

This Kingdom contains two groups of organisms having very primitive cells that lack a nuclear membrane, mitochondria, endoplasmic reticulum, Golgi apparatus, and lysosomes. Such cells are said to be procaryotic, in contrast to the eucaryotic cells of other organisms. The bacteria are divided into several classes, but the characters on which they are based have little meaning except for experts and they will not be listed here. The blue-green algae are all placed in a single class. A few representative genera are listed for each group.

Division I. Schizophyta (from Greek *schizein,* "to split," and *phyton,* "plant"). The bacteria. Minute forms, existing as single cells or chains of cells. Chlorophyll lacking. Reproduction generally by fission as the name indicates. About 1,600 species are known but there are surely many more. *Bacillus, Cytophaga, Streptococcus.*

Division II. Cyanophyta (from Greek, *kyanos,* "blue," and *phyton,* "plant"). The blue-green algae. Pigments present but not organized into plastids. Color produced by pigments chlorophyll and phycocyanin; yellow, orange, brown, and red pigments may also be present. Plant body a single cell or clusters or filaments of cells, often covered with a gelatinous sheath. See Figure C–2. Some 1,500 known species, on damp objects above ground, in the soil, and in both fresh and salt water. *Nostoc, Oscillatoria.*

Kingdom Plantae

The plants are here divided into thirteen divisions. The first six contain the relatively simple, primarily aquatic plants known collectively as *algae.* The others are land plants, which are more or less well adapted to terrestrial life.

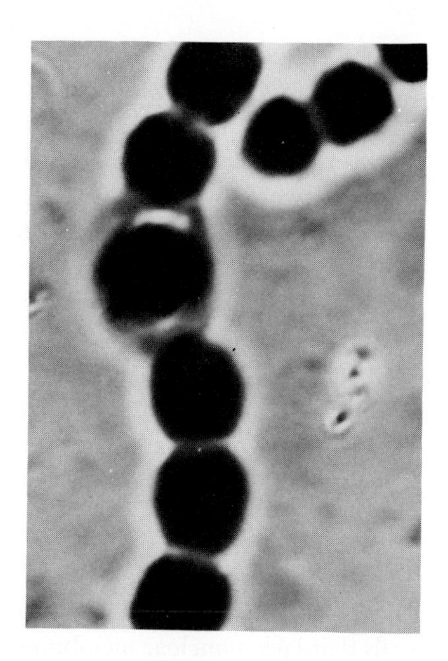

Figure C–2. *Nostoc*, a blue-green alga. The individual cells are approximately 5 micra (about 0.0002 inch) in diameter. (Photo by Paul H. Smith from a microscope slide mount courtesy of Ward's Natural Science Establishment, Inc., Rochester N.Y.)

Division I. Euglenophyta (from Greek, *eu*, "good," *glene*, "pupil of eye," and *phyton*, "plant"—a reference to the eye spot found in some forms). Flagellated algae. Body a single cell, usually without true cell walls, and bearing one or more whiplike flagella. See Figure C–3. Some have chlorophyll, which gives them a grassy green color; others lack pigments. Pigments in definite plastids. Both autotrophs and heterotrophs. About 450 species common in fresh water in which decaying organic material is present. A few are parasitic. *Euglena, Leishmania.*

Division II. Chlorophyta (from Greek, *chloros*, "green," and *phyton*, "plant"). Green algae. Plant body unicellular or multicellular, motile or nonmotile, no true tissues or organs. Chlorophyll present, along with some yellow pigments, carotenoids; pigments in well-organized plastids. Nearly 6,000 species, widely distributed in fresh water and the sea. *Oedogonium, Spirogyra, Volvox.*

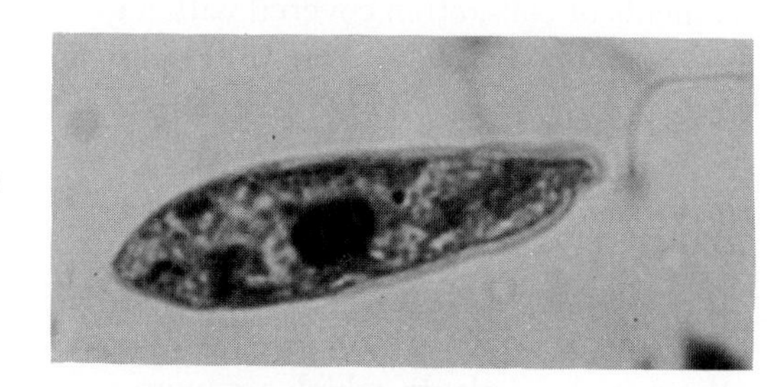

Figure C–3. *Euglena*, a flagellated alga. (Courtesy Ward's Natural Science Establishment, Inc., Rochester, N.Y.)

Division III. Chrysophyta (from Greek, *chrysos,* "gold," and *phyton,* "plant"). Yellow-green algae and diatoms. Single-celled or colonial algae, yellowish to brown in color because of the presence of yellow and brown carotenoid pigments, although chlorophyll is sometimes present. Diatoms have cell walls containing silica. About 6,000 species, distributed in soil, fresh water, and the seas. *Botrydium, Mallomonas.*

Division IV. Phaeophyta (from Greek, *phaios,* "dusky," and *phyton,* "plant"). Brown algae. Colonial algae, brown in color because of pigment fucoxanthin. Little or no tissue differentiation, but cells forming colonies ranging from tiny, delicate filaments to massive, bladelike or leathery fronds about sixty meters (over 200 feet) in length. See Figure C–4. Many attached by hold-fast, but others free-floating. Almost entirely marine. Some 1,000 species known. *Fucus, Sargassum.*

Division V. Pyrrophyta (from Greek, *pyr,* "fire," and *phyton,* "plant"). Dinoflagellates and their allies. Unicellular, flagellated algae with yellow-green or golden-brown plastids and usually with heavy cell walls divided into plates. About 1,000 known species, mostly marine. *Ceratium, Gymnodinium.*

Division VI. Rhodophyta (from Greek, *rhodon,* "rose," and *phyton,* "plant"). The red algae. Red color produced chiefly by red pigment phycoerythrin. Small, filamentous, cylindrical, sheetlike, or ribbon-like colonies attached to sea floor. About 2,500 species known. *Chondrus, Dasya.*

Figure C–4. A brown alga (*Fucus*) and a red alga. James Bay, Canada. (Courtesy J. Kenneth Doutt.)

Division VII. Bryophyta (from Greek, *bryos,* "moss," and *phyton,* "plant"). The liverworts and mosses. See Figure C–5. Sex organs (antheridia and archegonia) multicellular, with an outer layer of sterile cells. Zygote begins development while within female organ or within embryo sac. Tissue differentiation, but no roots nor vascular tissue. Main plant body the gametophyte generation, sporophyte an unbranched structure that remains attached to gametophyte. About 22,000 species distributed throughout the world. *Marchantia, Sphagnum.*

Division VIII. Lycopodophyta (from Greek, *lykos,* "wolf," *podo,* "foot," and *phyton* "plant," from the genus *Lycopodium,* shaped like a wolf's foot). The club mosses. See Figure C–6. Vascular tissue present. Roots present, stems simple and unjointed, leaves simple. Reproduction by spores. About 1,000 species. *Lycopodium, Selaginella.*

Division IX. Arthrophyta (from Greek, *arthron,* "joint," and *phyton,* "plant"). Horsetails. See Figure C–7. Simple roots, stems, and leaves, stems jointed. Reproduction by spores. Vascular tissue present. Generally small, but some may reach nearly a meter (three feet) in height. Twenty-five living species, all members of a single genus, *Equisetum.*

Division X. Pterophyta (from Greek, *pteris,* "fern," and *phyton,* "plant"). The ferns. See Figure C–8. Typically with well-developed roots, stems, and generally large, complex leaves. Sporophyte generation predominant. Reproduction by spores. The stem usually reclining on or under ground, but sometimes erect, as in tree ferns. Nearly 10,000 species known. *Dryopteris, Osmunda, Pteridium.*

Division XI. Cycadophyta (from *cycad,* the name of a plant and *phyton,* "plant"). Cycads. Roots, stems, and large leaves. Naked seeds, usually borne on

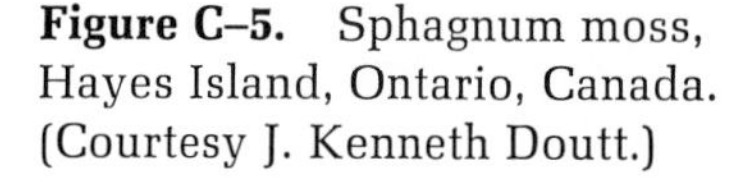

Figure C–5. Sphagnum moss, Hayes Island, Ontario, Canada. (Courtesy J. Kenneth Doutt.)

Figure C–6. A club moss, *Lycopodium flabelliforme*, growing on Carnegie Museum's Powdermill Nature Reserve, Westmoreland Co., Pa. (Carnegie Museum—M. Graham Netting photo.)

Figure C–7. Horsetail, *Equisetum arvense.* (Carnegie Museum—O. E. Jennings photo.)

Figure C–8. Royal fern, *Osmunda regalis*, growing on Carnegie Museum's Powdermill Nature Reserve, Westmoreland Co., Pa. (Carnegie Museum—M. Graham Netting photo.)

cones. Ciliated, swimming sperm. About ninety species in warm climates. *Cycas, Zamia.*

Division XII. Coniferophyta (from Latin, *conus,* "cone," *ferre,* "to bear," and Greek *phyton,* "plant"). The conifers. Vascular plants that bear seeds but not flowers. About 550 conifers or "softwoods," the ginkgo, and a few dozen species of atypical forms that do not fit into any of these categories. *Ginkgo, Pinus, Sequoia, Taxus, Welwitschia.*

Division XIII. Anthophyta (from Greek, *anthos,* "flower," and *phyton,* "plant"). The flowering plants. The predominant plants in the world today, with about 250,000 species. Seed enclosed by carpels, the non-motile sperm reaching the egg through a pollen tube after pollen grain has been deposited on stigma. Leaves typically broad.

CLASS DICOTYLEDONEAE. The dicotyledons. Plants in which embryo has two cotyledons, floral parts mostly in fours or fives, the vascular tissue in the stem a cylinder, cambium present, net-veined leaves. See Figure C–9. About 200,000 species. Typical families with their numbers of species are buttercups (Ranunculaceae), 1,500; mustards (Cruciferae), 3,000; roses (Rosaceae), 3,000; and legumes (Leguminaceae), 12,000.

CLASS MONOCOTYLEDONEAE. The monocotyledons. Embryo with a single

Figure C–9. Phlox, a dicot (Burpee Seeds photo.)

cotyledon, floral parts usually in threes, vascular tissue in stem usually in scattered bundles, cambium usually absent in adult plants, leaves with parallel veins. Mostly herbaceous plants, although some trees. See Figure C–10. About 50,000 species. Some representative families with the number of included species are lilies (Liliaceae), 3,000; palms (Palmaceae), 2,000; grasses (Gramineae), 7,500; and orchids (Orchidaceae), 17,500.

Kingdom Fungi

The fungi are eucaryotic organisms that lack chlorophyll and absorb nutrients from the environment.

Figure C–10. Trillium, a monocot. (Carnegie Museum, M. Graham Netting photo.)

Division I. Myxomycophyta (from Greek, *myxa,* "mucus," *mykes,* "fungus," and *phyton,* "plant"). Slime molds. Body either an amorphous mass of multinucleate protoplasm not subdivided by internal septa or formed by a fusion of separate, ameboid cells. Asexual reproduction by means of nonmotile spores, sexual reproduction by the fusion of ameboid gametes. Probably less than 500 species. *Physarum, Spongospora.*

Division II. Eumycophyta (from Greek, *eu,* "good," *mykes,* "fungus," and *phyton,* "plant"). The true fungi, including the mushrooms, yeasts, and related forms. See Figure C–11. Body a single cell or composed of elongated filaments, the hyphae. Hyphae may or may not be divided by septa into definitive nucleate cells. All either saprophytic or parasitic. One group lives as commensals with green or blue-green algae to form lichens. See Figure C–12. A very widespread group, with nearly 40,000 species. *Amanita, Penicillium.*

Figure C–11. Oyster mushroom. (Carnegie Museum—A. C. Lloyd photo.)

Figure C–12. A lichen (*Lecanora*) on a rock at Carnegie Museum's Powdermill Nature Reserve, Westmoreland Co., Pa. (Carnegie Museum—LeRoy K. Henry photo.)

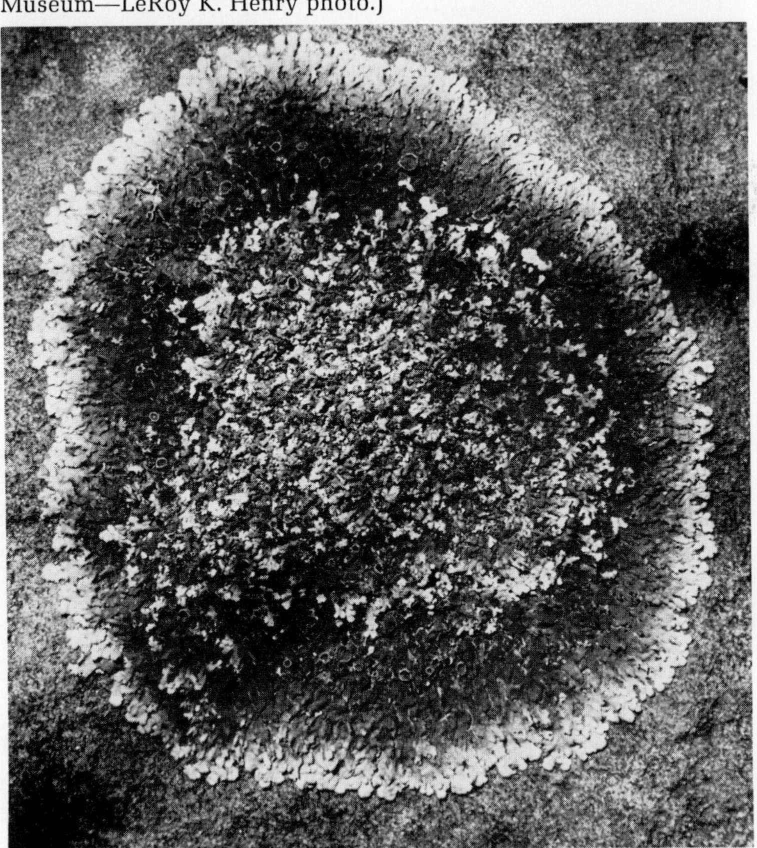

Kingdom Animalia

There are well over a million different kinds of animals. They are all heterotrophs and they do not show the alternation of diploid and haploid generations characteristic of plants; otherwise they are a remarkably diverse group.

The number of phyla recognized by different workers varies. Some may feel that the difference between two groups is significant enough to justify placing them in separate phyla, others are more impressed by the resemblance and lump the two in a single phylum. All are agreed that there are about a dozen major phyla, and it is only in the recognition of small, not very well known groups that there are differences of opinion. The number of recognized phyla may vary from eighteen to twenty-six. Twenty are included here.

For all the major invertebrate phyla this list carries the classification down to classes. (Some minor classes are omitted.) For the phylum Chordata brief statements are included about the orders in the classes of tetrapods, i.e., the amphibians, reptiles, birds, and mammals.

Phylum I. Protozoa (from Greek, *protos,* "first," and *zoon,* "animal"). Mostly microscopic, unicellular organisms, although some are arranged in loosely organized colonies. The word *protozoa* serves as both a common name and a technical name for the group. Protozoa are classified primarily on the basis of the method of locomotion. Marine, freshwater, terrestrial (in moist places), or parasitic. About 30,000 species. Three classes are recognized (four if the heterotrophic euglenoid flagellates are classed as animals rather than plants).

CLASS SARCODINA. Most move by means of pseudopods, although some may have flagella. Usually free-living, but a few are parasitic. *Amoeba, Entamoeba, Actinophrys.*

CLASS SPOROZOA. All parasitic and immotile as adults. Life histories complex and involve sexual reproductive stages. *Plasmodium, Monocystis.*

CLASS CILIATA. Protozoa that move by ciliary action. Most are free-living, aquatic forms, but a few are parasitic. In one group, the adults are stalked and use tentacles in capturing food, although the immature stages are ciliated and free-swimming. Frequently a sexual-type reproduction involving conjugation. See Figure C–13. *Didinium, Paramecium, Stentor, Podophrys.*

Phylum II. Mesozoa (from Greek *meso,* "middle," and *zoon,* "animal"). A small group of minute animals parasitic in the bodies of other invertebrates. The animal composed of a single layer of cells surrounding one, several, or many reproductive cells. Reproductive habits complex with alternate sexual and asexual stages. About fifty species. The largest barely six millimeters (a quarter of an inch) long. *Pseudicyema.*

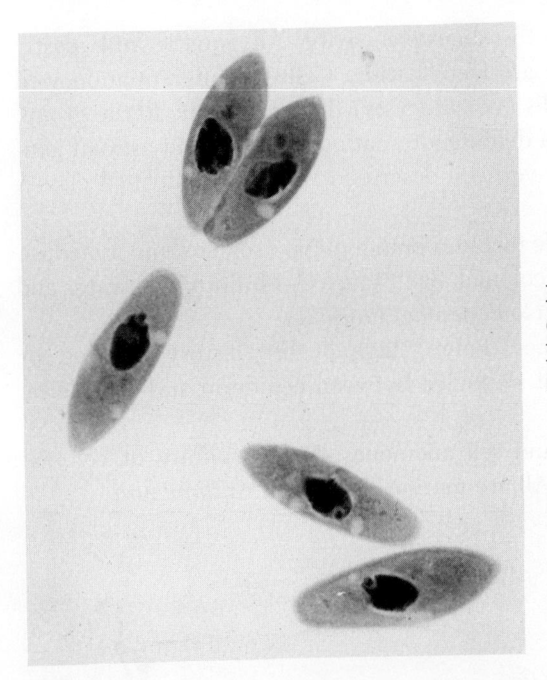

Figure C–13. *Paramecium* in conjugation. (Courtesy Ward's Natural Science Establishment, Inc., Rochester, N.Y.)

Phylum III. Porifera (from Latin, *porus*, "pore," and *ferre*, "to bear"). Sponges. Primitive aquatic metazoans, mostly marine but with a few freshwater forms. Invariably sessile. Body made up of an outer dermal epithelium and an inner gastric epithelium between which there is usually a noncellular gelatinous substance known as *mesoglea*. The outer and inner layers of cells are not homologous to ectoderm and endoderm of higher animals. A skeleton of calcareous or siliceous spicules and/or fibrous spongin lies in the mesoglea together with the reproductive cells. Numerous openings or pores in the body wall lead into a simple or complex series of canals through which water flows to bring food and oxygen to the cells and carry away waste products. About 4,200 species known at present. Three classes are recognized.

CLASS CALCAREA. Marine species in which the skeleton is made up of calcium carbonate. *Sycon, Scypha.*

CLASS HEXACTINELLIDA. Deepwater marine species with the skeleton formed of silica, giving them the common name of *glass sponges. Euplectella, Hyalonema.*

CLASS DESMOSPONGIAE. Skeleton of siliceous spicules and/or spongin fibers. Marine and freshwater. The commercial sponge is included here. *Hippospongia, Spongia, Spongilla.*

Phylum IV. Cnidaria (from Greek, *knide*, "nettle"). Jellyfishes, corals, sea anemones, and hydroids. Simple, radially symmetrical, true diploblastic metazoans with a single opening to the digestive cavity. All aquatic and mostly marine. Some are sessile, others are free-floating. Cellular differentiation well marked, with special stinging cells (nematocysts) that are unique to the group. Life history often complex with a definite alternation of sexual and asexual generations. About 11,000 species have been described. Three well-defined classes are included in the phylum.

CLASS HYDROZOA. Life cycle includes both a polyp (sessile) and a medusa (free-floating) stage. Sometimes colonial. See Figure C–14. Both freshwater and saltwater forms. *Hydra, Obelia, Gonionemus, Physalia.*

CLASS SCYPHOZOA. Jellyfishes. Polyp stage of life history reduced or absent. Extensive development of mesoglea between ectoderm and endoderm. All are marine. *Cyanea, Pelagia.*

CLASS ANTHOZOA. Corals and sea anemones. Either solitary or colonial polyps, with no medusoid stage. All are marine. *Gorgonia, Anthopleura.*

Figure C–14. *Obelia*, a colonial hydrozoan. (Courtesy Ward's Natural Science Establishment, Inc., Rochester, N.Y.)

Phylum V. Ctenophora (from Greek *ktenos*, "comb," and *phora*, "carrying"). Relatively simple marine metazoans, familiarly known as sea walnuts or comb jellies. Adult body wall basically diploblastic but with some muscle cells in mesoglea between ectoderm and endoderm that seem to represent a primitive-type mesoderm. The group is sometimes considered the most primitive triploblastic phylum. Radially symmetrical. Tentacles equipped with adhesive cells rather than stinging cells. Body bearing eight rows of swimming plates or combs. Body nearly transparent and combs iridescent with reflected light during the day. About eighty species known. *Pleurobrachia.*

Phylum VI. Platyhelminthes (from Greek *platys*, "wide," and *helminthes*, "worm"). Flatworms. True triploblastic animals with well-developed mesoderm and bilateral symmetry. Gut cavity opens to outside by a single opening. No coelom or pseudocoel. Both free-living and parasitic forms. Some 15,000 species have been named. The phylum is divided into three classes.

CLASS TURBELLARIA. Free-living flatworms. All have a ventral mouth and are ciliated externally. Marine, freshwater, and terrestrial. *Dugesia, Bipalium.*

CLASS TREMATODA. Flukes. Suckers are present, the mouth is usually anterior. Covered externally with a cuticle. All are parasitic. *Fasciola, Schistosoma.*

CLASS CESTODA. Tapeworms. External cuticle present, digestive system absent. Body composed of few to many segments called *proglottids.* Anterior segment (scolex) with suckers and (generally) hooks for attachment to host. See Figure C—15. Entirely parasitic. *Echinococcus, Taenia.*

Phylum VII. Rhynchocoela (from Greek, *rhynchos*, "snout," and *koilos*, "hollow"). Nemertine worms. An anus present at posterior end of gut, but no pseudocoel or coelom. Small, slender, flattish worms, with a remarkable proboscis apparatus used in the capture of food. It gives the name *proboscis worms* to the group. About 600 species known. All marine except one genus that lives in fresh water and another that is terrestrial in the tropics. *Lincus.*

Phylum VIII. Aschelminthes (from Greek, *ascus*, "a bladder," and *helminthes*, "worm"). Cavity worms and their allies. Generally small animals characterized by the presence of an anus and a pseudocoel. Digestive tract simple except for a specialized, muscular pharynx. Over 80,000 species. The phylum includes two quite different major classes (sometimes considered separate phyla) and several minor ones (not listed here).

CLASS ROTIFERA. Wheel animalcules. Aquatic, usually in fresh water. Bilaterally symmetrical animals with the pharynx modified into a grinding organ (mastax) and with a crown of cilia on the anterior end. The beating cilia look like a rotating wheel in many forms. Usually solitary, free-moving animals,

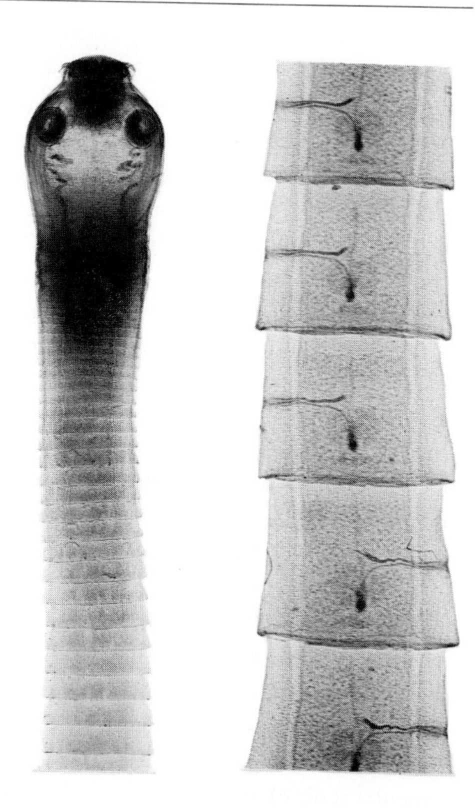

Figure C–15. Tapeworm, *Taenia pisiformis,* showing the head (scolex) and several immature body segments (proglottids). (Courtesy Ward's Natural Science Establishment, Inc., Rochester, N.Y.)

but some are sessile and some colonial. Most are freshwater forms, some are parasitic. Sexes separate. *Philodina, Rotaria, Conochilus.*

CLASS NEMATODA. Roundworms. See Figure C–16. Bilaterally symmetrical, slender, elongate worms, usually tapered at both ends. Mouth surrounded by lips and sensory papillae or bristles. No cilia present. Sexes usually separate. One of the most numerous and widespread of the metazoan groups, occurring

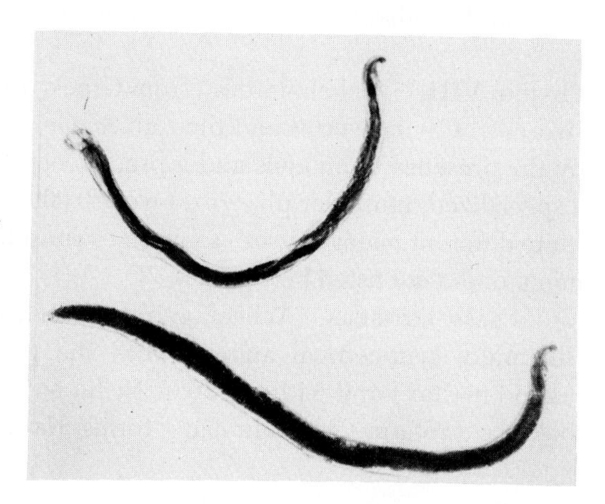

Figure C–16. Male and female American hookworms. (Courtesy Ward's Natural Science Establishment, Inc., Rochester, N.Y.)

on land, in the sea, in fresh water, from the polar regions to the tropics, and from great depths in the oceans to mountain tops, deserts, and hot springs. Free-living and parasitic forms. Ninety thousand individuals have been taken from a rotting apple on the ground in an apple orchard. *Ascaris, Trichinella, Wuchereria.*

Phylum IX. Acanthocephala (from Greek, *acantheis,* "spiny," and *cephalicus,* "relating to the head"). The spiny-headed worms. Parasitic worms with a spiny proboscis. Digestive tract complete and pseudocoel present. Only some 800 species known in this small phylum. *Acanthogyrus, Acanthocephalus.*

Phylum X. Entoprocta (from Greek, *entos,* "within," and *proctus,* "the anus"). Small, sessile animals, mostly marine, with a pseudocoel and a wheel of cilia within which both mouth and anus open, hence the name. A few are solitary but the great majority of the some sixty species are colonial. *Pedicellina.*

Phylum XI. Ectoprocta (from Greek, *ectos,* "outside," and *proctus,* "the anus"). Coelomate animals with a lophophore, a circular fold of the body wall bearing ciliated tentacles that surrounds the mouth. With few exceptions, sessile, colonial animals. Some 4,000 species are known, about 500 in fresh water, all others are marine. *Bugula, Plumatella.*

Phylum XII. Brachiopoda (from Greek, *brachium,* "armlike branch," and *podo,* "foot." An allusion to the lophophore, which is made up of two looped, ciliated branches). Lampshells. These animals superficially resemble bivalve mollusks with their two shells, but the brachiopod shells are not paired right and left but are dorsal and ventral. The two-looped ciliated arms of the lophophore lie coiled within the shell. Only about 300 living species, all marine. *Lingula.*

Phylum XIII. Mollusca (from Latin, *molluscus,* "soft"). The familiar snails, clams, squids, and octopods. Triploblastic, coelomate protostomes with well-developed respiratory, circulatory, excretory, nervous, and reproductive systems. Bilaterally symmetrical, but one group (the snails) secondarily modified toward asymmetry. All with a fleshy muscular organ called the *foot,* and most with a calcareous shell. The phylum is well known because the shells are popular with collectors, they make excellent fossils, and man has eaten the animals from time immemorial. Mostly marine but also widespread in fresh water and on land. Approximately 110,000 living species, divided into five major classes and several minor ones.

CLASS AMPHINEURA. Chitons. Ovate, flattish mollusks that live on rocks around the seacoasts. Generally the back is covered with eight transverse plates. *Chiton, Chaetopleura.*

CLASS GASTROPODA. Snails, slugs, and marine whelks. All have a filelike feeding apparatus, the radula. Marine, freshwater, and terrestrial. Many economically important. *Haliotis, Helix, Littorina.*

CLASS PELECYPODA. Clams and oysters. Bivalved aquatic mollusks, many with economic value. *Anodonta, Ostrea, Venus.*

CLASS SCAPHOPODA. Toothshells. Shell tusk-shaped and open at both ends. Marine bottom dwellers. *Dentalium.*

CLASS CEPHALOPODA. Squids, *Nautilus,* and octopods. The foot is divided into a number of arms or tentacles used in locomotion and feeding. Except in *Nautilus,* shell reduced and internal (squids) or absent (octopods). All are marine. *Loligo, Nautilus, Sepia.*

Phylum XIV. Annelida (from Latin, *annulus,* "a ring"). The segmented worms, including the familiar angleworms and leeches. Highly developed, triploblastic, coelomate protostomes with metamerism well developed; each segment is essentially a duplicate of the others. Excretory organs, blood vessels, and nervous system all segmentally arranged. Most are marine, a number live in fresh water or damp soil. The majority are free-living, but a few are parasitic. About 9,000 species known, belonging to three major classes.

CLASS POLYCHAETA. Marine worms. Fleshy, nonjointed appendages (parapodia) on each of the body segments. *Aphrodite, Chaetopterus, Nereis.*

CLASS OLIGOCHAETA. Earthworms. Without parapodia or suckers. Found in damp soil and fresh water. *Lumbricus, Tubifex.*

CLASS HIRUDINEA. The leeches. Parapodia absent but with anterior and posterior suckers with which they attach. Many are predaceous rather than bloodsucking. Marine, freshwater, and terrestrial. *Haemopsis, Hirudo, Placobdella.*

Phylum XV. Onychophora (from Greek, *onyx,* "claw," and *phoros,* "bearing"). A small group somewhat intermediate between the annelids and the arthropods. Like the former they are segmented and have paired excretory nephridia in most body segments; like the latter they have paired appendages with claws, the appendages showing incipient segmentation. Terrestrial in tropical regions. About eighty species known. *Peripatus.*

Phylum XVI. Arthropoda (from Greek, *arthron,* "joint," and *podos,* "foot"). Insects, spiders, centipedes, crustaceans, and their allies. Complex, triploblastic, bilaterally symmetrical, metameric, coelomate protostomes. The only invertebrates that have external skeletons and completely jointed appendages. Further differing from the annelids and onychophorans in the tendency of the body segments to fuse into larger units. At least 80 per cent of all living species belong to this phylum. Over 900,000 living species are known, of which about 800,000 are insects. This large phylum is divided into three subphyla, one of which includes only the extinct trilobites.

SUBPHYLUM CHELICERATA. Body divided into cephalothorax and abdomen, antennae lacking, first pair of appendages modified into pincer-like appendages called *chelicerae*.

CLASS MEROSTOMATA. Abdominal appendages modified to form gills, a spikelike "tail." The living marine horseshoe crabs or king crabs and the extinct eurypterids. *Limulus*.

CLASS PYCNOGONIDA. Sea Spiders. Marine forms with a narrow body and very long legs. *Nymphon*.

CLASS ARACHNIDA. Spiders, scorpions, ticks, and their allies. See Figure C–17. In addition to the chelicerae, they have a second pair of appendages called *pedipalps*, which may be variously modified, and four pairs of appendages modified as walking legs. Mostly terrestrial, although some have become secondarily aquatic. *Centruroides, Dermacentor, Latrodectus*.

Figure C–17. An arachnid, Daddy Longlegs, on a mallow flower. (Courtesy Ward's Natural Science Establishment, Inc., Rochester, N.Y.)

SUBPHYLUM MANDIBULATA. Sensory appendages (antennae) present; appendages near mouth modified to form chewing structures. (Many authorities believe that the crustaceans evolved separately from the trilobites and should not be included in the subphylum.)

CLASS CRUSTACEA. Crabs, lobsters, barnacles, sow bugs, and others. Two pairs of antennae. Adults breathe by means of gills. Most marine, but many freshwater and some terrestrial. *Balanus, Callinectes, Cambarus.*

CLASS DIPLOPODA. The millipedes. See Figure C–18. One pair of short antennae. Many body segments each with two pairs of legs. *Julus, Spirobolus.*

CLASS CHILOPODA. The centipedes. Antennae large and prominent. Many body segments, each bearing a single pair of legs. *Scutigera, Scolopendra.*

CLASS INSECTA. Body divided into three parts: head, thorax, and abdomen. Adults breathe by means of spiracles. One pair of antennae and three pairs of walking legs, wings usually present. There are many orders in this class. Some of the more familiar ones are the following:

Figure C–18. A millipede photographed at night on Carnegie Museum's Powdermill Nature Reserve, Westmoreland Co., Pa. (Carnegie Museum—M. Graham Netting photo.)

Order Neuroptera. Ant lions, lace wings, doodle bugs. *Chrysops, Myrmeleon.*

Order Orthoptera. Grasshoppers, locusts, crickets, cockroaches and others. *Mantis, Romalea, Blatta.*

Order Odonata. Dragonflies and damselflies. *Anax. Gomphus.*

Order Ephemeroptera. Mayflies. *Ephemera.*

Order Hemiptera. True bugs. *Cimex, Ranatra.*

Order Homoptera. Cicadas and aphids. *Aphis, Cicada.*

Order Coleoptera. Beetles, weevils, and others. The largest order with at least a quarter of a million species. *Tenebrio, Dytiscus.*

Order Lepidoptera. Butterflies and moths. See Figure C–19. *Tinea, Sphinx, Papilio.*

Order Siphonaptera. Fleas. *Ctenocephalides, Pulex.*

Order Hymenoptera. Bees, ants, wasps, and others. *Apis, Formica, Vespa.*

Order Diptera. Flies, mosquitoes, and their relatives. *Culex, Musca.*

Phylum XVII. Echinodermata (from Greek, *echinos,* "hedgehog," and *derma,* "skin," referring to the spines with which most are covered). The starfishes, sea

Figure C–19. A newly emerged Monarch butterfly at Carnegie Museum's Powdermill Nature Reserve, Westmoreland Co., Pa. (Carnegie Museum—M. Graham Netting photo.)

urchins, sand dollars, and others. Like the chordates, they have an internal skeleton derived from the mesoderm and deuterostomic development. They also have a unique system of fluid-filled canals and appendages derived from the coelom, the water vascular system. Adults usually radially symmetrical and built on a pentamerous (five-part) body plan. Probably the most closely related to the chordates of all the invertebrates. They are entirely marine. Less than 6,000 living species, in five classes.

CLASS CRINOIDEA. Sea lilies and feather stars. Either stalked forms that attach to the ocean floor, or free-swimming with reduced stalk. Body drawn out into arms. *Antedon, Pentacrinus.*

CLASS ASTEROIDEA. Starfishes. See Figure C–20. Free-moving bottom crawlers. The arms are not sharply set off from the central body. *Asterias, Solaster.*

CLASS OPHIUROIDEA. Brittle stars. Arms markedly slender and set off from the central body. *Aphiothrix, Ophioderma.*

CLASS ECHINOIDEA. Sea urchins, sand dollars, and others. Body circular or oval in shape, flattened. Arms lacking, spines well developed. *Arbacia, Strongylocentrotus.*

CLASS HOLOTHURIOIDEA. Sea cucumbers. Skeleton reduced so that body is soft, cylindrical. No arms are present. *Cucumaria, Thyone.*

Phylum XVIII. Chaetognatha (from Greek, *chaeton,* "bristle," and *gnathos,* "jaw"). Arrow-worms. A small phylum, apparently not closely related to any other group. Triploblastic coelomate deuterostomes with well-developed digestive system but no circulatory or excretory organs. The tail extends beyond the anus, as in no other group except the chordates. May represent a very early offshoot of deuterostomic line. Size range from about six to 120 millimeters (a quarter of an inch to nearly five inches). Active swimmers, darting after prey like tiny arrows. About sixty species, all marine. *Sagitta.*

Figure C–20. A starfish, *Orthasterias columbiana.* (Courtesy Ward's Natural Science Establishment, Inc., Rochester, N.Y.)

Phylum XIX. Hemichordata (from Greek, *hemi,* "half," and *chorda,* "string"). Acorn worms or tongue worms. Wormlike creatures that range from twenty-five to 450 millimeters (one to eighteen inches) in length. A proboscis that projects anterior to mouth like a tongue. Coelomate deuterostomes. Because some have pharyngeal slits and dorsal nervous tissue they were formerly placed with the chordates. It seems more probable that they represent an early offshoot of the line that gave rise to both echinoderms and chordates. All are marine, burrowing in soft mud of ocean floor. About 100 species known. *Rhabdopleura, Dolichoglossus.*

Phylum XX. Chordata (from Greek, *chorda,* "string," in allusion to the string- or rodlike notochord in the dorsal region). The chordates all have an elongated notochord, hollow dorsal nervous system, and pharyngeal pouches at some stage in their life cycle. Three subphyla. About 45,000 described species, most belonging to subphylum Vertebrata.

SUBPHYLUM UROCHORDATA. Sea squirts or tunicates. Adults have little resemblance to other chordates, but larvae show their true relationships. Latter are free-swimming forms with well-developed nervous system, notochord, and pharyngeal pouches that break through to outside to form slits. Adults of most species attach to piling or rock and metamorphose (degenerate?) into sessile, rounded, soft-bodied creatures. Some are colonial. Nearly microscopic to over ten centimeters (about four inches) long. All are marine. *Appendicularia, Ciona, Molgula.*

SUBPHYLUM CEPHALOCHORDATA. Amphioxus, the lancelets. Small, fishlike marine animals, usually found buried in sand at bottom not far from shore. Well-developed notochord, hollow dorsal nervous system, and numerous gill slits. Metamerism strikingly evident in V-shaped muscle segments of body trunk. No true head or brain. *Branchiostoma.*

SUBPHYLUM VERTEBRATA. Spinal column of vertebral elements made of cartilage or bone. Evidence of metamerism in arrangement of muscles, nerves, and blood vessels as well as vertebrae. A true brain, heart, and liver are always present. Seven living classes; three are fishlike (pisciform) and four are tetrapods or four-footed vertebrates.

CLASS AGNATHA. Lampreys and hagfish. See Figure C–21. Jawless vertebrates, without scales or paired appendages. Mostly external parasites on bodies of bony fishes. Marine and freshwater. *Petromyzon, Bdellostoma.*

CLASS CHONDRICHTHYES. Sharks, rays, skates, and chimeras. The cartilaginous fishes. Skeleton of cartilage and gill slits usually opening separately to outside. All have jaws. Mostly marine but a few freshwater forms. *Squalus, Raja.*

CLASS OSTEICHTHYES. The bony fishes. Jawed fishes with bony skeletons. Gill slits covered by a bony plate, the operculum. Marine and freshwater. Two subclasses.

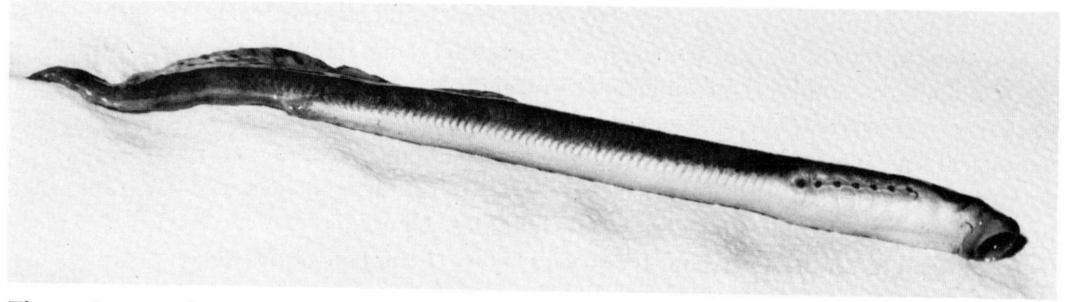

Figure C–21. The brook lamprey, *Lampetra*, taken from a stream on Carnegie Museum's Powdermill Nature Reserve, Westmoreland Co., Pa. (Carnegie Museum, M. Graham Netting Photo.)

SUBCLASS ACTINOPTERYGII. The ray-finned fishes. Nostrils not connected with mouth cavity. Fins without fleshy lobe at base or bony axis. A flourishing group, occupying the seas and fresh waters of the world. *Amia, Perca, Tarpon.*

SUBCLASS SARCOPTERYGII. Lobe-finned fishes, lung fishes, and brachiopterygians. Typically with lungs, a fleshy lobe at base of fin, and with a passage from the nasal cavity opening into mouth. *Latimeria, Protopterus, Polypterus.*

CLASS AMPHIBIA. Frogs, salamanders, sirens, and caecilians. The first class of vertebrates to develop the tetrapod limb for life on land. Larvae usually aquatic, adults usually terrestrial. Four orders.

Order Gymnophiona. The caecilians. Wormlike tropical forms without legs. Small dermal scales present in some. *Ichthyophis.*

Order Trachystomata. The sirens. Elongated, aquatic amphibians with a single pair of tiny, anteriorly placed legs. Living forms known only from southern United States and northern Mexico. *Siren, Pseudobranchus.*

Order Caudata. Salamanders, newts, mudpuppies. See Figure C–22. Four-legged, lizard-shaped amphibians with well-developed tails. Primarily northern hemisphere in distribution. Representatives of genus *Andrias* in eastern Asia are the largest living amphibians, reaching over one and a half meters (about five feet) in length. *Desmognathus, Plethodon.*

Order Anura. Frogs and toads. Tail lacking, hind legs elongated. See Figure C–23. A breeding call well developed in males of most species. *Bufo, Hyla, Rana.*

CLASS REPTILIA. Snakes, lizards, turtles, and others. Ectothermic vertebrates that reproduce by means of amniote egg. Scaled and air-breathing throughout posthatching life. Four orders.

Figure C–22. Long-tailed Salamander, *Eurycea longicauda*, taken on Carnegie Museum's Powdermill Nature Reserve, Westmoreland Co., Pa. (Carnegie Museum—M. Graham Netting photo.)

Order Squamata. Lizards and snakes. Reptiles with teeth and paired copulatory organs. Body elongate, with or without legs. Primarily terrestrial but some secondarily aquatic. *Anolis, Natrix, Python.*

Order Testudinata. Turtles. See Figure C–24. Toothless reptiles with a protective shell. Marine, freshwater, terrestrial. *Chelonia, Clemmys, Testudo.*

Figure C–23. American toad, *Bufo americanus*, calling male. (Carnegie Museum—M. Graham Netting photo.)

Figure C–24. Gopher turtle, *Gopherus polyphemus*, taken at Pineland Plantation, Baker Co., Georgia. (Carnegie Museum—M. Graham Netting photo.)

Order Crocodilia. Alligators and their allies. Lizard-like in general appearance but differing in a number of internal details. Copulatory organ of male single. *Alligator, Crocodylus.*

Order Rhynchocephalia. Tuatara. The "living fossil" of New Zealand. Lizard-like in appearance but not closely related. Male lacks copulatory organ. Teeth on roof of mouth as well as on jaws. *Sphenodon.*

CLASS AVES. Birds. See Figure C–25. Feathers are unique to the group. Endotherms that lay amniote eggs. Class is divided into two subclasses, one of which includes only the extinct, toothed birds of the Jurassic, *Archeopteryx.* All living forms are toothless and are included in the single subclass Neornithes. A

Figure C–25. Mute Swan in the moat of Kronborg Castle, Elsinore, Denmark. (Carnegie Museum, M. Graham Netting photo.)

large and successful group with about 8,600 living species. Modern birds are divided into about twenty-eight orders. Among the more familiar are the perching birds, order Passeriformes; the grain-eating birds, order Galliformes; and the birds of prey, order Falconiformes. *Passer, Gallus, Falco.*

CLASS MAMMALIA. Rats, bats, dogs, horses, men. Vertebrates with hair that nourish their young by means of mammary glands. About 3,500 living species. Two subclasses.

SUBCLASS PROTOTHERIA. Egg-laying mammals. The spiny anteaters and duck-billed platypus of the Australian region. *Tachyglossus, Ornithorhynchus.*

SUBCLASS THERIA. Live-bearing mammals.

INFRACLASS METATHERIA. The pouched mammals or marsupials. Largely but not entirely confined to Australia. *Didelphis, Macropus.*

INFRACLASS EUTHERIA. Placental mammals. The largest and most successful group of mammals, worldwide in distribution. Sixteen orders of living forms.

Order Insectivora. Moles, shrews, hedgehogs. Small animals with numerous, sharp-pointed teeth. *Scalopus, Sorex.*

Order Dermoptera. The "flying lemurs." Neither flying nor lemurs. A well-developed fold of skin on each side that serves as a gliding membrane. One living genus, confined to southeastern Asia. *Galeopithecus.*

Order Chiroptera. Bats. Forelimb modified into a flying organ. *Desmodus, Myotis.*

Order Primates. Lemurs, monkeys, great apes, man. High intelligence. *Ateles, Gorilla, Homo.*

Order Edentata. New World anteaters, sloths, armadillos. Specialized mammals with teeth poorly developed or absent. *Dasypus, Myrmecophaga.*

Order Pholidota. Pangolins. The scaly anteaters of the Old World. Without teeth, body covered with heavy scales. *Manis.*

Order Lagomorpha. Rabbits, hares, pikas. Incisors modified for gnawing as in rodents but differing in many details of internal anatomy. *Lepus, Sylvilagus.*

Order Rodentia. Rats, mice, squirrels, beavers. Most of the mammals of the world are rodents. Only two incisors in each jaw, highly modified for gnawing. *Mus, Sciurus, Castor.*

Order Cetacea. Whales, dolphins, porpoises. See Figure C–26. Marine mammals, highly modified for aquatic existence. *Mesoplodon, Physeter, Tursiops.*

Order Carnivora. The flesh-eating mammals. Cats, bears, weasels, seals. See Figure C–27. Canine teeth well developed, digits with claws. Some, such as seals and walrus, modified for aquatic existence. Any

Figure C–26. Bottlenose Porpoise, *Tursiops truncatus*, at Aquatorium, St. Petersburg Beach, Fla. (Carnegie Museum—M. Graham Netting photo.)

animal that eats meat is a carnivore, but not necessarily a member of the order Carnivora. *Canis, Felis, Phoca.*

Order Tubulidentata. Aardvark. The "earth pig" of Africa. Long-snouted and long-eared with poorly developed teeth. *Orycteropus.*

Order Proboscidea. Elephants. Upper lip and nose elongated into a trunk, incisors elongated to form tusks. *Elephas.*

Figure C–27. Cheetah, *Acinonyx jubatus*, Amboseli Game Reserve, Kenya. (Carnegie Museum—M. Graham Netting photo.)

Order Hyracoidea. The conies of Africa and Arabia. Small herbivores, digits with hoof-like nails. *Hyrax.*

Order Sirenia. Manatee or sea cow. Large marine mammals in which the pelvic girdle and hind limbs have entirely disappeared. *Manatus.*

Order Perissodactyla. The odd-toed ungulates. Horse, rhinoceros, tapir, and others. See Figure C–28. Hoofed mammals with an odd number of toes. *Equus, Rhinoceros.*

Order Artiodactyla. Camels, cows, pigs, and others. Hoofed mammals with an even number of toes. *Bos, Camelus, Sus.*

Figure C–28. East African zebras, *Equus burchelli*, on the Amboseli Game Reserve, Kenya. (Carnegie Museum—M. Graham Netting photo.)

APPENDIX **D**

Glossary

Accommodate: To bring the light rays from sources at various distances to a focal point on the retina; in man this is accomplished by modifying the shape of the lens.

Acetylcholine: A substance released by the axons of nerve cells by which impulses are transmitted from one neuron to another or to a muscle.

Actin: A muscle protein that forms filaments.

Active transport: Transport of a substance through a membrane by other than osmotic activity or diffusion.

Actomyosin: A complex of actin and myosin.

Addison's disease: A disease caused by the hypofunction of the adrenal cortex, with symptoms of muscular weakness, apathy, and pigmentation of the skin; generally fatal if not treated.

ADP: Adenosine diphosphate; a substance involved in energy exchange within the cell.

Adrenal: Endocrine gland located adjacent to the kidney.

Adrenalin: See epinephrine.

Aerobic respiration: Metabolic breakdown of foods involving oxygen and resulting in the formation of ATP.

Agnatha: A class of jawless vertebrates.

Allantois: An extraembryonic membrane in the amniote embryo involved in excretion.

Allele: Alternate form of a gene.

Alveolus: One of the little sacs in the lungs formed by the termination of the bronchioles.

Amino acid: One of the compounds containing NH_2 from which proteins are formed.

Ammonia: A colorless gas, NH_3, soluble in water, with a pungent odor.

Amnion: An extraembryonic membrane.

AMP: Adenosine monophosphate, usually as cyclic AMP, a substance that mediates the effects of hormones in cells.

Amphibians: A class of vertebrates; they have limbs rather than fins and do not have an amnion associated with the embryo.

Amplexus: A clasping of one sex by the other for external fertilization.

Anaerobic respiration: A metabolic breakdown of foods not involving oxygen, resulting in the formation of ATP.

Anaphase: The stage of mitosis in which the daughter chromosomes migrate toward opposite poles.

Androgens: Male hormones produced by the testes and by the adrenal cortex.

Aneuploidy: Having an unbalanced set of chromosomes.

Angiosperms: See Anthophytes.

Anthophytes: The flowering plants, formerly called angiosperms.

Antibody: A serum globulin synthesized by the body in response to an antigenic stimulus.

Antidiuretic hormone: A hormone that suppresses the excretion of urine.

Antigen: A substance that induces the production of antibodies when introduced into the body.

Apocrine: Pertaining to glands in which the cells are partially disintegrated in the process of secretion.

Aqueous chamber: The chamber of the eye in front of the lens.

Archenteron: The embryonic primitive gut.

Arteries: Muscular blood vessels that generally convey blood away from the heart.

Arthropoda: A phylum of animals with external skeletons and jointed appendages.

Ascorbic acid: Vitamin C, $C_6H_8O_6$.

Asexual: Literally, without sex.

Asexual reproduction: Reproduction not involving gamete formation.

ATP: Adenosine triphosphate, a substance involved in energy exchange within the cell.

Autonomic system: A part of the peripheral nervous system concerned with involuntary actions.

Autotroph: Self-feeder; an organism capable of manufacturing its own food.

Axon: Neural fiber that conveys impulses away from the nerve cell body.

Backcross: A cross between the heterozygous F_1 and the homozygous recessive parent.

B-cells: Cells of the immune system that form antibodies against certain bacteria.

Bile: Fluid secreted by the liver.

Biocompound (Biochemical compound): Organic compounds, often large and complex, found only in organisms.

Birth canal: The inner or true pelvis, the cervix of the uterus, and all of the soft tissues that line these structures.

Bisexual: Having two sexes.

Bladder: A storage bag, usually referring to the sac in which urine is stored.

Blastocoel: The cavity of the blastula.

Blastomeres: Cells formed in the primary divisions of the egg.

Blastopore: The opening formed when the blastula invaginates.

Blastula: A hollow, ball-like structure formed of blastomeres.

Blood: A fluid tissue that circulates in the body and functions in transport.

Bone: The mineralized portion of a vertebrate skeleton.

Bony labyrinth: Bony, fluid-filled tubes that surround the membranous part of the inner ear.

Bowman's capsule: Invaginated end of a kidney tubule that surrounds a glomerulus.

Brain stem: That part of the brain comprising the midbrain, pons, and medulla.

Breathing: The physical process of pumping air into and out of the lungs.

Bronchiole: A small tubule in the lung that terminates in an alveolus.

Bronchus: Either one of the two main branches of the trachea.

Bryophyta: The liverworts and mosses.

Canaliculus: A tiny passage in bone through which interchange takes place between cells.

Canine: One of the four elongate stabbing teeth in dogs and cats; may be absent or otherwise modified in other animals. Also a member of the family Canidae, the dogs.

Capillary: One of the smallest blood vessels. Continuous with the smallest arteries and/or veins.

Carbohydrate: A compound of carbon, hydrogen, and oxygen in which the proportion of hydrogen to oxygen is about the same as in water, H_2O.

Cardiac muscle: Heart muscle.

Cartilage: The tough connective tissue found in vertebrate skeletons.

Cecum: A blind pouch or saclike diverticulum.

Central nervous system: That part of the nervous system made up of the brain and spinal cord.

Centriole: Small rodlike structure concerned with cell division.

Centromere: See kinetochore.

Cerebellum: The part of the brain that functions in the coordination and integration of postural and voluntary movements.

Cerebrum: The part of the brain that is the center of consciousness and memory.

Chiasma: A crossing, e.g., a chromosome chiasma.

Cholesterol: One of the steroids; an important component of cell membranes.

Cholinesterase: An enzyme that splits acetylcholine.

Chondrichthyes: Fishlike vertebrates that have jaws but lack a bony skeleton.

Chordata: The phylum of animals that at some time during their lives have a notochord develop below the neural cord.

Chorion: An extraembryonic membrane that surrounds the amniote embryo.

Chromosome: Body in the nucleus composed of DNA and protein.

Chyme: The fluid, partially digested, stomach contents.

Cilia: Tiny, threadlike structures on the surface of a cell that beat back and forth.

Clavicle: The collar bone.

Cleavage: The mitotic segmentation of the zygote into blastomeres.

Clitoris: A female genital structure homologous with the penis of the male.

Cnidaria: A phylum of invertebrate animals that have specialized stinging cells (formerly called Coelenterata).

Cochlea: A portion of the inner ear used in the perception of sound; in man it is coiled like a snail, hence the name cochlea (meaning "snail shell").

Cochlear duct: A fluid-filled tube in the inner ear that conveys vibrations.

Coelenterata: See Cnidaria.

Coelom: The cavity formed within the mesoderm.

Coenzyme: A nonprotein substance that functions in conjunction with an enzyme.

Collagen: Protein fiber found in connective tissues.

Colon: The main part of the large intestine.

Commensalism: A living together.

Cone: A cell in the retina of the eye used in color perception *or* a reproductive structure in nonflowering seed plants.

Coniferophyte: A seed-bearing plant that lacks flowers (formerly called a gymnosperm).

Connective tissue: Tissue that sustains, supports, or binds other tissues.

Consanguinity: Inbreeding, particularly in human populations; close relationship.

Contractile tissue: Muscle tissue.

Copulation: Insemination of the female by the male, usually with the aid of a special structure; in mammals this is the penis.

Coronary artery: Artery that supplies blood to the heart muscle.

Cowper's gland: Either of two glands near the base of the penis that supplies a fluid medium for the sperm.

Creatine: A nitrogenous compound in muscle.

Cryptorchidism: The condition in males of having the gonads in the abdominal cavity instead of the scrotum.

Dehydration: Removal of water.

Dendrite: Neural fiber that conveys impulses toward the nerve cell body.

Deuterostome: Organism in which cleavage is indeterminate.

Diad: The double chromatid body resulting from the first meiotic division.

Diaphragm: A muscular partition that separates the abdominal cavity from the thoracic cavity.

Diastole: The stage of dilation of the heart.

Dicotyledon: Flowering plant in which two food-storing leaves or cotyledons are formed in embryonic development.

Diffusion: Movement by random molecular activity.

Dihybrid: A genetic cross involving two pairs of heterozygous genes.

Disaccharide: Any of the class of sugars that yields two monosaccharides upon hydrolysis.

Division: In botany, one of the major categories into which plants are classified.

DNA: Deoxyribonucleic acid; self-duplicating molecule by which hereditary characters are transmitted.

Dominant: A gene that suppresses the phenotypic expression of its allele.

Dorsal: The back side of an animal.

Ductless gland: Same as endocrine gland.

Ductus arteriosus: A fetal artery that lets blood bypass the lungs.

Eccrine: A gland that secretes its product without injury or destruction of the secreting cells.

Echinodermata: A phylum of animals containing the starfishes and their relatives.

Elastic fiber: Yellowish fiber, elastic in quality, found in the intercellular substance of connective tissue.

Endemic goiter: Enlargement of the thyroid gland in compensation for a lack of enough iodine in the food and water.

Endocrine gland: Ductless glard; a hormone-secreting gland.

Endoderm: The innermost layer of tissue in a diploblastic or triploblastic embryo.

Endometrium: The glandular lining of the uterus.

Endoplasmic reticulum: A series of membranes in the cytoplasm of the cell on which ribosomes may be located.

Engram: The place where a memory is stored in the brain.

Enteron: A true gut, one with both an anterior and posterior opening.

Environment: The aggregate of all external conditions affecting the life of an organism.

Enzyme: A protein that regulates chemical reactions but is not used up in those reactions.

Epiblast: The outer layer of cells in an embryo two cell layers thick.

Epidermis: The outer layer of the integument.

Epididymis: An oblong body attached (in humans) to the upper part of each testis; it contains the ducts leading from the testis.

Epinephrine (Adrenalin): A hormone of the adrenal medulla that controls carbohydrate metabolism; also secreted by axons of the sympathetic nervous system.

Epithelial tissue: Covering or lining tissue.

Equatorial plate: The plane on which the chromosomes arrange themselves prior to separating.

Erythroblast: The nucleated cell from which a red blood cell, an erythrocyte, is formed.

Erythrocyte: Red blood cell.

Estrogens: A general term for a number of female sex hormones.

Eunuchism: The condition in males of being without functional gonads.

Euploidy: The condition of having the entire set of chromosomes arranged other than in pairs.

Eustachian tube: A tube that connects the middle ear chamber with the pharynx.

Eutheria: The placental mammals.

Exocrine gland: A gland that secretes its product through a duct.

Exophthalmic goiter: A condition brought about by overproduction of the thyroid gland, characterized by protruding eyeballs.

Extraembryonic membranes: Membranes that are not tissues of the embryo itself, e.g., amnion, chorion.

F_1: The first filial generation.

F_2: The second filial generation.

FAD: Flavin adenine dinucleotide, a molecule that acts as a hydrogen acceptor-donor.

Fallopian tube: The oviduct of mammals.

Fascia: A connective tissue sheath covering a muscle or other structure.

Fatty acid: A hydrocarbon chain with a terminal COOH group.

Fertilization: The fusion of the nuclei of gametes.

Fetus: Specifically a human embryo any time after two months' development. The term may be applied to the unborn young of any viviparous mammal.

Fibrinogen: A soluble protein in the blood plasma involved in the formation of clots.

Fission: A type of asexual reproduction involving simple division of a cell by mitosis.

Fixed Action Pattern: A coordinated muscular movement that is innate and species specific.

Flagella: The long slender process in some cells that is utilized in locomotion.

Follicle: A very small secretory or excretory sac or gland.

Fontanel: An opening.

Foods: The fats, carbohydrates, proteins, and other substances from which the body derives nourishment.

Foramen ovale: An oval window; specifically, an opening located in the atrial partition of the heart of a fetus or one in the cochlea of the ear.

Gamete: A haploid cell that fuses with another to form a zygote.

Gamma globulin: A class of proteins found in blood; most antibodies are gamma globulins.

Gastrin: A hormone produced by the stomach mucosa.

Gastrocoel: The cavity of the archenteron.

Gastrula: The embryonic stage that consists of two cell layers surrounding a cavity.

Gastrulation: The process in which a young embryo develops its primary germ layers.

Gene: A determiner of a genetic trait, a specific linear segment of nucleotides in DNA.

Genotype: A list of the genes in an organism.

Germ cell: Gamete, a cell capable of participating in sexual reproduction.

Glomerulus: A tiny network of capillaries in the kidney through which filtration takes place.

Glucagon: A hormone produced by the pancreas that aids in the conversion of glycogen to glucose in the liver.

Glucocorticoid: One of a group of hormones produced by the adrenal cortex involved primarily in the regulation of carbohydrate metabolism.

Glycogen: A carbohydrate storage material in animals.

Glycolysis: The anaerobic breakdown of sugars in metabolism.

Golgi apparatus: A small network of membranes generally located near the nucleus and concerned with secretory activities.

Gonadotrophic hormone: A hormone that stimulates the gonads.

Gonads: Organs in animals capable of producing gametes; the testes and ovaries.

Gymnosperm: See Coniferophyte.

Habituation: The simplest type of learning; an animal learns to ignore stimuli that are meaningless.

Haversian canal: A canal in bone surrounded by concentric rings of osteocytes for the passage of an artery, vein, and nerve fibers.

Heart: A muscular organ of the circulatory system that functions in pumping blood through the system.

Hemopoietic activity: Pertaining to the formation of blood cells.

Heterotroph: An organism that feeds on other organisms.

Heterozygous: The condition of having the genes arranged in allelic pairs.

Hibernation: A genetically fixed, seasonal waning of the metabolic processes.

Homeostatic control: The maintenance of relatively stabilized internal conditions.

Homozygous: Having the genes arranged in identical pairs.

Hydroxyapatite: A crystalline mineral forming a major constituent of bones and teeth.

Hymen: A membranous fold that partially occludes the external opening of the vagina.

Hyperparathyroidism: An overfunctioning of the parathyroid glands that brings about muscular weakness, bone decalcification, and an increase in calcium in the blood and urine.

Hyperthyroidism: Overactivity of the thyroid gland that increases the body metabolism.

Hypoblast: The inner layer of cells in an embryo made up of two cell layers.

Hypothalamus: Part of the brain stem concerned with emotions and homeostatic activities such as eating and drinking.

Hypothyroidism: Underfunctioning of the thyroid gland. May cause such pathological conditions as cretinism and myxedema.

Immune System: The system that protects the body against harmful invaders; includes thymus gland, lymph nodes, and white blood cells.

Impulse: The wave of excitation that passes along a nerve fiber.

Incisor: One of the most anterior teeth in a mammal; often, but not invariably, modified for nipping or gnawing as in the rodents.

Integument: The outer, cellular layer of an animal, including appendages such as hair and nails.

Interkinesis: The period between the first and second meiotic divisions.

Interphase: The period between the end of one cell division and the initiation of the next.

Intrauterine Device (IUD): A plastic or metal device inserted in the uterus as a contraceptive.

Isometric: Of muscles having variable tension without change in length.

Isostacy: The tendency of the earth's crust to maintain its relative height by adjusting its position in the denser rock on which it rests.

Isotonic: (1) Of muscles having a uniform tension and change in length on contraction. (2) Of solutions having the same concentrations.

Kidney: Organ involved in the excretion of urine.

Kinetochore: The locus on a chromosome to which spindle fibers attach, formerly called centromere.

Lacteal: Intestinal lymphatic that takes up emulsified fat from the gut.

Lacuna: A small hollow or depression, specifically in bone or cartilage the site of the living cell.

Larynx: The voice box.

Leukocyte: White blood cell.

Lichen: Plant formed by an alga and a fungus living together in a mutualistic relationship.

Lipid: One of a group of biochemical substances, insoluble in water, but soluble in fat solvents, e.g., fatty acids, steroids.

Lymph: The fluid in the lymphatic vessels; it is similar to blood plasma and contains leukocytes.

Lymphatic: Thin-walled, valved vessel that conveys lymph from the tissues back into the bloodstream.

Lysosome: Cellular organelle concerned with intracellular digestion.

Malpighian body: The functional structure in the kidney.

Mammal: A member of the class of vertebrates in which the female suckles the young.

Mammary gland: A milk-producing gland.

Matrix: The substance in which something is embedded; in animals, the intercellular substance of connective tissues.

Meiosis: A type of cell division that in animals results in haploid gametes and in plants results in haploid spores.

Melanin: A yellow to black pigment in the skin.

Membranous labyrinth: A series of membranous tubes in the inner ear that contain the sensory receptors.

Meninges: The membranes that surround the central nervous system.

Menstruation: The cyclic, physiologic uterine bleeding in the females of some primates.

Meristem: An undifferentiated plant tissue.

Mesenchyme: An undifferentiated animal tissue.

Mesoderm: The third or middle germ layer formed in a developing embryo.

Metabolism: The chemical processes by which a living organism is produced and maintained.

Metaphase: The stage of cell division in which the chromosomes are arranged on an equatorial plate.

Metatheria: That group of mammals in which the young complete development in a pouch or marsupium.

Microfilament: A cytoplasmic fiber of protein.

Microtubule: A cytoplasmic tubule formed of microfilaments.

Mineralocorticoid: One of a group of hormones produced by the adrenal cortex involved primarily in mineral metabolism.

Mitochondria: Small, rounded, or oval bodies in the cytoplasm that are centers of ATP formation by chemical respiration.

Mitosis: A type of cell division in which the nuclear material of a mother cell is duplicated in each of two daughter cells.

Mitotic cycle: A complete cell division from interphase to interphase.

Molar: One of the grinding teeth; in humans these are the nondeciduous teeth.

Mollusca: A phylum of soft-bodied invertebrates characterized by a muscular foot; many of them bear shells, e.g., clams, snails.

Monocotyledon: A flowering plant in which only a single food-storing leaf is formed in the seed, e.g., corn, wheat, lily.

Monohybrid: A genetic cross involving only one pair of alleles.

Monosaccharide: A simple sugar, one

that cannot be decomposed by hydrolysis.

Mucus: A lubricating substance secreted by cells in an epithelial tissue.

Mutation: Any sudden, heritable change in the genetic material.

Mutualism: Living together for the mutual benefit of both parties.

Myelin: The fatty substance that forms a sheath around many nerve fibers.

Myofibril: An elongate structure within a muscle cell; it contains the contractile elements.

Myoglobin: An oxygen-storing protein found in muscle.

Myosin: The most abundant protein in muscle, forming fibers.

NAD: Nicotinamide adenine dinucleotide, an organic molecule that serves as a hydrogen acceptor-donor.

NADP: Nicotinamide adenine dinucleotide phosphate, an organic molecule that serves as a hydrogen acceptor-donor.

Nematoda: Any of the worms without a true coelom but with a complete gut, or enteron.

Nervous tissue: Tissue made of neurons or parts of neurons together with protective material.

Neural tube: Nervous tissue initially formed from ectoderm in the form of a tube.

Neurectoderm: The portion of the ectoderm that gives rise to the neural tube.

Neuron: A nerve cell.

Niacin: Nicotinic acid, one of the B vitamins.

Nipple: The conic organ that gives outlet to the milk; differs from a teat in that it does not contain a cistern for the accumulation of milk.

Noradrenalin: See Norepinephrine.

Norepinephrine (Noradrenalin): A secretion of the adrenal medulla that operates primarily in the control of vasoconstriction; also secreted by axons of sympathetic nervous system.

Notochord: A moderately stiff rod of cartilaginous-like material lying between the gut and nerve cord; characteristic of the phylum of animals that includes the vertebrates.

NPN: Nonprotein nitrogen-containing compounds.

Nucleic acids: Polymers of nucleotides, DNA and RNA.

Nucleolus: A round little body within the nucleus of a cell, rich in proteins and RNA.

Nucleotide: One of the compounds that goes into the formation of nucleic acid.

Nucleus: A rounded body in the cell that is the controlling center of the cell.

Oögenesis: Meiosis in the female; egg formation.

Oögonium: The primordial cell from which a primary oöcyte arises.

Organ: A part of the body, made up of tissues, that has a special function, e.g., the heart, the brain.

Organism: A living body.

Organogenesis: The embryonic formation of an organ.

Osmosis: The passage of a solvent (water) through a membrane toward a greater concentration of a solute.

Osteichthyes: A class of vertebrates commonly known as the bony fishes.

Osteoblast: Bone-forming cell.

Osteoclast: Bone-destroying cell.

Osteocyte: A bone cell.

Osteoprogenitor cell: A cell that gives rise to osteoblasts and osteoclasts.

Oval window: (1) An opening in the median wall of the heart. (2) An opening in the large end of the cochlea.

Ovary: The female gonad; organ that produces eggs.

Ovulation: The discharge of an ovum by the ovary.

Ovum: An egg cell; the female gamete.

Palate: The roof of the mouth.

Pancreas: A large internal gland both endocrine and exocrine in function.

Pancreatic juice: The fluid containing the digestive enzymes produced by the pancreas.

Parasitism: Symbiosis in which one kind of organism is benefited and the other injured.

Parathormone: A hormone of the parathyroid glands that regulates both the calcium level of the blood and phosphorus metabolism.

Parathyroids: Small endocrine glands lying adjacent to the thyroid gland; they control calcium and phosphorus levels in tissues.

Penis: The special copulatory organ in the males of mammals and some reptiles and birds.

Pepsin: An enzyme of the gastric juice that acts on proteins.

Pepsinogen: A material produced by gastric cells, converted into pepsin by hydrochloric acid.

Peptides: Compounds containing amino acid, formed by the hydrolysis of proteins.

Peripheral nervous system: That part of the nervous system exclusive of the brain and spinal cord.

Peristalsis: Contraction waves of the gut and other muscular tubes.

Peritoneum: The thin sheet of mesodermal tissue lining the coelom.

Peroxisome: An organelle in the cell, resembling a lysosome but differing in function.

Phagocytosis: The engulfing of organisms, other cells, and foreign particles by a cell.

Pharynx: The first part of the endodermal portion of the alimentary canal.

Phenotype: The physical expression of genic effects.

Phloem: A living, conducting, vascular tissue in the higher plants that transports food.

Photosynthesis: The process by which the energy of sunlight is converted into chemical bond energy.

Phylum: Any one of the major categories into which animals are classified.

"Pill": Any of a number of kinds of pills containing sex hormones that act as contraceptives.

Pinocytosis: The engulfing of liquids by cells.

Pituitary: A multiple endocrine gland lying below the floor of the brain.

Placenta: An extraembryonic structure that in most mammals attaches the embryo to the uterus.

Placoderms: An extinct class of primitive, jawed, fishlike vertebrates.

Plasma: The fluid portion of the blood in which the cells are suspended.

Plasma membrane: The thin, modified layer of cytoplasm that forms the surface of the cell.

Plastid: One of a number of organelles in the cytoplasm of plant cells. The most familiar ones are the chloroplasts that contain chlorophyll.

Platelets: Tiny discs found in the blood, concerned with the coagulation of the blood.

Platyhelminthes: The phylum of flatworms.

Pleiotropic: Genes that affect several different traits.

Pleura: The peritoneum lining that portion of the chest cavity that contains the lungs.

Ploidy: The number of sets of chromosomes in the cell. Used also as a suffix denoting the degree of multiplication of the chromosome sets.

Polar body: A nonfunctional daughter cell resulting from meiosis in the female.

Polygenes: Genes having similar, cumulative effects; the same as multiple factors.

Polymer: A chain of similar molecules bound together.

Polysaccharide: Any of a group of carbohydrates that contains more than four simple carbohydrates bonded together, e.g., the starches, glycogen.

Porifera: The phylum of animals to which the sponges belong.

Premolar: One of the most posterior deciduous teeth; in humans premolars are crushing teeth but in other mammals they may be modified for other purposes, as for shearing in cats and dogs.

Primary oöcyte: A female germ cell ready to undergo meiosis.

Primary spermatocyte: A male germ cell ready to undergo meiosis.

Primates: The order of mammals to which man belongs.

Progesterone: The hormone produced by the corpus luteum that functions in preparing the uterus for the implantation of the ovum.

Prophase: The stage in cell division in which the chromosomes shorten and thicken and move to the equatorial plate.

Proprioceptor: One of the sensory receptors that receives stimuli concerning the state of contraction of muscles and degree of flexion of the joints.

Prostate: A gland in the male that surrounds the neck of the bladder and the urethra.

Protein: Any one of a group of long polymers of amino acids.

Protoplasm: The matter of which living cells are made.

Protostome: An animal in which the mouth develops at the same end of the embryo as the blastopore.

Prototheria: The egg-laying mammals.

Protozoa: Unicellular animals.

Pure line: A strain of plants or animals homozygous for a particular complement of genes.

Pylorus: The constriction between the stomach and small intestine.

Recessive: A gene or trait whose expression is inhibited by its allele.

Reflex arc: That part of the nervous system used in a simple, fixed, unlearned reaction.

Reproduction: The phenomenon of giving rise to another generation of like individuals.

Reptilia: The class of ectothermic amniotes that are covered with epidermal scales.

Reticular Formation: Fiber tracts in the brain that connect one part to another.

Retina: The layer at the back of the eye containing the sensory receptor cells.

Rhodopsin: A photosensitive substance in the rods of the retina.

Ribosome: A small body on the membrane of the endoplasmic reticulum or in the cytoplasm; the site of protein synthesis.

Ritualization: The process by which a fixed action pattern that is characteristic of one behavior sequence becomes part of another sequence in which it serves as a sign stimulus.

RNA: Ribonucleic acid; functions in the translation of the genetic code contained in DNA.

Rod: A cell in the retina of the eye that functions in light-dark discrimination.

Sacrum: The structure made up of the vertebrae that connect with the pelvic girdle.

Saliva: The watery, enzymatic fluid produced by the salivary glands.

Saprophyte: A plant, fungus or bacterium that takes its nourishment from dead organisms.

Sarcolemma: The plasma membrane of a muscle fiber.

Sarcomere: A segment of a myofibril.

Scapula: The shoulder blade.

Sclera: The outer, tough coat of the eyeball.

Scrotum: A sac of skin that houses the gonads in the male of many mammals.

Sebaceous gland: Oil-producing gland.

Secondary oöcyte: The large cell resulting from the first meiotic division in the female.

Secondary spermatocyte: One of the cells resulting from the first meiotic division in the male.

Secretion: A specific substance produced and discharged by a gland.

Semen: The fluid in which the sperm are conveyed.

Seminal vesicle: The sac in which semen is held prior to ejaculation.

Sere: A succession of ecological communities.

Sertoli cell: Elongated cell in the tubule of the testis. The spermatids become attached to the end of these cells, apparently for the purpose of nutrition.

Sexual reproduction: Reproduction involving specialized germ cells or gametes.

Sign stimulus: A stimulus that initiates a fixed action sequence.

Skeletal muscle: Striated, voluntary muscle.

Social hierarchy: The organization of an animal society in levels of dominance-subordination.

Solute: A substance in solution in a solvent.

Solvent: The fluid in which a solute is dissolved.

Soma: The tissues that are specialized for the maintenance of the body rather than reproduction.

Sperm: The male gamete.

Spermatid: One of the cells resulting from meiosis in the male; it develops without further division into a flagellated sperm.

Spermatogenesis: Meiosis in the male that produces sperm.

Spermatogonium: Male germ cell prior to the time of meiosis.

Spermatophore: The gelatinous structure containing sperm deposited by a male salamander.

Sphincter: A constricting muscle.

Spindle fibers: Protein fibers (really microtubules) connecting the centrioles with the kinetochores of chromosomes during cell division.

Starch: A polysaccharide produced by plants.

Steroid: One of a group of compounds that resembles fats in solubility properties but not in structure.

Stratum corneum: The outermost, horny layer of the epidermis.

Stratum germinativum: The lowest layer of the epidermis and the one from which the other layers are derived.

Symbiosis: Living together.

Synapse: (1) The place at which an impulse passes from one nerve cell to another. (2) The coming together of homologous chromosomes during the first meiotic division.

Systole: The period of the heart's contraction or the contraction itself.

T-cells: Cells of the immune system that protect against viruses and some bacteria.

Telophase: The last, or reconstruction, phase of mitosis.

Territory: Part of an animal's home range that is defended against invaders of the same species.

Testis: The male gonad.

Testosterone: The male hormone produced by the testis.

Tetany: Continuous, steady contraction of a muscle without twitching.

Theria: The placental mammals.

Thiamine: Vitamin B_1.

Thrombocyte: A blood platelet.

Thymus: An organ that forms part of the immune system.

Thyroid: A ductless gland situated in the neck region in man.

Thyroxine: A hormone of the thyroid gland.

Tonus: The slight, continuous contraction of a skeletal muscle that aids in the maintenance of posture.

Trachea: The tube that conveys air between the pharynx and a bronchus.

Tracheophyte: A plant that has conductive tissues.

Transpiration: The loss of water vapor from a plant.

Trihybrid: Involving three pair of genes.

Triploblastic: Having three germ layers—ectoderm, endoderm, and mesoderm.

Tropomyosin: A protein in muscle.

Turbinate: Scroll-shaped bone in the nasal chamber.

Tympanic membrane: A thin membrane that separates the middle ear chamber from the outer ear chamber.

Urea: A relatively nontoxic, water-soluble form of nitrogenous waste.

Ureter: The tube conveying urine from the kidney to the bladder.

Urethra: The tube conveying urine from the bladder to the outside.

Uric acid: A nontoxic, nonsoluble form of nitrogenous waste.

Urine: The excretory substance produced by the kidneys.

Uterus: An expanded portion of the oviducts in which the embryo develops.

Vacuole: A space in a cell.

Vagina: That part of a female reproductive tract that receives the penis during copulation.

Vas deferens: The duct leading from the epididymis to the urethra.

Vasopressin: A hormone of the posterior lobe of the pituitary that increases water reabsorption by the kidney.

Vein: A noncontractile blood vessel that generally conveys blood toward the heart.

Ventral: The front or belly side; opposite to dorsal.

Vestibular membrane: A membrane in the inner ear.

Visceral muscle: Smooth, involuntary muscles.

Virion: The inert, extracellular form of a virus.

Vitamin: A small, organic molecule, functioning as a coenzyme, that is not synthesized by the organism.

Vitreous chamber: The chamber of the eye behind the lens.

Wax: Compound of fatty acids with an alcohol larger than glycerol.

Xylem: The vascular tissue in plants that transports water and minerals.

Yolk sac: A sac of stored food utilized by a developing embryo.

Z-line: The line that separates one sarcomere from another.

Zygote: A cell formed by the union of two gametes.

Index